ADVANCES IN CHEMICAL PHYSICS

VOLUME LIX

Advances in
CHEMICAL PHYSICS

VOLUME LIX

INDEX: VOLUMES I–LV

PREPARED BY

Patricia L. Radloff
The James Franck Institute
The University of Chicago
Chicago, Illinois

SERIES EDITORS

I. PRIGOGINE

University of Brussels
Brussels, Belgium
and
University of Texas
Austin, Texas

AND

STUART A. RICE

Department of Chemistry
and
The James Franck Institute
The University of Chicago
Chicago, Illinois

AN INTERSCIENCE® PUBLICATION
JOHN WILEY & SONS
NEW YORK • CHICHESTER • BRISBANE • TORONTO • SINGAPORE

An Interscience® Publication

Copyright © 1985 by John Wiley & Sons, Inc.

Library of Congress Catalog Number: 58-9935

ISBN 0-471-80427-4

Printed in the United States of America

10 9 8 7 6 5 4 3 2 1

CONTENTS

ADVANCES IN CHEMICAL PHYSICS

VOLUME LIX

ARTICLE INDEX

A

Alcohol narcosis

Johnson, F. H., "Hydrostatic Pressure Reversal of Alcohol Narcosis in Aquatic Animals," **21**:647.

Atmospheric chemistry

McGowan, J. W., Kummler, R. H., and Gilmore, F. R., "Excitation De-excitation Processes Relevant to the Upper Atmosphere," **28**:379.

Michels, H. H., "Electronic Excited States of Selected Atmospheric Systems," **45**:225.

Nicolet, M., "Atmospheric Chemistry," **55**:63.

Zander, R., "Commentary: Observational Aspects Related to the Chemical Evolution of Our Atmosphere," **55**:79.

B

Bénard convection

see Convection

Benzenes

Goodman, L. and Rava, R. P., "Two-Photon Spectroscopy of Perturbed Benzenes," **54**:177.

Parmenter, C. S., "Radiative and Nonradiative Processes in Benzene," **22**:365.

see also Electronic spectroscopy

Bifurcations

Kauffman, S. A., "Bifurcations in Insect Morphogenesis," **55**:219.

Nicolis, G., "Bifurcations and Symmetry Breaking in Far-from-Equilibrium Systems: Toward a Dynamics of Complexity," **55**:177.

Biomolecule conformation

Urry, D. W., "Biomolecular Conformation and Biological Activity," **21**:581.

Boron

Houtman, J. A. and Duffey, G. H., "A Free-Electron Study of Common Boron Frameworks," **21**:7.

Brownian motion

Dagonnier, R., "Theory of Quantum Brownian Motion," **16**:1.

Guth, E., "Brownian Motion and Indeterminacy Relations," **15**:363.

1

C

Carcinogenesis

Daudel, R., "The Relation Between Structure and Chemical Reactivity of Aromatic Hydrocarbons with Particular Reference to Carcinogenic Properties," **1**:165.

Pullman, B., "Steroids, Purine-Pyrimidine Pairs and Polycyclic Aromatic Carcinogens," **8**:163.

Catalysis

Gomer, R., "Electron Spectroscopy of Chemisorption on Metals," **27**:211.

Bernasek, S. L., "Heterogeneous Reaction Dynamics," **41**:477.

Hammes, G. G., "Control of the Catalytic Activity of Enzymes by the Near and Remote Environment of a Polyatomic Framework," **39**:177.

Higuchi, I., "The Pore Structure of Catalysts and Catalytic Reaction Rates," **21**:791.

Klotz, I. M., "Synzymes: Synthetic Polymers with Enzymelike Catalytic Activities," **39**:109.

Laforgue, A., Rousseau, J., and Imelik, B., "Chemisorption of Ethylene on Metals and Catalytic Hydrogenation," **8**:141.

Chaos

Brumer, P., "Intramolecular Energy Transfer: Theories for the Onset of Statistical Behavior," **47**:Part 1:201.

Ford, J., "The Transition from Analytic Dynamics to Statistical Mechanics," **24**:155.

Kauffman, S. A., "Bifurcations in Insect Morphogenesis," **55**:219.

Percival, I. C., "Semiclassical Theory of Bound States," **36**:1.

Rice, S. A., "An Overview of the Dynamics of Intramolecular Transfer of Vibrational Energy," **47**:Part 1:117.

Tabor, M., "The Onset of Chaotic Motion in Dynamical Systems," **46**:73.

Charge transfer

Baede, A. P. M., "Charge Transfer Between Neutrals at Hyperthermal Energies," **30**:463.

Englman, R., "Optical Absorptions and Charge Transfer States," **8**:13.

Lichten, W., "Resonant Charge Exchange in Atomic Collisions," **13**:41.

Stebbings, R. F., "Charge Transfer," **10**:195.

see also Collisional processes

Chemical bonding

see Valence theory

Chemical Dynamics

Adelman, S. A., "Chemical Reaction Dynamics in Liquid Solution," **53**:61.

Adelman, S. A., "Generalized Langevin Equations and Many-body Problems in Chemical Dynamics," **44**:143.

Anderson, J. B., "The Reaction F + $H_2 \rightarrow$ HF + H," **41**:229.

Aroeste, H., "Toward an Analytic Theory of Chemical Reactions," **6**:1.

Atkins, P. W. and Evans, G. T., "Theories of Chemically Induced Electron Spin Polarization," **35**:1.

Baede, A. P. M., "Charge Transfer Between Neutrals at Hyperthermal Energies," **30**:463.

Baer, M., "A Review of Quantum-Mechanical Approximate Treat-

ments of Three-Body Reactive Systems," **49**:191.

Balint-Kurti, G. G., "Potential Energy Surfaces for Chemical Reaction," **30**:137.

Baronavski, A., Umstead, M. E., and Lin, M. C., "Laser Diagnostics of Reaction Product Energy Distributions," **47**:Part 2:85.

Barriol, J. and Regnier, J., "Calculation of Transition Energies from the Geometry of the System," **8**:5.

Basilevsky, M. V., "Transition State Stabilization Energy as a Measure of Chemical Reactivity," **33**:345.

Bass, L. B. and Moore, W. J., "Theory of Rate Processes Applied to Release of Acetylcholine at Synapses," **21**:619.

Ben-Shaul, A., "Chemical Laser Kinetics," **47**:Part 2:55.

Bernasek, S. L., "Heterogeneous Reaction Dynamics," **41**:477.

Beswick, J. A. and Jortner, J., "Intramolecular Dynamics of van der Waals Molecules," **47**:Part 1:363.

Bigeleisen, J. and Wolfsberg, M., "Theoretical and Experimental Aspects of Isotope Effects in Chemical Kinetics," **1**:15.

Careri, G., "Some Physical Aspects of Gaseous Chemical Kinetics," **1**:119.

Dewar, M. J. S., "Chemical Reactivity," **8**:65.

Eschenroeder, A. Q., "Reaction Kinetics in Hypersonic Flows," **13**:19.

Eyring, E. M., Rich, L. D., McCoy, L. L., Graham, R. C., and Taylor, N., "Activation Parameters and Deprotonation Kinetics of Intramolecularly Hydrogen Bonded Acids," **21**:327.

Grice, R., "Reactive Scattering," **30**:247.

Haas, Y., "Electronically Excited Fragments Formed by Unimolecular Multiple Photon Dissociation," **47**:Part 1:713.

Haas, Y. and Asscher, M., "Two-Photon Excitation as a Kinetic Tool: Application to Nitric Oxide Fluorescence Quenching," **47**:Part 2:17.

Herschbach, D., "Reactive Scattering in Molecular Beams," **10**:319.

Johnston, H. S, "Large Tunnelling Corrections in Chemical Reaction Rates," **3**:131.

Kapral, R., "Kinetic Theory of Chemical Reactions in Liquids," **48**:71.

Keck, J. C., "Variational Theory of Reaction Rates," **13**:85.

Kreevoy, M. M. and Konasewich, D. E., "The Brønsted α and the Primary Hydrogen Isotope Effect. A Test of the Marcus Theory," **21**:243.

Laidler, K. J. and Tweedale, A., "The Current Status of Eyring's Rate Theory," **21**:113.

Leone, S. R., "Photofragment Dynamics," **50**:255.

Levine, R. D., "The Information Theoretic Approach to Intramolecular Dynamics," **47**:Part 1:239.

Light, J. C., "Quantum Theories of Chemical Kinetics," **19**:1.

Lin, M. C., "Dynamics of Oxygen Atom Reactions," **42**:113.

Mahan, B. H., "Recombination of Gaseous Ions," **23**:1.

McGowan, J. W., Kummler, R. H., and Gilmore, F. R., "Excitation De-excitation Processes Relevant to the Upper Atmosphere,"

28:379.

McQuarrie, D. A., "Stochastic Theory of Chemical Rate Processes," 15:149.

Menziger, M., "Electronic Chemiluminescence in Gases," 42:1.

Micha, D. A., "Quantum Theory of Reactive Molecular Collisions," 30:7.

Montroll, E. W. and Shuler, K. E., "The Application of the Theory of Stochastic Processes to Chemical Kinetics," 1:361.

Mortensen, E. M., "Permeabilities for Reactions of the Type $H + H_2 = H_2 + H$ Treated as a Linear Encounter Using Variational and Distorted Wave Techniques," 21:127.

Nicolet, M., "Atmospheric Chemistry," 55:63.

Smith, I. W. M., "The Production of Excited Species in Simple Chemical Reactions," 28:1.

Welge, K. H. and Schmiedl, R., "Doppler Spectroscopy of Photofragments," 47:Part 2:133.

Widom, B., "Collision Theory of Chemical Reaction Rates," 5:355.

Chemical kinetics

see Chemical dynamics

Chemiluminescence

Menziger, M., "Electronic Chemiluminescence in Gases," 42:1.

Chemisorption

Gomer, R., "Electron Spectroscopy of Chemisorption on Metals," 27:211.

Laforgue, A., Rousseau, J., and Imelik, B., "Chemisorption of Ethylene on Metals and Catalytic Hydrogenation," 8:141.

Chirality

Hansen, A. H. and Bouman, T. D., "Natural Chiroptical Spectroscopy: Theory and Computations," 44:545.

Noyes, R. M., "Commentary: On Selection of Chirality," 55:111.

Chlorophyll

Kaufmann, K. J. and Wasielewski, M. R., "Studies of Chlorophyll *in Vitro*," 47:Part 2:579.

Clathrates

van der Waals, J. H. and Platteeuw, J. C., "Clathrate Solutions," 2:1.

Clusters

Hoare, M. R., "Structure and Dynamics of Simple Microclusters," 40:49.

Ng, C. Y., "Molecular Beam Photoionization Studies of Molecules and Clusters," 52:263.

Coherence effects

Cantrell, C. D., Makarov, A. A., and Louisell, W. H., "Laser Excitation of SF_6: Spectroscopy and Coherent Pulse Propagation Effects," 47:Part 1:583.

Wiersma, D. A., "Coherent Optical Transient Studies of Dephasing and Relaxation in Electronic Transitions of Large Molecules in Condensed Media," 47:Part 2:421.

Collisional processes

Amme, R. C., "Vibrational and Rotational Excitation in Gaseous Collisions," 28:171.

Baede, A. P. M., "Charge Transfer Between Neutrals at Hyperthermal Energies," 30:463.

Bak, T. A. and Sørensen, G. P., "Vibrational Relaxation of a Gas of Diatomic Molecules," 15:219.

Birnbaum, G., Guillot, B., and Bra-

tos, S., "Theory of Collision—Induced Line Shapes—Absorption and Light Scattering at Low Density," **51**:49.

Breckenridge, W. H. and Umemoto, H., "Collisional Quenching of Electronically Excited Metal Atoms," **50**:325.

Brunner, T. A. and Pritchard, D., "Fitting Laws for Rotationally Inelastic Collisions," **50**:589.

Clark, A. P., Dickinson, A. S. and Richards, D., "The Correspondence Principle in Heavy-Particle Collisions," **36**:63.

DePristo, A.E. and Rabitz, H., "Vibrational and Rotational Collision Processes," **42**:271.

Englman, R., "Optical Absorptions and Charge Transfer States," **8**:13.

Freed, K. F., "Collision Induced Intersystem Crossing," **47**:Part 2:291.

Freed, K. F., "Collisional Effects on Electronic Relaxation Processes," **42**:207.

Frommhold, L., "Collision-Induced Scattering of Light and the Diatom Polarizabilities," **46**:1.

Haas, Y., "Electronically Excited Fragments Formed by Unimolecular Multiple Photon Dissociation," **47**:Part 1:713.

Hertel, I. V., "Collisional Energy-Transfer Spectroscopy with Laser-Excited Atoms in Crossed Atom Beams: A New Method for Investigating the Quenching of Electronically Excited Atoms by Molecules," **45**:341.

Kempter, V., "Electronic Excitation in Collisions between Neutrals," **30**:417.

Kodaira, M. and Watanabe, T., "Collisional Transfer of Triplet Excitations Between Helium Atoms," **21**:167.

Lacmann, K., "Collisional Ionization," **42**:513.

Lichten, W., "Resonant Charge Exchange in Atomic Collisions," **13**:41.

Micha, D. A., "Quantum Theory of Reactive Molecular Collisions," **30**:7.

Miller, W. H., "Classical-Limit Quantum Mechanics and the Theory of Molecular Collisions," **25**:69.

Miller, W. H., "Classical S-Matrix in Molecular Collisions," **30**:77.

Muschlitz, E. E., Jr., "Collisions of Electronically Excited Atoms and Molecules," **10**:171.

Niehaus, A., "Spontaneous Ionization in Slow Collisions," **45**:399.

Nikitin, E. E., "Theory of Nonadiabatic Collision Processes Including Excited Alkali Atoms," **28**:317.

Rice, S. A., "Collision Induced Intramolecular Energy Transfer in Electronically Excited Polyatomic Molecules," **47**:Part 2:237.

Stebbings, R. F., "Charge Transfer," **10**:195.

Tramer, A. and Nitzan, A., "Collisional Effects in Electronic Relaxation," **47**:Part 2:337.

Widom, B., "Collision Theory of Chemical Reaction Rates," **5**:355.

see also Energy transfer; Nonradiative processes; Vibrational relaxation

Convection

Hurle, D. T. J. and Jakeman, E., "On the Nature of Oscillatory Convection in Two-Component Fluids,"

32:277.

Koschmieder, E. L., "Bénard Convection," 26:177.

Koschmieder, E. L., "Stability of Supercritical Bénard Convection and Taylor Vortex Flow," 32:109.

Moore, D. R., "Numerical Models for Convection," 32:65.

Morgan, J. C., "The Effect of Prandtl Number on Finite Amplitude Bénard Convection," 32:77.

Platten, J. K. and Chavepeyer, G., "Finite Amplitude Instability in the Two-Component Bénard Problem," 32:281.

Schlecter, R. S., Velarde, M. G., and Platten, J. K., "Two-Component Bénard Problem," 26:265.

Thomaes, G., "The Bénard Instability in Liquid Mixtures," 32:269.

Weiss, N. O., "Magnetic Fields and Convection," 32:101.

see also Dissipative structures

Correlation functions

Berne, B. J. and Harp, G. D., "On the Calculation of Time Correlation Functions," 17:63.

Evans, M., Evans, G., and Davies, R., "Experimental and Theoretical Studies of Rototranslational Correlation Functions," 44:255.

Verdier, P. H., "Fluctuations in Autocorrelation Functions in Diffusing Systems," 15:137.

Critical Phenomena

Chay, T. R. and Ree, F. H., "Inequalities for Critical Indices near Gas-Liquid Critical Point," 21:431.

Coopersmith, M. H., "The Thermodynamic Description of Phase Transitions," 17:43.

Fixman, M., "The Critical Region," 6:175.

Kozak, J. J., "Nonlinear Problems in the Theory of Phase Transitions," 40:229.

see also Phase transitions

Crystal growth

Tensmeyer, L. G., Landis, P. W., and Coffey, H. F., "A Continuous Fall Method of Growing Crystals from Solution," 21:767.

Van Hook, A., "Application of Transition State Ideas to Crystal Growth Problems," 21:715.

Crystals

Karle, J., "The Phase Problem in Structural Analysis," 16:131.

Zandler, M. E. and Thomson, T. R., "Application of the Significant Structure Theories to Plastic Crystals," 21:517.

see also Molecular solids; Solids

D

Deoxyribonucleic acid

Douzou, P. and Sadron, C., "The Electronic Properties of Deoxyribonucleic Acid," 7:339.

Hoffmann, T. A. and Ladik, J., "Quantum Mechanical Considerations on Some Properties of DNA," 7:84.

Löwdin, P., "Some Aspects of the Biological Problems of Heredity, Mutations, Aging, and Tumors in View of the the Quantum Theory of the DNA Molecule," 8:177.

Pullman, B., "Steroids, Purine-Pyrimidine Pairs and Polycyclic Aromatic Carcinogens," 8:163.

Shimanouchi, T., Tsuboi, M., and Kyogoku, Y., "Infrared Spectra of Nucleic Acids and Related Com-

pounds," 7:435.

Wartell, R. M. and Montroll, E. W., "Equilibrium Denaturation of Natural and of Periodic Synthetic DNA Molecules," 22:129.

Detonations

Cook, M. A., "Contributions of Professor Henry Eyring to the Theory of Detonation," 21:779.

Hirschfelder, J. O. and Curtiss, C. F., "Propagation of Flames and Detonations," 3:59.

Williams, F., "Electronic States of Solid Explosives and Their Probable Role in Detonations," 21:289.

Diamagnetism

Pacault, A., Hoarau, J., and Marchand, A., "Aspects Récents du Diamagnétisme," 3:171.

Diamond

Hall, H. T., "The Synthesis of Diamond," 21:721.

Wentorf, R. H., Jr., "Diamond Synthesis," 9:365.

Dissipative structures

Bak, T. A., "Nonlinear Problems in Thermodynamics of Irreversible Processes," 3:33.

Ball, J. G. and Himmelblau, D. M., "The Local Potential Applied to Instability Problems," 13:267.

Biot, M. S., "On a Unified Thermodynamic Approach to a Large Class of Instabilities of Dissipative Continua," 32:13.

Boon, J. P., "Theoretical Models for Bacterial Motion and Chemotaxis," 29:169.

Clarke, B. L., "Stability of Complex Reaction Networks," 43:1.

Hanusse, P., Ross, J., and Ortoleva, P., "Instability and Far-from-Equilibrium States of Chemically Reacting Systems," 38:317.

Hess, B., Boiteux, A., Busse, H., and Gerisch, G., "Spatiotemporal Organization in Chemical and Cellular Systems," 29:137.

Hess, B., Goldbeter, A., and Lefever, R., "Temporal, Spatial, and Functional Order in Regulated Biochemical and Cellular Systems," 38:363.

Hurle, D. T. J. and Jakeman, E., "On the Nature of Oscillatory Convection in Two-Component Fluids," 32:277.

Kahrig, E. and Besserdich, H., "A Short Remark About Various Dissipative Structures," 29:133.

Koschmieder, E. L., "Bénard Convection," 26:177.

Koschmieder, E. L., "Stability of Supercritical Bénard Convection and Taylor Vortex Flow," 32:109.

Moore, D. R., "Numerical Models for Convection," 32:65.

Morgan, J. C., "The Effect of Prandtl Number on Finite Amplitude Bénard Convection," 32:77.

Nazarea, A. D., "Electric Fields and Self-Coherent Patterns and Structures in Chemical Systems: Large-Scale Effects and Biological Implications," 38:415.

Nicolis, G., "Bifurcations and Symmetry Breaking in Far-from-Equilibrium Systems: Toward a Dynamics of Complexity," 55:177.

Nicolis, G., "Dissipative Instabilities, Structure, and Evolution," 29:29.

Nicolis, G., "Stability and Dissipative Structures in Open Systems Far from Equilbrium," 19:209.

Nicolis, G., Erneux, T., and Herschkowitz-Kaufman, M., "Pat-

tern Formation in Reacting and Diffusing Systems," **38**:263.

Nicolis, G., Prigogine, I., and Glansdorff, P., "On the Mechanism of Instabilities in Nonlinear Systems," **32**:1.

Novak, B., "Periodical Signals in the Spatial Differentiation of Plant Cells," **29**:281.

Ortoleva, P. and Ross, J., "Studies in Dissipative Phenomena with Biological Applications," **29**:49.

Pieranski, P. and Guyon, E., "Cylindrical Couette Flow Instabilities in Nematic Liquid Crystals," **32**:151.

Platten, J. K. and Chavepeyer, G., "Finite Amplitude Instability in the Two-Component Bénard Problem," **32**:281.

Prigogine, I., "Quantum States and Dissipative Systems," **15**:11.

Prigogine, I. and Lefever, R., "Coupling Between Diffusion and Chemical Reactions," **39**:1.

Prigogine, I. and Lefever, R., "Stability and Self-Organization in Open Systems," **29**:1.

Richter, P. H., Procaccia, I., and Ross, J., "Chemical Instabilities," **43**:217.

Roberts, P. H., "Concepts in Hydrodynamic Stability Theory," **32**:17.

Rohde, S. M., "Some Remarks on Variational Methods, the Local Potential, and Finite Element Methods with Application to Certain Continuum Mechanics Problems," **32**:37.

Ross, J. and Richter, P. H., "Commentary: Dissipation Regulation in Oscillatory Reactions. Application to Glycolysis," **55**:169.

Schlecter, R. S., Velarde, M. G., and Platten, J. K., "Two-Component Bénard Problem," **26**:265.

Schlögl, F., "Glansdorff-Prigogine Criterion and Statistical Theory," **32**:61.

Schoffeniels, E., "Commentary: A Complex Biological Oscillator," **55**:171.

Thomaes, G., "The Bénard Instability in Liquid Mixtures," **32**:269.

Thomas, D., "Diffusion-Reaction in Structured Media and Membranes Bearing Enzymes," **39**:229.

Thomas, D., "Functional Organization in Artificial Enzyme Membranes: Accomplishments and Prospects," **29**:113.

Thomas, R., "Logical Description, Analysis, and Synthesis of Biological and Other Networks Comprising Feedback Loops," **55**:247.

Turner, J. S., "Complex Periodic and Nonperiodic Behavior in the Belousov-Zhabotinski Reaction," **55**:205.

Turner, J. S., "Finite Fluctuations, Nonlinear Thermodynamics, and Far-From-Equilibrium Transitions Between Multiple Steady States," **29**:63.

Turner, J. S., "Laboratory Experiments on Double-Diffusive Instabilities," **32**:135.

Walter, C., "The Global Stability of Prey-Predator Systems with Second-Order Dissipation," **29**:125.

Weiss, N. O., "Magnetic Fields and Convection," **32**:101.

Wolpert, L., "The Development Pattern: Mechanisms Based on Positional Information," **29**:253.

DNA

see Deoxyribonucleic acid

Dynamics of Life

Eigen, M., "The Origin and Evolution of Life at the Molecular Level," **55**:119.

Stover, B., "The Dynamics of Life. IV. Lenticular Cataracts Induced by Aging and Irradiation," **21**:661.

E

Electrochromism

Labhart, H., "Electrochromism," **13**:179.

Electrode chemistry

Elving, P. J. and Pullman, B., "Mechanisms of Organic Electrode Reactions," **3**:1.

Tomassi, W., "Power Electrodes and Their Applications," **3**:239.

Electrolyte solutions

Cernuschi, F., Giambiagi, M., and Segre, M., "Application of a Short-Range Ordered Model to Strong Electrolytes," **21**:455.

Résibois, P. and Hasselle-Schuermans, N., "Microscopic Approach to Equilibrium and Non-Equilibrium Properties of Electrolytes," **11**:159.

Stecki, J., "Ionic Solvation," **6**:413.

Electron correlation

Brueckner, K. A., "The Correlation Energy of a Non-Uniform Electron Gas," **14**:215.

Čižek, J., "On the Use of the Cluster Expansion and the Technique of Diagrams in Calculations of Correlation Effects in Atoms and Molecules," **14**:35.

Grimaldi, F., "Correlation Effects in Diatomic Molecules Obtained from Configuration Interaction Using Hartree-Fock Orbitals. Effects on Energy and Monoelectronic Operators," **14**:341.

Löwdin, P., "Correlation Problem in Many-Electron Quantum Mechanics. I. Review of Different Approaches and Discussion of Some Current Ideas," **2**:207.

Löwdin, P., "Some Aspects of the Correlation Problem and Possible Extensions of the Independent-Particle Method," **14**:283.

Löwdin, P., "Some Recent Developments in the Quantum Theory of Many-Electron Systems and the Correlation Problem," **8**:3.

Nesbet, R. K., "Electronic Correlation in Atoms and Molecules," **9**:321.

Sinanoğlu, O., "Electron Correlation in Atoms and Molecules," **14**:237.

Yoshizumi, H., "Correlation Problem in Many-Electron Quantum Mechanics. II. Bibliographical Survey of the Historical Developments with Comments," **2**:323.

Electron gas

Bellemans, A. and de Leener, M., "Electron Gas in a Lattice of Positive Charges," **6**:85.

Brueckner, K. A., "The Correlation Energy of a Non-Uniform Electron Gas," **14**:215.

Electron spectroscopy

Brion, C. E. and Hamnett, A., "Continuum Optical Oscillator-Strength Measurements by Electron Spectroscopy in the Gas Phase," **45**:2.

Christophorou, L. G., Grant, M., and McCorkle, D. L., "Interactions of Slow electrons with Benzene and Benzene Derivatives," **36**:413.

Gomer, R., "Electron Spectroscopy of Chemisorption on Metals," 27:211.

Shirley, D. A., "ESCA," 23:85.

Somorjai, G. A. and Farrell, H. H., "Low-Energy Electron Diffraction," 20:215.

Taylor, H. S., "Models, Interpretations, and Calculations Concerning Resonant Electron Scattering Processes in Atoms and Molecules," 18:91.

Trajmar, S., Rice, J. K., and Kupperman, A., "Electron-Impact Spectroscopy," 18:15.

Electron spin resonance

Atkins, P. W. and Evans, G. T., "Theories of Chemically Induced Electron Spin Polarization," 35:1.

Carrington, A., Levy, D. H., and Miller, T. A., "Electron Resonance of Gaseous Diatomic Molecules," 18:149.

Ehrenberg, A., "Some Recent Developments in the Study of Paramagnetic Species of Biological Interest," 7:602.

Kaufmann, K. J. and Wasielewski, M. R., "Studies of Chlorophyll *in Vitro*," 47:Part 2:579.

Nicolau, C. S., "ESR Investigations on Different Plant Systems," 7:628.

Smaller, B., "Recent Advances in EPR Spectroscopy," 7:532.

Electronic spectroscopy

Goodman, L. and Rava, R. P., "Two-Photon Spectroscopy of Perturbed Benzenes," 54:177.

Parmenter, C. S., "Radiative and Nonradiative Processes in Benzene," 22:365.

Ross, I. G., "High-Resolution Electronic Spectra of Large Polya-tomic Molecules," 20:341.

Electronic structure

Čížek, J., "On the Use of the Cluster Expansion and the Technique of Diagrams in Calculations of Correlation Effects in Atoms and Molecules," 14:35.

Douzou, P. and Sadron, C., "The Electronic Properties of Deoxyribonucleic Acid," 7:339.

Drickamer, H. G. and Zahner, J. C., "The Effect of Pressure on Electronic Structure," 4:161.

Engelbrecht, L. and Hinze, J., "Molecular Properties Observed and Computed," 44:1.

Englman, R., "Optical Absorptions and Charge Transfer States," 8:13.

Fernández-Alonso, J. I., "Electronic Structures in Quantum Biochemistry," 7:3.

Green, S., "Sources of Error and Expected Accuracy in *Ab Initio* One-Electron Operator Properties: The Molecular Dipole Moment," 25:179.

Grimaldi, F., "Correlation Effects in Diatomic Molecules Obtained from Configuration Interaction Using Hartree-Fock Orbitals. Effects on Energy and Monoelectronic Operators," 14:341.

Halperin, B. I., "Properties of a Particle in a One-dimensional Random Potential," 13:123.

Harris, F. E. and Michels, H. H., "The Evaluation of Molecular Integrals for Slater-type Orbitals," 13:205.

Herrick, D. R., "New Symmetry Properties of Atoms and Molecules," 52:1.

Hinze, J., "An Overview of Computational Methods for Large

Molecules," **26**:213.

Hoffmann, T. A. and Ladik, J., "Quantum Mechanical Considerations on Some Properties of DNA," **7**:84.

Jucys, A. P., "On the Hartree-Fock Method in Multi-Configuration Approximation," **14**:191.

Jucys, A. P. and Kaminskas, V. A., "On the Application of the Extended Method of Calculation to the Atomic Electrons," **14**:207.

Judd, B. R., "Effective Operators for Configurations of Equivalent Electrons," **14**:91.

Kelly, H. P., "Applications of Many-Body Diagram Techniques in Atomic Physics," **14**:129.

Kiode, S. and Oguchi, T., "Theories on the Magnetic Properties of Compounds," **5**:189.

Kotani, M., "Electronic Structure and Magnetic Properties of Hemoproteins, Particularly Hemoglobins," **7**:159.

Kvasnička, V., "Application of Diagrammatic Quasidegenerate RSPT in Quantum Molecular Physics," **36**:345.

Kvasnička, V., Laurinc, V., Biskupič, S., and Haring, M., "Coupled-Cluster Approach in the Electronic Structure Theory of Molecules," **52**:181.

Kwun, S. and Ree, T., "Ionization Potentials and Electron Affinities of Some Polyphenyl Molecules," **21**:37.

Löwdin, P., "Correlation Problem in Many-Electron Quantum Mechanics. I. Review of Different Approaches and Discussion of Some Current Ideas," **2**:207.

Löwdin, P., "Some Aspects of the Correlation Problem and Possible Extensions of the Independent-Particle Method," **14**:283.

Löwdin, P., "Some Recent Developments in the Quantum Theory of Many-Electron Systems and the Correlation Problem," **8**:3.

Matsen, F. A., "Helium Molecular States," **21**:1.

Michels, H. H., "Electronic Excited States of Selected Atmospheric Systems," **45**:225.

Nesbet, R. K., "Atomic Bethe-Goldstone Equations," **14**:1.

Nesbet, R. K., "Electronic Correlation in Atoms and Molecules," **9**:321.

Olsen, J., Yeager, D. L., and Jørgensen, P., "Optimization and Characterization of a Multiconfigurational Self-Consistent Field (MCSCF) State," **54**:1.

Percival, I. C., "Semiclassical Theory of Bound States," **36**:1.

Rajagopal, A. K., "Theory of Inhomogeneous Electron Systems: Spin-Density-Functional Formalism," **41**:59.

Ramana, M. V. and Rajagopal, A. K., "Inhomogeneous Relativistic Electron Systems: A Density Functional Formalism," **54**:231.

Sandars, P. G. H., "A Linked Diagram Treatment of Configuration Interaction in Open-Shell Atoms," **14**:365.

Sinanoğlu, O., "Electron Correlation in Atoms and Molecules," **14**:237.

Sinanoğlu, O., "Many-Electron Theory of Atoms, Molecules, and Their Interactions," **6**:315.

Tolmachev, V. V., "The Field-Theoretic Form of the Perturbation Theory for Many-Electron

Atoms. I. Abstract Theory," **14**:421.

Tolmachev, V. V., "The Field-Theoretic Form of the Perturbation Theory for Many-Electron Atoms. II. Atomic Systems," **14**:471.

Weeks, J. D., Hazi, A., and Rice, S. A., "On the Use of Pseudopotentials in the Quantum Theory of Atoms and Molecules," **16**:283.

Wooley, R. G., "The Electrodynamics of Atoms and Molecules," **33**:153.

Yoshizumi, H., "Correlation Problem in Many-Electron Quantum Mechanics. II. Bibliographical Survey of the Historical Developments with Comments," **2**:323.

Energy conversion

Stucki, J. W., "Optimization of Mitochondrial Energy Conversions," **55**:141.

Energy transfer

Brumer, P., "Intramolecular Energy Transfer: Theories for the Onset of Statistical Behavior," **47**:Part 1:201.

Hertel, I. V., "Collisional Energy-Transfer Spectroscopy with Laser-Excited Atoms in Crossed Atom Beams: A New Method for Investigating the Quenching of Electronically Excited Atoms by Molecules," **45**:341.

Hertel, I. V., "Progress in Electronic-to-Vibrational Energy Transfer," **50**:475.

Houston, P. L., "Electronic to Vibrational Energy Transfer from Excited Halogen Atoms," **47**:Part 2:381.

McGowan, J. W., Kummler, R. H., and Gilmore, F. R., "Excitation

De-excitation Processes Relevant to the Upper Atmosphere," **28**:379.

Moore, C. B., "Vibration → Vibration Energy Transfer," **23**:41.

Rice, S. A., "An Overview of the Dynamics of Intramolecular Transfer of Vibrational Energy," **47**:Part 1:117.

Rice, S. A., "Collision Induced Intramolecular Energy Transfer in Electronically Excited Polyatomic Molecules," **47**:Part 2:237.

Weitz, E. and Flynn, G., "Vibrational Energy Flow in the Ground Electronic States of Polyatomic Molecules," **47**:Part 2:185.

Yang, K., "The Mechanism of Electronic Energy Transfer Between Excited Mercury ($^3P_1$) Atoms and Gaseous Paraffins," **21**:187.

see also Collisional processes; Nonradiative processes; Vibrational relaxation

Entropy

Mead, C. A., "Generalizations of the Entropy Concept: Principles, Limitations, and Prospects for Applications," **51**:113.

Pfeifer, P., "A Deductive Approach to Reduced Dynamics and Entropy in Finite Quantum Systems," **53**:341.

Enzyme kinetics

Walter, C., "The Use of Product Inhibition and Other Kinetic Methods in the Determination of Mechanisms of Enzyme Action," **7**:645.

see also Catalysis

Evolution

Eigen, M., "The Origin and Evolution of Life at the Molecular Level," **55**:119.

Margoliash, E., "The Molecular

Variations of Cytochrome *c* as a Function of the Evolution Species," **29**:191.

Miller, S. L., "The Prebiotic Synthesis of Organic Molecules and Polymers," **55**:85.

Nicolis, G., "Dissipative Instabilities, Structure, and Evolution," **29**:29.

Prigogine, I., "Nonequilibrium Thermodynamics and Chemical Evolution: An Overview," **55**:43.

Zander, R., "Commentary: Observational Aspects Related to the Chemical Evolution of Our Atmosphere," **55**:79.

Excitons

Aslangul, C. and Kottis, P., "Density Operator Description of Excitonic Absorption and Motion in Finite Molecular Systems," **41**:321.

Barriol, J. and Regnier, J., "Calculation of Transition Energies from the Geometry of the System," **8**:5.

Craig, D. P., "Model Calculations in the Theory of Mixed-Crystal Spectra," **8**:27.

Drickamer, H. G. and Zahner, J. C., "The Effect of Pressure on Electronic Structure," **4**:161.

Kearns, D. R., "Electronic Conduction in Organic Molecular Solids," **7**:282.

Philpott, M. R., "Some Aspects of Exciton Theory," **23**:227.

Schnepp, O. and Jacobi, N., "The Lattice Vibrations of Molecular Solids," **22**:205.

Turlet, J. M., Kottis, Ph., and Philpott, M. R., "Polariton and Surface Exciton State Effects in the Photodynamics of Organic Molecular Crystals," **54**:303.

Explosives

see Detonations

F

Flames

Hirschfelder, J. O. and Curtiss, C. F., "Propagation of Flames and Detonations," **3**:59.

Fluorescence

Chance, R. R., Prock, A., and Silbey, R., "Molecular Fluorescence and Energy Transfer Near Interfaces," **37**:1.

Clyne, M. A. A. and McDermid, I. S., "Laser-Induced Fluorescence: Electronically Excited States of Small Molecules," **50**:1.

Haas, Y. and Asscher, M., "Two-Photon Excitation as a Kinetic Tool: Application to Nitric Oxide Fluorescence Quenching," **47**:Part 2:17.

Krause, L., "Sensitized Fluorescence and Quenching," **28**:267.

see also Electronic spectroscopy; Quenching; Radiative processes

Forbidden transitions

Liehr, A. D., "Forbidden Transitions in Organic and Inorganic Systems," **5**:241.

G

Glasses

Angell, C. A., Clarke, J. H. R., and Woodcock, L. V., "Interaction Potentials and Glass Formation: A Survey of Computer Experiments," **48**:397.

Grest, G. S. and Cohen, M. H., "Liquids, Glasses, and the Glass Transition: A Free Volume Approach," **48**:455.

Nagel, S. R., "Metallic Glasses,"

51:227.

Gravel

Winterkorn, H. F., "Analogies Between Macromeritic and Molecular Systems and the Mechanical Properties of Sand and Gravel Assemblies," **21**:751.

Gravitational systems

Haggerty, M. J. and Severne, G., "Kinetic Theory of Gravitational Systems," **35**:119.

H

Hydrogen bonding

Eyring, E. M., Rich, L. D., McCoy, L. L., Graham, R. C., and Taylor, N., "Activation Parameters and Deprotonation Kinetics of Intramolecularly Hydrogen Bonded Acids," **21**:327.

I

Information theory

Levine, R. D., "The Information Theoretic Approach to Intramolecular Dynamics," **47**:Part 1:239.

Infrared spectroscopy

Lecomte, J., "Prospects for the Use of Infrared Spectroscopy in Biology," **7**:408.

Instability

Biot, M. S., "On a Unifed Thermodynamic Approach to a Large Class of Instabilities of Dissipative Continua," **32**:13.

Nicolis, G., "Dissipative Instabilities, Structure, and Evolution," **29**:29.

Pieranski, P. and Guyon, E., "Cylindrical Couette Flow Instabilities

in Nematic Liquid Crystals," **32**:151.

Platten, J. K. and Chavepeyer, G., "Finite Amplitude Instability in the Two-Component Bénard Problem," **32**:281.

Richter, P. H., Procaccia, I., and Ross, J., "Chemical Instabilities," **43**:217.

Thomaes, G., "The Bénard Instability in Liquid Mixtures," **32**:269.

Walgraef, D., Dewel, G., and Borckmans, P., "Nonequilibrium Phase Transitions and Chemical Instabilities," **49**:311.

see also Dissipative structures; Stability

Intercalation compounds

Solin, S. A., "The Nature and Structural Properties of Graphite Intercalation Compounds," **49**:455.

Interfaces

Chance, R. R., Prock, A., and Silbey, R., "Molecular Fluorescence and Energy Transfer Near Interfaces," **37**:1.

Davis, H. T. and Scriven, L. E., "Stress and Structure in Fluid Interfaces," **49**:357.

Harasima, A., "Molecular Theory of Surface Tension," **1**:203.

Jhon, M. S., Dahler, J. S., and Desai, R. C., "Theory of the Liquid-Vapor Interface," **46**:279.

Rice, S. A., Guidotti, D., Lemberg, H. L., Murphy, W. C., and Bloch, A. N., "Some Comments on the Electronic Properties of Liquid Metal Surfaces," **27**:543.

Viaud, P. R., "Theoretical and Experimental Study of Stationary Profiles of a Water-Ice Mobile Solidification Interface," **32**:163.

Vrij, A., Joosten, J. G. H., and

Fijnaut, H. M., "Light Scattering from Thin Liquid Films," **48**:329.

Wiegel, F. W. and Kox, A. J., "Theories of Lipid Monolayers," **41**:195.

Intermolecular forces

Bernstein, R. B. and Muckerman, J. T., "Determination of Intermolecular Forces via Low-Energy Molecular Beam Scattering," **12**:389.

Birnbaum, G., "Microwave Pressure Broadening and Its Application to Intermolecular Forces," **12**:487.

Bloom, M. and Oppenheim, I., "Intermolecular Forces Determined by Nuclear Magnetic Resonance," **12**:549.

Buckingham, A. D., "Permanent and Induced Molecular Moments and Long-Range Intermolecular Forces," **12**:107.

Dalgarno, A., "New Methods for Calculating Long-Range Intermolecular Forces," **12**:143.

Hirschfelder, J. O. and Meath, W. J., "The Nature of Intermolecular Forces," **12**:3.

Kihara, T., "Intermolecular Forces and Equation of State of Gases," **1**:267.

Kihara, T., "Multipolar Interactions in Molecular Crystals," **20**:1.

Kihara, T. and Koide, A., "Intermolecular Forces and Crystal Structures for D_2, N_2, O_2, F_2, and CO_2," **33**:51.

Linder, B., "Reaction Field Techniques and Their Applications to Intermolecular Forces," **12**:225.

Mason, E. A. and Monchick, L., "Methods for the Determination of Intermolecular Forces," **12**:329.

Pitzer, K. S., "Inter- and Intramolecular Forces and Molecular Polarizability," **2**:59.

Power, E. A., "Very Long-Range (Retardation Effect) Intermolecular Forces," **12**:167.

Sinanoğlu, O., "Intermolecular Forces in Liquids," **12**:283.

see also Potential energy surfaces

Ion-cyclotron resonance

Gray, G. A., "Ion Cyclotron Resonance," **19**:141.

Ion-molecule reactions

Dugan, J. V., Jr. and Magee, J. L., "Dynamics of Ion-Molecule Collisions," **21**:207.

Friedman, L. and Reuben, B. G., "A Review of Ion-Molecule Reactions," **19**:33.

Giese, C., "Ion-Neutral Reactions," **10**:247.

Koski, W. S., "Scattering of Positive Ions by Molecules," **30**:185.

Isotope effects

Bigeleisen, J. and Wolfsberg, M., "Theoretical and Experimental Aspects of Isotope Effects in Chemical Kinetics," **1**:15.

Johnston, H. S, "Large Tunnelling Corrections in Chemical Reaction Rates," **3**:131.

Kreevoy, M. M. and Konasewich, D. E., "The Brønsted α and the Primary Hydrogen Isotope Effect. A Test of the Marcus Theory," **21**:243.

K

Kinetic theory

see Statistical mechanics of irreversible processes

Kinetics

see Chemical dynamics

L

Laser induced fluorescence

see Fluorescence; Radiative processes

Lasers

Ben-Shaul, A., "Chemical Laser Kinetics," **47**:Part 2:55.

Bullis, R. H., "Applications to Lasers," **28**:423.

Cantrell, C. D., Makarov, A. A., and Louisell, W. H., "Laser Excitation of SF_6: Spectroscopy and Coherent Pulse Propagation Effects," **47**:Part 1:583.

Dion, D. R. and Hirschfelder, J. O., "Time-Dependent Perturbation of a Two-State Quantum System by a Sinusoidal Field," **35**:265.

Fleming, G. R., "Applications of Continuously Operating, Synchronously Mode-Locked Lasers," **49**:1.

Light scattering

Birnbaum, G., Guillot, B., and Bratos, S., "Theory of Collision - Induced Line Shapes—Absorption and Light Scattering at Low Density," **51**:49.

Boon, J. P., "Light Scattering from Nonequilibrium Fluid Systems," **32**:87.

Fleury, P. A. and Boon, J. P., "Laser Light Scattering in Fluid Systems," **24**:1.

Frommhold, L., "Collision-Induced Scattering of Light and the Diatom Polarizabilities," **46**:1.

Gelbart, W. M., "Depolarized Light

Scattering by Simple Fluids," **26**:1.

Vrij, A., Joosten, J. G. H., and Fijnaut, H. M., "Light Scattering from Thin Liquid Films," **48**:329.

Line shapes

Ben-Reuven, A., "Spectral Lineshapes in Gases in the Binary-Collision Approximation," **33**:235.

Ben-Reuven, A. and Rabin, Y., "*N*-Level Multiple Resonance," **47**:Part 1:555.

Birnbaum, G., "Microwave Pressure Broadening and Its Application to Intermolecular Forces," **12**:487.

Birnbaum, G., Guillot, B., and Bratos, S., "Theory of Collision - Induced Line Shapes—Absorption and Light Scattering at Low Density," **51**:49.

Greer, W. L. and Rice, S. A., "On the Theory of Optical Absorption Profiles in Condensed Systems," **17**:229.

Kubo, R., "A Stochastic Theory of Line Shape," **15**:101.

McGurk, J. C., Schmalz, T. G., and Flygare, W. H., "A Density Matrix, Bloch Equation Description of Infrared and Microwave Transient Phenomena," **25**:1.

Morrey, J. R. and Morgan, L. G., "T_n Frequency Functions as Energy Contours for Photon Absorbance in Condensed Systems," **21**:317.

Liquid crystals

Forster, D., "The Theory of Liquid Crystals," **31**:231.

Pieranski, P. and Guyon, E., "Cylindrical Couette Flow Instabilities in Nematic Liquid Crystals," **32**:151.

Tobolsky, A. V. and Samulski, E. T.,

"Solid 'Liquid-Crystalline' Films of Synthetic Polypeptides: A New State of Matter," **21**:529.

Liquid metals

Ballentine, L. E., "Theory of Electron States in Liquid Metals," **31**:263.

Bellemans, A. and de Leener, M., "Electron Gas in a Lattice of Positive Charges," **6**:85.

McLachlan, D., Jr., " The Gaseous Fraction in Liquid Metals," **21**:499.

Nachtrieb, N. H., "Solutions of Metals in Molten Salts," **31**:465.

Oriani, R. A. "Thermodynamics of Metallic Solutions," **2**:119.

Rice, S. A., Guidotti, D., Lemberg, H. L., Murphy, W. C., and Bloch, A. N., "Some Comments on the Electronic Properties of Liquid Metal Surfaces," **27**:543.

Liquids

Adelman, S. A. and Deutch, J. M., "The Structure of Polar Fluids," **31**:103.

Andersen, H. C., Chandler, D., and Weeks, J. D., "Role of Repulsive and Attractive Forces in Liquids: The Equilibrium Theory of Classical Fluids," **34**:105.

Blum, L. and Narten, A. H., "Diffraction by Molecular Liquids," **34**:203.

Boon, J. P., "Light Scattering from Nonequilibrium Fluid Systems," **32**:87.

Dahler, J. S. and Theodosopulu, M., "The Kinetic Theory of Dense Polyatomic Fluids," **31**:155.

Davis, H. T., "Kinetic Theory of Dense Fluids and Liquids Revisited," **24**:257.

Desai, R. C. and Yip, S., "A Stochastic Model for Neutron Scattering by Simple Liquids," **15**:129.

Egelstaff, P. A., "New Experimental Studies of the Structure of Fluids," **53**:1.

Fleury, P. A. and Boon, J. P., "Laser Light Scattering in Fluid Systems," **24**:1.

Frisch, H. L., "The Equation of State of the Classical Hard Sphere Fluid," **6**:229.

Gelbart, W. M., "Depolarized Light Scattering by Simple Fluids," **26**:1.

Jhon, M. S., "Sound Velocity and van der Waals Force in Liquids According to Significant Structure Theory," **21**:481.

Karnicky, J. F. and Pings, C. J., "Recent Advances in the Study of Liquids by X-ray Diffraction," **34**:157.

Lee, Y., Ree, F. H., and Ree, T., "Distribution Function of Classical Fluids of Hard Spheres. I," **21**:393.

Luks, K. D. and Kozak, J. J., "The Statistical Mechanics of Square-Well Fluids," **37**:139.

Pak, H. and Chang, S., "Transient State Theory of Significant Liquid Structure Applied to Water," **21**:487.

Prigogine, I., Nicolis, G., Allen, P. M., "Eyring's Theory of Viscosity of Dense Media and Nonequilibrium Statistical Mechanics," **21**:473.

Reiss, H., "Scaled Particle Methods in the Statistical Thermodynamics of Fluids," **9**:1.

Schwarz, K. W., "Mobilities of Charge Carriers in Superfluid Helium," **33**:1.

Stillinger, F. H., "Theory and Molecular Models for Water," **31**:1.

Utsugi, H. and Ree, T., "Application of the Absolute Reaction-Rate Theory to Non-Newtonian Flow," **21**:273.

Watts, R. O., Henderson, D., and Baxter, R. J., "Hard Spheres with Surface Adhesion: The Percus-Yevick Approximation and the Energy Equation," **21**:421.

M

Macromolecules

see Polymers

Magnetic circular dichroism

Stephens, P. J., "Magnetic Circular Dichroism," **35**:197.

Magnetic properties

Ehrenberg, A., "Some Recent Developments in the Study of Paramagnetic Species of Biological Interest," **7**:602.

Erber, T., Latal, H. G., and Harmon, B. N., "The Origin of Hysteresis in Simple Magnetic Systems," **20**:71.

Kiode, S. and Oguchi, T., "Theories on the Magnetic Properties of Compounds," **5**:189.

Kotani, M., "Electronic Structure and Magnetic Properties of Hemoproteins, Particularly Hemoglobins," **7**:159.

Philippot, J., "Nuclear Paramagnetic Relaxation in Solids," **11**:289.

Schoffa, G., "Magnetic Susceptibilities and the Chemical Bond in Hemoproteins," **7**:182.

Stephens, P. J., "Magnetic Circular Dichroism," **35**:197.

Weiss, N. O., "Magnetic Fields and Convection," **32**:101.

Many-body problem

Adelman, S. A., "Generalized Langevin Equations and Many-body Problems in Chemical Dynamics," **44**:143.

Kelly, H. P., "Applications of Many-Body Diagram Techniques in Atomic Physics," **14**:129.

Löwdin, P., "Correlation Problem in Many-Electron Quantum Mechanics. I. Review of Different Approaches and Discussion of Some Current Ideas," **2**:207.

Löwdin, P., "Some Recent Developments in the Quantum Theory of Many-Electron Systems and the Correlation Problem," **8**:3.

Sinanoğlu, O., "Many-Electron Theory of Atoms, Molecules, and Their Interactions," **6**:315.

Tolmachev, V. V., "The Field-Theoretic Form of the Perturbation Theory for Many-Electron Atoms. I. Abstract Theory," **14**:421.

Tolmachev, V. V., "The Field-Theoretic Form of the Perturbation Theory for Many-Electron Atoms. II. Atomic Systems," **14**:471.

Yoshizumi, H., "Correlation Problem in Many-Electron Quantum Mechanics. II. Bibliographical Survey of the Historical Developments with Comments," **2**:323.

see also Chemical dynamics

Measurement theory

Kosloff, R., "Thermodynamic Aspects of the Quantum-Mechanical Measurement Process," **46**:153.

Primas, H. and Müller-Herold, U.,

"Quantum-Mechanical System Theory: A Unifying Framework for Observations and Stochastic Processes in Quantum Mechanics," **38**:1.

Membranes

Bolis, L., "Structure and Transport in Biomembranes," **29**:301.

Chen Y., "Noise Analysis of Kinetic Systems and Its Applications to Membrane Channels," **37**:67.

Clayton, R. K., "Photochemical Reaction Centers and Photosynthetic Membranes," **19**:353.

de Levie, R., "Mathematical Modeling of Transport of Lipid-Soluble Ions and Ion-Carrier Complexes Through Lipid Bilayer Membranes," **37**:99.

Delmotte, M., Julien, J., Charlemagne, J., and Chanu, J., "Thermodynamic Considerations of the Excitable Membranes Behavior," **29**:343.

Goodwin, B. C., "A Membrane Model for Polar Transport and Gradient Formation," **29**:269.

Kobatake, Y., "Physiochemical Problems in Excitable Membranes," **29**:319.

Läuger, P., "Ion Transport Through Artificial Lipid Membranes," **29**:309.

Lefever, R. and Deneubourg, J. L., "Membrane Excitation," **29**:349.

McConnell, H., "Dynamic Properties of Membranes: Membrane Immunochemistry," **39**:275.

Monnier, A. M., "Excitability and Ionic Selectivity, Common Properties of Many Lipidic Derivatives," **29**:341.

Simon, W., "Selective Transport Processes in Artificial Membranes," **39**:287.

Thomas, D., "Diffusion-Reaction in Structured Media and Membranes Bearing Enzymes," **39**:229.

Thomas, D., "Functional Organization in Artificial Enzyme Membranes: Accomplishments and Prospects," **29**:113.

Woodbury, J. W., "Eyring Rate Theory Model of the Current-Voltage Relationships of Ion Channels in Excitable Membranes," **21**:601.

Mixtures

Bearman, R. J., Kirkwood, J. G., and Fixman, M., "Statistical-Mechanical Theory of Transport Processes. X. The Heat of Transport in Binary Liquid Solutions," **1**:1.

Bellemans, A., Mathot, V., and Simon, M., "Statistical Mechanics of Mixtures—The Average Potential Model," **11**:117.

Blander, M., "Dimensional Methods in the Statistical Mechanics of Ionic Systems," **11**:83.

Burton, L. L. and Van Artsdalen, E. R., "On the Nature of Solutions of Organic Compounds in Fused Salts," **21**:523.

Hijmans, J. and Holleman, Th., "The Principle of Corresponding States for Chain-molecule Liquids and their Mixtures," **16**:223.

Lamm, O., "Studies in the Kinematics of Isothermal Diffusion. A Macrodynamical Theory of Multicomponent Fluid Diffusion," **6**:295.

Nachtrieb, N. H., "Solutions of Metals in Molten Salts," **31**:465.

Oriani, R. A. "Thermodynamics of Metallic Solutions," **2**:119.

Rowlinson, J. S. and Richardson, M.

J., "The Solubility of Solids in Compressed Gases," **2**:85.

Schneider, G. M., "Phase Equilibria in Fluid Mixtures at High Pressures," **17**:1.

Thomaes, G., "The Bénard Instability in Liquid Mixtures," **32**:269.

van der Waals, J. H. and Platteeuw, J. C., "Clathrate Solutions," **2**:1.

see also Solutions

Molecular beams

Andersen, J. B., Andres, R. P., and Fenn, J. B., "Supersonic Nozzle Beams," **10**:275.

Bederson, B. and Robinson, E. J., "Beam Measurements of Atomic Polarizabilities," **10**:1.

Bernstein, R. B. and Muckerman, J. T., "Determination of Intermolecular Forces via Low-Energy Molecular Beam Scattering," **12**:389.

Bersohn, R. and Lin, S. H., "Orientation of Targets by Beam Excitation," **16**:67.

Eschenroeder, A. Q., "Reaction Kinetics in Hypersonic Flows," **13**:19.

Haberland, H., Lee, Y. T., and Siska, P. E., "Scattering of Noble-Gas Metastable Atoms in Molecular Beams," **45**:487.

Herschbach, D., "Reactive Scattering in Molecular Beams," **10**:319.

Lawley, K. P., "New Directions in Molecular Beams," **30**:1.

Ng, C. Y., "Molecular Beam Photoionization Studies of Molecules and Clusters," **52**:263.

Rice, S. A., "Collision Induced Intramolecular Energy Transfer in Electronically Excited Polyatomic Molecules," **47**:Part 2:237.

Molecular solids

Duchesne, J., "Nuclear Quadrupole Resonance in Irradiated Crystals," **2**:187.

Kearns, D. R., "Electronic Conduction in Organic Molecular Solids," **7**:282.

Meyer, L., "Phase Transitions in van der Waal's Lattices," **16**:343.

Schnepp, O. and Jacobi, N., "The Lattice Vibrations of Molecular Solids," **22**:205.

see also Excitons; Solids

Molecular structure

Bastiansen, O. and Skancke, P. N., "Electron Diffraction in Gases and Molecular Structure," **3**:323.

Brown, J. M. and Jørgensen, F., "Vibronic Energy Levels of a Linear Triatomic Molecule in a Degenerate Electronic State: A Unified Tratment of the Renner-Teller Effect," **52**:117.

Browne, J. C. and Matsen, F. A., "*Ab Initio* Calculations on Small Molecules," **23**:161.

Burdett, J. K., "Predictions of the Structure of Complex Solids," **49**:47.

Carlson, C. M., "A Linear Sum-over-Points Approach for Computing Electronic Energies: Application to He and H_2 Using Conroy Diophantine Points," **21**:49.

Lathouwers, L. and Van Leuven, P., "Generator Coordinate Theory of Nuclear Motion in Molecules," **49**:115.

Somorjai, G. A. and Farrell, H. H., "Low-Energy Electron Diffraction," **20**:215.

Wilson, E. B., Jr., "The Problem of Barriers to Internal Rotation in Molecules," **2**:367.

Molten salts

Burton, L. L. and Van Artsdalen, E. R., "On the Nature of Solutions of Organic Compounds in Fused Salts," **21**:523.

Nachtrieb, N. H., "Solutions of Metals in Molten Salts," **31**:465.

see also Mixtures

Monolayers

Wiegel, F. W. and Kox, A. J., "Theories of Lipid Monolayers," **41**:195.

see also Interfaces

Multiphoton processes

Goodman, L. and Rava, R. P., "Two-Photon Spectroscopy of Perturbed Benzenes," **54**:177.

Haas, Y., "Electronically Excited Fragments Formed by Unimolecular Multiple Photon Dissociation," **47**:Part 1:713.

Haas, Y. and Asscher, M., "Two-Photon Excitation as a Kinetic Tool: Application to Nitric Oxide Fluorescence Quenching," **47**:Part 2:17.

Hall, R. B., Kaldor, A., Cox, D. M., Horsley, A., Rabinowitz, P., Kramer, G. M., Bray, R. G., and Maas, E. T., Jr., "Infrared Laser Chemistry of Complex Molecules," **47**:Part 1:639.

Houston, P. L., "Initiation of Atom-Molecule Reactions by Infrared Multiphoton Dissociation," **47**:Part 1:625.

King, D. S., "Infrared Multiphoton Excitation and Dissociation," **50**:105.

Mukamel, S., "Reduced Equations of Motion for Collisionless Molecular Multiphoton Processes," **47**:Part 1:509.

Quack, M., "Reaction Dynamics and Statistical Mechanics of the Preparation of Highly Excited States by Intense Infrared Radiation," **50**:395.

Reisler, H. and Wittig, C., "Electronic Luminescence Resulting from Infrared Multiple Photon Excitation," **47**:Part 1:679.

Ronn, A. M., "Luminescence of Parent Molecule Induced by Multiphoton Infrared Excitation," **47**:Part 1:661.

Woodin, R. L. and Kaldor, A., "Enhancement of Chemical Reactions by Infrared Lasers," **47**:Part 2:3.

N

Nonequilibrium statistical mechanics

see Statistical mechanics of irreversible processes

Nonequilibrium thermodynamics

see Thermodynamics of irreversible processes

Nonradiative processes

Bondybey, V. E., "Experimental Studies of Nonradiative Processes in Low Temperature Matrices," **47**:Part 2:521.

Bondybey, V. E. and Brus, L. E., "Nonradiative Processes in Small Molecules in Low Temperature Solids," **41**:269.

Freed, K. F., "Collision Induced Intersystem Crossing," **47**:Part 2:291.

Freed, K. F., "Collisional Effects on Electronic Relaxation Processes," **42**:207.

Grecos, A. P., "Solvable Models for Unstable States in Quantum Physics," **38**:143.

Haas, Y. and Asscher, M., "Two-Photon Excitation as a Kinetic Tool: Application to Nitric Oxide Fluorescence Quenching," **47**:Part 2:17.

Krause, L., "Sensitized Fluorescence and Quenching," **28**:267.

Parmenter, C. S., "Radiative and Nonradiative Processes in Benzene," **22**:365.

see also Collisional processes; Energy transfer

Nuclear magnetic resonance

Bloom, M. and Oppenheim, I., "Intermolecular Forces Determined by Nuclear Magnetic Resonance," **12**:549.

Jardetzky, O., "The Study of Specific Molecular Interactions by Nuclear Magnetic Relaxation Methods," **7**:499.

Philippot, J., "Nuclear Paramagnetic Relaxation in Solids," **11**:289.

Nuclear quadrupole resonance

Scrocco, E., "Quantum Mechanical Interpretation of Nuclear Quadrupole Coupling," **5**:319.

Nucleation

Eyal, M., Agam, U., and Grabiner, F. R., "Infrared Laser-Enhanced Diffusion Cloud Reactions," **47**:Part 2:43.

Katz, J. L. and Donohue, M. D., "A Kinetic Approach to Homogeneous Nucleation Theory," **40**:137.

O

Optical activity

Moscowitz, A., "Theoretical Aspects of Optical Activity - Part One:

Small Molecules," **4**:67.

Schuster, P., "Commentary: On the Origin of Optical Activity," **55**:109.

Tinoco, I., Jr., "Theoretical Aspects of Optical Activity - Part Two: Polymers," **4**:113.

see also Chirality

Oxygen atom reactions

Lin, M. C., "Dynamics of Oxygen Atom Reactions," **42**:113.

P

Phase transitions

Coopersmith, M. H., "The Thermodynamic Description of Phase Transitions," **17**:43.

Grest, G. S. and Cohen, M. H., "Liquids, Glasses, and the Glass Transition: A Free Volume Approach," **48**:455.

Kozak, J. J., "Nonlinear Problems in the Theory of Phase Transitions," **40**:229.

Ubbelohde, A. R., "Melting Mechanisms of Crystals," **6**:459.

Viaud, P. R., "Theoretical and Experimental Study of Stationary Profiles of a Water-Ice Mobile Solidification Interface," **32**:163.

Walgraef, D., Dewel, G., and Borckmans, P., "Nonequilibrium Phase Transitions and Chemical Instabilities," **49**:311.

see also Critical phenomena

Photochemistry

Jortner, J. and Levine, R. D., "Photoselective Chemistry," **47**:Part 1:1.

Lau, A. M. F., "The Photon-as-Catalyst Effect in Laser Induced

Predissociation and Autoionization," **50**:191.

Leone, S. R., "Photofragment Dynamics," **50**:255.

Rice, S. A., "Some Comments on the Theory of Photochemical Reactions," **21**:153.

Welge, K. H. and Schmiedl, R., "Doppler Spectroscopy of Photofragments," **47**:Part 2:133.

Photoelectron spectroscopy

Cederbaum, L. S. and Domcke, W., "Theoretical Aspects of Ionization Potentials and Photoelectron Spectroscopy: A Green's Function Approach," **36**:205.

Peshkin, M., "Angular Distributions of Photoelectrons: Consequences of Symmetry," **18**:1.

Plummer, E. W. and Eberhardt, W., "Angle-Resolved Photoemission as a Tool for the Study of Surfaces," **49**:533.

Photosynthesis

Sybesma, C., "Commentary: A Prebiotic Origin of Photosynthesis?," **55**:113.

Picosecond spectroscopy

Fleming, G. R., "Applications of Continuously Operating, Synchronously Mode-Locked Lasers," **49**:1.

Kenney-Wallace, G. A., "Picosecond Spectroscopy and Dynamics of Electron Relaxation Processes in Liquids," **47**:Part 2:535.

Rentzepis, P. M., "Picosecond Spectroscopy and Molecular Relaxation," **23**:189.

Rentzepis, P. M. and Barbara, P. F., "Proton Transfer: A Primary Picosecond Event," **47**:Part 2:627.

Wiersma, D. A., "Coherent Optical Transient Studies of Dephasing and Relaxation in Electronic Transitions of Large Molecules in Condensed Media," **47**:Part 2:421.

Plasmas

Klimontovich, Yu. L., "Kinetic Theory of Plasmas," **38**:193.

Prager, S., "The One-Dimensional Plasma," **4**:201.

Plastic crystals

Zandler, M. E. and Thomson, T. R., "Application of the Significant Structure Theories to Plastic Crystals," **21**:517.

Polymers

Curtiss, C. F., Bird, R. B., and Hassager, O., "Kinetic Theory and Rheology of Macromolecular Solutions," **35**:31.

de Brouckère, L. and Mandel, M., "Dielectric Properties of Dilute Polymer Solutions," **1**:77.

Domb, C., "Self-Avoiding Walks on Lattices," **15**:229.

Freed, K. F., "Functional Integrals and Polymer Statistics," **22**:1.

Gibbs, J. H., "The Kinetics of Biopolymerization on Nucleic Acid Templates," **15**:185.

Klotz, I. M., "Synzymes: Synthetic Polymers with Enzymelike Catalytic Activities," **39**:109.

Mazur, J., "Non-Self Intersecting Random Walks on Lattices with Near-Neighbor Interactions," **15**:261.

Miller, S. L., "The Prebiotic Synthesis of Organic Molecules and Polymers," **55**:85.

Moncrieff-Yeates, M. and White, H. J., Jr., "The Interaction of Cellulose with Simple Salt Solutions and with Dyebaths," **21**:685.

Orwoll, R. A. and Stockmayer, W. H., "Stochastic Models for Chain Dynamics," **15**:305.

Rabinowitz, P., Silberberg, A., Simha, R., and Loftus, E., "Reversible Kinetics on Linear Lattices with Neighbor Effects," **15**:281.

Szwarc, M., "Recent Advances in Polymer Chemistry," **2**:147.

Tinoco, I., Jr., "Theoretical Aspects of Optical Activity - Part Two: Polymers," **4**:113.

Whittington, S. G., "Statistical Mechanics of Polymer Solutions and Polymer Adsorption," **51**:1.

Zwanzig, R., "Langevin Theory of Polymer Dynamics in Dilute Solution," **15**:325.

Porphyrins

Braterman, P. S., Davies, R. C., and Williams, R. J. P., "The Properties of Metal Porphyrin and Similar Complexes," **7**:359.

Potential energy surfaces

Bottcher, C., "Excited-State Potential Surfaces and Their Applications," **42**:169.

Frost, A. A., "The Potential Energy Surface of the H_3 System Using Floating Gaussian Orbitals," **21**:65.

Hirst, D. M., "The Calculation of Potential Energy Surfaces for Excited States," **50**:517.

Kaufman, J. J., "Potential Energy Surface Considerations for Excited State Reactions," **28**:113.

Le Roy, R. J. and Carley, J. S., "Spectroscopy and Potential Energy Surfaces of van der Waals Molecules," **42**:353.

Mueller, C. R., Smith, B., McGuire, P., Williams, W., Chakraborti, P.,

and Penta, J., "Intermolecular Potentials and Macroscopic Properties of Argon and Neon from Differential Cross Sections," **21**:369.

Truhlar, D. G. and Wyatt, R. E., "H + H_2: Potential-Energy Surfaces and Elastic and Inelastic Scattering," **36**:141.

Tully, J. C., "Semiempirical Diatomics-in-Molecules Potential Energy Surfaces," **42**:63.

see also Intermolecular forces

Predator-prey systems

Walter, C., "The Global Stability of Prey-Predator Systems with Second-Order Dissipation," **29**:125.

Pressure gradients

Tamayama, M., "On the Pressure Gradient in a Cylindrical Sample Container under High Pressure," **21**:737.

Prophecy

Hirschfelder, J. O., "A Forecast for Theoretical Chemistry," **21**:73.

Proteins

Braams, R. and Van Herpen, G., "The Effects of Ionizing Radiations on Some Fibrous Proteins," **7**:259.

Campbell, I. D., Dobson, C. M., and Williams, R. J. P., "Structures and Energetics of Proteins and Their Active Sites," **39**:55.

Cerf, R., "Cooperative Conformational Kinetics of Synthetic and Biological Chain Molecules," **33**:73.

Curtis, D. H. and Spikes, J. D., "Kinetics of the Interactions of Formaldehyde, Acetaldehyde, and Acrolein with Rattail Tendon," **21**:557.

Dougherty, T. F. and Dolowitz, D.

A., "Physiologic Actions of Heparin not Related to Blood Clotting," **21**:633.

Eley, D. D. and Leslie, R. B., "Adsorption of Water on Solid Proteins with Special Reference to Haemoglobin," **7**:238.

Ikegami, A., "Statistical Thermodynamics of Proteins and Protein Denaturation," **46**:363.

Kotani, M., "Electronic Structure and Magnetic Properties of Hemoproteins, Particularly Hemoglobins," **7**:159.

Lumfry, R., "Protein Conformations, 'Rack' Mechanisms and Water," **21**:567.

Margoliash, E., "The Molecular Variations of Cytochrome c as a Function of the Evolution Species," **29**:191.

Pollard, E. C., "Thermal Effects on Protein, Nucleic Acid, and Viruses," **7**:201.

Rigby, B. J., "The Thermal Stability of Collagen: Its Significance in Biology and Physiology," **21**:537.

Schoffa, G., "Magnetic Susceptibilities and the Chemical Bond in Hemoproteins," **7**:182.

Proton transfer

Huppert, D., Gutman, M., and Kaufmann, K. J., "Laser Studies of Proton Transfer," **47**:Part 2:643.

Rentzepis, P. M. and Barbara, P. F., "Proton Transfer: A Primary Picosecond Event," **47**:Part 2:627.

Q

Quenching

Breckenridge, W. H. and Umemoto, H., "Collisional Quenching of Electronically Excited Metal Atoms," **50**:325.

Haas, Y. and Asscher, M., "Two-Photon Excitation as a Kinetic Tool: Application to Nitric Oxide Fluorescence Quenching," **47**:Part 2:17.

Hertel, I. V., "Collisional Energy-Transfer Spectroscopy with Laser-Excited Atoms in Crossed Atom Beams: A New Method for Investigating the Quenching of Electronically Excited Atoms by Molecules," **45**:341.

Krause, L., "Sensitized Fluorescence and Quenching," **28**:267.

R

Radiation chemistry

Braams, R. and Van Herpen, G., "The Effects of Ionizing Radiations on Some Fibrous Proteins," **7**:259.

Jagger, J., "Photoprotection from Far Ultraviolet Effects in Cells," **7**:584.

Radiative processes

Chance, R. R., Prock, A., and Silbey, R., "Molecular Fluorescence and Energy Transfer Near Interfaces," **37**:1.

Clyne, M. A. A. and McDermid, I. S., "Laser-Induced Fluorescence: Electronically Excited States of Small Molecules," **50**:1.

Menziger, M., "Electronic Chemi-

luminescence in Gases," **42**:1.

Parmenter, C. S., "Radiative and Nonradiative Processes in Benzene," **22**:365.

see also Fluorescence; Multiphoton processes; Quenching

Reaction networks

Clarke, B. L., "Stability of Complex Reaction Networks," **43**:1.

Reactivity

see Chemical dynamics

Recombination dynamics

Mahan, B. H., "Recombination of Gaseous Ions," **23**:1.

Renner-Teller effect

Brown, J. M. and Jorgensen, F., "Vibronic Energy Levels of a Linear Triatomic Molecule in a Degenerate Electronic State: A Unified Tratment of the Renner-Teller Effect," **52**:117.

S

Sand

Winterkorn, H. F., "Analogies Between Macromeritic and Molecular Systems and the Mechanical Properties of Sand and Gravel Assemblies," **21**:751.

Scattering

Amdur, I. and Jordan, J. E., "Elastic Scattering of High-Energy Beams: Repulsive Forces," **10**:29.

Bernstein, R. B., "Quantum Effects in Elastic Molecular Scattering," **10**:75.

Bernstein, R. B. and Muckerman, J. T., "Determination of Intermolecular Forces via Low-Energy Molecular Beam Scattering," **12**:389.

Buck, U., "Elastic Scattering," **30**:313.

Desai, R. C. and Yip, S., "A Stochastic Model for Neutron Scattering by Simple Liquids," **15**:129.

Eno, L. and Rabitz, H., "Sensitivity Analysis and Its Role in Quantum Scattering," **51**:177.

Greene, E. F., Moursund, A. L., and Ross, J., "Elastic Scattering in Chemically Reactive Systems," **10**:135.

Grice, R., "Reactive Scattering," **30**:247.

Haas, Y., "Electronically Excited Fragments Formed by Unimolecular Multiple Photon Dissociation," **47**:Part 1:713.

Haberland, H., Lee, Y. T., and Siska, P. E., "Scattering of Noble-Gas Metastable Atoms in Molecular Beams," **45**:487.

Herschbach, D., "Reactive Scattering in Molecular Beams," **10**:319.

Koski, W. S., "Scattering of Positive Ions by Molecules," **30**:185.

Loesch, H. J., "Scattering of Non-Spherical Molecules," **42**:421.

Micha, D. A., "Quantum Theory of Reactive Molecular Collisions," **30**:7.

Mortensen, E. M., "Permeabilities for Reactions of the Type $H + H_2 = H_2 + H$ Treated as a Linear Encounter Using Variational and Distorted Wave Techniques," **21**:127.

Reuss, J., "Scattering from Oriented Molecules," **30**:389.

Taylor, H. S., "Models, Interpretations, and Calculations Concerning Resonant Electron Scattering Processes in Atoms and Molecules," **18**:91.

Truhlar, D. G., Abdallah, J., Jr., and Smith, R. G., "Algebraic Variational Methods in Scattering Theory," **25**:211.

Truhlar, D. G., Mead, C. A., and Brandt, M. A., "Time-Reversal Invariance, Representations for Scattering Wave Functions, Symmetry of the Scattering Matrix, and Differential Cross Sections," **33**:295.

Truhlar, D. G. and Wyatt, R. E., "H + H_2: Potential-Energy Surfaces and Elastic and Inelastic Scattering," **36**:141.

see also Collisional processes; Light scattering; Molecular beams

Solids

Allnatt, A. R., "Statistical Mechanics of Point-Defect Interactions in Solids," **11**:1.

Burdett, J. K., "Predictions of the Structure of Complex Solids," **49**:47.

McAllister, P. V. and Cutler, I. B., "The Kinetics of the Thermal Grooving of Solids," **21**:669.

Solin, S. A., "The Nature and Structural Properties of Graphite Intercalation Compounds," **49**:455.

Weber, G. R. and Eyring, L., "Self-Diffusion of Oxygen in Iota Phase Praseodymium Oxide," **21**:253.

see also Crystals; Excitons; Molecular solids

Solitons

Collins, M. A., "Solitons in Chemical Physics," **53**:225.

Solutions

Adelman, S. A., "Chemical Reaction Dynamics in Liquid Solution," **53**:61.

Curtiss, C. F., Bird, R. B., and Hassager, O., "Kinetic Theory and Rheology of Macromolecular Solutions," **35**:31.

Das, T. P., "Structure and Properties of Metal-Ammonia Solutions," **4**:303.

de Brouckère, L. and Mandel, M., "Dielectric Properties of Dilute Polymer Solutions," **1**:77.

Moncrieff-Yeates, M. and White, H. J., Jr., "The Interaction of Cellulose with Simple Salt Solutions and with Dyebaths," **21**:685.

Oriani, R. A. "Thermodynamics of Metallic Solutions," **2**:119.

van der Waals, J. H. and Platteeuw, J. C., "Clathrate Solutions," **2**:1.

Whittington, S. G., "Statistical Mechanics of Polymer Solutions and Polymer Adsorption," **51**:1.

Zwanzig, R., "Langevin Theory of Polymer Dynamics in Dilute Solution," **15**:325.

see also Mixtures

Solvated electrons

Brodsky, A. M. and Tsarevsky, A. V., "On a Theoretical Description of Solvated Electrons," **44**:483.

Das, T. P., "Structure and Properties of Metal-Ammonia Solutions," **4**:303.

Davis, H. T. and Brown, R. G., "Low-Energy Electrons in Nonpolar Fluid Films," **31**:329.

Kenney-Wallace, G. A., "Picosecond Spectroscopy and Dynamics of Electron Relaxation Processes in Liquids," **47**:Part 2:535.

Schwarz, K. W., "Mobilities of Charge Carriers in Superfluid Helium," **33**:1.

Solvay conferences

Amaldi, E., "The Solvay Conferences in Physics," **55**:7.

Solvay, J., "The Solvay Conferences and the International Cooperation in Science," **55**:1.

Ubbelohde, A. R., "The Solvay Conferences in Chemistry, 1922–1976," **55**:37.

Spectral density

Gordon, R. G., "Error Bounds and Spectral Densities," **15**:79.

Stability

Nicolis, G., "Stability and Dissipative Structures in Open Systems Far from Equilbrium," **19**:209.

Perdang, J., "Stellar Evolutionary Stability in Relation to Spectral Theory," **32**:207.

Roberts, P. H., "Concepts in Hydrodynamic Stability Theory," **32**:17.

Walter, C., "The Global Stability of Prey-Predator Systems with Second-Order Dissipation," **29**:125.

see also Dissipative structures; Instability

Statistical mechanics

Adelman, S. A., "Generalized Langevin Equations and Many-body Problems in Chemical Dynamics," **44**:143.

Bellemans, A., Mathot, V., and Simon, M., "Statistical Mechanics of Mixtures—The Average Potential Model," **11**:117.

Blander, M., "Dimensional Methods in the Statistical Mechanics of Ionic Systems," **11**:83.

Burke, T. and Siegel, A., "A New Series Expansion for Path Integrals in Statistical Mechanics," **15**:355.

Dagonnier, R., "Theory of Quantum Brownian Motion," **16**:1.

Emch, G. G., "An Introduction to C*-Algebraic Methods in Physics," **22**:315.

Ford, J., "The Transition from Analytic Dynamics to Statistical Mechanics," **24**:155.

Friedmann, H., "The Quantum Mechanical Distribution Function of Molecular Systems: Translational and Rotational Motions," **4**:225.

Girardeau, M. D. and Mazo, R. M., "Variational Methods in Statistical Mechanics," **24**:187.

Guth, E., "Brownian Motion and Indeterminacy Relations," **15**:363.

Haggerty, M. J. and Severne, G., "Kinetic Theory of Gravitational Systems," **35**:119.

Hijmans, J. and Holleman, Th., "The Principle of Corresponding States for Chain-molecule Liquids and their Mixtures," **16**:223.

Hoare, M. R., "The Linear Gas," **20**:135.

Kerner, E. H., "Statistical-Mechanical Theories in Biology," **19**:325.

Kihara, T., "Convex Molecules in Gaseous and Crystalline States," **5**:147.

Kilpatrick, J. E., "The Computation of Virial Coefficients," **20**:39.

Kitahara, K., "The Hamilton-Jacobi-Equation Approach to Fluctuation Phenomena," **29**:85.

Lin, S. H. and Lin Ma, C. Y., "Calculation of Statistical Complexions of Polyatomic Molecules and Ions," **21**:143.

Mazur, J., "Non-Self Intersecting Random Walks on Lattices with Near-Neighbor Interactions," **15**:261.

Mazur, P., "On Statistical Mechanics and Electromagnetic Properties of

Matter," **1**:309.

Miller, R. H., "Equilibrium in Stellar Systems," **26**:107.

Perdang, J., "Stellar Evolutionary Stability in Relation to Spectral Theory," **32**:207.

Reiss, H., "Scaled Particle Methods in the Statistical Thermodynamics of Fluids," **9**:1.

Rowlinson, J. S., "Penetrable Sphere Models of Liquid-Vapor Equilibrium," **41**:1.

Steele, W. A., "The Rotation of Molecules in Dense Phases," **34**:1.

Stell, G., Patey, G. N., and Høye, J. S., "Dielectric Constants of Fluid Models: Statistical Mechanical Theory and Its Quantitative Implementation," **48**:183.

Stillinger, F. H., "Theory and Molecular Models for Water," **31**:1.

van Kampen, N. G., "Thermal Fluctuations in Nonlinear Systems," **15**:65.

Verdier, P. H., "Fluctuations in Autocorrelation Functions in Diffusing Systems," **15**:137.

Statistical mechanics of irreversible processes

Balescu, R. and Résibois, P., "Aspects of Kinetic Theory," **38**:173.

Bearman, R. J., Kirkwood, J. G., and Fixman, M., "Statistical-Mechanical Theory of Transport Processes. X. The Heat of Transport in Binary Liquid Solutions," **1**:1.

Berne, B. J. and Harp, G. D., "On the Calculation of Time Correlation Functions," **17**:63.

Boon, J. P., "Light Scattering from Nonequilibrium Fluid Systems," **32**:87.

Brocas, J., " On the Comparison between Generalized Boltzmann Equations," **11**:317.

Collins, F. C. and Raffel, H., "Transport Processes in Liquids," **1**:135.

Dagonnier, R., "Theory of Quantum Brownian Motion," **16**:1.

Dahler, J. S. and Theodosopulu, M., "The Kinetic Theory of Dense Polyatomic Fluids," **31**:155.

Davis, H. T., "Kinetic Theory of Dense Fluids and Liquids Revisited," **24**:257.

Erber, T., Latal, H. G., and Harmon, B. N., "The Origin of Hysteresis in Simple Magnetic Systems," **20**:71.

Evans, M., Evans, G., and Davies, R., "Experimental and Theoretical Studies of Rototranslational Correlation Functions," **44**:255.

Frisch, H. L., "Commentary: Stochastic Models of Synthesis of Asymmetric Forms," **55**:201.

Kapral, R., "Kinetic Theory of Chemical Reactions in Liquids," **48**:71.

Koopman, B. O., "Relaxed Motion in Irreversible Molecular Statistics," **15**:37.

McLennan, J. A., Jr., "The Formal Statistical Theory of Transport Processes," **5**:261.

Mead, C. A., "Generalizations of the Entropy Concept: Principles, Limitations, and Prospects for Applications," **51**:113.

Nicolis, G., "Local Potential Methods in the Study of Kinetic Equations," **13**:299.

Ono, S., "Variational Principles in Thermodynamics and Statistical Mechanics of Irreversible Processes," **3**:267.

Prigogine, I., "Quantum States and

Dissipative Systems," **15**:11.

Schlögl, F., "Glansdorff-Prigogine Criterion and Statistical Theory," **32**:61.

Weber, G. R. and Eyring, L., "Self-Diffusion of Oxygen in Iota Phase Praseodymium Oxide," **21**:253.

Stellar systems

Miller, R. H., "Equilibrium in Stellar Systems," **26**:107.

Perdang, J., "Stellar Evolutionary Stability in Relation to Spectral Theory," **32**:207.

Thomas, R. N., "Stellar Atmospheres, Nonequilibrium Thermodynamics, and Irreversibility," **32**:259.

Stochastic processes

Desai, R. C. and Yip, S., "A Stochastic Model for Neutron Scattering by Simple Liquids," **15**:129.

McQuarrie, D. A., "Stochastic Theory of Chemical Rate Processes," **15**:149.

Montroll, E. W., "Enzyme Cascades and Their Control in Blood Plasma," **26**:145.

Montroll, E. W. and Shuler, K. E., "The Application of the Theory of Stochastic Processes to Chemical Kinetics," **1**:361.

Oppenheim, I., "Probabilistic and Dynamic Approaches to the Theory of Stochastic Processes," **15**:1.

Ree, F. H., Ree, T. S., Ree, T., and Eyring, H., "Random Walk and Related Physical Problems," **4**:1.

van Kampen, N. G., "The Expansion of the Master Equation," **34**:245.

Weiss, G. H., "First Passage Time Problems in Chemical Physics," **13**:1.

Weiss, G. H., "Some Models for the Decay of Initial Correlations in Dynamical Systems," **15**:199.

Weiss, G. H. and Rubin, R. J., "Random Walks: Theory and Selected Applications," **52**:363.

Surface tension

Harasima, A., "Molecular Theory of Surface Tension," **1**:203.

Surfaces

Bodé, D. D., Jr., "The Calculated Heat of Adsorption of Water on Mercury, Silver, Gold, and Platinum," **21**:361.

Duke, C. B., "Determination of the Structure and Properties of Solid Surfaces by Electron Diffraction and Emission," **27**:1.

Economou, E. N. and Ngai, K. L., "Surface Plasma Oscillations and Related Surface Effects in Solids," **27**:265.

Eriksson, J. C., "On the Thermodynamics of Surface Systems," **6**:145.

Kliewer, K. L. and Fuchs, R., "Theory of Dynamical Properties of Dielectric Surfaces," **27**:355.

Koutecký, J., "Quantum Chemistry of Crystal Surfaces," **9**:85.

Laforgue, A., Rousseau, J., and Imelik, B., "Chemisorption of Ethylene on Metals and Catalytic Hydrogenation," **8**:141.

Plummer, E. W. and Eberhardt, W., "Angle-Resolved Photoemission as a Tool for the Study of Surfaces," **49**:533.

Rajagopal, A. K., "Theory of Inhomogeneous Electron Systems: Spin-Density-Functional Formalism," **41**:59.

Ramana, M. V. and Rajagopal, A. K., "Inhomogeneous Relativistic Electron Systems: A Density

Functional Formalism," **54**:231.

Rice, S. A., Guidotti, D., Lemberg, H. L., Murphy, W. C., and Bloch, A. N., "Some Comments on the Electronic Properties of Liquid Metal Surfaces," **27**:543.

Rowlinson, J. S., "Penetrable Sphere Models of Liquid-Vapor Equilibrium," **41**:1.

Somorjai, G. A. and Farrell, H. H., "Low-Energy Electron Diffraction," **20**:215.

see also Interfaces

T

Thermodynamics of irreversible processes

Bak, T. A., "Nonlinear Problems in Thermodynamics of Irreversible Processes," **3**:33.

Baranowski, B., de Vries, A. E., Haring, A., and Paul, R., "Thermal Diffusion in Systems with some Transformable Components," **16**:101.

Bolis, L., "Structure and Transport in Biomembranes," **29**:301.

Chanu, J., "Thermal Diffusion of Halides in Aqueous Solution," **13**:349.

Coleman, B. D., "Thermodynamics of Discrete Mechanical Systems with Memory," **24**:95.

Edelen, D. G. B., "The Thermodynamics of Evolving Chemical Systems and the Approach to Equilibrium," **33**:399.

Eigen, M., "How Does Information Originate? Principles of Biological Self-Organization," **38**:211.

Eigen, M., "The Origin and Evolution of Life at the Molecular

Level," **55**:119.

Lamm, O., "Studies in the Kinematics of Isothermal Diffusion. A Macrodynamical Theory of Multicomponent Fluid Diffusion," **6**:295.

Ono, S., "Variational Principles in Thermodynamics and Statistical Mechanics of Irreversible Processes," **3**:267.

Prigogine, I., "Nonequilibrium Thermodynamics and Chemical Evolution: An Overview," **55**:43.

Spohn, H. and Lebowitz, J. L., "Irreversible Thermodynamics for Quantum Systems Weakly Coupled to Thermal Reservoirs," **38**:109.

Suzuki, M., "Passage from an Initial Unstable State to a Final Stable State," **46**:195.

Thomas, R. N., "Stellar Atmospheres, Nonequilibrium Thermodynamics, and Irreversibility," **32**:259.

Yao, S. J. and Zwolinski, B. J., "Studies on Rates of Nonequilibrium Processes," **21**:91.

Toeplitz determinants

Fisher, M. E. and Hartwig, R. E., "Toeplitz Determinants: Some Applications, Theorems, and Conjectures," **15**:333.

Transport processes

Bearman, R. J., Kirkwood, J. G., and Fixman, M., "Statistical-Mechanical Theory of Transport Processes. X. The Heat of Transport in Binary Liquid Solutions," **1**:1.

Bolis, L., "Structure and Transport in Biomembranes," **29**:301.

Collins, F. C. and Raffel, H., "Transport Processes in Liquids," **1**:135.

de Levie, R., "Mathematical Model-

ing of Transport of Lipid-Soluble Ions and Ion-Carrier Complexes Through Lipid Bilayer Membranes," **37**:99.

Goodwin, B. C., "A Membrane Model for Polar Transport and Gradient Formation," **29**:269.

Läuger, P., "Ion Transport Through Artificial Lipid Membranes," **29**:309.

McLennan, J. A., Jr., "The Formal Statistical Theory of Transport Processes," **5**:261.

Simon, W., "Selective Transport Processes in Artificial Membranes," **39**:287.

see also Statistical mechanics of irreversible processes

V

Vacuum ultraviolet spectroscopy

Wallace, S. C., "Nonlinear Optics and Laser Spectroscopy in the Vacuum Ultraviolet," **47**:Part 2:153.

Valence theory

Blyholder, G. and Coulson, C. A., "Basis of Extended Hückel Formalism," **21**:25.

Carlson, C. M., "A Linear Sum-over-Points Approach for Computing Electronic Energies: Application to He and H_2 Using Conroy Diophantine Points," **21**:49.

Coulson, C. A. and MacLagan, R. G. A. R., "Notes on the Two-Particle Density Matrix in π-Electron Theory," **21**:303.

de Brouckere, G., "Calculations of Observables in Metallic Complexes by the Molecular Orbital Theory," **37**:203.

Ferreira, R., "Electronegativity and Chemical Bonding," **13**:55.

Hartmann, H., "New Developments in the One-Electron Theory of π-Electron Systems," **5**:1.

Hecht, H., and Lewis, W. B., and Eastman, M. P., "Covalency Effects in Octahedral $5f^1$ Complexes," **21**:351.

Herman, M. F., Freed, K. F., and Yeager, D. L., "Analysis and Evaluation of Ionization Potentials, Electron Affinities, and Excitation Energies by the Equations of Motion—Green's Function Method," **48**:1.

Houtman, J. A. and Duffey, G. H., "A Free-Electron Study of Common Boron Frameworks," **21**:7.

Jørgensen, C. K., "Applications of Ligand-Field Theory to Complexes," **8**:47.

Jørgensen, C. K., "Spectroscopy of Transition-Group Complexes," **5**:33.

Kende, A., "A Self-Consistent Field Molecular Orbital Treatment of Carbonyl Base Strength," **8**:133.

Longuet-Higgins, H. C., "Recent Developments in Molecular Orbital Theory," **1**:239.

Nicholson, B. J., "Approximate Molecular Orbital Theories," **18**:249.

van der Waals molecules

Beswick, J. A. and Jortner, J., "Intramolecular Dynamics of van der Waals Molecules," **47**:Part 1:363.

Le Roy, R. J. and Carley, J. S., "Spectroscopy and Potential Energy Surfaces of van der Waals Molecules," **42**:353.

Levy, D. H., "van der Waals Molecules," **47**:Part 1:323.

Vibrational relaxation

Bak, T. A. and Sørensen, G. P., "Vibrational Relaxation of a Gas of Diatomic Molecules," **15**:219.

Diestler, D., "Theoretical Studies of Vibrational Relaxation of Small Molecules in Dense Media," **42**:305.

Oxtoby, D. W., "Dephasing of Molecular Vibrations in Liquids," **40**:1.

Oxtoby, D. W., "Vibrational Population Relaxation in Liquids," **47**:Part 2:487.

Sage, M. L. and Jortner, J., "Bond Modes," **47**:Part 1:293.

see also Collisional processes; Energy transfer

Vibrational spectroscopy

Carney, G. D., Sprandel, L. L., and Kern, C. W., "Variational Approaches to Vibration-Rotation Spectroscopy for Polyatomic Molecules," **37**:305.

Weinstock, B. and Goodman, G. L., "Vibrational Properties of Hexafluoride Molecules," **9**:169.

Virial coefficients

Kilpatrick, J. E., "The Computation of Virial Coefficients," **20**:39.

W

Water

Pak, H. and Chang, S., "Transient State Theory of Significant Liquid Structure Applied to Water," **21**:487.

Stillinger, F. H., "Theory and Molecular Models for Water," **31**:1.

AUTHOR INDEX

A

Abdallah, J., Jr., *see* Truhlar, D. G., 25:211.

Adelman, S. A., "Chemical Reaction Dynamics in Liquid Solution," 53:61.

Adelman, S. A., "Generalized Langevin Equations and Many-body Problems in Chemical Dynamics," 44:143.

Adelman, S. A. and Deutch, J. M., "The Structure of Polar Fluids," 31:103.

Agam, U., *see* Eyal, M., 47:Part 2:43.

Allen, P. M., *see* Prigogine, I., 21:473.

Allnatt, A. R., "Statistical Mechanics of Point-Defect Interactions in Solids," 11:1.

Amaldi, E., "The Solvay Conferences in Physics," 55:7.

Amdur, I. and Jordan, J. E., "Elastic Scattering of High-Energy Beams: Repulsive Forces," 10:29.

Amme, R. C., "Vibrational and Rotational Excitation in Gaseous Collisions," 28:171.

Andersen, H. C., Chandler, D., and Weeks, J. D., "Role of Repulsive and Attractive Forces in Liquids: The Equilibrium Theory of Clas-

sical Fluids," 34:105.

Andersen, J. B., Andres, R. P., and Fenn, J. B., "Supersonic Nozzle Beams," 10:275.

Anderson, J. B., "The Reaction $F + H_2 \rightarrow HF + H$," 41:229.

Andres, R. P., *see* Andersen, J. B., 10:275.

Angell, C. A., Clarke, J. H. R., and Woodcock, L. V., "Interaction Potentials and Glass Formation: A Survey of Computer Experiments," 48:397.

Aroeste, H., "Toward an Analytic Theory of Chemical Reactions," 6:1.

Aslangul, C. and Kottis, P., "Density Operator Description of Excitonic Absorption and Motion in Finite Molecular Systems," 41:321.

Asscher, M., *see* Haas, Y., 47:Part 2:17.

Atkins, P. W. and Evans, G. T., "Theories of Chemically Induced Electron Spin Polarization," 35:1.

Baede, A. P. M., "Charge Transfer Between Neutrals at Hyperthermal Energies," 30:463.

Baer, M., "A Review of Quantum-Mechanical Approximate Treatments of Three-Body Reactive

35

Systems," **49**:191.

Bak, T. A., "Nonlinear Problems in Thermodynamics of Irreversible Processes," **3**:33.

Bak, T. A. and Sørensen, G. P., "Vibrational Relaxation of a Gas of Diatomic Molecules," **15**:219.

Balescu, R. and Résibois, P., "Aspects of Kinetic Theory," **38**:173.

Balint-Kurti, G. G., "Potential Energy Surfaces for Chemical Reaction," **30**:137.

Ball, J. G. and Himmelblau, D. M., "The Local Potential Applied to Instability Problems," **13**:267.

Ballentine, L. E., "Theory of Electron States in Liquid Metals," **31**:263.

Baranowski, B., de Vries, A. E., Haring, A., and Paul, R., "Thermal Diffusion in Systems with some Transformable Components," **16**:101.

Barbara, P. F., *see* **Rentzepis, P. M.,** **47**:Part 2:627.

Baronavski, A., Umstead, M. E., and Lin, M. C., "Laser Diagnostics of Reaction Product Energy Distributions," **47**:Part 2:85.

Barriol, J. and Regnier, J., "Calculation of Transition Energies from the Geometry of the System," **8**:5.

Basilevsky, M. V., "Transition State Stabilization Energy as a Measure of Chemical Reactivity," **33**:345.

Bass, L. B. and Moore, W. J., "Theory of Rate Processes Applied to Release of Acetylcholine at Synapses," **21**:619.

Bastiansen, O. and Skancke, P. N., "Electron Diffraction in Gases and Molecular Structure," **3**:323.

Baxter, R. J., *see* **Watts, R. O.,** **21**:421.

Bearman, R. J., Kirkwood, J. G., and Fixman, M., "Statistical-Mechanical Theory of Transport Processes. X. The Heat of Transport in Binary Liquid Solutions," **1**:1.

Bederson, B. and Robinson, E. J., "Beam Measurements of Atomic Polarizabilities," **10**:1.

Bellemans, A. and de Leener, M., "Electron Gas in a Lattice of Positive Charges," **6**:85.

Bellemans, A., Mathot, V., and Simon, M., "Statistical Mechanics of Mixtures—The Average Potential Model," **11**:117.

Ben-Reuven, A., "Spectral Lineshapes in Gases in the Binary-Collision Approximation," **33**:235.

Ben-Reuven, A. and Rabin, Y., "N-Level Multiple Resonance," **47**:Part 1:555.

Ben-Shaul, A., "Chemical Laser Kinetics," **47**:Part 2:55.

Bernasek, S. L., "Heterogeneous Reaction Dynamics," **41**:477.

Berne, B. J. and Harp, G. D., "On the Calculation of Time Correlation Functions," **17**:63.

Bernstein, R. B., "Quantum Effects in Elastic Molecular Scattering," **10**:75.

Bernstein, R. B. and Muckerman, J. T., "Determination of Intermolecular Forces via Low-Energy Molecular Beam Scattering," **12**:389.

Bersohn, R. and Lin, S. H., "Orientation of Targets by Beam Excitation," **16**:67.

Besserdich, H., *see* **Kahrig, E.,** **29**:133.

Beswick, J. A. and Jortner, J., "Intramolecular Dynamics of van der Waals Molecules," **47**:Part 1:363.

Bigeleisen, J. and Wolfsberg, M., "Theoretical and Experimental Aspects of Isotope Effects in Chemical Kinetics," **1**:15.

Biot, M. S., "On a Unified Thermodynamic Approach to a Large Class of Instabilities of Dissipative Continua," **32**:13.

Bird, R. B., *see* **Curtiss, C. F., 35**:31.

Birnbaum, G., "Microwave Pressure Broadening and Its Application to Intermolecular Forces," **12**:487.

Birnbaum, G., Guillot, B., and Bratos, S., "Theory of Collision - Induced Line Shapes—Absorption and Light Scattering at Low Density," **51**:49.

Biskupič, S., *see* **Kvasnička, V., 52**:181.

Blander, M., "Dimensional Methods in the Statistical Mechanics of Ionic Systems," **11**:83.

Bloch, A. N., *see* **Rice, S. A., 27**:543.

Bloom, M. and Oppenheim, I., "Intermolecular Forces Determined by Nuclear Magnetic Resonance," **12**:549.

Blum, L. and Narten, A. H., "Diffraction by Molecular Liquids," **34**:203.

Blyholder, G. and Coulson, C. A., "Basis of Extended Hückel Formalism," **21**:25.

Bock, R. M., "Commentary: On Paper by Manfred Eigen," **55**:139

Bodé, D. D., Jr., "The Calculated Heat of Adsorption of Water on Mercury, Silver, Gold, and Platinum," **21**:361.

Boiteux, A., *see* **Hess, B., 29**:137.

Bolis, L., "Structure and Transport in Biomembranes," **29**:301.

Bondybey, V. E., "Experimental Studies of Nonradiative Processes in Low Temperature Matrices," **47**:Part 2:521.

Bondybey, V. E. and Brus, L. E., "Nonradiative Processes in Small Molecules in Low Temperature Solids," **41**:269.

Boon, J. P., "Light Scattering from Nonequilibrium Fluid Systems," **32**:87.

Boon, J. P., "Theoretical Models for Bacterial Motion and Chemotaxis," **29**:169.

Boon, J. P., *see* **Fleury, P. A., 24**:1.

Borckmans, P., *see* **Walgraef, D., 49**:311.

Bottcher, C., "Excited-State Potential Surfaces and Their Applications," **42**:169.

Bouman, T. D., *see* **Hansen, A. E., 44**:545.

Braams, R. and Van Herpen, G., "The Effects of Ionizing Radiations on Some Fibrous Proteins," **7**:259.

Brandt, M. A., *see* **Truhlar, D. G., 33**:295.

Braterman, P. S., Davies, R. C., and Williams, R. J. P., "The Properties of Metal Porphyrin and Similar Complexes," **7**:359.

Bratos, S., *see* **Birnbaum, G., 51**:49.

Bray, R. G., *see* **Hall, R. B., 47**:Part 1:639.

Breckenridge, W. H. and Umemoto, H., "Collisional Quenching of Electronically Excited Metal Atoms," **50**:325.

Brion, C. E. and Hamnett, A., "Continuum Optical Oscillator-Strength Measurements by Electron Spectroscopy in the Gas Phase," **45**:2.

Brocas, J., " On the Comparison between Generalized Boltzmann Equations," **11**:317.

Brodsky, A. M. and Tsarevsky, A. V., "On a Theoretical Description of Solvated Electrons," **44**:483.

Brown, J. M. and Jorgensen, F., "Vibronic Energy Levels of a Lin-

ear Triatomic Molecule in a Degenerate Electronic State: A Unified Tratment of the Renner-Teller Effect," **52**:117.

Brown, R. G., *see* **Davis, H. T.,** **31**:329.

Browne, J. C. and Matsen, F. A., "*Ab Initio* Calculations on Small Molecules," **23**:161.

Brueckner, K. A., "The Correlation Energy of a Non-Uniform Electron Gas," **14**:215.

Brumer, P., "Intramolecular Energy Transfer: Theories for the Onset of Statistical Behavior," **47**:Part 1:201.

Brunner, T. A. and Pritchard, D., "Fitting Laws for Rotationally Inelastic Collisions," **50**:589.

Brus, L. E., *see* **Bondybey, V. E.,** **41**:269.

Buck, U., "Elastic Scattering," **30**:313.

Buckingham, A. D., "Permanent and Induced Molecular Moments and Long-Range Intermolecular Forces," **12**:107.

Bullis, R. H., "Applications to Lasers," **28**:423.

Burdett, J. K., "Predictions of the Structure of Complex Solids," **49**:47.

Burke, T. and Siegel, A., "A New Series Expansion for Path Integrals in Statistical Mechanics," **15**:355.

Burland, D. M. and Zewail, A. H., "Coherent Processes in Molecular Crystals," **40**:369.

Burton, L. L. and Van Artsdalen, E. R., "On the Nature of Solutions of Organic Compounds in Fused Salts," **21**:523.

Busse, H., *see* **Hess, B.,** **29**:137.

Campbell, I. D., Dobson, C. M., and Williams, R. J. P., "Structures and Energetics of Proteins and Their Active Sites," **39**:55.

Cantrell, C. D., Makarov, A. A., and Louisell, W. H., "Laser Excitation of SF_6: Spectroscopy and Coherent Pulse Propagation Effects," **47**:Part 1:583.

Careri, G., "Some Physical Aspects of Gaseous Chemical Kinetics," **1**:119.

Carley, J. S., *see* **Le Roy, R. J.,** **42**:353.

Carlson, C. M., "A Linear Sum-over-Points Approach for Computing Electronic Energies: Application to He and H_2 Using Conroy Diophantine Points," **21**:49.

Carney, G. D., Sprandel, L. L., and Kern, C. W., "Variational Approaches to Vibration-Rotation Spectroscopy for Polyatomic Molecules," **37**:305.

Carrington, A., Levy, D. H., and Miller, T. A., "Electron Resonance of Gaseous Diatomic Molecules," **18**:149.

Cederbaum, L. S. and Domcke, W., "Theoretical Aspects of Ionization Potentials and Photoelectron Spectroscopy: A Green's Function Approach," **36**:205.

Cerf, R., "Cooperative Conformational Kinetics of Synthetic and Biological Chain Molecules," **33**:73.

Cernuschi, F., Giambiagi, M., and Segre, M., "Application of a Short-Range Ordered Model to Strong Electrolytes," **21**:455.

Chakraborti, P., *see* **Mueller, C. R.,** **21**:369.

Chance, R. R., Prock, A., and Silbey, R., "Molecular Fluorescence and Energy Transfer Near Interfaces," **37**:1.

Chandler, D., *see* **Andersen, H. C.,** **34**:105.

Chang, S., *see* **Pak, H.,** **21**:487.

Chanu, J., "Thermal Diffusion of Halides in Aqueous Solution," 13:349.

Chanu, J., see Delmotte, M., 29:343.

Charlemagne, J., see Delmotte, M., 29:343.

Chavepeyer, G., see Platten, J. K., 32:281.

Chay, T. R. and Ree, F. H., "Inequalities for Critical Indices near Gas-Liquid Critical Point," 21:431.

Chen Y., "Noise Analysis of Kinetic Systems and Its Applications to Membrane Channels," 37:67.

Christophorou, L. G., Grant, M., and McCorkle, D. L., "Interactions of Slow electrons with Benzene and Benzene Derivatives," 36:413.

Čížek, J., "On the Use of the Cluster Expansion and the Technique of Diagrams in Calculations of Correlation Effects in Atoms and Molecules," 14:35.

Clark, A. P., Dickinson, A. S. and Richards, D., "The Correspondence Principle in Heavy-Particle Collisions," 36:63.

Clarke, B. L., "Stability of Complex Reaction Networks," 43:1.

Clarke, J. H. R., see Angell, C. A., 48:397.

Clayton, R. K., "Photochemical Reaction Centers and Photosynthetic Membranes," 19:353.

Clyne, M. A. A. and McDermid, I. S., "Laser-Induced Fluorescence: Electronically Excited States of Small Molecules," 50:1.

Coffey, H. F., see Tensmeyer, L. G., 21:767.

Cohen, M. H., see Grest, G. S., 48:455.

Coleman, B. D., "Thermodynamics of Discrete Mechanical Systems with Memory," 24:95.

Collins, F. C. and Raffel, H., "Transport Processes in Liquids," 1:135.

Collins, M. A., "Solitons in Chemical Physics," 53:225.

Cook, M. A., "Contributions of Professor Henry Eyring to the Theory of Detonation," 21:779.

Coopersmith, M. H., "The Thermodynamic Description of Phase Transitions," 17:43.

Coulson, C. A. and MacLagan, R. G. A. R., "Notes on the Two-Particle Density Matrix in π-Electron Theory," 21:303.

Coulson, C. A., see Blyholder, G., 21:25.

Cox, D. M., see Hall, R. B., 47:Part 1:639.

Craig, D. P., "Model Calculations in the Theory of Mixed-Crystal Spectra," 8:27.

Curtis, D. H. and Spikes, J. D., "Kinetics of the Interactions of Formaldehyde, Acetaldehyde, and Acrolein with Rattail Tendon," 21:557.

Curtiss, C. F., Bird, R. B., and Hassager, O., "Kinetic Theory and Rheology of Macromolecular Solutions," 35:31.

Curtiss, C. F., see Hirschfelder, J. O., 3:59.

Cutler, I. B., see McAllister, P. V., 21:669.

Dagonnier, R., "Theory of Quantum Brownian Motion," 16:1.

Dahler, J. S. and Theodosopulu, M., "The Kinetic Theory of Dense Polyatomic Fluids," 31:155.

Dahler, J. S., see Jhon, M. S., 46:279.

Dalgarno, A., "New Methods for Calculating Long-Range Intermolecular Forces," 12:143.

Das, T. P., "Structure and Properties of

Metal-Ammonia Solutions," 4:303.

Daudel, R., "The Relation Between Structure and Chemical Reactivity of Aromatic Hydrocarbons with Particular Reference to Carcinogenic Properties," 1:165.

Davies, R., see Evans, M., 44:255.

Davies, R. C., see Braterman, P. S., 7:359.

Davis, H. T., "Kinetic Theory of Dense Fluids and Liquids Revisited," 24:257.

Davis, H. T. and Brown, R. G., "Low-Energy Electrons in Nonpolar Fluid Films," 31:329.

Davis, H. T. and Scriven, L. E., "Stress and Structure in Fluid Interfaces," 49:357.

de Brouckere, G., "Calculations of Observables in Metallic Complexes by the Molecular Orbital Theory," 37:203.

de Brouckère, L. and Mandel, M., "Dielectric Properties of Dilute Polymer Solutions," 1:77.

de Leener, M., see Bellemans, A., 6:85.

de Levie, R., "Mathematical Modeling of Transport of Lipid-Soluble Ions and Ion-Carrier Complexes Through Lipid Bilayer Membranes," 37:99.

Delmotte, M., Julien, J., Charlemagne, J., and Chanu, J., "Thermodynamic Considerations of the Excitable Membranes Behavior," 29:343.

Deneubourg, J. L., see Lefever, R., 29:349.

DePristo, A.E. and Rabitz, H., "Vibrational and Rotational Collision Processes," 42:271.

Desai, R. C. and Yip, S., "A Stochastic Model for Neutron Scattering by Simple Liquids," 15:129.

Desai, R. C., see Jhon, M. S., 46:279.

Deutch, J. M., see Adelman, S. A., 31:103.

de Vries, A. E., see Baranowski, B., 16:101.

Dewar, M. J. S., "Chemical Reactivity," 8:65.

Dewel, G., see Walgraef, D., 49:311.

Dickinson, A. S., see Clark, A. P., 36:63.

Diestler, D., "Theoretical Studies of Vibrational Relaxation of Small Molecules in Dense Media," 42:305.

Dion, D. R. and Hirschfelder, J. O., "Time-Dependent Perturbation of a Two-State Quantum System by a Sinusoidal Field," 35:265.

Dobson, C. M., see Campbell, I. D., 39:55.

Dolowitz, D. A., see Dougherty, T. F., 21:633.

Domb, C., "Self-Avoiding Walks on Lattices," 15:229.

Domcke, W., see Cederbaum, L. S., 36:205.

Donohue, M. D., see Katz, J. L., 40:137.

Dougherty, T. F. and Dolowitz, D. A., "Physiologic Actions of Heparin not Related to Blood Clotting," 21:633.

Douzou, P. and Sadron, C., "The Electronic Properties of Deoxyribonucleic Acid," 7:339.

Drickamer, H. G. and Zahner, J. C., "The Effect of Pressure on Electronic Structure," 4:161.

Duchesne, J., "Nuclear Quadrupole Resonance in Irradiated Crystals," 2:187.

Duffey, G. H., see Houtman, J. A., 21:7.

Dugan, J. V., Jr. and Magee, J. L., "Dynamics of Ion-Molecule Col-

lisions," 21:207.

Duke, C. B., "Determination of the Structure and Properties of Solid Surfaces by Electron Diffraction and Emission," 27:1.

Eastman, M. P., see Hecht, H. G., 21:351.

Eberhardt, W., see Plummer, E. W., 49:533.

Economou, E. N. and Ngai, K. L., "Surface Plasma Oscillations and Related Surface Effects in Solids," 27:265.

Edelen, D. G. B., "The Thermodynamics of Evolving Chemical Systems and the Approach to Equilibrium," 33:399.

Egelstaff, P. A., "New Experimental Studies of the Structure of Fluids," 53:1.

Ehrenberg, A., "Some Recent Developments in the Study of Paramagnetic Species of Biological Interest," 7:602.

Eigen, M., "How Does Information Originate? Principles of Biological Self-Organization," 38:211.

Eigen, M., "The Origin and Evolution of Life at the Molecular Level," 55:119.

Eley, D. D. and Leslie, R. B., "Adsorption of Water on Solid Proteins with Special Reference to Haemoglobin," 7:238.

Elving, P. J. and Pullman, B., "Mechanisms of Organic Electrode Reactions," 3:1.

Emch, G. G., "An Introduction to C*-Algebraic Methods in Physics," 22:315.

Engelbrecht, L. and Hinze, J., "Molecular Properties Observed and Computed," 44:1.

Englman, R., "Optical Absorptions and Charge Transfer States," 8:13.

Eno, L. and Rabitz, H., "Sensitivity Analysis and Its Role in Quantum Scattering," 51:177.

Erber, T., Latal, H. G., and Harmon, B. N., "The Origin of Hysteresis in Simple Magnetic Systems," 20:71.

Eriksson, J. C., "On the Thermodynamics of Surface Systems," 6:145.

Erneux, T., see Nicolis, G., 38:263.

Eschenroeder, A. Q., "Reaction Kinetics in Hypersonic Flows," 13:19.

Evans, G., see Evans, M., 44:255.

Evans, G. T., see Atkins, P. W., 35:1.

Evans, M., Evans, G., and Davies, R., "Experimental and Theoretical Studies of Rototranslational Correlation Functions," 44:255.

Eyal, M., Agam, U., and Grabiner, F. R., "Infrared Laser-Enhanced Diffusion Cloud Reactions," 47:Part 2:43.

Eyring, E. M., Rich, L. D., McCoy, L. L., Graham, R. C., and Taylor, N., "Activation Parameters and Deprotonation Kinetics of Intramolecularly Hydrogen Bonded Acids," 21:327.

Eyring, H., see Ree, F. H., 4:1.

Eyring, L., see Weber, G. R., 21:253.

Farrell, H. H., see Somorjai, G. A., 20:215.

Fenn, J. B., see Andersen, J. B., 10:275.

Fernández-Alonso, J. I., "Electronic Structures in Quantum Biochemistry," 7:3.

Ferreira, R., "Electronegativity and Chemical Bonding," 13:55.

Fijnaut, H. M., see Vrij, A., 48:329.

Fisher, M. E. and Hartwig, R. E., "Toeplitz Determinants: Some Appli-

cations, Theorems, and Conjectures," **15**:333.

Fixman, M., "The Critical Region," **6**:175.

Fixman, M., *see* **Bearman, R. J., 1**:1.

Fleming, G. R., "Applications of Continuously Operating, Synchronously Mode-Locked Lasers," **49**:1.

Fleury, P. A. and Boon, J. P., "Laser Light Scattering in Fluid Systems," **24**:1.

Flygare, W. H., *see* **McGurk, J. C., 25**:1.

Flynn, G., *see* **Weitz, E., 47**:Part 2:185.

Ford, J., "The Transition from Analytic Dynamics to Statistical Mechanics," **24**:155.

Forster, D., "The Theory of Liquid Crystals," **31**:231.

Freed, K. F., "Collision Induced Intersystem Crossing," **47**:Part 2:291.

Freed, K. F., "Collisional Effects on Electronic Relaxation Processes," **42**:207.

Freed, K. F., "Functional Integrals and Polymer Statistics," **22**:1.

Freed, K. F., *see* **Herman, M. F., 48**:1.

Friedman, L. and Reuben, B. G., "A Review of Ion-Molecule Reactions," **19**:33.

Friedmann, H., "The Quantum Mechanical Distribution Function of Molecular Systems: Translational and Rotational Motions," **4**:225.

Frisch, H. L., "Commentary: Stochastic Models of Synthesis of Asymmetric Forms," **55**:201.

Frisch, H. L., "The Equation of State of the Classical Hard Sphere Fluid," **6**:229.

Frommhold, L., "Collision-Induced Scattering of Light and the Diatom Polarizabilities," **46**:1.

Frost, A. A., "The Potential Energy Surface of the H_3 System Using Floating Gaussian Orbitals," **21**:65.

Fuchs, R., *see* **Kliewer, K. L., 27**:355.

Gelbart, W. M., "Depolarized Light Scattering by Simple Fluids," **26**:1.

Gerisch, G., *see* **Hess, B., 29**:137.

Giambiagi, M., *see* **Cernuschi, F., 21**:455.

Gibbs, J. H., "The Kinetics of Biopolymerization on Nucleic Acid Templates," **15**:185.

Giese, C., "Ion-Neutral Reactions," **10**:247.

Gilmer, G. H., *see* **Weeks, J. D., 40**:157.

Gilmore, F. R., *see* **McGowan, J. W., 28**:379.

Girardeau, M. D. and Mazo, R. M., "Variational Methods in Statistical Mechanics," **24**:187.

Glansdorff, P., *see* **Nicolis, G., 32**:1.

Goldbeter, A., *see* **Hess, B., 38**:363.

Gomer, R., "Electron Spectroscopy of Chemisorption on Metals," **27**:211.

Goodman, G. L., *see* **Weinstock, B., 9**:169.

Goodman, L. and Rava, R. P., "Two-Photon Spectroscopy of Perturbed Benzenes," **54**:177.

Goodwin, B. C., "A Membrane Model for Polar Transport and Gradient Formation," **29**:269.

Gordon, R. G., "Error Bounds and Spectral Densities," **15**:79.

Grabiner, F. R., *see* **Eyal, M., 47**:Part 2:43.

Graham, R. C., *see* **Eyring, E. M., 21**:237.

Grant, M., *see* **Christophorou, L. G., 36**:413.

Gray, G. A., "Ion Cyclotron Resonance," **19**:141.

Grecos, A. P., "Solvable Models for Unstable States in Quantum Physics," **38**:143.

Green, S., "Sources of Error and Expected Accuracy in *Ab Initio* One-Electron Operator Properties: The Molecular Dipole Moment," **25**:179.

Greene, E. F., Moursund, A. L., and Ross, J., "Elastic Scattering in Chemically Reactive Systems," **10**:135.

Greer, W. L. and Rice, S. A., "On the Theory of Optical Absorption Profiles in Condensed Systems," **17**:229.

Grest, G. S. and Cohen, M. H., "Liquids, Glasses, and the Glass Transition: A Free Volume Approach," **48**:455.

Grice, R., "Reactive Scattering," **30**:247.

Grimaldi, F., "Correlation Effects in Diatomic Molecules Obtained from Configuration Interaction Using Hartree-Fock Orbitals. Effects on Energy and Monoelectronic Operators," **14**:341.

Guidotti, D., *see* Rice, S. A., **27**:543.

Guillot, B., *see* Birnbaum, G., **51**:49.

Guth, E., "Brownian Motion and Indeterminacy Relations," **15**:363.

Gutman, M., *see* Huppert, D., **47**:Part 2:643.

Guyon, E., *see* Pieranski, P., **32**:151.

Haas, Y., "Electronically Excited Fragments Formed by Unimolecular Multiple Photon Dissociation," **47**:Part 1:713.

Haas, Y. and Asscher, M., "Two-Photon Excitation as a Kinetic Tool: Application to Nitric Oxide Fluorescence Quenching," **47**:Part 2:17.

Haberland, H., Lee, Y. T., and Siska, P. E., "Scattering of Noble-Gas Metastable Atoms in Molecular Beams," **45**:487.

Haggerty, M. J. and Severne, G., "Kinetic Theory of Gravitational Systems," **35**:119.

Hall, H. T., "The Synthesis of Diamond," **21**:721.

Hall, R. B., Kaldor, A., Cox, D. M., Horsley, A., Rabinowitz, P., Kramer, G. M., Bray, R. G., and Maas, E. T., Jr., "Infrared Laser Chemistry of Complex Molecules," **47**:Part 1:639.

Halperin, B. I., "Properties of a Particle in a One-dimensional Random Potential," **13**:123.

Hammes, G. G., "Control of the Catalytic Activity of Enzymes by the Near and Remote Environment of a Polyatomic Framework," **39**:177.

Hamnett, A., *see* Brion, C. E., **45**:2.

Hansen, A. H. and Bouman, T. D., "Natural Chiroptical Spectroscopy: Theory and Computations," **44**:545.

Hanusse, P., Ross, J., and Ortoleva, P., "Instability and Far-from-Equilibrium States of Chemically Reacting Systems," **38**:317.

Harasima, A., "Molecular Theory of Surface Tension," **1**:203.

Haring, A., *see* Baranowski, B., **16**:101.

Haring, M., *see* Kvasnička, V., **52**:181.

Harmon, B. N., *see* Erber, T., **20**:71.

Harp, G. D., *see* Berne, B. J., **17**:63.

Harris, F. E. and Michels, H. H., "The Evaluation of Molecular Integrals for Slater-type Orbitals," **13**:205.

Hartmann, H., "New Developments in the One-Electron Theory of π-Electron Systems," **5**:1.

Hartwig, R. E., *see* Fisher, M. E., **15**:333.

Hassager, O., *see* Curtiss, C. F., **35**:31.

Hasselle-Schuermans, N., *see* Résibois, P., **11**:159.

Hazi, A., *see* Weeks, J. D., **16**:283.

Hecht, H., and Lewis, W. B., and Eastman, M. P., "Covalency Effects in Octahedral $5f^1$ Complexes," **21**:351.

Henderson, D., *see* Watts, R. O., **21**:421.

Herman, M. F., Freed, K. F., and Yeager, D. L., "Analysis and Evaluation of Ionization Potentials, Electron Affinities, and Excitation Energies by the Equations of Motion—Green's Function Method," **48**:1.

Herrick, D. R., "New Symmetry Properties of Atoms and Molecules," **52**:1.

Herschbach, D., "Reactive Scattering in Molecular Beams," **10**:319.

Herschkowitz-Kaufman, M., *see* Nicolis, G., **38**:263.

Hertel, I. V., "Collisional Energy-Transfer Spectroscopy with Laser-Excited Atoms in Crossed Atom Beams: A New Method for Investigating the Quenching of Electronically Excited Atoms by Molecules," **45**:341.

Hertel, I. V., "Progress in Electronic-to-Vibrational Energy Transfer," **50**:475.

Hess, B., Boiteux, A., Busse, H., and Gerisch, G., "Spatiotemporal Organization in Chemical and Cellular Systems," **29**:137.

Hess, B., Goldbeter, A., and Lefever, R., "Temporal, Spatial, and Functional Order in Regulated Biochemical and Cellular Systems," **38**:363.

Higuchi, I., "The Pore Structure of Catalysts and Catalytic Reaction Rates," **21**:791.

Hijmans, J. and Holleman, Th., "The Principle of Corresponding States for Chain-molecule Liquids and their Mixtures," **16**:223.

Himmelblau, D. M., *see* Ball, J. C., **13**:267.

Hinze, J., "An Overview of Computational Methods for Large Molecules," **26**:213.

Hinze, J., *see* Engelbrecht, L., **44**:1.

Hirschfelder, J. O., "A Forecast for Theoretical Chemistry," **21**:73.

Hirschfelder, J. O. and Curtiss, C. F., "Propagation of Flames and Detonations," **3**:59.

Hirschfelder, J. O. and Meath, W. J., "The Nature of Intermolecular Forces," **12**:3.

Hirschfelder, J. O., *see* Dion, D. R., **35**:265.

Hirst, D. M., "The Calculation of Potential Energy Surfaces for Excited States," **50**:517.

Hoarau, J., *see* Pacault, A., **3**:171.

Hoare, M. R., "Structure and Dynamics of Simple Microclusters," **40**:49.

Hoare, M. R., "The Linear Gas," **20**:135.

Hoffmann, T. A. and Ladik, J., "Quantum Mechanical Considerations on Some Properties of DNA," **7**:84.

Holleman, Th., *see* Hijmans, J., **16**:223.

Horsley, J. A., *see* Bray, R. B., **47**:Part 1:639.

Houston, P. L., "Electronic to Vibrational Energy Transfer from

Excited Halogen Atoms," **47**:Part 2:381.

Houston, P. L., "Initiation of Atom-Molecule Reactions by Infrared Multiphoton Dissociation," **47**:Part 1:625.

Houtman, J. A. and Duffey, G. H., "A Free-Electron Study of Common Boron Frameworks," **21**:7.

Høye, J. S., see **Stell, G.**, **48**:183.

Huppert, D., Gutman, M., and Kaufmann, K. J., "Laser Studies of Proton Transfer," **47**:Part 2:643.

Hurle, D. T. J. and Jakeman, E., "On the Nature of Oscillatory Convection in Two-Component Fluids," **32**:277.

Ikegami, A., "Statistical Thermodynamics of Proteins and Protein Denaturation," **46**:363.

Imelik, B., see **Laforgue, A.**, **8**:141.

Jacobi, N., see **Schnepp, O.**, **22**:205.

Jagger, J., "Photoprotection from Far Ultraviolet Effects in Cells," **7**:584.

Jakeman, E., see **Hurle, D. T. J.**, **32**:277.

Jardetzky, O., "The Study of Specific Molecular Interactions by Nuclear Magnetic Relaxation Methods," **7**:499.

Jhon, M. S., "Sound Velocity and van der Waals Force in Liquids According to Significant Structure Theory," **21**:481.

Jhon, M. S., Dahler, J. S., and Desai, R. C., "Theory of the Liquid-Vapor Interface," **46**:279.

Johnson, F. H., "Hydrostatic Pressure Reversal of Alcohol Narcosis in Aquatic Animals," **21**:647.

Johnston, H. S, "Large Tunnelling Corrections in Chemical Reaction Rates," **3**:131.

Joosten, J. G. H., see **Vrij, A.**, **48**:329.

Jordan, J. E., see **Amdur, I.**, **10**:29.

Jørgensen, C. K., "Applications of Ligand-Field Theory to Complexes," **8**:47.

Jørgensen, C. K., "Spectroscopy of Transition-Group Complexes," **5**:33.

Jørgensen, F., see **Brown, J. M.**, **52**:117.

Jørgensen, P., see **Olsen, J.**, **54**:1.

Jortner, J. and Levine, R. D., "Photoselective Chemistry," **47**:Part 1:1.

Jortner, J., see **Beswick, J. A.**, **47**:Part 1:293.

Jortner, J., see **Sage, M. L.**, **47**:Part 1:293.

Jucys, A. P., "On the Hartree-Fock Method in Multi-Configuration Approximation," **14**:191.

Jucys, A. P. and Kaminskas, V. A., "On the Application of the Extended Method of Calculation to the Atomic Electrons," **14**:207.

Judd, B. R., "Effective Operators for Configurations of Equivalent Electrons," **14**:91.

Julien, J., see **Delmotte, M.**, **29**:343.

Kahrig, E. and Besserdich, H., "A Short Remark About Various Dissipative Structures," **29**:133.

Kaldor, A., see **Hall, R. B.**, **47**:Part 1:639.

Kaldor, A., see **Woodin, R. L.**, **47**:Part 2:3.

Kaminskas, V. A., see **Jucys, A. P.**, **14**:207.

Kapral, R., "Kinetic Theory of Chemical Reactions in Liquids," **48**:71.

Karle, J., "The Phase Problem in Structural Analysis," **16**:131.

Karnicky, J. F. and Pings, C. J., "Recent Advances in the Study of Liquids

by X-ray Diffraction," **34**:157.

Katz, J. L. and Donohue, M. D., "A Kinetic Approach to Homogeneous Nucleation Theory," **40**:137.

Kauffman, S. A., "Bifurcations in Insect Morphogenesis," **55**:219.

Kaufman, J. J., "Potential Energy Surface Considerations for Excited State Reactions," **28**:113.

Kaufmann, K. J. and Wasielewski, M. R., "Studies of Chlorophyll *in Vitro*," **47**:Part 2:579.

Kaufmann, K. J., *see* **Huppert, D.**, **47**:Part 2:643.

Kearns, D. R., "Electronic Conduction in Organic Molecular Solids," **7**:282.

Keck, J. C., "Variational Theory of Reaction Rates," **13**:85.

Kelly, H. P., "Applications of Many-Body Diagram Techniques in Atomic Physics," **14**:129.

Kempter, V., "Electronic Excitation in Collisions between Neutrals," **30**:417.

Kende, A., "A Self-Consistent Field Molecular Orbital Treatment of Carbonyl Base Strength," **8**:133.

Kenney-Wallace, G. A., "Picosecond Spectroscopy and Dynamics of Electron Relaxation Processes in Liquids," **47**:Part 2:535.

Kern, C. W., *see* **Carney, G. D.**, **37**:305.

Kerner, E. H., "Statistical-Mechanical Theories in Biology," **19**:325.

Kihara, T., "Convex Molecules in Gaseous and Crystalline States," **5**:147.

Kihara, T., "Intermolecular Forces and Equation of State of Gases," **1**:267.

Kihara, T., "Multipolar Interactions in Molecular Crystals," **20**:1.

Kihara, T. and Koide, A., "Intermolecular Forces and Crystal Structures for D_2, N_2, O_2, F_2, and CO_2," **33**:51.

Kilpatrick, J. E., "The Computation of Virial Coefficients," **20**:39.

King, D. S., "Infrared Multiphoton Excitation and Dissociation," **50**:105.

Kiode, A., *see* **Kihara, T.**, **33**:51.

Kiode, S. and Oguchi, T., "Theories on the Magnetic Properties of Compounds," **5**:189.

Kirkwood, J. G., *see* **Bearman, R. J.**, **1**:1.

Kitahara, K., "The Hamilton-Jacobi-Equation Approach to Fluctuation Phenomena," **29**:85.

Kliewer, K. L. and Fuchs, R., "Theory of Dynamical Properties of Dielectric Surfaces," **27**:355.

Klimontovich, Yu. L., "Kinetic Theory of Plasmas," **38**:193.

Klotz, I. M., "Synzymes: Synthetic Polymers with Enzymelike Catalytic Activities," **39**:109.

Kobatake, Y., "Physiochemical Problems in Excitable Membranes," **29**:319.

Kodaira, M. and Watanabe, T., "Collisional Transfer of Triplet Excitations Between Helium Atoms," **21**:167.

Konasewich, D. E., *see* **Kreevoy, M. M.**, **21**:243.

Koopman, B. O., "Relaxed Motion in Irreversible Molecular Statistics," **15**:37.

Koschmieder, E. L., "Bénard Convection," **26**:177.

Koschmieder, E. L., "Stability of Supercritical Bénard Convection and Taylor Vortex Flow," **32**:109.

Koski, W. S., "Scattering of Positive

Ions by Molecules," **30**:185.

Kosloff, R., "Thermodynamic Aspects of the Quantum-Mechanical Measurement Process," **46**:153.

Kotani, M., "Electronic Structure and Magnetic Properties of Hemoproteins, Particularly Hemoglobins," **7**:159.

Kottis, P., *see* **Aslangul, C., 41**:321.

Kottis, P., *see* **Turlet, J. M., 54**:303.

Koutecký, J., "Quantum Chemistry of Crystal Surfaces," **9**:85.

Kox, A. J., *see* **Wiegel, F. W., 41**:195.

Kozak, J. J., "Nonlinear Problems in the Theory of Phase Transitions," **40**:229.

Kozak, J. J., *see* **Luks, K. D., 37**:139.

Kramer, G. M., *see* **Hall, R. B., 47**:Part 1:639.

Krause, L., "Sensitized Fluorescence and Quenching," **28**:267.

Kreevoy, M. M. and Konasewich, D. E., "The Brønsted α and the Primary Hydrogen Isotope Effect. A Test of the Marcus Theory," **21**:243.

Kubo, R., "A Stochastic Theory of Line Shape," **15**:101.

Kummler, R. H., *see* **McGowan, J. W., 28**:379.

Kuppermann, A., *see* **Trajmar, S., 18**:15.

Kvasnička, V., "Application of Diagrammatic Quasidegenerate RSPT in Quantum Molecular Physics," **36**:345.

Kvasnička, V., Laurinc, V., Biskupič, S., and Haring, M., "Coupled-Cluster Approach in the Electronic Structure Theory of Molecules," **52**:181.

Kwun, S. and Ree, T., "Ionization Potentials and Electron Affinities of Some Polyphenyl Molecules," **21**:37.

Kyogoku, Y., *see* **Shimanouchi, T., 7**:435.

Labhart, H., "Electrochromism," **13**:179.

Lacmann, K., "Collisional Ionization," **42**:513.

Ladik, J., *see* **Hoffmann, T., 7**:84.

Laforgue, A., Rousseau, J., and Imelik, B., "Chemisorption of Ethylene on Metals and Catalytic Hydrogenation," **8**:141.

Laidler, K. J. and Tweedale, A., "The Current Status of Eyring's Rate Theory," **21**:113.

Lamm, O., "Studies in the Kinematics of Isothermal Diffusion. A Macrodynamical Theory of Multicomponent Fluid Diffusion," **6**:295.

Landis, P. W., *see* **Tensmeyer, L. G., 21**:767.

Latal, H. G., *see* **Erber, T., 20**:71.

Lathouwers, L. and Van Leuven, P., "Generator Coordinate Theory of Nuclear Motion in Molecules," **49**:115.

Lau, A. M. F., "The Photon-as-Catalyst Effect in Laser Induced Predissociation and Autoionization," **50**:191.

Läuger, P., "Ion Transport Through Artificial Lipid Membranes," **29**:309.

Laurinc, V., *see* **Kvasnička, V., 52**:181.

Lawley, K. P., "New Directions in Molecular Beams," **30**:1.

Lebowitz, J. L., *see* **Spohn, H., 38**:109.

Lecomte, J., "Prospects for the Use of Infrared Spectroscopy in Biology," **7**:408.

Lee, Y., Ree, F. H., and Ree, T., "Distribution Function of Classical Fluids of Hard Spheres. I," **21**:393.

Lee, Y. T., *see* Haberland, H., **45**:487.

Lefever, R. and Deneubourg, J. L., "Membrane Excitation," **29**:349.

Lefever, R., *see* Prigogine, I., **29**:1; **39**:1.

Lefever. R., *see* Hess, B., **38**:363.

Lemberg, H. L., *see* Rice, S. A., **27**:543.

Leone, S. R., "Photofragment Dynamics," **50**:255.

Le Roy, R. J. and Carley, J. S., "Spectroscopy and Potential Energy Surfaces of van der Waals Molecules," **42**:353.

Leslie, R. B., *see* Eley, D. D., **7**:238.

Levine, R. D., "The Information Theoretic Approach to Intramolecular Dynamics," **47**:Part 1:239.

Levine, R. D., *see* Jortner, J., **47**:Part 1:1.

Levy, D. H., "van der Waals Molecules," **47**:Part 1:323.

Levy, D. H., *see* Carrington, A., **18**:149.

Lewis, W. B., *see* Hecht, H. G., **21**:351.

Lichten, W., "Resonant Charge Exchange in Atomic Collisions," **13**:41.

Liehr, A. D., "Forbidden Transitions in Organic and Inorganic Systems," **5**:241.

Lifschitz, C., *see* Tiernan, T. O., **45**:82.

Light, J. C., "Quantum Theories of Chemical Kinetics," **19**:1.

Lin, M. C., "Dynamics of Oxygen Atom Reactions," **42**:113.

Lin, M. C., *see* Baronavski, A., **47**:Part 2:85.

Lin, S. H. and Lin Ma, C. Y., "Calculation of Statistical Complexions of Polyatomic Molecules and Ions," **21**:143.

Lin, S. H., *see* Bersohn, R., **16**:67.

Lin Ma, C. Y., *see* Lin, S. H., **21**:143.

Linder, B., "Reaction Field Techniques and Their Applications to Intermolecular Forces," **12**:225.

Loesch, H. J., "Scattering of Non-Spherical Molecules," **42**:421.

Loftus, E., *see* Rabinowitz, P., **15**:281.

Longuet-Higgins, H. C., "Recent Developments in Molecular Orbital Theory," **1**:239.

Louisell, W. H., *see* Cantrell, C. D., **47**:Part 1:583.

Löwdin, P., "Correlation Problem in Many-Electron Quantum Mechanics. I. Review of Different Approaches and Discussion of Some Current Ideas," **2**:207.

Löwdin, P., "Some Aspects of the Biological Problems of Heredity, Mutations, Aging, and Tumors in View of the the Quantum Theory of the DNA Molecule," **8**:177.

Löwdin, P., "Some Aspects of the Correlation Problem and Possible Extensions of the Independent-Particle Method," **14**:283.

Löwdin, P., "Some Recent Developments in the Quantum Theory of Many-Electron Systems and the Correlation Problem," **8**:3.

Luks, K. D. and Kozak, J. J., "The Statistical Mechanics of Square-Well Fluids," **37**:139.

Lumfry, R., "Protein Conformations, 'Rack' Mechanisms and Water," **21**:567.

Maas, E. T., Jr., *see* Hall, R. B., **47**:Part 1:639.

MacLagan, R. G. A. R., *see* Coulson, C. A., **21**:303.

Magee, J. L., *see* Dugan, J. V., Jr., **21**:207.

Mahan, B. H., "Recombination of Gaseous Ions," **23**:1.

Makarov, A. A., *see* Cantrell, C. D., **47**:Part 1:583.

Mandel, M., *see* de Brouckère, L., **1**:77.

Marchand, A., *see* Pacault, A., **3**:171.

Margoliash, E., "The Molecular Variations of Cytochrome *c* as a Function of the Evolution Species," **29**:191.

Mason, E. A. and Monchick, L., "Methods for the Determination of Intermolecular Forces," **12**:329.

Mathot, V., *see* Bellemans, A., **11**:117.

Matsen, F. A., "Helium Molecular States," **21**:1.

Matsen, F. A., *see* Browne, J. C., **23**:161.

Mazo, R. M., *see* Girardeau, M. D., **24**:187.

Mazur, J., "Non-Self Intersecting Random Walks on Lattices with Near-Neighbor Interactions," **15**:261.

Mazur, P., "On Statistical Mechanics and Electromagnetic Properties of Matter," **1**:309.

McAllister, P. V. and Cutler, I. B., "The Kinetics of the Thermal Grooving of Solids," **21**:669.

McConnell, H., "Dynamic Properties of Membranes: Membrane Immunochemistry," **39**:275.

McCorkle, D. L., *see* Christophorou, L. G., **36**:413.

McCoy, L. L., *see* Eyring, E. M., **21**:237.

McDermid, I. S., *see* Clyne, M. A. A., **50**:1.

McGowan, J. W., Kummler, R. H., and Gilmore, F. R., "Excitation De-excitation Processes Relevant to the Upper Atmosphere," **28**:379.

McGuire, P., *see* Mueller, C. R., **21**:369.

McGurk, J. C., Schmalz, T. G., and Flygare, W. H., "A Density Matrix, Bloch Equation Description of Infrared and Microwave Transient Phenomena," **25**:1.

McLachlan, D., Jr., "The Gaseous Fraction in Liquid Metals," **21**:499.

McLennan, J. A., Jr., "The Formal Statistical Theory of Transport Processes," **5**:261.

McQuarrie, D. A., "Stochastic Theory of Chemical Rate Processes," **15**:149.

Mead, C. A., "Generalizations of the Entropy Concept: Principles, Limitations, and Prospects for Applications," **51**:113.

Mead, C. A., *see* Truhlar, D. G., **33**:295.

Meath, W. J., *see* Hirschfelder, J. O., **12**:3.

Menziger, M., "Electronic Chemiluminescence in Gases," **42**:1.

Meyer, L., "Phase Transitions in van der Waal's Lattices," **16**:343.

Micha, D. A., "Quantum Theory of Reactive Molecular Collisions," **30**:7.

Michels, H. H., "Electronic Excited States of Selected Atmospheric Systems," **45**:225.

Michels, H. H., *see* Harris, F. E., **13**:205.

Miller, R. H., "Equilibrium in Stellar Systems," **26**:107.

Miller, S. L., "The Prebiotic Synthesis of Organic Molecules and Polymers," **55**:85.

Miller, T. A., *see* Carrington, A., **18**:149.

Miller, W. H., "Classical-Limit Quantum Mechanics and the Theory of Molecular Collisions," **25**:69.

Miller, W. H., "Classical *S*-Matrix in Molecular Collisions," **30**:77.

Monchick, L., *see* Mason, E. A., **12**:329.

Moncrieff-Yeates, M. and White, H. J., Jr., "The Interaction of Cellulose with Simple Salt Solutions and with Dyebaths," **21**:685.

Monnier, A. M., "Excitability and Ionic Selectivity, Common Properties of Many Lipidic Derivatives," 29:341.

Montroll, E. W., "Enzyme Cascades and Their Control in Blood Plasma," 26:145.

Montroll, E. W. and Shuler, K. E., "The Application of the Theory of Stochastic Processes to Chemical Kinetics," 1:361.

Montroll, E. W., see Wartell, R. M., 22:129.

Moore, C. B., "Vibration → Vibration Energy Transfer," 23:41.

Moore, D. R., "Numerical Models for Convection," 32:65.

Moore, W. J., see Bass, L., 21:619.

Morgan, J. C., "The Effect of Prandtl Number on Finite Amplitude Bénard Convection," 32:77.

Morgan, L. G., see Morrey, J. R., 21:317.

Morrey, J. R. and Morgan, L. G., "T_n Frequency Functions as Energy Contours for Photon Absorbance in Condensed Systems," 21:317.

Mortensen, E. M., "Permeabilities for Reactions of the Type $H + H_2 = H_2 + H$ Treated as a Linear Encounter Using Variational and Distorted Wave Techniques," 21:127.

Moscowitz, A., "Theoretical Aspects of Optical Activity - Part One: Small Molecules," 4:67.

Moursund, A. L., see Greene, E. F., 10:135.

Muckerman, J. T., see Bernstein, R. B., 12:389.

Mueller, C. R., Smith, B., McGuire, P., Williams, W., Chakraborti, P., and Penta, J., "Intermolecular Potentials and Macroscopic Properties of Argon and Neon from Differential Cross Sections," 21:369.

Mukamel, S., "Reduced Equations of Motion for Collisionless Molecular Multiphoton Processes," 47:Part 1:509.

Müller-Herold, U., see Primas, H., 38:1.

Murphy, W. C., see Rice, S. A., 27:543.

Muschlitz, E. E., Jr., "Collisions of Electronically Excited Atoms and Molecules," 10:171.

Nachtrieb, N. H., "Solutions of Metals in Molten Salts," 31:465.

Nagel, S. R., "Metallic Glasses," 51:227.

Narten, A. H., see Blum, L., 34:203.

Nazarea, A. D., "Electric Fields and Self-Coherent Patterns and Structures in Chemical Systems: Large-Scale Effects and Biological Implications," 38:415.

Nesbet, R. K., "Atomic Bethe-Goldstone Equations," 14:1.

Nesbet, R. K., "Electronic Correlation in Atoms and Molecules," 9:321.

Ng, C. Y., "Molecular Beam Photoionization Studies of Molecules and Clusters," 52:263.

Ngai, K. L., see Economou, E. N., 27:265.

Nicholson, B. J., "Approximate Molecular Orbital Theories," 18:249.

Nicolau, C. S., "ESR Investigations on Different Plant Systems," 7:628.

Nicolet, M., "Atmospheric Chemistry," 55:63.

Nicolis, G., "Bifurcations and Symmetry Breaking in Far-from-Equilibrium Systems: Toward a Dynamics of Complexity," 55:177.

Nicolis, G., "Dissipative Instabilities, Structure, and Evolution," 29:29.

Nicolis, G., "Local Potential Methods in

the Study of Kinetic Equations," **13**:299.

Nicolis, G., "Stability and Dissipative Structures in Open Systems Far from Equilbrium," **19**:209.

Nicolis, G., Erneux, T., and Herschkowitz-Kaufman, M., "Pattern Formation in Reacting and Diffusing Systems," **38**:263.

Nicolis, G., Prigogine, I., and Glansdorff, P., "On the Mechanism of Instabilities in Nonlinear Systems," **32**:1.

Nicolis, G., see Prigogine, I., **21**:473.

Niehaus, A., "Spontaneous Ionization in Slow Collisions," **45**:399.

Nikitin, E. E., "Theory of Nonadiabatic Collision Processes Including Excited Alkali Atoms," **28**:317.

Nitzan, A., see Tramer, A., **47**:Part 2:337.

Novak, B., "Periodical Signals in the Spatial Differentiation of Plant Cells," **29**:281.

Noyes, R. M., "Commentary: On Selection of Chirality," **55**:111.

Oguchi, T., see Koide, S., **5**:189.

Olsen, J., Yeager, D. L., and Jørgensen, P., "Optimization and Characterization of a Multiconfigurational Self-Consistent Field (MCSCF) State," **54**:1.

Ono, S., "Variational Principles in Thermodynamics and Statistical Mechanics of Irreversible Processes," **3**:267.

Oppenheim, I., "Probabilistic and Dynamic Approaches to the Theory of Stochastic Processes," **15**:1.

Oppenheim, I., see Bloom, M., **12**:549.

Oriani, R. A. "Thermodynamics of Metallic Solutions," **2**:119.

Ortoleva, P. and Ross, J., "Studies in Dissipative Phenomena with Biological Applications," **29**:49.

Ortoleva, P., see Hanusse, P., **38**:317.

Orwoll, R. A. and Stockmayer, W. H., "Stochastic Models for Chain Dynamics," **15**:305.

Oxtoby, D. W., "Dephasing of Molecular Vibrations in Liquids," **40**:1.

Oxtoby, D. W., "Vibrational Population Relaxation in Liquids," **47**:Part 2:487.

Pacault, A., Hoarau, J., and Marchand, A., "Aspects Récents du Diamagnétisme," **3**:171.

Pak, H. and Chang, S., "Transient State Theory of Significant Liquid Structure Applied to Water," **21**:487.

Parmenter, C. S., "Radiative and Nonradiative Processes in Benzene," **22**:365.

Patey, G. N., see Stell, G., **48**:183.

Paul, R., see Baranowski, B., **16**:101.

Penta, J., see Mueller, C. R., **21**:369.

Percival, I. C., " Semiclassical Theory of Bound States," **36**:1.

Perdang, J., "Stellar Evolutionary Stability in Relation to Spectral Theory," **32**:207.

Peshkin, M., "Angular Distributions of Photoelectrons: Consequences of Symmetry," **18**:1.

Pfeifer, P. "A Deductive Approach to Reduced Dynamics and Entropy in Finite Quantum Systems," **53**:341.

Philippot, J., "Nuclear Paramagnetic Relaxation in Solids," **11**:289.

Philpott, M. R., "Some Aspects of Exciton Theory," **23**:227.

Philpott, M. R., see Turlet, J. M., **54**:303.

Pieranski, P. and Guyon, E., "Cylindri-

cal Couette Flow Instabilities in Nematic Liquid Crystals," 32:151.

Pings, C. J., *see* Karnicky, J. F., 34:157.

Pitzer, K. S., "Inter- and Intramolecular Forces and Molecular Polarizability," 2:59.

Platteeuw, J. C., *see* van der Waals, J. H., 2:1.

Platten, J. K. and Chavepeyer, G., "Finite Amplitude Instability in the Two-Component Bénard Problem," 32:281.

Platten, J. K., *see* Schlecter, R. S., 26:265.

Plummer, E. W. and Eberhardt, W., "Angle-Resolved Photoemission as a Tool for the Study of Surfaces," 49:533.

Pollard, E. C., "Thermal Effects on Protein, Nucleic Acid, and Viruses," 7:201.

Power, E. A., "Very Long-Range (Retardation Effect) Intermolecular Forces," 12:167.

Prager, S., "The One-Dimensional Plasma," 4:201.

Prigogine, I., "Nonequilibrium Thermodynamics and Chemical Evolution: An Overview," 55:43.

Prigogine, I., "Quantum States and Dissipative Systems," 15:11.

Prigogine, I. and Lefever, R., "Coupling Between Diffusion and Chemical Reactions," 39:1.

Prigogine, I. and Lefever, R., "Stability and Self-Organization in Open Systems," 29:1.

Prigogine, I., Nicolis, G., Allen, P. M., "Eyring's Theory of Viscosity of Dense Media and Nonequilibrium Statistical Mechanics," 21:473.

Prigogine, I., *see* Nicolis, G., 32:1.

Primas, H. and Müller-Herold, U., "Quantum-Mechanical System Theory: A Unifying Framework for Observations and Stochastic Processes in Quantum Mechanics," 38:1.

Pritchard, D., *see* Brunner, T. A., 50:589.

Procaccia, I., *see* Richter, P. H., 43:217.

Prock, A., *see* Chance, R. R., 37:1.

Pullman, B., "Steroids, Purine-Pyrimidine Pairs and Polycyclic Aromatic Carcinogens," 8:163.

Pullman, B., *see* Elving, P. J., 3:1.

Quack, M., "Reaction Dynamics and Statistical Mechanics of the Preparation of Highly Excited States by Intense Infrared Radiation," 50:395.

Rabin, Y., *see* Ben-Reuven, A., 47:Part 1:555.

Rabinowitz, P., Silberberg, A., Simha, R., and Loftus, E., "Reversible Kinetics on Linear Lattices with Neighbor Effects," 15:281.

Rabinowitz, P., *see* Hall, R. B., 47:Part 1:639.

Rabitz, H., *see* DePristo, A. E., 42:271; Eno, L., 51:177.

Raffel, H., *see* Collins, F. C., 1:135.

Rajagopal, A. K., "Theory of Inhomogeneous Electron Systems: Spin-Density-Functional Formalism," 41:59.

Rajagopal, A. K., *see* Ramana, M. V., 54:231.

Ramana, M. V. and Rajagopal, A. K., "Inhomogeneous Relativistic Electron Systems: A Density Functional Formalism," 54:231.

Rava, R. P., *see* Goodman, L., 54:177.

Ree, F. H., Ree, T. S., Ree, T., and Eyring, H., "Random Walk and

Related Physical Problems," 4:1.

Ree, F. H., *see* Chay, T. R., 21:431.; Lee, Y., 21:393.

Ree, T., *see* Kwun, S., 21:37.; Lee, Y., 21:393.; Ree, F. H., 4:1, *see* Utsugi, H., 21:273.

Ree, T. S., *see* Ree, F. H., 4:1.

Regnier, J., *see* Barriol, J., 8:5.

Reisler, H. and Wittig, C., "Electronic Luminescence Resulting from Infrared Multiple Photon Excitation," 47:Part 1:679.

Reiss, H., "Scaled Particle Methods in the Statistical Thermodynamics of Fluids," 9:1.

Rentzepis, P. M., "Picosecond Spectroscopy and Molecular Relaxation," 23:189.

Rentzepis, P. M. and Barbara, P. F., "Proton Transfer: A Primary Picosecond Event," 47:Part 2:627.

Résibois, P. and Hasselle-Schuermans, N., "Microscopic Approach to Equilibrium and Non-Equilibrium Properties of Electrolytes," 11:159.

Résibois, P., *see* Balescu, R., 38:173.

Reuben, B. G., *see* Friedman, L., 19:33.

Reuss, J., "Scattering from Oriented Molecules," 30:389.

Rice, J. K., *see* Trajmar, S., 18:15.

Rice, S. A., "Collision Induced Intramolecular Energy Transfer in Electronically Excited Polyatomic Molecules," 47:Part 2:237.

Rice, S. A., "An Overview of the Dynamics of Intramolecular Transfer of Vibrational Energy," 47:Part 1:117.

Rice, S. A., "Some Comments on the Theory of Photochemical Reactions," 21:153.

Rice, S. A., Guidotti, D., Lemberg, H. L., Murphy, W. C., and Bloch, A. N., "Some Comments on the Electronic Properties of Liquid Metal Surfaces," 27:543.

Rice, S. A., *see* Greer, W. L., 17:229.; Weeks, J. D., 16:283.

Rich, L. D., *see* Eyring, E. M., 21:237.

Richards, D., *see* Clark, A. P., 36:63.

Richardson, M. J., *see* Rowlinson, J. S., 2:85.

Richter, P. H., Procaccia, I., and Ross, J., "Chemical Instabilities," 43:217.

Richter, P. H., *see* Ross, J., 55:169.

Rigby, B. J., "The Thermal Stability of Collagen: Its Significance in Biology and Physiology," 21:537.

Roberts, P. H., "Concepts in Hydrodynamic Stability Theory," 32:17.

Robinson, E. J., *see* Bederson, B., 10:1.

Rohde, S. M., "Some Remarks on Variational Methods, the Local Potential, and Finite Element Methods with Application to Certain Continuum Mechanics Problems," 32:37.

Ronn, A. M., "Luminescence of Parent Molecule Induced by Multiphoton Infrared Excitation," 47:Part 1:661.

Ross, I. G., "High-Resolution Electronic Spectra of Large Polyatomic Molecules," 20:341.

Ross, J. and Richter, P. H., "Commentary: Dissipation Regulation in Oscillatory Reactions. Application to Glycolysis," 55:169.

Ross, J., *see* Greene, E. F., 10:135.; Hanusse, P., 38:317.; Ortoleva, P., 29:49.; Richter, P. H., 43:217.

Rousseau, Mme. J., *see* Laforgue, A., 8:141.

Rowlinson, J. S., "Penetrable Sphere Models of Liquid-Vapor Equilibrium," **41**:1.

Rowlinson, J. S. and Richardson, M. J., "The Solubility of Solids in Compressed Gases," **2**:85.

Rubin, R. J., *see* Weiss, G. H., **52**:363.

Sadron, C., *see* Douzou, P., **7**:339.

Sage, M. L. and Jortner, J., "Bond Modes," **47**:Part 1:293.

Samulski, E. T., *see* Tobolsky, A. V., **21**:529.

Sandars, P. G. H., "A Linked Diagram Treatment of Configuration Interaction in Open-Shell Atoms," **14**:365.

Schlecter, R. S., Velarde, M. G., and Platten, J. K., "Two-Component Bénard Problem," **26**:265.

Schlögl, F., "Glansdorff-Prigogine Criterion and Statistical Theory," **32**:61.

Schmalz, T. G., *see* McGurk, J. C., **25**:1.

Schmiedl, R., *see* Welge, K. H., **47**:Part 2:133.

Schneider, G. M., "Phase Equilibria in Fluid Mixtures at High Pressures," **17**:1.

Schnepp, O. and Jacobi, N., "The Lattice Vibrations of Molecular Solids," **22**:205.

Schoffa, G., "Magnetic Susceptibilities and the Chemical Bond in Hemoproteins," **7**:182.

Schoffeniels, E., "Commentary: A Complex Biological Oscillator," **55**:171.

Schuster, P., "Commentary: On the Origin of Optical Activity," **55**:109.

Schwarz, K. W., "Mobilities of Charge Carriers in Superfluid Helium," **33**:1.

Scriven, L. E., *see* Davis, H. T., **49**:357.

Scrocco, E., "Quantum Mechanical Interpretation of Nuclear Quadrupole Coupling," **5**:319.

Segre, M., *see* Cernuschi, M., **21**:455.

Severne, G., *see* Haggerty, M. J., **35**:119.

Shimanouchi, T., Tsuboi, M., and Kyogoku, Y., "Infrared Spectra of Nucleic Acids and Related Compounds," **7**:435.

Shirley, D. A., "ESCA," **23**:85.

Shuler, K. E., *see* Montroll, E. W., **1**:361.

Siegel, A., *see* Burke, T., **15**:355.

Silberberg, A., *see* Rabinowitz, P., **15**:281.

Silbey, R., *see* Chance, R. R., **37**:1.

Simha, R., *see* Rabinowitz, P., **15**:281.

Simon, M., *see* Bellemans, A., **11**:117.

Simon, W., "Selective Transport Processes in Artificial Membranes," **39**:287.

Sinanoğlu, O., "Electron Correlation in Atoms and Molecules," **14**:237.

Sinanoğlu, O., "Intermolecular Forces in Liquids," **12**:283.

Sinanoğlu, O., "Many-Electron Theory of Atoms, Molecules, and Their Interactions," **6**:315.

Siska, P. E., *see* Haberland, H., **45**:487.

Skancke, P. N., *see* Bastiansen, O., **3**:323.

Smaller, B., "Recent Advances in EPR Spectroscopy," **7**:532.

Smith, B., *see* Mueller, C. R., **21**:369.

Smith, I. W. M., "The Production of Excited Species in Simple Chemical Reactions," **28**:1.

Smith, R. L., *see* Truhlar, D. G., **25**:211.

Solin, S. A., "The Nature and Structural Properties of Graphite Intercala-

tion Compounds," **49**:455.

Solvay, J., "The Solvay Conferences and the International Cooperation in Science," **55**:1.

Somorjai, G. A. and Farrell, H. H., "Low-Energy Electron Diffraction," **20**:215.

Sørensen, G. P., *see* **Bak, T. A.**, **15**:219.

Spikes, J. D., *see* **Curtis, D. H.**, **21**:557.

Spohn, H. and Lebowitz, J. L., "Irreversible Thermodynamics for Quantum Systems Weakly Coupled to Thermal Reservoirs," **38**:109.

Sprandel, L. L., *see* **Carney, G. D.**, **37**:305.

Stebbings, R. F., "Charge Transfer," **10**:195.

Stecki, J., "Ionic Solvation," **6**:413.

Steele, W. A., "The Rotation of Molecules in Dense Phases," **34**:1.

Stell, G., Patey, G. N., and Høye, J. S., "Dielectric Constants of Fluid Models: Statistical Mechanical Theory and Its Quantitative Implementation," **48**:183.

Stephens, P. J., "Magnetic Circular Dichroism," **35**:197.

Stillinger, F. H., "Theory and Molecular Models for Water," **31**:1.

Stockmayer, W. H., *see* **Orwoll, R. A.**, **15**:305.

Stover, B., "The Dynamics of Life. IV. Lenticular Cataracts Induced by Aging and Irradiation," **21**:661.

Stucki, J. W., "Optimization of Mitochondrial Energy Conversions," **55**:141.

Suzuki, M., "Passage from an Initial Unstable State to a Final Stable State," **46**:195.

Sybesma, C., "Commentary: A Prebiotic Origin of Photosynthesis?," **55**:113.

Szwarc, M., "Recent Advances in Polymer Chemistry," **2**:147.

Tabor, M., "The Onset of Chaotic Motion in Dynamical Systems," **46**:73.

Tamayama, M., "On the Pressure Gradient in a Cylindrical Sample Container under High Pressure," **21**:737.

Taylor, H. S., "Models, Interpretations, and Calculations Concerning Resonant Electron Scattering Processes in Atoms and Molecules," **18**:91.

Taylor, N., *see* **Eyring, E. M.**, **21**:237.

Tensmeyer, L. G., Landis, P. W., and Coffey, H. F., "A Continuous Fall Method of Growing Crystals from Solution," **21**:767.

Theodosopulu, M., *see* **Dahler, J. S.**, **31**:155.

Thomaes, G., "The Bénard Instability in Liquid Mixtures," **32**:269.

Thomas, D., "Diffusion-Reaction in Structured Media and Membranes Bearing Enzymes," **39**:229

Thomas, D., "Functional Organization in Artificial Enzyme Membranes: Accomplishments and Prospects," **29**:113.

Thomas, R., "Logical Description, Analysis, and Synthesis of Biological and Other Networks Comprising Feedback Loops," **55**:247.

Thomas, R. N., "Stellar Atmospheres, Nonequilibrium Thermodynamics, and Irreversibility," **32**:259.

Thomson, T. R., *see* **Zandler, M. E.**, **21**:517.

Tiernan, T. O. and Lifschitz, C., "Role of Excited States in Ion-Neutral Collisions," **45**:82.

Tinoco, I., Jr., "Theoretical Aspects of

Optical Activity - Part Two: Polymers," **4**:113.

Tobolsky, A. V. and Samulski, E. T., "Solid 'Liquid-Crystalline' Films of Synthetic Polypeptides: A New State of Matter," **21**:529.

Tolmachev, V. V., "The Field-Theoretic Form of the Perturbation Theory for Many-Electron Atoms. I. Abstract Theory," **14**:421.

Tolmachev, V. V., "The Field-Theoretic Form of the Perturbation Theory for Many-Electron Atoms. II. Atomic Systems," **14**:471.

Tomassi, W., "Power Electrodes and Their Applications," **3**:239.

Trajmar, S., Rice, J. K., and Kupperman, A., "Electron-Impact Spectroscopy," **18**:15.

Tramer, A. and Nitzan, A., "Collisional Effects in Electronic Relaxation," **47**:Part 2:337.

Truhlar, D. G., Abdallah, J., Jr., and Smith, R. G., "Algebraic Variational Methods in Scattering Theory," **25**:211.

Truhlar, D. G., Mead, C. A., and Brandt, M. A., "Time-Reversal Invariance, Representations for Scattering Wave Functions, Symmetry of the Scattering Matrix, and Differential Cross Sections," **33**:295.

Truhlar, D. G. and Wyatt, R. E., "H + H_2: Potential-Energy Surfaces and Elastic and Inelastic Scattering," **36**:141.

Tsarevsky, A. V., *see* **Brodsky, A. M.**, **44**:483.

Tsuboi, M., *see* **Shimanouchi, T.**, **7**:435.

Tully, J. C., "Semiempirical Diatomics-in-Molecules Potential Energy Surfaces," **42**:63.

Turlet, J. M., Kottis, Ph., and Philpott, M. R., "Polariton and Surface Exciton State Effects in the Photodynamics of Organic Molecular Crystals," **54**:303.

Turner, J. S., "Complex Periodic and Nonperiodic Behavior in the Belousov-Zhabotinski Reaction," **55**:205.

Turner, J. S., "Finite Fluctuations, Nonlinear Thermodynamics, and Far-From-Equilibrium Transitions Between Multiple Steady States," **29**:63.

Turner, J. S., "Laboratory Experiments on Double-Diffusive Instabilities," **32**:135.

Tweedale, A., *see* **Laidler, K. J.**, **21**:113.

Ubbelohde, A. R., "Melting Mechanisms of Crystals," **6**:459.

Ubbelohde, A. R., "The Solvay Conferences in Chemistry, 1922–1976," **55**:37.

Umstead, M. E., *see* **Baronavski, A.**, **47**:Part 2:85.

Urry, D. W., "Biomolecular Conformation and Biological Activity," **21**:581.

Usemoto, H., *see* **Breckenridge, W. H.**, **50**:325.

Utsugi, H. and Ree, T., "Application of the Absolute Reaction-Rate Theory to Non-Newtonian Flow," **21**:273.

Van Artsdalen, E. R., *see* **Burton, L. L.**, **21**:523.

van der Waals, J. H. and Platteeuw, J. C., "Clathrate Solutions," **2**:1.

Van Herpen, G., *see* **Braams, R.**, **7**:259.

Van Hook, A., "Application of Transition State Ideas to Crystal Growth Problems," **21**:715.

van Kampen, N. G., "The Expansion of the Master Equation," **34**:245.

van Kampen, N. G., "Thermal Fluctua-

tions in Nonlinear Systems," **15**:65.

Van Leuven, P., *see* **Lathouwers, L.,** **49**:115.

Verdier, P. H., "Fluctuations in Autocorrelation Functions in Diffusing Systems," **15**:137.

Verlade, M. G., *see* **Schlecter, R. S.,** **26**:265.

Viaud, P. R., "Theoretical and Experimental Study of Stationary Profiles of a Water-Ice Mobile Solidification Interface," **32**:163.

Vrij, A., Joosten, J. G. H., and Fijnaut, H. M., "Light Scattering from Thin Liquid Films," **48**:329.

Walgraef, D., Dewel, G., and Borckmans, P., "Nonequilibrium Phase Transitions and Chemical Instabilities," **49**:311.

Wallace, S. C., "Nonlinear Optics and Laser Spectroscopy in the Vacuum Ultraviolet," **47**:Part 2:153.

Walter, C., "The Global Stability of Prey-Predator Systems with Second-Order Dissipation," **29**:125.

Walter, C., "The Use of Product Inhibition and Other Kinetic Methods in the Determination of Mechanisms of Enzyme Action," **7**:645.

Wartell, R. M. and Montroll, E. W., "Equilibrium Denaturation of Natural and of Periodic Synthetic DNA Molecules," **22**:129.

Wasielewski, M. R., *see* **Kaufmann, K. J.,** **47**:Part 2:579.

Watanabe, T., *see* **Kodaira, M.,** **21**:167.

Watts, R. O., Henderson, D., and Baxter, R. J., "Hard Spheres with Surface Adhesion: The Percus-Yevick Approximation and the Energy Equation," **21**:421.

Weber, G. R. and Eyring, L., "Self-Diffusion of Oxygen in Iota Phase Praseodymium Oxide," **21**:253.

Weeks, J. D. and Gilmer, G. H., "Dynamics of Crystal Growth," **40**:157.

Weeks, J. D., Hazi, A., and Rice, S. A., "On the Use of Pseudopotentials in the Quantum Theory of Atoms and Molecules," **16**:283.

Weeks, J. D., *see* **Andersen, H. C.,** **34**:105.

Weinstock, B. and Goodman, G. L., "Vibrational Properties of Hexafluoride Molecules," **9**:169.

Weiss, G. H., "First Passage Time Problems in Chemical Physics," **13**:1.

Weiss, G. H., "Some Models for the Decay of Initial Correlations in Dynamical Systems," **15**:199.

Weiss, G. H. and Rubin, R. J., "Random Walks: Theory and Selected Applications," **52**:363.

Weiss, N. O., "Magnetic Fields and Convection," **32**:101.

Weitz, E. and Flynn, G., "Vibrational Energy Flow in the Ground Electronic States of Polyatomic Molecules," **47**:Part 2:185.

Welge, K. H. and Schmiedl, R., "Doppler Spectroscopy of Photofragments," **47**:Part 2:133.

Wentorf, R. H., Jr., "Diamond Synthesis," **9**:365.

White, H. J., Jr., *see* **Moncrieff-Yeates, M.,** **21**:685.

Whittington, S. G., "Statistical Mechanics of Polymer Solutions and Polymer Adsorption," **51**:1.

Widom, B., "Collision Theory of Chemical Reaction Rates," **5**:355.

Wiegel, F. W. and Kox, A. J., "Theories of Lipid Monolayers," **41**:195.

Wiersma, D. A., "Coherent Optical Transient Studies of Dephasing

and Relaxation in Electronic Transitions of Large Molecules in Condensed Media," **47**:Part 2:421.

Williams, F., "Electronic States of Solid Explosives and Their Probable Role in Detonations," **21**:289.

Williams, R. J. P., *see* **Braterman, P. S.,** 7:359.

Williams, R. J. P., *see* **Campbell, I. D.,** **39**:55.

Williams, W., *see* **Mueller, C. R.,** **21**:369.

Wilson, E. B., Jr., "The Problem of Barriers to Internal Rotation in Molecules," 2:367.

Winterkorn, H. F., "Analogies Between Macromeritic and Molecular Systems and the Mechanical Properties of Sand and Gravel Assemblies," **21**:751.

Wittig, C., *see* **Reisler, H.,** **47**:Part 1:679.

Wolfsberg, M., *see* **Bigeleisen, J.,** 1:15.

Wolpert, L., "The Development Pattern: Mechanisms Based on Positional Information," **29**:253.

Woodbury, J. W., "Eyring Rate Theory Model of the Current-Voltage Relationships of Ion Channels in Excitable Membranes," **21**:601.

Woodcock, L. V., *see* **Angell, C. A.,** **48**:397.

Woodin, R. L. and Kaldor, A., "Enhancement of Chemical Reactions by Infrared Lasers," **47**:Part 2:3.

Wooley, R. G., "The Electrodynamics of Atoms and Molecules," **33**:153.

Wyatt, R. E., *see* **Truhlar, D. G., 36**:141.

Yang, K., "The Mechanism of Electronic Energy Transfer Between Excited Mercury ($^3P_1$) Atoms and Gaseous Paraffins," **21**:187.

Yao, S. J. and Zwolinski, B. J., "Studies on Rates of Nonequilibrium Processes," **21**:91.

Yeager, D. L., *see* **Herman, M. F., 48**:1.

Yeager, D. L., *see* **Olsen, J., 54**:1.

Yip, S., *see* **Desai, R. C., 15**:129.

Yoshizumi, H., "Correlation Problem in Many-Electron Quantum Mechanics. II. Bibliographical Survey of the Historical Developments with Comments," 2:323.

Zahner, J. C., *see* **Drickamer, H. G.,** 4:161.

Zander, R., "Commentary: Observational Aspects Related to the Chemical Evolution of Our Atmosphere," **55**:79.

Zandler, M. E. and Thomson, T. R., "Application of the Significant Structure Theories to Plastic Crystals," **21**:517.

Zewail, A. H., *see* **Burland, D. M., 40**:369.

Zwanzig, R., "Langevin Theory of Polymer Dynamics in Dilute Solution," **15**:325.

Zwolinski, B. J., *see* **Yao, S. J., 21**:73.

SUBJECT INDEX

A

α and β bands, 7:382

α process, 44:399

α-β transformation, 16:267–268, 380

Ab initio calculations, 12:332; 42:427, 522

Ab initio pair potential, 53:21, 33ff, 40, 46

Abelian integral, 30:338, 353

Abelian theorem, for Laplace transforms, 52:386, 388, 394–395

Abrikosov diagram, 36:224

Absolute reaction rate theory, 1:30, 124, 362, 391; 10:265–266

applied to non-Newtonian flow, 21:273

Absorbing barrier, 52:410, 425, 480

Absorbing boundary, 52:407ff, 449, 455, 480, 481

Absorbing point, 52:411–413

Absorbing sphere model, 50:345

Absorptance, 27:436, 437, 457, 461, 501–502

Absorption, anisotropic, 49:17, 31–33

of carbon dioxide in water, 13:274

collision-induced, 51:49–109

correction, for molecular fluids, 53:20–30

in dilute metal-ammonia solutions,

4:312, 313, 315

double resonance, 25:19, 53, 56

electric dipole, 33:248

Lamb dip, 25:16

molecular gas, 50:71–84

multiphoton, *see* Multiphoton excitation, Two-photon excitation

multipole, 33:238

nonresonant, 12:491, 539, 540; *see also* Nonresonant absorption

optical, inert gas liquid and solid, 31:375, 376

irradiated glasses, 31:371

irradiated liquid propane, 31:369

proper modes, 44:337

quadrupolar, 51:79

rotation-translation, 51:74

rotational band, 51:74

saturation dips, 25:21

steady state, 25:15

double resonance, 25:19

surface induced, 27:508

transient, 25:22, 23; *see also* Transient absorption

two-photon, monitoring techniques, 47:Part 2:20–22; *see also* Multiphoton excitation; Two-photon excitation

velocity dependence, 25:12

see also Far infrared; Induced absorption; Power absorption; Translational absorption

Absorption coefficient, 4:143; 12:121; 25:7; 27:507; 29:59; 44:347, 549ff
of a barrier, 1:382
in electron diffraction, 53:5
integrated, 51:57

Absorption cross section, 27:441; 30:48; 44:309

Absorption intensity, 4:149

Absorption line shape, bond modes and, 47:Part 1:315–321; *see also* Line shape

Absorption spectra, 33:238, 248, 290
of actinides, 5:135
of alkali halides, 5:132
of aromatic compounds, 5:17ff
of charge transfer complexes, 8:13ff
dip in, 51:58
of ionic melts, 6:477
optical, 4:68, 71, 83, 89, 93, 94, 107, 109, 128, 136, 161, 165, 182, 186, 187
line shape, 13:126
see also Electron transfer spectra; Electronic transitions
satellite bands in, 33:289
of transition metal complexes, 5:256

AC conductivity, 52:461, 477

Accelerator, 36:37

Accessible domain, 43:17, 40

Accessible state, physical limitations to, 43:40–43

Acentric factor, 12:292

Acetabularia, current oscillations in, 29:291, 292
current-voltage relation in, 29:293
potential gradient in, 29:289
potential oscillations in, 29:290, 291

Acetaldehyde, barrier height of internal rotation, 2:378, 382, 383, 388

Acetoacetate decarboxylase, 39:153, 156

Acetone, 47:Part 1:641
clusters, 52:345–347

Acetonitrile, 34:190
in clathrate, 2:20
clathrate in hydroquinone, 2:7, 11
isotropic Raman linewidth, 40:39

Acetyl acetonate, 7:376

Acetyl fluoride, barrier height of internal rotation, 2:383

Acetyl phosphate, 7:66

Acetylcholine, 21:621; 39:325

Acetylene, 52:347–350
clathrate in hydroquinone, 2:7
electron impact spectra, 18:56, 57
interatomic distances, 5:156
as quencher, 50:366
solid, 22:260
librations of, 22:221

Acetylene hydrate, thermodynamic data and lattice constants, 2:8

N-Acetylglycine, 7:547

Acetylsalicylate, deacylation of, 39:116, 119

Acid-bases, 1:62

Acid dissociation constants of oxime complexes, 7:374

Acidity, gas phase, of alcohols, 19:192, 204
of amines, 19:192

Acids, 1:67–69

Acoustical absorption, in crystals, 42:320

Acrylamides, polymerization of, 2:181

Acrylonitrile, 2:155

Actinides, 41:185
absorption spectra, 5:135
hexafluorides, 9:185ff

Landé parameter, **5**:134

Actinomycin C, 8:173

Action, 33:169

classical, **30**:87, 321

Action-angle variables, 25:105, 111, 123, 129, 133, 134, 153; **30**:86, 91, 110, 129; **31**:203; **36**:68; **46**:75; **53**:315

Action function, 36:6, 8, 10, 23, 52, 67; **41**:128

Action integral, 30:105; **36**:10, 26, 30

Action potential, 29:269, 321, 331, 346, 349

electrical stimulation, **29**:331

pressure stimulation, **29**:332, 333

space-clamped, **29**:368

Action space, 5:356–358, 370

Activated complex, 28:145

Activated complex theory, 1:23, 29–31, 66, 73, 127; **10**:190, 261, 263, 264–265, 266–267; **12**:284; **21**:115

applied to H + H$_2$ reaction, **21**:120

Activated state, 1:18, 56, 61, 119, 124, 125, 129, 130, 389; **7**:209

Activation, 29:50, 260

of membrane-bound enzymes, **29**:114, 119

sodium, **29**:361

Activation energy, 1:33, 87, 120–122, 126, 128, 132, 162, 363, 365, 390; **7**:147, 212; **25**:145, 146; **33**:105; **36**:143, 159, 163; **41**:235, 239–241; **42**:10

algebraic methods, **25**:212, 214, 266–268, 279

effect of impurities, **7**:307ff

for methyl rotation, **44**:381

Activation enthalpy, 33:99, 100, 107, 116, 117

negative, **33**:100, 117, 118, 120

Activation entropy, 33:105, 107

negative, **33**:118, 120

Active center, 7:124

Active medium, 27:360

Active molecule, 1:31

Active nitrogen, 28:387

Active site, of enzyme, **39**:170, 172

of lyzozyme, **39**:70

of protein, **39**:55–108, 339

Active transport, *see* Transport

Activity coefficient, 2:125; **11**:8, 30, 31, 36, 42, 44–46, 57, 59, 69, 70, 72

of copper-gold (Cu$_{75}$Au$_{25}$, Cu$_5$Au$_5$) alloys, **2**:125

Debye-Hückel values of, **11**:42, 44

Activity, absolute, **2**:12, 13; **20**:40

Adamantane (sym-tricyclodecane), multipole interaction, **5**:180

Adams and Wilde approximations, 3:91

Addition polymerization, 2:148

Addition reaction, 19:101, 105

Additivity, of bond energies, **9**:352

of intermolecular potential, **1**:267, 294

of magnetic properties, **3**:202

Adenine, 7:47ff, 52, 85

internuclear distance, **7**:87

molecular diagrams, **7**:94

transition moment, **7**:95

Adenine-thymine pairs, 7:54ff

Adenosine, infrared spectra, **7**:445ff, 452, 457ff

substrates, **7**:716ff

Adenosine diphosphate, 7:63ff

Adenosine hydrochloride, 7:447

Adenosine monophosphate, 7:66

Adenosine triphosphate, 7:63, 64; **19**:363; **29**:272, 303

in glycolysis, **29**:138, 144

hydrolysis of, **29**:345

Adenylate cyclase, 38:367, 385; **39**:206–

208
  allosteric model of, **38**:390
Adenylic acid, 7:349
Adiabatic approximation, 1:342; **10**:201,
    212; **12**:9, 12, 13, 16, 24; **14**:425,
    429; **16**:297; **28**:324, 353; **30**:20;
    36:278, 280, 290; **44**:5, 24, 45,
    491; **49**:117, 130–159, 181, 183,
    195; **52**:122, 123
  application to, $B^1\Sigma_u^+$ H$_2$, **12**:24
    ground state D$_2$, **12**:22
    ground state H$_2$, **12**:22
    ground state HD, **12**:22
  vibrational, **30**:31
Adiabatic breakdown, 49:143
Adiabatic collision, 28:179, 324, 353
Adiabatic compressibility, 16:236–239
  reduced, **16**:242–243
Adiabatic correction, 44:104; **49**:138,
    140, 148, 152
Adiabatic correction potential, 52:129
Adiabatic coupling, 28:327, 350
Adiabatic-diabatic transformation,
    49:278–283
    in three dimensions, **49**:283–287
Adiabatic eigenvalue problem, 49:139
Adiabatic elimination, 49:318, 323, 326
Adiabatic level crossing, 35:22
Adiabatic limit, 49:141
  Langevin model, **53**:94–96
  truncated chain model, **53**:100
Adiabatic modes, 32:228
  classification of, **32**:228, 242, 253
  continuous eigenvalues of, **32**:228,
    242, 253
  convection criterion related to,
    32:228, 253
  dynamical stability, **32**:228, 253
  p- and g-, **32**:228, 241, 252
Adiabatic nuclear motion function,
    49:138, 140, 181

Adiabatic potential, 10:66–67; **18**:138
Adiabatic principle, 28:176, 177
Adiabatic process, 24:107
Adiabatic region, 28:324, 325, 355
Adiabatic representation, 49:176–178
Adiabatic separation, 44:3, 104
Adiabatic spectroscopic constants,
    49:171
Adiabatic state, 23:164; **30**:305, 465,
    469, 512
Adiabatic transition, 10:198, 254, 257,
    267, 269
Adiabatic turn-on, 35:274, 316, 317
Adipic acid radical, 7:546
Adjoint tableaux, 14:100
ADP, *see* Adenosine diphosphate
Adsorption, 27:218
  activated, **41**:498
  electronegative, **27**:218
  electropositive, **27**:218
  of gases, amorphous, **20**:324
    chemisorption, **20**:314, 315, 321
    coadsorption, **20**:323
    correlations, **20**:327
    desorption, **20**:327, 328
    physical, **20**:314, 315
  Gibbs law, **6**:151–153
  Langmuir formula for, **6**:170
  lattice model of, **52**:449
  at liquid-gas interface, **6**:169–170
  at liquid-liquid interface, **6**:169–170
  potentiometric investigations of
    adsorbents, **3**:258
  related to catalysis, **21**:793
  thermodynamics, **7**:244
  of water on metals, **21**:363
Adsorption isotherm, 7:239ff; **37**:125
Adsorption overlayer, 27:146
  diffusion of impurities, **27**:145
  inequivalent domains, **27**:148

nonuniform, **27**:145

out of phase domains, **27**:148

reflection registry degeneracy, **27**:149

registry degeneracy, **27**:148

rotational registry degeneracy, **27**:149

translational registry degeneracy, **27**:148

uniform, **27**:145

Advective terms, 32:78, 79, 80, 82

Aerodynamic methods, 28:213

After-effect function, 17:71, 77, 81

Afterglow, flowing, **45**:113–115, 427

Aggregates, in stellar systems, **26**:137, 140, 141, 142

Aggregation, and dissipative structures, **29**:138, 154

of amphiphilic molecules, **29**:309

models of, **29**:295

of slime molds, **29**:152, 171, 295

Aging, 7:85

of colloid, **43**:238

Agrofonov, Martinov and Sarkisov theory, 48:244–245

Agrostis stolonifera, **7**:625

free radicals from irradiation, **7**:640

Air, + carbon dioxide, **2**:96, 107, 111

second virial coefficient, **2**:111

+ ice system, **2**:98

Airglow, twilight, **42**:149

Airy function, 10:89–90, 104, 167

boundary method, **42**:388

Alametycin, 29:358

Alanine radical, 7:550

Alcohol narcosis, 21:647

Aldehydes, aromatic, **8**:88

base strength of, **8**:133, 136

Alfrey-Price correlation, 33:347, 381

Algebra, Abelian, **38**:25

B*, **38**:38

of bounded linear operators, **38**:21

of observables, **38**:24

duals of, **38**:32

representations of, **38**:37

type II and type III, **38**:13

W*, **38**:37

Algebraic mapping, 46:79, 134

quantization of, **46**:135

Alhassid-Levine method, 53:367

Aliphatic substitution, nucleophilic, **8**:92

Alkali-halogen systems, 42:11, 250

Alkali metal, 27:526

atom, autoionization, **50**:200, 234

excitation of, **45**:391–393

highly excited, **42**:12

quenching of, **50**:333

crystals, magnetic susceptibility of, **3**:220

dimer, **42**:7

photodissociation of, **50**:268, 290

predissociation of, **50**:227

reactions of, **30**:259, 280, 430

electronic correlation energy, **2**:252

ground state energy of, **6**:125, 129

polarizability of, **10**:9–20

Alkali metal halides, 34:193

absorption spectra, **5**:132

association in vapors, **11**:94, 96, 98

color centers in, **4**:162, 194–198

corresponding states law, **11**:95

dimerization, **11**:94

enthalpy of, **11**:97

entropy of, **11**:97

heats of fusion, **9**:70

heavy metals in, **4**:162, 188–194

phosphors, **4**:188–194

reduced equation of state, **11**:93

scattering from, **42**:479

thermodynamic data for vapors,

11:96

trimers, **11**:94

Alkali metal hydrides, scattering from, **42**:482

Alkaline earth chloride solutions, **34**:195

Alkaline earth crystals, magnetic susceptibility of, **3**:220

Alkaloids, infrared spectra, **7**:421

Alkanes, as quenchers, **50**:364ff

normal, **16**:228–230, 234, 252, 263, 270, 278

mixtures, **16**:229, 247–248, 252, 257–262, 276

principle of corresponding states for, **16**:235–245

Allen and Cole kinetic theory, **24**:292

All-or-none approximation, **33**:100, 101, 105, 116

All-or-none conformational change, **33**:106

All-or-none phenomenon, **29**:43, 337, 346, 350

All-or-none reaction, **33**:99, 108

Allosteric activator, **46**:380, 382

Allosteric effect, **46**:380, 382

Allosteric effector, **39**:192; **46**:377, 382

Allosteric enzyme, **38**:364; **39**:6, 177

Allosteric equilibrium, **39**:91

Allosteric inhibitor, **46**:382

Allosteric mechanisms, **39**:59, 92

Allosteric model, **29**:34, 146, 303; **39**:30, 31, 193–195

Allosteric protein, **46**:378, 382

Allosteric regulation, **39**:193, 196, 202

Alloxazine ring, **7**:70

Alloys, density of states of binary, **13**:154, 155

order-disorder, **11**:40, 41

Alternancy symmetry, **52**:74–81

Alternant hydrocarbons, **7**:37, 41

substitution in, **8**:75ff

Alternant molecular orbital method, **6**:335; **8**:4; **14**:330

Aluminum, **27**:526; **41**:141, 166, 171

bulk, **49**:561

electronic correlation energy (Al^{3-}), **2**:240

liquid, electron states, **31**:281

local surface field, **49**:457–470

sulfur adsorption, **49**:602

surface state, **49**:573

symmetry properties, **49**:575

Amide group, resonance coupling, **53**:287

Aminidases, **7**:74

Amino acid, **7**:265, 546

N-carboxyanhydride, **2**:170

constants, **7**:548

internal rotation of, **39**:73, 80

isoenergy contours, **39**:97

prebiotic synthesis, **55**:91–94

radicals, **7**:546ff

Ammonia, **34**:185, 193

adsorbed on iridium (111), **49**:634–635

adsorbed on nickel (111), **49**:635

aqueous, **34**:193

clathrates, **2**:5

clusters, **52**:347

inversion microwave spectrum, **12**:133

superposition of configurations, **2**:296

Ammonium dihydrogen phosphate crystal, **49**:21

Amnesia kernel, **20**:153, 191

Amorphous packing, **40**:72, 82

Amorphous phase, **48**:457

Amorphous semiconductor, energy transport of, **52**:475

AMP, *see* Adenosine monophosphate

Amphiphatic molecule, **41**:195

Amphiphilic molecule, **41**:195

Amplitude grating, **49**:17

Amplitude modulation, **44**:324, 327

Analytic continuation, **25**:70, 137, 139; **30**:114, 337

Analytic direct correlation approximation, **48**:217

for polarizable molecules, **48**:288

Analyticity, **25**:254

Analyzer cells, **19**:147

trapped ion, **19**:150

Andersen and Chandler approximation, **48**:224

Anderson localization model, **31**:336, 451; **40**:457, 458

Anderson's theory, **12**:493, 504, 509, 533

dipole-dipole broadening, **12**:512, 516, 525–529, 531, 534

frequency shift, **12**:497

phase shift, **12**:496, 503

Anemonin, **16**:177

Angle function, **51**:122, 150–153

Angular acceleration, **44**:402

Angular displacement, **44**:278

higher derivatives of, **44**:402

Angular distribution, **12**:399, 400, 481; **36**:212; **41**:249; **44**:268

of emitted molecular fragments, **16**:80–81

in harpooning, **10**:377

of photofragments, **50**:276ff

quantum interferences in, **12**:389

in reactive scattering, **10**:328–340, 360–361

in violent charge transfer, **10**:212–215

see also Cross section, Interference pattern

Angular frequency, **44**:303

Angular momentum, **12**:14, 15; **18**:183; **28**:20–22; **35**:124, 157; **44**:273, 276, 357; **49**:123–125, 161–162, 169, 181

autocorrelation function, **17**:163, 171

barrier resonance, **18**:96

consumption, **42**:16, 19

coupling, **30**:372

of disk, **44**:279

eigenstates, **49**:180–181

mechanical, **7**:164

nuclear, **49**:162–163, 171, 183

operators, **9**:223

orbital, **7**:162

projected kernel, **49**:165, 172, 181

quantum numbers, **7**:10; **25**:215, 256

of stellar systems, **26**:118, 122, 129, 135

total, **7**:163; **49**:159

Angular overlap model, **8**:60

Angular resolution, **18**:23, 86

Angular speed autocorrelation function, **17**:153

Angular velocity, **44**:402

autocorrelation function, **44**:391, 400

correlation function, **44**:279, 303

disk, **44**:268

Euler space, **44**:360

sphere, **44**:268

Anharmonic converter, **52**:149–150

Anharmonic correction, **52**:153

Anharmonic effects, **27**:404–414, 476, 480, 483

Anharmonic force, **36**:32

Anharmonic oscillator, V-V transfer, **23**:41, 55

Anharmonicity, **52**:149–151

atomic lattice vibration, **53**:294

classical motion, **53**:36–317

molecular vibration, **53**:318ff
periodic vibration, **53**:314
vibrational, **40**:23, 28, 29
xenon-xenon bond, **53**:316
1, 8-Anilino-naphthalene, 49:34
Anionic polymerization, 2:176, 177, 181
Anisotropic excitation, 16:82–90
distribution function, **16**:85–88
selection rules, **16**:89–90
Anisotropic interaction, 2:33
dipolar, **7**:540ff
intermolecular, **12**:555, 557, 564, 571, 572, 581, 585, 597
Anisotropic molecular fluids, depolarized light scattering by, **26**:55–61
Anisotropic molecules, 24:43
Anisotropic velocity distribution, 35:121, 124, 154
Anisotropy, 36:167, 175, 176, 179, 182
hard core, **44**:434
Anisotropy coefficient, 53:264
Anisotropy ratio, 44:555, 561, 567, 575, 597, 601, 629
Anisotropy terms, 44:68, 453
Annealing, 7:221
Annihilation operators, 14:41, 95
Anomalous dispersion, phase determination by, **16**:205–215
Anomalous fluctuation theorem, 46:200
Anomalous scatterer, 16:206, 208–209, 213
Anomalous skin effect, 27:278, 281, 282, 290, 291
Anthracene, 2:102, 153; **40**:446, 447, 463ff, 472ff
core parameters, **5**:137
crystal, **54**:372–376
a-polarized reflectivity system, **54**:396–398
b-polarized reflectivity system, **54**:385–390

dispersion formulae, **54**:340
dispersion of site shift excitons, **54**:453
exciton surface polaritons, **54**:451–455
first singlet transition, surface excitons, **54**:322–325
fluorescence, **54**:322, 398–406
first singlet transition, **54**:320–325
measurement of thickness, **54**:374–375
mounting and orientation, **54**:376
refining and crystal growth, **54**:373–374
reflectivity, **54**:321–322
influence of surface modifications, **54**:396–398
and fluorescence, **54**:381–406
semiclassical theory of, **54**:340–353
dispersion due to one transition, **54**:348–352
dispersion equation of isotropic, **54**:340–346
effect of several transitions, **54**:352–353
structure, **54**:372–373
surface excitons, **54**:322–325
transmission and Davydov splitting, **54**:320–321
Davydov splitting, **23**:311
diffusion in air, **5**:163
electron and hole mobility, **7**:294, 304
exciton bands, **23**:309–327
fluorescence, first singlet transition, free molecule, **54**:320
monocrystal, **54**:320–325
fluorescent decay, **42**:260
ionization potential, **7**:302

reflection spectra, **23**:312, 317

rigid convex body constants, **5**:163

solid, **22**:226, 241, 259

9-Anthraldehyde, 8:134, 136

Anthraquinones, polarization of electronic transitions, **13**:193

Antibiotics, structure of, **7**:422

Antibodies, 7:262

Antibonding band, semiconductor, **53**:292–293

Antibonding molecular orbital, 7:19ff, 92, 144, 168

Anticommutation relations, 14:42, 96

Antiferromagnet, continuum limit Hamiltonian, **53**:234

domain walls, **53**:261

dynamic structure factor, **53**:256

kinks and, **53**:250–251

magnetic and elastic dynamics, **53**:264

solitary waves, **53**:238

spin Hamiltonian, **53**:232

spin vectors, **53**:231

Antiferromagnetic element, 51:241, 253

Antiferromagnetism, 1:303; **2**:313; **20**:77, 89, 91

Antigenic effect, 7:262

Antigraph relations, 36:234, 271

Anti-heroin agent, 16:199

Antikinks, *see* Kinks-antikinks

Antimony pentachloride, 20:33

Antimony sulfide, domain walls, **53**:275

Antiparallel domains, 53:274

Anti-radiation agent, 16:199

Antiresonance, 25:213, 275, 291

Antiserum, 7:223

Antisolitons, density, **53**:245

Anti-Stokes vibrational transition, 42:307

Antisymmetric functions, 2:260, 261;

8:36

Antisymmetrized products, 6:321, 339

Antitumor activity, 7:60ff

Aperiodic crystals, 7:202

Aperiodic motion, 46:33

Appearance energy, 52:265

Appearance potential, 10:175, 191, 227–228, 252; **19**:43, 57, 123, 126

Appearance potential spectroscopy, 27:26, 27

Approximate density functional model, 49:414, 416, 419, 423

Approximate methods, validity of, **9**:347

Apurinic acid, infrared spectra, **7**:466

Aqueous electrolyte theory, 11:108

Arc discontinuity, of glycolytic oscillator, **38**:378

Arc jet nozzle beams, 10:310

Area-preserving mapping, 24:160, 164, 174, 177, 179

Arenomium ions, 8:70ff

L-Arginine dihydrate, 16:187

ArH⁺, ArD⁺, 19:39, 42, 44, 47, 52, 53, 59, 84, 85, 97

Argon, 1:138, 151, 152, 160, 276, 282, 292, 293; **16**:111–112, 115–117, 120–121, 125, 128, 381–382; **27**:476, 479, 486; **34**:179; **37**:175; **41**:140; **49**:295

clathrate, **2**:20, 21, 30, 37, 41

clusters, **52**:281

collision cross sections, **21**:373

complexes of iodine and, **47**:Part 1:358–360

compressibility of, **9**:55

correlation function calculation, **53**:192, 195–196, 197

crystalline, **21**:384

dimer, **41**:157; **46**:52

electron impact spectra, **18**:21

equivalent chain model, **53**:102

fluid structure, **53**:13–15

hexagonal close-packed, **16**:345–347, 350, 352–354, 360, 381–382

hydrate, **2**:28, 31, 32

  thermodynamic data and lattice constants, **2**:8

  vapor pressure, **2**:31

light scattering, **24**:40, 50, 52, 53

liquid, **21**:385

London coefficient for, **6**:399

metastable, polarizability of, **10**:20–25

  quenching of, **50**:310, 381, 510

microclusters, **40**:55

polarizability, **46**:52

  magnetic susceptibility, and ionization energy, **2**:71

  potential energy, **2**:71

second virial coefficient, **21**:379

stacking fault energy, **16**:346

structural and thermodynamic data, **16**:346–348

see also Scattering; van der Waals molecules

Argon-carbon monoxide, phase diagram, **16**:350, 352, 354, 356

Argon-fluorine, phase diagram, **16**:350, 356, 359

Argon-helium mixture, **16**:127

  viscosity of, **13**:363

Argon-nitrogen, phase diagram, **16**:350–351, 354–355

Argon-nitrogen dioxide-dinitrogen tetroxide mixture, **16**:103, 111

Argon-oxygen, phase diagram, **16**:350, 356, 358–359

Aromatic hydrocarbons, **2**:200; **7**:152, 305

  absorption spectra, **5**:17ff

  carbonyls, **35**:4

  carcinogens, **8**:119, 163ff

coordinate transformations, **5**:253

diamagnetic anisotropy of, **3**:190

electrical properties, **7**:305

forbidden transitions, **5**:241

heteroanalogues, **9**:139ff

homogeneous complexes, **7**:321ff

magnetic properties, **7**:323ff

nitrogen containing, **7**:91

  basicity of, **7**:59

photoconduction, **7**:310

protonation of, **8**:70ff

symmetry operations, **5**:253

vapors, mobility of thermal electrons in, **36**:505, 506

vibronic coupling constants, **5**:254, 257

see also Hydrocarbons

Arrested relaxation, method of, **28**:56, 57, 58, 92

Arrhenius equation, **1**:17, 65, 120, 388; **48**:407, 459, 468

Arrhenius kinetics, detonations, **3**:121

  flames, **3**:76

Arrhenius plot, **25**:144

Arrhenius-Thovert effect, **6**:310

Arrival rate ratio, **40**:140, 141, 143

Arsenic bromide, **34**:187

Arsenic pentafluoride, **49**:459

Arsenic structure, **49**:67, 96–97, 100, 104

Articulation point, **11**:337, 345, 346, 347, 357, 362, 378; **34**:150

Articulation star, **20**:49

Ashiem-Zondek reaction, **7**:433

Aspartate transcarbamylase, **39**:195–198

Aspartic acid, radicals, **7**:551

Aspect ratio, **26**:185, 194, 196, 200

Aspirin, aminolysis of, **39**:116, 117, 119

Association complexes, in ionic melts,

6:469–476

Association reaction, 28:33–38

Associative electron detachment, 30:277, 527

Associative ionization, 36:186, 188, 190, 198; 42:514, 578; 52:303, 317

Aston bands, 10:209; 45:92

Astronomy, 26:115

Asymmetric charge transfer, 10:197, 240–241

Asymmetric forms, stochastic models of synthesis, 55:201–204

Asymmetric random walk, 52:483

Asymmetric resonance, 10:197–198, 242–243, 254–256

Asymmetric rotor, 20:346; 44:265, 276, 357, 359

  Brownian angular diffusion, 44:266

  Coriolis coupling, 20:352

  energies and line strengths, 20:346, 353

  inertial defect, 20:352

  Langevin equation, 44:374

  nonrigid, 20:351

  Stark effect in electronic spectra, 20:351

Asymptotic expression, for correlation, 48:198

Asymptotic form, of wave function, 25:216

Asymptotic master equation, 53:349, 365

Asymptotic phase shift, soliton collisions, 53:239

Asymptotic polarization, 35:20

Asymptotic properties, central limit theorem, 52:377–381

  discrete results, 52:391–407

Asymptotic ray, 43:175

Asymptotic series, 46:210

Asymptotic stability, 33:434–436

tests for, 43:84, 90, 91, 206

Asynchronous logical description of feedback loops, 55:252–266

  conditions of various sequences, 55:259–264

  equations, 55:255–256

  final states, 55:264–266

  graph and sequences of states, 55:258–259

  involvement of time, 55:252–253

  state tables, 55:256–257

  variables and functions used, 55:253–255

Atmosphere, chemical evolution of, 55:79–94

  pollution, 55:80–84

  trends, 55:80

  variability, 55:79–80

Atmospheric chemistry, 50:262; 55:63–78

  action of minor constituents, 55:67–71

  hydrogen and mesospheric chemistry, 55:71–72

  hydrogen, nitrogen, and halogen compounds in troposphere, 55:74–76

  pure oxygen atmosphere, 55:64–67

  solar radiation absorption, 55:64

  stratospheric chemistry, halogen compounds, 55:73–74

    oxygen and nitrogen compounds, 55:72–73

Atmospheric systems, excited states, 45:237–340

Atom-atom potential, 44:259, 374, 401, 436

Atom-atom scattering, 25:238, 282

Atom-chain collsions, 53:101–106

Atom-exchange reaction, 28:38–62, 388, 394, 397, 412

Atomic chains, 9:101

Atomic charges, correlation of, 23:119

Atomic coordinates, dynamics of, 53:280

Atomic deformation polarization, 1:77

Atomic diameter, 50:331

Atomic interaction, atomic lattice, 53:294–297
  molecular crystal, 53:278
  recombination dynamics, 53:106–111
  solitons, 53:320

Atomic magnetic susceptibility, 3:204

Atomic many-electron theory, 6:376–383

Atomic orbital basis, 42:172

Atomic orbitals, 1:176, 241, 245, 248, 252, 254, 256, 258, 259; 7:10ff; 9:88
  orthogonal, 1:259

Atomic polarizability, 10:1–27
  of alkali atoms, 10:8–20
  anisotropy, 10:5–6, 21
  beam measurements of, 10:9–26
  of metastable $^3P_2$ argon, 10:20–25
  tensor, 10:3–4, 20–21, 25

Atomic radial distribution function, 34:168

Atomic recombination, 42:2

Atomic relaxation, 41:308–310

Atomic resonance fluorescence, 50:3

Atomic scattering factor, 16:137, 205, 207–210

Atom-molecule reaction, initiation by MPD, 47:Part 1:625–638

Atom-molecule scattering, 25:282, 283

Atoms, crossed beams, 45:342–394
  collisional quenching of, 50:325–394
  laser excited, in crossed beams, 45:342–394
  quenching of electronically excited states, 45:342–394

reaction with hydrogen halides, 47:Part 2:6–7

Atom-size effect, 2:141

Atom-surface scattering, 30:105

ATP, see Adenosine triphosphate

ATPase, 29:138, 344; 39:205, 326

Attenuated total reflection, 27:339, 445–449, 519–523, 525–526

Attractant, 29:170

Attractive force, 41:205, 206
  intermolecular, 2:60

Attractor state, 38:337, 341

Audiofrequency range, 44:259

Aufbau process, 7:27

Auger effect, 10:21, 174–175

Auger electron emission yields, from Mo, 45:92

Auger processes, 27:18

Auger spectroscopy, 27:26; 41:487

Autocatalysis, 29:33, 50, 67, 125, 345; 38:235, 364
  in Belousov-Zhabotinski reaction, 29:157
  and cell aggregation, 29:295
  and evolution, 29:20, 26
  in the trimolecular model, 29:33, 159

Autocorrelation function, 15:66, 75, 130, 137ff; 17:85, 108; 20:198, 208; 29:176; 44:345, 347, 400, 425, 429; 51:52; 52:375
  angular momentum, 17:153, 163, 171
  angular velocity, 44:391, 400
  coupled, 44:284
  derivative, 44:428
  dipolar, 44:262, 338, 342, 360, 380
  dipole, 17:142
  fluctuations in, 15:137ff
  kink-position, 53:273–274
  momentum, 17:163
  rotational kinetic energy, 17:155

translational kinetic energy, **17**:155

velocity, **17**:141, 152

see also Correlation function; Fluctuations; specific autocorrelation functions

Autocorrelation technique, 49:5–6, 8–11

Autoionization, 10:178; **30**:5, 419, 448, 449, 455, 458; **42**:195; **50**:191–254; **52**:33, 264, 281, 288, 294

electronic, **52**:287

mechanism, **52**:285

of negative ions, **19**:152

vibrational, **52**:287

Automorphism, 38:16, 49, 72, 73

of orthomodular lattice, **38**:16

of propositional system, **38**:16

on center, **38**:72, 73

Autonomous center, 38:345, 348

Available energy rate, 13:96, 105

Avena sativa, **7**:634

Average field approximation, 42:200

Average potential model, 2:134; **11**:117, 119, 121, 123, **16**:226

Averages, notation for, **35**:62

over Euler angles, **35**:47

over internal configurations, **35**:100

phase space, **35**:56

Avoided crossing, 42:2, 79, 104, 190

field induced, **50**:239

Avrami equation, 15:180

Avrami-Erofeev equation, 15:180

Axilrod-Teller-Muto forces, 30:364

Axilrod-Teller triple dipole term, 53:13, 18

Axon, 19:339

Azauracil, 7:53

Azulene, deuterated, **23**:206

emission of, **23**:202

gas phase, **23**:207

ground state, **23**:204

rate of repopulation, **23**:205

intersystem crossing, **23**:206

isolated molecule, **23**:203

quantum yield, **23**:207

relaxation time, **23**:203, 207

Azulene-1-aldehyde, 8:135

Azulene-1,3-dipropionic acid, 16:181–183

B

β **process, 44**:399

β-**filter, 34**:162

B⁺ algebra, 40:303

Bacillus subtilis, **7**:220

Backfolding, 41:202, 218, 219

Background scattering, 10:41–42, 50–53, 62

Bäcklund transformation, 53:239, 301, 328–332

Backscattering, 10:44, 46, 106

Backward evolution operator, 36:40

Bacteria, motility of, **29**:174

photosynthetic, **47**:Part 2:584–586

random walk of, **52**:488

Bacterial viruses, photoreactivation, **7**:586

Bacteriochlorophyll, 19:354

light-harvesting, **19**:354, 361

oxidized radical, **19**:355

Bacteriophage, 7:223, 233

Bacteriopheophytin, 19:355

Bacteriorhodopsin, 39:224, 227; **49**:34

Baer-Kouri-Levin-Tobocman approach, 49:263–269

Baker-Hausdorff formula, 33:205, 213

Baker's transformation, 24:177

Balanced filter, 34:162

Balance equation, **46**:313

Balescu-Lenard operator, **13**:338

Banach space, **32**:216; **53**:348

Band structure, **2**:248; **4**:197; **16**:378; **41**:164; **51**:261

  aluminum, **49**:561

  calculations, **7**:293

  copper, **49**:556–558

  GaAs, **49**:560

  gold, **49**:562

  in pure crystals, **8**:29

  TaS$_2$, **49**:545

Band-to-band transition, **40**:392, 414, 415

Banfield and Kenyon's radical, **7**:319

Barium, metastable, **42**:10, 14

  quenching of, **50**:384

  reaction with hydrogen halides, **50**:95

  see also Chemiluminescent reactions

Barium bromide, Soret effect, **13**:361

Barium chloride, Soret effect, **13**:361

Barium diethylphosphate, absorption spectra, **7**:488

Barium dimethylphosphate, **7**:488

Barium ferrite crystal structure, **5**:182

Barium fluoride, laser-induced fluorescence, **50**:92

Barium oxide, electron pair correlations of, **6**:370

Barium titanate, **27**:44, 505

Barker potential, **53**:14, 15, 19, 20

Barnett-Coulson expansion, **13**:262

Barometric distribution, **35**:148

  generalized, **35**:151

Barotropic phenomenon, **17**:2, 12

Barrier, **36**:143, 145, 151, 153, 155, 162, 163, 165

  absorbing, **13**:10

  rate constant, **13**:112, 117

  reflecting, **13**:10

  rotational, **13**:105, 112, 117

Barrier height, **36**:160, 162; **41**:253, 265, 266; **50**:566

  to internal rotation, **44**:375

  vibrationally adiabatic, **36**:160, 163, 165

Barrier profile, **37**:123

Bartell-Glidewell approach, **49**:56–58

Barycentric velocity, **16**:105–106

Base-pairing in oligonucleotides, **33**:79, 98, 99

Base pairs, anomalous and normal, **7**:111, 115

  donor-acceptor, **7**:148ff

  electronic structure, **7**:108

  stereochemical investigations, **7**:109

  Watson-Crick, **7**:148

Base strength, of aldehydes, **8**:133, 136

  of carbonyl compounds, **8**:133ff

Bases, basicity order, **7**:366

  superimposed, **7**:122ff

  total π-electron energy of, **7**:93

Basis functions, **9**:328ff; **26**:219, 243

  Hartree-Fock condition, **9**:329

Batchinski-Hildebrand equation, **48**:430

Batchinski's equation, **48**:410

Bauer-Fisher-Gilmore model, **49**:295, 300–302 **50**:345, 348, 350, 353, 360, 371, 384, 481, 488

Bayes' theorem, **51**:170

BBGKY hierarchy, **26**:125; **35**:137; **37**:140; **38**:195; **40**:233, 300, 302, 305, 312, 364

BCS model, **22**:319

Beam detector, **10**:58, 61–62, 139–141, 249–252, 270, 324–328

  for charge transfer, **10**:200–218, 228–234

  differential surface ionization, **30**:248, 270

electron bombardment, **30**:274, 281, 430

for excited species, **10**:174–177, 228–234

geometry of, **10**:38–39, 46–47

negative surface ionization, **30**:494

unstable gases, **10**:219–222, 225

Beam divider, **44**:321

Beam expander, **50**:9

Beam resonance method, **30**:258

Beam scattering, **12**:85, 358

bibliography of data, **12**:445ff

Beam size, **36**:175

Beam source, **10**:57–60, 69, 139, 248–249, 323–324; **42**:443ff

discharge, **30**:284, 287, 289, 292, 318

dissociative, **30**:285, 289

excited species, **10**:172–174, 225–227

excited state, **30**:318

hyperthermal, **30**:464

ion neutralization, **30**:427, 446

nozzle, **30**:248, 259, 401

seeded, **30**:256, 295, 427, 430

sputtering, **30**:429, 494

state selected, alternating gradient, **30**:396

four pole field, **30**:395

six pole field, **30**:396

unstable gases, **10**:219, 221–222

unstable species, **30**:318, 373

vibrationally hot, **30**:514

see also Nozzle beams

Beams, atom, **45**:342–394

ion-neutral, **45**:108, 113–118

oriented, **30**:258, 389

scattering of metastable noble gas atoms, **45**:487–585

BEBO model, **50**:364

Becker-Doering theory, **40**:177

Beer-Lambert law, **42**:6

Behaviour surface, **38**:340

Belleman's equation, **6**:435

Belousov-Zhabotinskii reaction, **29**:156, 157; **38**:319, 320, 327, 334, 348, 408; **43**:142, 147, 153, 211, 223, 224, 228; **55**:50–54, 190, 197

"buffered", **55**:206–211

Field-Noyes model, **55**:206–207

nonequilibrium phase transitions, **55**:207–209

quasiharmonic and relaxation oscillations, **55**:209–210

transitions in closed reaction, **55**:210–211

experimental arrangement, **55**:51

mixed mode oscillations, **55**:53

nonequilibrium behavior, **55**:55

open system, **55**:211–215

bursts of oscillations, **55**:211–212

mixed mode oscillations, **55**:212–214

period-doubling bifurcations, **55**:214–215

simple model, **55**:211

power spectrum of oscillations, **55**:52–53

see also Zhabotinskii reaction

Bénard convection, **26**:177; **29**:133; **32**:18, 68, 77ff, 92, 109ff, 153, 273

in the atmosphere, **26**:178

balance theorem, **26**:202

in bounded containers, **26**:178, 202

bounded laterally, **26**:196

with centrifugal force, **26**:208

convective wavelength, **32**:110, 115, 116, 117, 123, 131

with Coriolis force, **26**:207, 208

finite amplitude theory, **26**:200, 201; **32**:177ff

heat transfer by, **26**:185, 189–191,

194, 201ff, 207–209
  higher modes of, **26**:184
  instability, **32**:269ff, 281ff
  internal heating, **26**:208, 209
  maximal heat transfer, **26**:201, 203
  nonuniform heating, **26**:207
  with rotation, **26**:207, 208
  Schmidt-Milverton plot, **32**:270, 278
  with shear, **26**:187, 206
  supercritical, **26**:202; **32**:109ff
  thermal diffusion effect on, **32**:269
  two-component, **32**:4, 269, 277, 281
  in two component fluids, **26**:178, 265ff
  wave length, **26**:185ff, 197–199, 202, 203, 207–209
  wave number, **26**:178, 180, 193
  see also Cells
Bénard instability, **24**:24; **49**:313, 320, 328–329, 346
  hexagonal cell structure, **49**:333
Benzaldehyde, **8**:134, 136
Benzene, **1**:62, 99, 100, 104, 138, 160, 178, 182ff, 247ff **8**:8ff; **22**:365ff; **34**:190; **40**:43, 456, 474 **47**:Part 1:6; **49**:30, 35, 48
  absorption spectrum, **22**:378, 385
  charge density function, **5**:244
  core parameters, **5**:157, 178
  crystal, **5**:171, 172, 178
  electron impact spectra, **18**:21
  electronic states, **5**:244
  electronic transitions, **5**:241
  electron pair correlations of, **6**:370
  electron scattering, **22**:373, 374
    at 1850Å, **22**:371
    at 2600Å, **22**:370
    at 3400Å, **22**:372
  excited states, **22**:368, 370
  exciton bands, **23**:329

fluorescence lifetime, **22**:396, 397, 404, 414
  in condensed phase, **22**:405
fluorescent decay, **42**:260
forbidden transitions, **5**:241
Franck-Condon factors, **22**:384
interactions with slow electrons, **36**:413–520
  core-excited shape resonances, **36**:434
  differential cross section, energy dependence, **36**:427
    for vibrational excitation, **36**:427
  dissociative excitation, **36**:462
  electron affinity, **36**:453
  electronic transitions, **36**:429
  electronic transmission spectrum, **36**:431
  energy functions for fluorescence, **36**:430
  energy loss spectra, **36**:422, 423
  fragmentation, **36**:504
  higher ionization potentials, **36**:502
  negative ion states of, **36**:439
  optical emission, **36**:430
  optical excitation spectrum, **36**:423
  shape resonance, **36**:442
  short-lived negative ions, **36**:452
  singlet-triplet transitions, **36**:423
  threshold electron excitation, **36**:426
interatomic distances, **5**:156
ion, uninegative, **5**:251
  unipositive, **5**:251
molecular vibrations, **53**:318–319
photochemistry, **2**:200; **22**:402
π-electron molecular orbitals, **22**:369
quantum yield of fluorescence,

22:373, 395, 396, 401

in condensed phase, 22:404

relaxation, from higher vibrational levels, 22:403

from isolated $^1B_{2u}$ molecules, 22:407

from levels excited with 2536Å line of Hg, 22:408

from single vibronic levels, 22:411, 414

from thermal vibrational levels, 22:392, 402

resonance fluorescence, 22:380, 415

rigid convex body constants, 5:163

Rydberg states of, 16:315–316

Rydberg transitions, 22:369

sensitized butene-2 isomerization, 22:398

sensitized ketene dissociation, 22:401

sensitized phosphorescence in biacetyl, 22:401

singlet states, 22:369

single vibronic level fluorescence, 22:337, 379, 381, 382, 383, 385

solid, 22:221, 226, 256

librations of, 22:221

Raman spectrum of, 22:207, 226

spectroscopic notation, 22:376

triplet formation, 22:397

triplet state, $^3B_{1u}$ (T_1), 22:372

$^3B_{2u}$ (T_3), 22:373

$^3E_{2u}$ (T_2), 22:373

energies, 22:372

triplet yields from higher S_1 levels, 22:406

two photon absorption, 22:371

vibrational frequencies, 22:374, 375

vibrations, 5:244

vibronic coupling constants, 5:257

vibronic perturbations, 5:244, 253

see also Benzene derivatives; Benzenes

Benzene derivatives, interactions with slow electrons, 36:413–520

CNDO calculations, 36:450

CNDO-2 calculations, 36:483

cross section for average momentum transfer, 36:510

dissociative electron attachment, 36:455

cross section, 36:457, 459

double negative ion resonances, 36:446

electron charge density distribution, 36:483

electron donating and withdrawing substituents, 36:448

electronic transitions, 36:429

intramolecular charge transfer, 36:485

intramolecular complexing, 36:438

ionization cross section, 36:487

long-lived parent negative ions, 36:464, 468, 469

multiple ionization, 36:503

negative ion resonances, 36:445

NO_2 containing derivatives, 36:478

nuclear excited Feshbach resonances, 36:468

photoelectron spectroscopy, 36:500

π charge density, 36:447

thermal attachment rates and electron affinities, 36:477

thermal electron mobility, 36:477

threshold-electron excitation spectra, 36:426, 429, 487

Benzenes, two-photon spectroscopy of perturbed, 54:117–230

B_{1u} state, 54:200–201

$^1B_{2u} \leftarrow {}^1A_{1g}$ transition, 54:179–180

B_{2u} state, 54:198–200

electrostatic effects, 54:201–211

inductive perturbations, **54**:202–204, 209–211

L_a state, **54**:209–216

L_b state, **54**:201–209

resonance perturbations, **54**:204–209, 211

mass perturbations, Duschinsky rotation, **54**:188–191

harmonic mode scrambling, **54**:181–188

mono-substituted benzenes, **54**:211

perturbation scheme, **54**:194–198

phenylacetylene, benzonitrile, and styrene, **54**:220–224

polysubstitution, **54**:216–220

regularities **54**:192–194

symbols used, **54**:226–227

toluene and fluorobenzene, **54**:191–192

vibronic mechanisms, **54**:198–201

Benzonitrile, 54:220–224

Benzophenone, 23:221, 222; **40**:465; **42**:217, 226

electronic relaxation, **23**:222

gaseous, **23**:222

intersystem crossing, **23**:221

nonexponential behavior, **23**:222

quantum interference, **23**:222

p-**Benzoquinone**, crystal structure, **5**:185

3, 4-Benzpyrene, relaxation of, **23**:208, 209

Benzvalene, 22:402

N-**Benzyl-1, 4-dihydronicotinamide, 16**:192–193

Bernal, sphere packing model of, **40**:73

Bernoulli equation, 19:307

Beryllium, 6:354; **14**:18, 154, 157, 158, 164, 165, 171, 203; **27**:526; **41**:141

electronic energy of, **15**:329–333, 339

ion (Be^{2+}), **2**:299

electron correlation in, **2**:324

electronic correlation energy, **2**:249

ground state energy of, **2**:302, 312

ion (Be^{3+}), electronic correlation energy, **2**:249

isoelectronic series, **6**:366, 380

Rydberg states of, **16**:322–324

symmetry properties, **6**:372

valence pair energy, **16**:336, 338–339

Beryllium fluoride, 34:190; **41**:152

Beryllium hydride, coupled-cluster calculation for, **52**:258

ion, electron correlation, **2**:324

Bessel functions, 13:254, 255; **32**:82, 84, 85; **44**:282

of the first kind, **44**:331

Beta-distribution, noisy, **19**:342

Betagalactosidase, 7:212

Bethe-Goldstone equation, 9:335, 341, 344ff, 354ff, 356; **14**:1ff, 11, 12;

Nesbet's hierarchy, **52**:198–200

Bethe-Guggenheim pair distribution approximation, 2:122

Bethe Salpeter's Hamiltonian operator, 2:240

Bethe splitting, 4:161

Bhatnager-Gross-Krook equation, 46:338; **48**:123–125

Biacetyl, 22:410; **42**:226

Biaxial crystals, 27:430–433, 435, 527

Bicentric coordinates, 12:299

Bifunctional monomeric unit, 2:149

Bifunctional polymer, 2:178

Bifurcation analysis, 49:335, 338

Bifurcation equation, 40:286

Bifurcation parameter, 38:272

Bifurcation point, 32:291; **37**:111, 143, 169, 170; **40**:240, 252, 276, 280, 283, 336, 348

Bifurcations, 24:238, 242, 244, 246;

38:109, 271, 274, 277, 292, 294, 303; **43**:218

complex systems and, **55**:177–199
  adaptive chemistry, **55**:197
  chiral forms, **55**:191–197
  global bifurcations, **55**:183–186
  local phenomena, **55**:179–181
  physicochemical examples, **55**:186–191
  primary equations, **55**:178–179
of exactly solvable models, **38**:296
first primary branch, **38**:272
fluctuations, **38**:114
  renormalization group formulation, **38**:311
group invariance, **40**:284, 287
hard, **40**:242, 246
Hopf, **43**:60
insect morphogenesis, **55**:219–245
  analysis of reaction-diffusion systems, **55**:239–242
  dynamical model, **55**:226–234
  epigenetic code, **55**:234–238
limit cycle, **38**:277
and long range correlation, **38**:300
on ring, **38**:294
secondary, **38**:278
at simple eigenvalue, **38**:272, 273
soft, **40**:242, 247, 252, 276
solvability conditions, **38**:274, 276
of space dependent solutions, **38**:271
of steady state solutions, **38**:273, 290
and strange attractors, **38**:297
of successive primary branches, **38**:277
tertiary and higher, **38**:297
of time periodic solution, **38**:271
of travelling wave, **38**:294
Bifurcation set, **40**:245, 252; **43**:9, 37, 60, 166–206

method of approximating, **43**:185
Bifurcation solutions, stability of, **40**:303
Bifurcation theory, **32**:1, 2, 18; **37**:195; **38**:324, 327, 343, 346, 357; **40**:240, 259, 265; **49**:335
Bijvoet differences, **16**:205–206
Bilayer, lipid, **29**:304, 309
Bile pigments, absorption spectra, **7**:398
Bilirubin, **7**:399
Bimetallic interface, **41**:171
Bimolecular annhilation, **40**:445
Bimolecular collision, **44**:323
Bimolecular reaction, **1**:31, 122; **48**:72, 94
  atom plus hydrogen halide, **47**:Part 2:6–7
  collisional deactivation, **47**:Part 2:12–13
  laser-enhanced, at high pressure, **47**:Part 2:10–12
  polyatomic, **47**:Part 2:7–10
  rate constant, **19**:157
  stochastic models of, **15**:157
  vibrational enhancement, **47**:Part 2:4–13
Binary collision, **19**:343; **26**:111, 123, 133; **44**:285
Binary collision approximation in line broadening, **33**:235–294
Binary collision model, **42**:321
  of liquids, **40**:28–31, 34, 36, 38
Binary collision operator, **48**:97
Binary diffusion coefficient, **15**:104, 109, 122, 128
Binary mixture, **11**:106
Binary stars, **26**:109, 116, 137, 140
  dynamical formation of, **26**:116, 132ff, 137, 138, 143
Binding, nonspecific, **46**:376, 385
  of anions, **39**:110, 112, 171, 336
  of calcium, **39**:88

of small molecule, **39**:163–165

specific, **46**:377, 382

of substrate, **39**:150

thermodynamics of, **39**:110

Binding energy, 35:167, 180; **36**:145, 151, 153, 164

of adsorbed species, CO, table of, **49**:623

N_2, table of, **49**:629

of microclusters, **40**:62

see also Bond dissociation energy

Binding mechanism, 2:3

Binding site, 29:271, 366; **39**:109, 114, 116, 171

Biocatalyzed reactions, 7:61ff

Biochemical kinetics, 19:325, 326, 327

multiple steady states, **19**:318

oscillatory processes, **19**:314, 315, 317

Biological coordinates, 19:327

Biological substances, structural determination, **7**:418

Biological systems, photoselective chemistry, **47**:Part 1:9

Bioluminescence, 21:649

Biophysics, photoselective chemistry, **47**:Part 1:98–102

Biopolymerization on nucleic acid templates, 15:185

Bipolar spherical expansion, 13:207

Biradical, 35:24

Biradical intermediate, 42:138

Birefringence, 44:456

of 7CB solution, **44**:388

Birth and death process, 29:69; **34**:251; **38**:298

Bis(o-dodecacarborane), 16:175–176

Bismuth, solution of noble metals in liquid, **2**:133

liquid, electron states, **31**:290

Bismuth-cadmium alloy (Bi_5Cd_5), ther-

modynamic quantities, **2**:136

Bistability, 38:331; **43**:223, 224

Bistochastic matrix, 51:120–121, 125

Bixon-Jortner model, 50:429

Black spot formation, 48:373

Black-sphere model, 10:128

Blip-function theory, 34:108, 113

exact at low density, **34**:115

Bloch and de Dominicis perturbation expansion, 6:88

Bloch condition, 7:138

Bloch equations, 1:284, 285; **11**:290; **12**:550; **15**:356; **22**:34 **25**:9; **35**:25, 26; **38**:2; **40**:389, 418, 419; **50**:109, 114

degenerate levels, **25**:13

general solution, **25**:57

nuclear magnetic resonance, **25**:9

steady-state solution, **25**:14

Bloch function, in pure crystals, **8**:8ff

Bloch-Redfield equations, 47:Part 1:557–559

Bloch-Rice model for liquid mercury, 27:572

Bloch-Siegert shift, 35:268, 270

Bloch sum, 49:73

Bloch theorem, 49:575

Bloch theory, 53:144

Bloch wall, 46:320

Bloch wave, 49:570

Bloch-Zunger approach, 49:108

Blood coagulation, 26:145

BO radical, A-X system, **50**:75

chemiluminescence from, **50**:78

collisional quenching of A state, **50**:82

as laser system, **50**:83

mechanism of formation, **50**:83

potential energy curves, **50**:77

preparation of, **50**:75

radiative lifetime of A state, **50**:80

Bochner's theorem, 17:115, 116

Body-centered cubic crystal, 49:647, 649

Body-fixed approximation, 36:111

centrifugal decoupling, **36**:123, 184

Boerdijk, sphere packing structure, **40**:73

Bogoliubov inequality, 46:361

Bohm-Aharanov effect, 33:156, 157

Bohn-Gross kernel, 20:151

Bohr magneton, 7:121, 166ff, 180; **53**:230

Bohr radius, 1:276

Bohr-Sommerfeld quantization rule, 36:68

Bolometer, 30:319, 367; **42**:364, 441

Boltzmann approximation for kinetic equations, 15:26

Boltzmann collision operator, 13:338; **24**:295

Boltzmann constant, 1:6, 19, 179, 279

Boltzmann distribution, 1:362, 380, 385–387; **4**:51; **13**:303; **35**:124, 148; **52**:290

chain atoms, **53**:105

soliton energy, **53**:317

Boltzmann equation, 1:362, 363; **5**:263; **13**:303; **15**:39, 199; **28**:429–436; **31**:157–159, 167; **35**:124, 151 **38**:173, 175, 182, 187; **48**:93

collision term, **11**:319

collisionless, **35**:124, 125, 139

for electrons, **31**:416

generalized, **11**:317ff, 319, 337, 356

linearized, **20**:137, 147, 159

in many particle dynamics, **6**:68

Boltzmann equipartition theorem, 53:171

Boltzmann factor, 1:280, 285, 296; **7**:502

Boltzmann H theorem, 11:333, 335; **38**:173

Boltzmann operator, 15:26, 31; **25**:164, 165; **38**:174

generalized, **11**:319, 320, 330, 353, 356, 359, 365

Boltzmann-Vlasov equation, 27:269, 274, 276

Bond dissociation, carbon-halogen, **3**:11, 14

in organic electrode reactions, **3**:3

Bond dissociation energy, 3:167

Ar_2^+, **52**:281, 283

carbon-halogen compounds, **42**:561

$(C_2H_4)_2^+$, **52**:351

$(CO)_2$, **52**:302

$(CO)_2^+-CO$, **52**:303

$(CO_2)_2^+$, **52**:331, 334

$(CO_2)_2^+-CO_2$, **52**:331, 334

$(CO_2)_3^+-CO_2$, **52**:331

$(CS_2)_2^+$, **52**:309, 311, 334

Kr_2^+, **52**:281, 283

$KrAr^+$, **52**:285

$(NH_3)_2^+$, **52**:347

$(N_2)_2^+$, **52**:302

$(N_2)_2^+-N_2$, **52**:303

$(N_2O)_2^+$, **52**:331, 334

$(NO)_2^+$, **52**:292, 302

NO^+-N_2O, **52**:331

$(O_2)_2^+$, **52**:293, 302

$(OCS)_2^+$, **52**:325, 326, 334

$(OCS-CS_2)^+$, **52**:327

$(SO_2)_2^+$, **52**:334, 343

SO^+-SO_2, **52**:344

Xe_2^+, **52**:281, 283

$XeAr^+$, **52**:285

$XeKr^+$, **52**:285

Bond distance, 41:159

Bond energy, 1:184; **6**:387, 408

of halogens, **2**:61

Bond length, and Placzek correction, **53**:51

atomic lattice, **53**:295, 306

molecular crystal, **53**:278

Morse lattice, **53**:303

in polyacetylene, **53**:289, 291

resonance interactions and , **53**:284

Toda lattice, **53**:300

vs. travelling wave coordinate, **53**:313

Bond modes, 47:Part 1:293–322

absorption intensity, **47**:Part 1:302–306

absorption lineshapes, **47**:Part 1:315–321

cooperative excitation, **47**:Part 1:309–313

energetics, **47**:Part 1:298–302

intramolecular dynamics, **47**:Part 1:313–315

Bond order, 1:178, 179, 184, 242, 250; **5**:23; **7**:41ff, 50, 53, 98

Bond percolation, 52:494

Bond polarizability, 12:136

Bond rotation, 39:91

Bond strength, 1:178; **21**:307

table of, **50**:354

Bond stretch attraction model, 50:373, 477, 500, 510

Bonding, 6:387–398; **36**:157, 164

chemical, relation to electronegativity, **13**:55, 64

effect of molecular environment, **6**:387, 395–396

hydrogen, **16**:163–164, 171–173, 179, 181, 186, 193

overlap, **16**:381–382

Bonding molecular orbital, 7:92, 168

Bone, 7:270

Borates, 16:185

Borel-measurable function, 38:57

Borel σ field, 38:57

Borel sum, 46:210

generalized, **46**:275

Born and Green, method of, 1:163, 218, 224

Born approximation, 10:92–93, 129; **20**:149; **25**:247, 279, 283; **28**:363; **31**:430; **42**:285; **44**:284

collision dynamics, **6**:42–43

distorted wave, **36**:181, 183

electron diffraction, **53**:5

electron scattering, **18**:20, 21, 31, 65, 72, 74, 75, 78, 80, 81

electronic excitation by electron impact, **36**:417, 420

Born-Bethe approximation, 28:203, 204

Born energy, 37:120, 121

Born equation, in ionic solvation, **6**:444–451

Born-Green equation, virial coefficients, **6**:265, 269

Born-Green-Kirkwood equation, 6:193

Born-Green-Yvon hierarchy, 21:400; **31**:33

superposition approximation, **21**:401

Born-Huang expansion, 49:156–158

Born-Huang method, 49:116, 139

Born-Kármán conditions, 9:88

Born-Meyer-Huggins potential, 48:438

Born-Ochkur-Rudge approximation, 18:80, 81

Born-Oppenheimer angular radial separation (BOARS) method, 42:385, 390, 412, 415

Born-Oppenheimer approximation, 6:9–22; **7**:8, 85; **10**:31–32, 34, 66, 369–370, 377; **11**:13, 14; **12**:6, 8, 9, 12, 15, 16; **13**:87; **17**:231, 237, 241, 259, 266; **25**:182, 201, 279; **26**:217; **28**:65, 118, 119, 147, 321; **30**:465; **35**:202; **36**:19, 45, 142, 167, 175, 184, 189, 418; **42**:17, 21, 66, 170, 187, 209, 367, 386, 424, 517; **44**:18, 23, 555, 557, 564, 597;

45:229, 403–405, 496; **49**:273–274; **50**:519; **53**:319

application to, $B^1\Sigma_u^+$ H_2, **12**:24

ground state D_2, **12**:22

ground state H_2, **12**:22

ground state HD, **12**:22

for a diatomic molecule, **6**:17

generalized, **35**:261

light scattering by simple fluids, **26**:12

mass polarization correction, **12**:6, 11, 27

Born-Oppenheimer exchange approximation, 18:31, 33

Born-Oppenheimer Hamiltonian, 48:8

Born-Oppenheimer perturbation theory, 49:117, 131–139, 142, 145, 156–159, 170, 193, 195–196

Born-Oppenheimer potential, 52:118, 122

Born-Oppenheimer separation, 45:231–232, 239

Boron, 1:172, 240

compounds of, free-electron energy levels, **21**:11–21

icosahedral fragments, **21**:10, 22

isoelectronic series, **6**:380

^{10}B resonance for the nucleus of, **2**:204

Boron (B_2), 41:152

ground state, quadrupole-quadrupole interaction, **12**:44

Boron fluoride, as initiator of polymerization, **2**:156

Boron hydride (BH), potential curves, **52**:254–258

Boron nitride, 41:159

Boron trichloride, 20:33

interatomic distances, **5**:156

Boron trifluoride, interatomic distances, **5**:156

Bose-Einstein condensation, 16:23–24

Bose gas, 22:334; **41**:181, 204, 218, 219, 220

Boson fluid, friction coefficient in, **16**:23–26, 44–46

static mobility in, **16**:30–31

Boson-operator algebra, 42:329

Boson states, 14:104ff

Bottleneck effect, 50:131

Bound modes, 27:283, 284, 293, 284, 302, 319

virtual, **27**:285, 293, 294, 319

Bound state, computation of wave function, **46**:131

induced absorption, **51**:56

induced light scattering, **51**:96

irregular spectrum, **46**:121

level spacing distribution, **46**:124

nodal structure, **46**:132

regular spectrum, **46**:121

second differences of, **46**:123

semiclassical limit, **46**:119

semiclassical wavefunctions, **46**:125

sensitivity to perturbation, **46**:123

Bound state calculations, 25:213

Bound state wave function, 36:30

Bound subsystem, 35:184

Boundary, artificial, **34**:258, 259

between ordered and disordered region, **33**:100

dead loaded, **33**:422

inclined, **32**:140, 141

natural, **34**:258, 259

of stability, **43**:9, 167

Boundary conditions, 29:34, 52, 133, 180, 255; **41**:203, 218; **48**:72

atomic lattice, **53**:295, 297

of chemical potentials and diffusion flux, **33**:429

of chemical systems, **38**:319, 334

complete absorption, **48**:78

for distribution function, **11**:321, 324

Ewald, **48**:249

periodic, **44**:402, 420, 441; **48**:246, 249–253; **53**:311

of place and traction, **33**:421

radiation, **48**:78

for random walks, **52**:407–430

    at absorbing surfaces, **52**:407ff, 449, 453ff, 480, 481

    at partially absorbing point, **52**:429

    at reflecting point, **52**:428, 429

for scattering problems, **25**:216ff, 249ff, 264, 268

of temperature and heat flux, **33**:429

Boundary forces, for relaxed motion, **15**:42ff, 51

Boundary layer, absence of, **32**:85

stability analysis, **13**:274, 281, 289

thermal, **32**:79, 80

Boundary value method, **30**:15

Bounding techniques, **32**:21

Boussinesq approximation, **26**:179, 194; **32**:18, 78

Boussinesq fluid, **32**:115

Bovine plasma albumin, **7**:254

Bovine serum albumin, **7**:243

Box normalization, **25**:278

Box toroid, **36**:21

Boyle's law, **40**:230

Bracketing theorem, **14**:315

Bragg envelope, **27**:174

Bragg reflection, **53**:27

Bragg scattering, **20**:232; **53**:57, 251, 253

Bragg-Williams approximation, **24**:243, 244; **41**:22

Branching diagram, **40**:240, 241, 247, 252

Branching ratio, **30**:39; **50**:140, 241

Branching rules, **14**:101ff, 108

α-**Brasses**, electronic specific heat of, **2**:131

Bravais lattice, **20**:9; **49**:84

Bravais net, **27**:50, 142, 152

Bray-Liebhafsky system, **43**:144, 148, 153, 211

Bray reaction, **38**:327

Brazovski generalized potential, **49**:325–326

BrCl, B-X system, **50**:49

quenching by Cl_2, **50**:49

BrCN, **50**:258

Breather solution, **53**:240–241, 263, 264

Bäcklund transformation, **53**:331–332

and high-temperature thermodynamics, **53**:247

Breathing sphere model, **50**:346

Breit corrections, **41**:85

Breit-Dirac equation, **53**:6, 50

Breit energies in mercury and fermium, **54**:272

Breit Hamiltonian, **12**:174, 210, 219

Breit interaction, local approximation for, **54**:296–297

Breit operator, **12**:209, 213, 215

Breit-Pauli Hamiltonian, **12**:72, 73; **23**:162

relativistic correction term, **12**:6

Breit-Wigner equation, **42**:376; **49**:230

Bremsstrahlung, **38**:208

Brewster angle, **27**:408–410, 412, 420, 436, 439

BrF, B-X system, **50**:58

predissociation of B state, **50**:58

quenching of, **50**:60

Bridgeman's zone of indifference, **21**:718

Brillouin relation, **44**:602, 603, 611, 612

Brillouin scattering, **17**:66; **24**:8, 19, 20, 23, 40, 42, 43, 48; **27**:463

Brillouin zone, **27**:455, 458, 470–471, 510, 524; **49**:48, 73, 86, 94
  bcc, **49**:647–649
  bulk, **49**:553, 559, 572, 614
  fcc, **49**:645–647
  hcp, **49**:649
  NaCl structure, **27**:488, 510
  reconstructed surface, **27**:512
  surface, **49**:553, 561, 572, 614
Brillouin's perturbation theory, **8**:31; **12**:46, 47
Brillouin's theorem, **2**:284; **9**:330; **14**:292; **25**:186; **26**:248
Brillouin-Wigner expansion, **14**:343
Brillouin-Wigner perturbation theory, **36**:348
Brink-Weiguny condition, **49**:123, 163
BrO, electron resonance spectra, **18**:216, 228, 229, 231, 233
Broad-band excitation, **40**:379–383
Broersma apparatus, **3**:173
Broken path model, **30**:26; **49**:205
Broken symmetry, **43**:218; **46**:310
Bromine, **34**:183
  atmospheric chemistry, **55**:74
  crystal structure, **5**:184
  hydrate, **2**:3, 22, 43, 47
  London energies, **2**:78
  molecule, **49**:459, 481
    A-X system, **50**:29
    B-X system, **50**:29
    hydrate, **2**:10
      thermodynamic data, lattice constants, **2**:71
    intramolecular London energies, **2**:78
    self-quenching of A state, **50**:29
  see also Chemiluminescent reactions
p-**Bromochlorobenzene-ethylene system**, **2**:102
Bromoethane, barrier height of internal rotation, **2**:382
  hydrate, **2**:10
    thermodynamic data, lattice energy, **2**:8
N-**Brosylmitomycin A**, **16**:207
Brot sum rule, **44**:309, 344
Brout-Carruthers expansion, **37**:142
  Brownian-dynamic model, **11**:246
Brownian model for vibrational relaxation of iodine, **53**:210
Brownian motion, **1**:87, 109, 130, 141, 143, 146, 153, 161, 365, 366; **4**:43; **5**:158; **7**:506, 509; **11**:202, 203, 209, 277, 290; **13**:14, 17; **17**:102; **29**:91, 170; **35**:38, 68, 70, 71, 87, 110; **38**:86; **44**:145, 257, 269; **52**:439
  angular, **44**:277
  disk, **44**:266
  domain wall, **53**:273, 274
  domain wall kinks, **53**:268
  equations of, **53**:64
  generalized transport equation, **16**:15
  indeterminacy relations for, **15**:363ff
  inverse, **20**:200, 209
  Langevin model, **53**:81–87
  linear response theory and, **53**:132
  nonlinear, **46**:211, 253
  polymer chains, **52**:448
  polymers, **22**:1, 6, 13, 23, 28, 32, 35, 38, 39, 50, 54, 70
  particle velocity, **53**:82, 83–84
  quantum, **16**:2–64
  rods, **52**:446
  solitons, **53**:228, 268
  spherical top, **44**:266
  translational, **44**:406, 416
Brownian oscillator, heatbath modeling, **53**:98
Brownian particle, **11**:254; **17**:147

angular momentum autocorrelation function, **17**:148

dipolar autocorrelation function, **17**:147

distribution function, **16**:34–37, 43–44, 49–52

velocity autocorrelation function, **17**:149

Brownian-static model, 11:240

Broyden asymmetric rank-1 update, 54:79–82

Broyden family (DFP and BFGS updates), 54:82–86

Brueckner condition, 9:339ff, 343ff

Brueckner-Goldstone perturbation series, 6:93, 123

Brueckner orbitals, 6:345; **36**:400;

Brueckner's approximation, 2:207, 273, 318, 319, 323

Brueckner theory, 6:321, 344, 345; **9**:342ff; **14**:309

Brunauer-Emmett-Teller equation, 7:240

Brunauer-Emmett-Teller monolayer, 7:249

Brusselator, 39:2, 8; **43**:231–236, 237, 240–244

simplified, **39**:3, 21

Bubbles, nucleation in absence of, **40**:138

Buckingham-Corner potential, 42:400, 402

Buckling of solids, 32:14

Buffering, thermodynamic, **55**:150–158

Bulk modulus, 41:162, 163, 185

Bulk properties, 12:370

Bulk relaxation time, 42:275

Bulk viscosity, 5:197; **24**:273, 277, 284, 287, 305, 310, 314, 329, 334;

see also Viscosity

Bunching, 36:53

Butadiene, configuration interaction of,

52:102

polymerization, **2**:160

trans-**Butadiene, 8**:8ff

n-**Butane + mercury system, 2**:103

Butterflies, 19:338

C

C*-algebras, *see* State

C-systems, **24**:159, 160, 171

C_6 constants, **12**:389, 414, 459, 481

Slater-Kirkwood approximation, **12**:414

Cadmium, 41:141

collision-induced emission, **30**:445

quenching of ^1P, **50**:351, 375, 382

quenching of ^3P, **50**:360, 375

reaction with H_2, **50**:98

Cadmium (Cd_2), dipole resonance interaction energy, **12**:41

Cadmium fluoride (CdF_2), 27:445

Cadmium halide structure, 49:104–105, 110

Cadmium-lead alloy (Cd_5Pb_5), thermodynamic quantities, **2**:136

Cadmium-magnesium alloy (Cd_5Mg_5), thermodynamic quantities, **2**:136, 142

Cadmium oxide (CdO), 27:414, 445

Cadmium-tin alloy (Cd_5Sn_5), thermodynamic quantities, **2**:136

Caffeine-riboflavin complex, 7:46

Cage, common cage effect, **53**:106ff

common cage formation, **53**:196

diffusing, **44**:405

of neighbors, **44**:277

Cage effects, 41:282–285; **42**:309; **48**:106

Cage motions, 53:100

coupling constant, **53**:107, 108

recombination dynamics, **53**:108

Cage structure, 53:69

Calcium, 41:141

quenching of ^1P, **50**:353

quenching of ^3P, **50**:357, 384

Calcium-binding protein, 39:83

Calcium fluoride (CaF$_2$), 27:445

Calcium microclusters, 40:58

Calculus of variations, 13:268; **25**:286

Calf thymus, 7:128

Calorimetric measurements of metal-ammonia solutions, 4:316, 357, 358

CaMoO$_4$, 27:447

Cancellation theorem, 41:154

Cancer, 7:85

antitumor activity, **7**:60ff

Cancer cells, 38:367, 394, 395

antigenic properties of, **38**:395

origin of, **38**:394

Canonical action, 36:10

Canonical commutation relations, 22:331

Canonical density operator, 46:166, 168, 169

Canonical distribution, 32:63

Canonical ensemble, 19:332; **22**:96, 99, 107; **38**:6; **44**:263, 339, 340

classical, **53**:166

distribution function, **53**:85, 219

Canonical form of dynamical equations, 24:137

Canonical formalism, 33:154, 159ff, 168, 171, 187ff, 195, 200

Canonical free energy, 24:108, 137

Canonical invariance, 20:198

Canonical Maslov index, 36:12

Canonical partition function, 53:242

Canonical transformation, 25:75, 76, 78, 95; **30**:79, 81, 89, 113; **33**:155, 203; **36**:12; **44**:554, 599, 601

dynamical variables, **53**:315

generator of, **25**:76–78, 83, 86ff, 99–101, 122; **30**:84, 113

point, **25**:78, 93

Poisson brackets, **53**:126

unitary, **33**:156, 206

Capillarity, Gibbs' theory of, **6**:145, 158

Capillary array, 42:529

Capillary wave, 46:304, 307

Caratheodory's theorem, 43:24

Carbanion, 2:180

Carbon, 1:49, 65, 70, 72, 89, 193, 252, 254

atoms, in graphite and diamond, **9**:366

dimer, **41**:152

isoelectronic series, **6**:381

phase diagram, **9**:367, 374

solubility test cell, **9**:383

solvents depositing graphite, **9**:382ff

Carbon blacks, magnetism of, **3**:226

Carbon-carbon bonds, 1:66, 67, 71, 114, 173ff, 247, 248

Carbon dioxide, 20:4, 12; **49**:439

absorption of by water, **13**:274

clathrate in hydroquinone, **2**:7, 11

clusters, **52**:330–334

critical behavior, **40**:230

crystal structure and intermolecular potential, **33**:51ff, 65

crystal, **5**:171, 172, 181

E → V energy transfer, **47**:Part 2:398–399

electrostatic interactions, **5**:174, 176, 181

as a "hilfsgas", **2**:18

hydrate, **2**:3

thermodynamic data and lattice constants, **2**:8

interatomic distances, **5**:156

light scattering, **24**:42, 57

multipole interactions, 5:181

photodissociation of, 50:302

as quencher, 50:358

rigid convex body constants, 5:163

solid, 22:207, 226, 239, 249, 289, 293, 303

    crystal structure, 22:248

    infrared spectra of, 22:207, 249, 250, 289

    lattice dynamics of, 22:225, 227, 228, 230, 250

    librations of, 22:221, 239, 241, 249

    transitional motions, 22:249

triple point, 2:97

vapor pressure, 2:96

V-V energy transfer, 23:4, 48, 49, 50, 60, 61, 62, 65, 66, 72, 77

Carbon dioxide + iodine system, 2:98

Carbon dioxide + *m*-nitrobenzene system, 2:101

Carbon dioxide + *o*-nitrochlorobenzene system, 2:101

Carbon dioxide + *o*-nitrophenol, 2:101

Carbon dioxide + *p*-toluidine, 2:101

Carbon dioxide + 3, 4 xylidine system, 2:101

Carbon disulfide, 52:306–309

    birefringence, 23:214

    clusters, 52:309–315

    E → V energy transfer, 47:Part 2:399

    interatomic distances, 5:156

    ion-molecule reaction, 52:315

    rigid convex body constants, 5:163

Carbon-halogen bond fission, 3:11, 14

Carbon-metal bonds, 8:147ff

Carbon monoxide, 12:45; 14:350; 41:152; 49:619–636

    α-β transformation in, 16:376

    adsorption of, 27:255, 256

    clusters, 52:301–304

Cu (100) adsorption, 49:621

differential cross section, 49:540

E → V energy transfer, 47:Part 2:397–398, 405–406

electron impact desorption of, 27:256

electron impact spectra, 18:48, 49, 60, 71

Fe (110) adsorption, 49:632–633

formation of vibrationally excited, 42:142

gas phase, 49:538–541

laser probing technique, 47:Part 2:86–90

lifetime of triplet states, 50:66

Ni (100) adsorption, 49:622–626, 642

Ni (111) adsorption, 49:621

Pd (111) adsorption, 49:522

potential curve of *a* state, 50:67

as quencher, 50:355, 365, 370

quenching of O(^1D), 42:149

resonances, 49:543–544

solid, 16:375–376; 22:221, 251

    librations of, 22:221, 291

structural and thermodynamic data, 16:346–348

surface binding energy, table of, 49:623

symmetry, 49:540

vibrational energy transfer in *d* state, 50:69

wave functions, 49:620

Carbon monoxide ion (CO⁺), laser-induced fluorescence, 50:66

Carbon oxide, as radiation source, 2:192

    gamma rays, 2:203

Carbon-oxygen bond, 1:93, 99

Carbon oxysulfide, E → V energy transfer, 47:Part 2:399

Carbon tetrachloride, 1:99ff, 138, 160,

185; **34**:189; **55**:76–77
  interatomic distances, **5**:156
  rigid convex body constants, **5**:163
Carbon tetrachloride-hydrogen sulfide -water ternary system, 2:49, 51, 52
Carbon tetrafluoride, 34:190
  interatomic distances, **5**:156
Carbonaceous chondrites, organic compounds in, **55**:99–100
Carbonium ion polymerization, 2:158
Carbonyl compounds, base strength of, **8**:133ff
  SCF-MO treatment of, **8**:133ff
Carbonyl cyanide phenylhydrazones, 19:368
Carboxyl phosphate, 7:63ff
Carboxylic groups initiatior, 2:174
Carboxypeptidase, 39:85
Carboxypeptidase A, 21:583
Carcinogenesis, 38:367, 394, 402
Carcinogens, 1:192, 199; **8**:179
  activity of, **1**:193, 198
  aromatic, **8**:163ff
  hydrocarbons, **7**:152ff
  properties of, **1**:166, 191, 192
Carleman-Verkua procedure, 20:194
Carotenoid pigments, 19:370
  absorption band shift, **19**:370, 371
  membrane potential, **19**:370
Carrier, 7:288; **39**:287, 294, 335, 336
  complex formation, **39**:295
  complex with cations, **39**:290
  concentration in membrane, **39**:295
  cyclic antibiotics, **29**:315, 358
  macrocyclic, **29**:311, 314
  mobilities, **7**:290ff
  neutral, **39**:291, 319
  photogeneration, **7**:288
  stability constant, **39**:294

  thermal generation, **7**:288
  vs. channel, **39**:324, 325, 326
Cartesian coordinates, in calculation of forbidden transitions, **5**:255
Cascade, electron, rate-limiting step, **13**:112
Casimir and Polder interaction energy, 12:82
Caspers theory of spin-spin relaxation, 11:305
Catalase, 7:160, 182, 615
Catalysis, 9:142, 165; **20**:321, 322
  acid-base, **39**:184, 186
  cooperative phenomena, **38**:354
  of decarboxylation, **39**:146
  energetics of, **39**:78
  of ester cleavage, **39**:120
  heterogeneous **38**:350; **39**:99
  microcluster theory of, **40**:51
  by synzymes, **39**:109, 147
Catalyst clathrates equilibrium, 2:35
Catalytic hydrogenation of ethylene, 8:141
Catalytic reaction rate, 21:791
  dependence on pore structure, **21**:793
Cataphoresis, 28:279
Cataracts, 21:661
Catastrophe, 26:134
  gravothermal, **26**:115, 122, 124
Catastrophe sets, 43:9, 167
Catastrophe theory, 32:2; **38**:339; **40**:242, 244, 251; **43**:207
Catecholamine, 7:522
Catochrome, 7:522
Cauchy integrals, 11:180
Cauchy kernel, 20:192
Cauchy-Kowaleski theorem, 19:347
Cauchy principal value, 20:158, 193; **53**:246
Cauchy's integral formula, 40:267

Causality principle, 4:89
Causation, backward, 38:91, 101
Caustics, 36:3, 23, 55; 46:127, 128, 136, 139
  clusters of, 46:139
Cavitation, 40:147
Cavity field, 1:80, 88; 12:269, 270
  fluctuating, 44:262, 263
Cavity formation in hard sphere fluid, 6:255, 259, 272ff, 282
Cavity function, 53:117, 120, 194, 208
Cavity potential, 48:121
Cayley Hamilton equation, 14:323
Cell, communication, 29:42
  contact, 29:254
  contraction, 29:254
  division, 29:43, 54, 126, 255
  of the eye, 29:343
  growth, 29:127
Cell-cluster expansion, 31:59
Cell mediated immune surveillance, 38:393
Cell model, 1:161, 203, 229, 233; 6:143, 238–240; 11:117, 118, 123; 16:226–228, 274
  of liquids, 40:28
  partition function, 16:227–228
    clathrates, 2:26
    metallic solutions, 2:135
  in vibrational relaxation, 42:348
Cell theory, Weres-Rice, 31:63
Cells, hexagonal, 32:77, 79
  in Bénard convection, 26:181, 190
    hexagonal, 26:183, 186–189, 190ff, 201, 204ff
    rectangular, 26:181
    ring, 26:181, 183, 187, 194, 203
    square, 26:181, 200, 201, 205
    supercritical Rayleigh number, 26:203

triangular, 26:181
  rolls, 32:77, 92
  ultraviolet photoprotection, 7:584
  see also Convection
Cellular division, 1:191
Cellular method, 2:226, 237, 252
Cellulose, interaction with dyebaths, 21:685, 696
  interaction with salts, 21:685, 691, 695
Center, of lattice, 38:17
  of Neumann algebra, 38:40
Center of interaction, 4:230, 231, 280, 286, 294
Center of inversion, in diatomic molecules, 5:48
  in ligand field theory, 5:43
  in tetrahedral symmetry, 5:43
Center of mass coordinates, 10:329; see also Coordinate systems
Center of mass motion, 44:286, 447
Center of resistance, 35:38, 43
Central decomposition, 38:40
Central field covalency, 5:68; 8:53
Central forces, elastic scattering by, 10:35–47
Central limit theorem, 22:11, 15, 16; 46:198; 52:365, 378, 432, 433, 441, 459; 53:83
Central measure, 38:40
Central motion, 36:52
Central nervous system, 19:341
Central potential, 36:26
Centrifugal barrier, 10:92, 162, 261, 263, 265, 371–372
Centrifugal decoupling, 42:378
Centrifugal potential, 12:18
Ceramics, 40:138
Cerebral samples, infrared spectroscopy, 7:414
Cerenkov radiation, 27:311, 318, 320;

38:208

Cerium, **41**:185

Cesium, **49**:485–486

  atom, correlations, **49**:504

  dimer, **41**:152

  see also Collisional ionization

Cesium chloride, **27**:528–529

CF, electron resonance spectra, **18**:228

CF$_3$NO, **49**:34

CH$_2$, geometry of, **42**:82, 84

  magnetic susceptibility of, **3**:204

CH$_3$, geometry of, **42**:84

CH$_5^+$, CD$_5^+$, **19**:38, 45, 51, 61, 86, 91, 92, 124

Chain, clamped, **53**:92, 98, 164–165

  dynamics, **53**:92

    inner product, **53**:169

  continuous, **22**:3, 5, 22ff, 32, 36ff, 54, 61, 62, 103

  double, **41**:402–407

  dynamics of, **15**:305ff

  effective, **22**:17

  equivalent, **53**:96–98

    heatbath modeling and, **53**:98

    statistical properties, **53**:166

    velcity autocorrelation function, **53**:102

  finite, **41**:327

  flexible, **22**:1, 11, 16–18, 20ff, 25, 32ff, 45, 61

  freely hinged, **22**:10, 17, 20, 101

  freely jointed, **15**:306

  freely rotating, **22**:18, 19, 21

  Gaussian, **22**:11, 15, 18ff, 37, 48, 69, 70, 74, 101, 116

  harmonic, **53**:300

    Einstein oscillations, **53**:99

    equivalent, **53**:96–98

    as model many-body system, **53**:74–76

  one-dimensional, **15**:311

  ordinary, **41**:383–407

  random flight, **22**:10, 13, 15, 20, 24, 37, 64, 111

  random, vibrational modes of, **13**:125, 126, 146, 149, 152, 156, 161, 162

  stiff, **22**:1, 11, 17, 35ff, 42ff

  wormlike, **22**:1, 21, 22, 36, 46–47, 53, 54

  see also Polymer; specific chains

Chain length, **16**:234, 247, 251, 253, 270, 278

  average, **16**:252–253, 257, 260, 270

Chain ligands, **7**:380

Chain molecules, principle of corresponding states for liquid of, **16**:226–235, 245, 279

Chain representation, **53**:68, 162–167

  atom-chain collisions, **53**:101–106

  clamped chain, **53**:164–165

  common cage effect, **53**:107

  coordinate variables, **53**:162–165

  equations of motion, **53**:179

  response functions, **53**:156–161, 178–179

  truncation, **53**:180

Chalcogen adsorption, **49**:605

Chalcogenide glasses, **44**:294

Chalcogen-ordered overlayer, **49**:594–606, 636

Chandrasekhar equation, **1**:146

Channel decoupling, **52**:28–38

Channel radius, **25**:249

Channels in membranes, **39**:324, 325, 326

Channeltron, **42**:532

Chaos, **38**:179; **43**:189

Chaotic motion, onset of, **46**:73–152

Chapman-Enskog method, **1**:362; **15**:50, 60; **31**:158

Chapman-Kolmogorov equation, **15**:67

Chapman-Smoluchowski-Kolmogorov equation, **15**:16, 31

Characteristic equations, **29**:88

Characteristic exponent, **35**:272, 274, 286, 304, 308, 381
  definition of, **35**:272
  theorems concerning, **35**:281

Characteristic flow, **29**:88, 95

Characteristic function, **17**:115; **38**:61, 69
  of π-electron systems, **5**:12, 13

Characteristic loss spectroscopy, **27**:25

Characteristic time, **46**:218

Charge carrier hopping, **44**:294, 396

Charge carrier production, **40**:439, 445

Charge density, **1**:314; **7**:98, 121; **33**:164, 165, 188, 194, 210, 215; **41**:166; **44**:450
  in chemical reactivity, **8**:102ff

Charge-dipole system, **48**:308

Charge distribution, **13**:211

Charge exchange, adiabatic path, **13**:46
  alkali atoms and ions, **13**:50
  Ar$^+$/Ar, **13**:51
  Cs/Rb$^+$, **13**:50
  damping, **13**:45, 46, 48
  diabatic path, **13**:47
  He$^+$/H, **13**:50
  He$^+$/He, **13**:46, 48
  He/H$^+$, **13**:50
  multiple ionization, **13**:51
  quality factor, **13**:50
  quasi-resonant, **13**:48
  resonant, **13**:41; **30**:479
  statistics, **13**:50
  see also Charge transfer

Charge-induced-dipole approximation, **52**:287

Charge stripping, **45**:141–142

Charge transfer, **7**:363, 368; **10**:195–246; **12**:47, 49, 68; **28**:394, 409, 410, 412, 413; **30**:463ff; **41**:277
  accidental (asymmetric) resonance, **10**:197–198, 242–243, 254–256
  asymmetric, **10**:197, 240–241
  atomic systems, **45**:177–189
  chemical forces in, **10**:367–387
  cross section, **10**:197–199, 200ff, 220ff, 225ff, 236–243
    total, **10**:201
  dissociative, **10**:202–206, 228, 250, 252–253
  excited products of, **10**:174, 228–234
  heavy particle transfer, **45**:193–196
  low-energy, **10**:248, 250, 252–257
  luminescent reactions, **45**:165–196
  with metastable primary beams, **10**:196, 225–228
  molecular systems, **45**:189–193
  neutral beam production by, **10**:57, 60–61, 174, 310–311
  quenching, **50**:345
  symmetric resonance, **10**:197–198, 239–240, 254
  threshold behavior, **10**:256–257
  with unstable gases, **10**:218–225
  see also Charge exchange; Electron transfer

Charge transfer bond, **7**:250

Charge transfer complex, **7**:70; **16**:175

Charge transfer method, **52**:354

Charge transfer source, **42**:532

Charge transfer spectra, **7**:368ff

Charge transfer states, covalency effects, **8**:15
  electronic radii, **8**:14ff
  exchange energy, **8**:15ff
  ligands, **8**:13ff
  optical absorption, **8**:13ff, 18ff

Charge transfer theory, **7**:44, 45ff

Charge transport, 4:323
  in amorphous solids, **52**:477
Charged particles, transport properties
  of, **5**:303
Chelates, 4:138, 182–188
Chelation, 7:426
Chemical activation, 20:140
Chemical affinity, 33:435
Chemical bond, 1:165, 170; **7**:11, 16ff,
  183; **51**:270
Chemical competitive technique, 1:36,
  37
Chemical dynamics, 33:346, 347, 395
  generalized Langevin equations in,
  44:143–254
Chemical equilibrium, approach to,
  6:55; *see also* Equilibrium
Chemical evolution, asymmetric forms,
  stochastic modes of synthesis,
  55:201–204
  atmosphere, **55**:63–84
  Belousov-Zhabotinskii reaction,
  55:205–217
  bifurcations, complex systems,
  55:177–199
    insect morphogenesis, **55**:219–
    245
  complex biological oscillator,
  55:171–176
  feedback loops, **55**:247–282
  glycolysis, **55**:169
  molecular level, **55**:119–137
  nonequilibrium thermodynamics,
  55:43–62
    dissipative structures, **55**:46–54
    sensitivity and selection, **55**:54–
    58
    stochastic aspects, **55**:58–60
  optical activity, **55**:109
  optimization of mitochondrial
    energy conversions, **55**:141–167

  prebiotic origin of photosynthesis,
  55:113–118
  prebiotic synthesis, **55**:85–107
  selection of chirality, **55**:111
  Solvay conferences, **55**:1–42
  in chemistry, **55**:37–42
  in physics, **55**:7–35
Chemical instability, 43:217–264;*see
  also* Instability; Stability
Chemical kinetics, 41:230
  on linear lattices, **15**:168
  role in stellar evolution, accumula-
    tion point in spectrum of, **32**:239
    chemical modes, **32**:232, 235, 242
    classification of spectra related
      with, **32**:242
    continuous modes of, **32**:232, 242
    effects on secular and vibrational
      spectra, **32**:240, 242
    unstable states in, **32**:233, 234
  stochastic theory of, **15**:149ff
Chemical laser, 41:237, 245; **42**:91, 122,
  440
Chemical laser kinetics, 47:Part 2:55–
  83
  kinetic scheme, **47**:Part 2:59–62
  model *vs.* experiment, **47**:Part 2:62–
  65
  rate equations, **47**:Part 2:57–59
  rotational effects, **47**:Part 2:65–72
  thermodynamics, lasing criteria,
  47:Part 2:72–75
    maximal work and constraints,
    47:Part 2:75–76
  time dependence, **47**:Part 2:77–80
Chemical networks, 29:1, 22; **43**:17
  entropy production in, **29**:22
Chemical potential, 1:38, 207, 279;
  5:264; **11**:5, 6, 11, 16, 17, 28–31;
  12:229, 232, 263, 264; **14**:220,
  224; **40**:141; **41**:74; **43**:30

associated with dispersion forces, 12:263

electrical part of, 12:232

excess, 12:262

hydrate lattice, 2:22

local equilibrium, 5:268, 262

orientation part of, 12:265

reduced, 33:416

Chemical reaction, 19:217; 24:2, 22, 59; 25:257, 259, 268; 34:284

affinity of, 19:221, 249

analytic theory of, collision dynamics, 6:35–55;see also Collision dynamics

interaction potential, 6:22–34

molecular orbital approximation, 6:23

semiempirical method, 6:25–31

valence bond method, 6:24

many particle dynamics, 6:55–80

species characterization, 6:9–22

diatomic molecules, 6:17

using Born-Oppenheimer approximation, 6:9–20

using Born-Oppenheimer separation, 6:20–22

species identification, 6:3–8

configuration criterion, 6:6

energy criterion, 6:5

energy-configuration criterion, 6:7

autocatalytic, 19:247, 258, 288, 291

bimolecular, 1:31, 122; 48:72, 94

coupling with diffusion, 39:1–54

dissociation, 1:121; 48:82

elementary, 37:71, 72

enhancement by infrared laser, 47:Part 2:3–16

bimolecular reaction, 47:Part 2:4–13

isomerization, 47:Part 2:13–14

excitation by, 28:3–96, 388, 412

fluorescence, 48:75

four center, 30:300

H + Cl$_2$, 19:17

H + H$_2$, 19:6, 9

ideal gases, 51:165–168

isomerization, 48:82

liquids, kinetic theory of, 48:71–182

orientation dependence, 30:390

quenching, 50:325ff

recombination, 48:86

stochiometric coefficients, 19:221

velocity, 19:220

see also Reaction; specific reactions

Chemical reaction mechanism, 3:5ff

carbon-halogen bond fission, 3:11

ketone reduction, 3:12

Chemical reaction rate, 3:139, 151

collision theory of, see Collision theory

nonequilibrium effects, 6:55

tunnelling corrections in, 3:131ff

Chemical reactivity, 1:165; 7:43ff; 8:65ff

Chemical relaxation process, dielectrically active, 4:379

Chemical relaxation time, 48:74

Chemical separation by thermal diffusion, 16:122–123

Chemical shift, 7:500; 12:75; 39:128, 129

Chemical short-range order, 51:238

Chemical trapping, 50:263, 308

Chemical twinning, 49:52

Chemical waves, 19:277, 287, 318; 38:334; 43:238

bifurcation of, 38:343

circular and spiral, 38:345

kinematic, 38:348

at localized states, 38:357

macroscopic theory, **38**:335
one-dimensional, **38**:335
propagation of, **38**:339
stability of, **38**:346
trigger, **38**:348
see also Waves
Chemically induced mutation, 7:85
Chemically reactive systems, nonlinear, 38:3–7
Chemically unstable gases, scattering in, 10:218–225
Chemiionization, 28:81; **30**:263, 269, 464, 528; **42**:13, 128; **52**:319
Chemiluminescence, 28:33ff, 57, 80ff; **30**:9, 263, 269, 276, 418, 454, 505; **41**:237, 242–248, 250; **47**:127; **50**:262, 301, 357
of chlorophylls, **19**:353, 372, 373
Chemiluminescent reactions, 42:1–62, 115, 130, 136
electronic, atomic, table of, **42**:41
Ba + CCl₄, **42**:31
Ba + N₂0, **42**:6, 14, 22, 32
Ba + OClO, **42**:31
Ba + S₂Cl₂, **42**:31
Ba + SO₂, **42**:29
Br + HI, **42**:27
Br + OClO, **42**:31
C⁺ + H₂, **42**:24
C⁺ + O₂, **42**:25
Ca + F₂, **42**:35
Ca + OClO, **42**:31
Ca + S₂Cl₂, **42**:21
Ca* + N₂O, **42**:6, 16
dioxetanes, **42**:47
dynamically induced, **42**:22
energy dependence of cross section, **42**:18
F + HI, **42**:27
Hg* + halogens, **42**:27
I₂* + F₂, **42**:22

In + I₂, **42**:18
ion-molecule, **42**:24
table of, **42**:42:48
La + O₂, **42**:32
metal + halogen, table of, **42**:38
metal + oxygen, table of, **42**:39
metals + N₂O, **42**:33
Mg + N₂O, **42**:16
Mg + OClO, **42**:31
N⁺ + hydrocarbons, **42**:24
Na*, **42**:12
Na₂ + Cl, **42**:12
NO + O₃, **42**:28
table of, **42**:45
NO + SO, table of, **42**:45
NO, fine structure components of, **42**:28
O + CS₂, **42**:122
O₂ + I, **42**:28
O₃, vibrational excitation in, **42**:28
Pb + N₂O, **42**:34
vibrational excitation, **42**:24
Sm + F₂, **42**:6
Sm + N₂O, **42**:6
Sn + N₂O, **42**:7, 22
Sr + OClO, **42**:31
Sr + S₂Cl₂, **42**:31
symmetry allowed, **42**:22
Tl + I₂, **42**:18
infrared, **42**:91, 117
ion-neutral collisions, **45**:165–196
He⁺ - Ar, **45**:178, 180
He⁺ - Xe, **45**:179
see also Luminescence
Chemisorption, 9:88, 105ff, 149, 161; **41**:165, 168–172
atom-jellium model, **41**:169
on crystal surface, **9**:146ff
ethylene on metals, **8**:141ff

heat of, **9**:90
irreversible, **9**:142
localized states, **9**:89ff, 91ff, 112, 140ff
  diamond-like, **9**:132
  one-dimensional, **9**:111
Shockley's and Tamm's states, **9**:107, 113, 141
theory of, **27**:212ff
  linear response, **27**:225
  Newns-Anderson model, **27**:214
  resonance, **27**:213
  valence bond, **27**:224
Chemoattractant, **52**:489, 490
Chemotactic coefficient, **29**:177
Chemotactic current, **29**:176
Chemotactic velocity, **29**:187
Chemotaxis, **52**:488
      one-dimensional migration, **29**:172, 173
      radial migration, **29**:172
    bacterial, **29**:169
    and band formation, **29**:172
    Nossal model, **29**:181
    Nossal-Weiss model, **29**:186
    in slime mold amoebae, **29**:152, 171, 296
    in spermatozoa, **29**:171
Chinese crab (*Eriocheir sinensis*), **55**:172
Chirality, **44**:546, 548, 564, 575ff, 620, 628; **55**:111
    functions, **44**:585
    homochirality, **44**:586
    spatiotemporal patterns, **55**:189–190
    see also Chiroptical spectroscopy; Optical rotation; Rotatory dispersion
Chiroptical spectroscopy, **44**:545–644
    aldehydes, **44**:628
    allenes, **44**:627

amides, **44**:621
benzene, **44**:627
butadiene, **44**:625
trans-2-butene, **44**:621
+-(D)-camphor, **44**:629
carboxylic acids, **44**:627
dienes, **44**:625
α-diketones, **44**:625
disulfides, **44**:621
ethylene, **44**:621
H_2O_2, **44**:621
lactones, **44**:627
5-methylenebicyclo [2, 2, 1] hept-2-ene, **44**:569, 633
nitramines, **44**:627
nitrosamines, **44**:627
1, 4-pentadiene, **44**:633
1-pentene, **44**:633, 634
peptides, **44**:627
β, γ-unsaturated ketones , **44**:632
Chloranil, as electron acceptor, **7**:254
Chlorella, electron spin resonance, **7**:637
Chlorine, **16**:127; **41**:169, 171
    adsorbed on Si (111), **49**:618
    crystal structure, **5**:184
    fluid structure, **53**:25, 26–29
    Fourier oscillations, **53**:55–56, 57
    frequency, **2**:289
    hydrate, **2**:3
      crystal structure of, **2**:4, 8
    interatomic distance, **5**:156
    London energy, **2**:78
    molecule, Franck-Condon factors, **50**:33
      intramolecular London energy, **2**:78
      ionization potential, **2**:71
      magnetic susceptibility, **2**:70, 71
      ortho and para forms, **50**:36

polarizability, 2:71

quenching of *B* state, 50:37

RKR curve for *B* state, 50:33

resonance line, 2:190

Chlorins, 7:386

N-Chloroacetyl-*p*-*O*-methyl-L-tyrosine, 16:196–197

Chlorocruorin, 7:386

Chloroethane, barrier height of internal rotation, 2:382

hydrate, thermodynamic data and lattice constants, 2:8

Chloroform hydrate, 2:3, 51

thermodynamic data and lattice constants, 2:8

Chloroform-hydrogen sulfide -water ternary system, 2:49, 50

Chloromethane hydrate, 2:50

thermodynamic data and lattice constants, 2:8

Chlorophyll, deuterated, 7:637

in vitro, 47:Part 2:579–626

aggregation of, 47:Part 2:587–589

covalently linked dimers, 47:Part 2:584–598

ESR linewidth, 47:Part 2:590–591

generation of dimers, 47:Part 2:593–594

photoactive dimer models, 47:Part 2:591–592

photochemistry of, 47:Part 2:589–590

primary processes, 47:Part 2:581–586

time resolved studies, 47:Part 2:598–623

electron-transfer reactions in cyclophanes, 47:Part 2:606–611

excited singlet state properties, 47:Part 2:611–621

models of reaction center, 47:Part 2:599–606

triplet dynamics of pyro-chlorophyll *a* dimer, 47:Part 2:622–623

light-harvesting, 19:354

luminescence, 19:353, 372, 373

Chloroplast, 7:629

electron spin resonance, 7:629

luminescence, 7:634

semiconductor action, 7:633

Cholesterol, 41:200

Choline phenyl ethers, 7:77

Christopherson's method, 32:55

Chromagranin A, 39:89, 91

Chroman complex, 2:4

Chromatium, 19:361, 363; 47:Part 1:98, 99

Chromatophores, 19:354

Chromium, 41:141, 161

Chromophore, 4:93ff, 108, 111; 7:586, 593ff; 8:48, 52

α-Chymotrypsin, 33:106, 107

Chymotrypsin, product inhibition, 7:668

Chymotrypsinogen, 33:109

CI, *see* Configuration interaction

Circadian rhythm, 19:316

Circular birefringence, 4:71, 72

Circular dichroism, 4:68, 70–72, 82, 83, 95, 105, 109; 12:122; 35:197; 44:546, 549ff, 589, 594

Cis-cis inversion, 44:380

Cis configuration, 2:158

of dipoles, 1:93, 95, 97

Cis effect, 7:374, 376

Cis-trans isomerization, 7:422; 44:380

Clamped nuclei approximation, 49:140, 147–148, 152–156

Clamped reagent, 53:182ff

Clamped solute, 53:77

Clamping approximation, 53:111–116, 179, 187–188, 206
    isolated solute atoms, 53:189–191
Clapeyron's equation, 1:299; 2:23, 43
Classical action, 25:98, 102, 132, 196
Classical barrier height, 36:143, 145, 151, 159
Classical deflection function, 10:77–81, 95, 100, 114, 136–137
    in semiclassical approximation, 10:95–98
Classical dynamics, 42:278, 434
    stochastic transition and, 47:Part 1:205–214
        divergence of trajectories, 47:Part 1:206
        integrable system, 47:Part 1:206–208
        K entropy, 47:Part 1:212–214
        mixing systems, 47:Part 1:208–210
        Poincaré surface of section, 47:Part 1:205–206
        typical systems, 47:Part 1:210–212
Classical equations of motion, 50:443
Classical limit, 17:138; 36:86, 122
    Egelstaff approximation, 17:199
    Schofield approximation, 17:141
Classical mechanics, criterion for applicability of, 10:53
Classical observable, 38:12
    definition, 38:40
Classical path approximation, 42:430, 432
Classical trajectory, 25:71, 98, 101, 109, 113, 123ff; 36:49; 42:65, 93, 101, 102, 126
    calculation, 50:313
    closed, 36:36, 49, 51
    complex-valued, 25:131–133, 140, 143, 146, 160, 168

    periodic, 25:168–171, 174
Classical turning point, 10:77, 85, 89, 92
Classically accessible region, 36:74, 96
Classically forbidden process, 42:281
Clathrate compounds, 2:2, 4, 14; 44:397
    heterogeneous equilibria, 2:5
    of hydrogen chloride, 2:4
    hydroquinone, 2:38
    ideal dilute solutions, 2:17
    molar heat, 2:21
    molar volume, 2:21
    solid solutions, 2:35, 38
    solutions, 2:36
    stability of, 2:5, 18
Clausius-Mosotti approximation, two-particle, 48:20
Clausius-Mosotti continuum, 48:210
Clausius-Mosotti equation, 1:80, 348, 349; 4:78; 6:284; 12:298
Clausius-Mosotti function, 46:13
Cleavamine, 16:207
Clebsch-Gordan coefficients, 13:253; 14:497; 25:91, 94, 95; 44:41, 123, 447ff
Cleft, role in enzyme action, 39:91, 170, 171, 172
ClF, B-X system, 50:62
    dissociation energy of B state, 50:62
    lasing action, 50:64
    photoionization of B state, 50:62
    predissociation, 50:62
    $X^1\Sigma^+$ state, 25:197
ClO, electron resonance spectra, 18:216, 228, 231, 232, 233
Close coupled equations, 19:19
Close coupling method, 18:72; 28:190; 36:104, 106, 181–183; 42:375, 389, 415, 429; 50:372
Closed Belousov-Zhabotinskii reaction, transitions in, 55:210–211
Closed-shell configuration, 7:45

Closed-shell molecules, 12:293
Closed system of equations, 38:197
Close-packing, 12:314
Close-packing density, 37:162
Closs, Kaptein, and Oosterhoff theory, 35:13
Cluster beams, 52:263–362
Cluster expansion, 11:318, 319, 325, 335, 375; 14:38
    for depolarized light scattering, 26:22
    for Helmholtz free energy, 11:19
    for polarizability of simple fluids, *ab initio* calculations, 26:55, 73, 74
        approximations for triplet component, 26:69–70
        for dielectric functions, 26:66
        for Kerr effect, 26:63
        superposition, 26:66, 67
    thermodynamic functions, 11:8, 19, 28, 77
Cluster functions, 14:239; 31:271, 279
    chain approximation, 31:310
    dimensionless, 31:271
    including coincidences, 31:275, 280
    Ursell, 11:321, 326
Cluster integrals, 1:278ff; 20:48, 49
    irreducible, 6:234, 235
    in radial distribution function, 6:192
    see also Diagrams
Cluster model, 44:501
    of metal-ammonia solutions, 4:346–350, 361–363
    of water, 31:80
Cluster operator, 52:186
Cluster series expansion, 48:237–238
Cluster sums, 6:192
Cluster theory, 9:2; 34:125, 131
    dense fluids, 34:131
    pair correlation, 34:125
Cluster-variation method, 31:54

Clustering reactions, 19:72, 109
Clusters, of atoms, in molten metals, 6:465
    of electrons, 6:331
        unlinked, 6:337–339
    of galaxies, 26:108, 138; 35:135, 156, 184
    in lipid bilayers, 39:264
    in membranes, 39:260
    microclusters, 40:49–136
    in nucleation theory, 40:137–156
    photoionization studies of, 52:263–362
        acetone, 52:345
        ammonia, 52:347
        argon, 52:281
        carbon dioxide, 52:330
        carbon disulfide, 52:306
        CO, 52:301–304
        ethlyene, 52:351
        H_2, 52:304
        HBr, 52:299
        HCl, 52:299
        HF, 52:299
        HI, 52:300
        N_2O, 52:330
        NO, 52:288
        O_2, 52:293
        rare gas, 52:280
        water, 52:344
    of stars, 26:109ff; 35:135, 154, 155, 184
    in synzymes, 39:131, 134
    in *o*-terphenyl, 6:473
    see also Microclusters
CN, 42:124, 226
Coarse-grained distribution, 35:125
Coarse-grained observable, 50:409ff, 429
Coarse-graining approximation, 22:185;

51:128, 134, 135; **53**:345, 362

Cobalamin, spectra, **7**:400

Cobalt, **41**:140, 141, 161

Cobalt-platinum, alloy, **2**:128
 enthalpy of formation, **2**:144
 solid solution, **2**:125
 solution, **2**:132

Cocarboxylase, **16**:162–164

Codimension two bifurcations, **55**:184

Coefficient of thermal expansion of liquid, **6**:281

Coenzymes, electronic structure, DPN, **7**:67ff
 FAD, **7**:67ff

Coexistence, **43**:245

Coexistence curve, **6**:176–181; **17**:31; **40**:334
 effect of gravity on, **6**:179
 Ising lattice theory, **6**:178, 194
 van der Waals gas, **6**:178, 187

Coexistence envelope, **37**:181

Coexistence of competitors, **38**:247, 325

Coexistence point, **43**:252

Coexistence region, **37**:197
 boundary locus, **37**:171

Cohen-Lekner theory, **31**:349

Coherence area, **24**:36

Coherence effects, energy acquisition and, **47**:Part 1:4

Coherence length, **20**:273; **40**:456

Coherence time, **40**:405, 417, 420.423, 424

Coherence zone, **27**:52, 53

Coherent excitation, in infrared, **50**:413ff

Coherent exciton, **40**:454, 457, 465

Coherent medium approximation, **52**:493

Coherent optical effects, **47**:Part 1:65–69
 experimental observables, **47**:Part 1:15–16

Coherent optical transient, **47**:Part 2:421–485
 dephasing and lineshape, **47**:Part 2:463–472
 hole-burning, **47**:Part 2:430, 434–444
 lineshape, **47**:Part 2:430–434
 nutation and free induction decay, **47**:Part 2:429–430, 457–461
 photon echo, **47**:Part 2:425–429
  in multilevel system, **47**:Part 2:472–481
  in organic mixed crystal, **47**:Part 2:444–457

Coherent picosecond excitation, **40**:7–10

Coherent potential method, **31**:312
 chain approximation, **31**:315
 quasi-crystalline approximation, **31**:315
 single site approximation, **31**:313

Coherent processes in molecular crystals, **40**:369–484

Coherent pulse propagation effects in SF$_6$, **47**:Part 1:583–624
 calculation of polarization density, **47**:Part 1:604–606
 electromagnetic field equations, **47**:Part 1:603–604
 radiated field, **47**:Part 1:609–611
 sidebands, **47**:Part 1:611–613
 spatial degeneracy, **47**:Part 1:613–617
 thin-sample approximation, **47**:Part 1:606–609

Coherent scattering, **17**:189

Coherent state, **38**:84; **40**:388; **44**:490

Coherent state representation, atomic lattice, **53**:314

Coherent transport, **40**:452

Cohesive energy, **5**:172; **41**:162, 163

Coincidence detection, **10**:215–218, 249, 270

Coincidence site structures, **27**:146

Collagen, **7**:264, 270ff
  thermal stability of, **21**:539, 560

Collapsed diatom limit, **42**:394

Collective correlation matrix, **44**:299

Collective effects, **11**:189, 200; **35**:140, 182, 183

Collective excitation, **38**:188; **41**:167

Collective fluctuation, **44**:300

Collective hydrodynamic interactions, **11**:271

Collective motion, **44**:436

Collective oscillation, **44**:296, 297

Collective properties, **8**:119

Collective rotation, **52**:124

Collective variables, **44**:295

Collective vibrations, **44**:488

Collinear elastic scattering, **53**:101, 103

Collinear oscillator problem, **42**:282

Collinear reactive scattering, **49**:193, 277, 281, 286
  resonances, **49**:230–232
  time delay, **49**:232–234

Collision, adiabatic, **36**:87
  Ar - N_2, **36**:114
  Ar - TlF, **36**:117
  atom-molecule, **36**:65
  atom + Morse oscillator, **36**:99
  atomic, effective range, **13**:49
    resonant charge exchange in, **13**:41
  between molecules or atoms, **40**:385
  binary, **19**:343; **31**:165–166, 223; **44**:285, 323, 467
  charged particle - hydrogen, **36**:66
  continuous, **42**:310
  coordinates, **36**:165
  correlated, **44**:355; **51**:58

detailed balance, **36**:77

duration of, **51**:50

elastic, **44**:353, 355

ellipsoids, **31**:180

experiments, **36**:14

H - CO, **36**:123

He - CO, **36**:123

He - H_2, **36**:123

He - HCN, **36**:124

He - N_2, **36**:114, 123

heavy particles, correspondence principle in, **36**:63–140

hyperthermal, **30**:256, 463

impulsive, **10**:268

incomplete, **35**:166

instantaneous, **44**:274

intimate, **10**:266–267

ion-molecule, **21**:209; **36**:77
  elastic, **30**:231–242
  inelastic, **30**:218–227
  reactive, **30**:9, 193–242
  see also Ion-molecule collisions

ion-neutral, **45**:82–224

Li + H_2, **36**:64, 122

loaded spheres, **31**:196

molecular, **45**:239–240

Ne - N_2, **36**:114

neutral-neutral, atom-diatom, **30**:394
  collinear, **30**:10, 13, 90, 92
  dissociative, **30**:493
  elastic, **30**:313ff
  electronic excitation, **30**:59, 193, 218, 445ff
  inelastic, **30**:218–227, 489
  ionizing, **30**:443, 449, 453, 458, 463ff

non-adiabatic, **30**:423ff

orbiting, **30**:305, 322, 328, 339, 369

polar molecules and electrons,

36:108

Rice-Allnatt model, **31**:199–201

rotationally inelastic, **50**:589–642

rough spheres, **31**:168

soliton, **53**:226, 239, 245, 268ff, 287, 301, 308ff, 328, 330

in stellar systems, **26**:108, 111, 112, 142

binary, **26**:111, 123, 133

sudden, **36**:97

three-body, **44**:445, 463

uncorrelated, **48**:107

violent, **10**:197–207

angular and energy distribution in, **10**:212–218

see also Scattering

Collision broadening, **50**:203

chemically reactive systems, **33**:239

classical-trajectory approximation in, **33**:276, 278, 285

density expansions in, **33**:262

direct scattering, **33**:260, 265

foreign gas, **33**:274

in gas mixtures, **33**:274

impact approximation in, **33**:267, 275

invariance under rotation, **33**:277

resonance-exchange, **33**:261, 265, 268

self, **33**:274

statistical-limit, **33**:287

velocity dependence in, **33**:279

Collision chamber techniques, for unstable gases, **10**:222–225

Collision complex, **10**:190, 261, 263, 266–267; **13**:98; **30**:68, 102, 105, 216, 302; **42**:14, 87, 108, 469

osculating, **30**:284

persistent, **10**:264–265

reactive, **30**:7ff, 247ff, 302

role in fluorescence quenching,

21:200

statistical, **30**:37, 47, 491

Collision cross section, of argon and neon; **21**:372

see also Cross section

Collision damping, **27**:275

Collision diameter, **1**:120, 121; **2**:106

Collision dynamics, **6**:35–55

Born approximation, **6**:42–43

classical, **28**:5–11

numerical solution for, **6**:53–55

rearrangement collisions, **6**:44–46

scattering cross section, **6**:40–42

separation of external motion, **6**:35–37

time-dependent perturbation theory, **6**:37–40

Collision effects, **47**:Part 2:185–418

electronic relaxation, **42**:207–270; **47**:Part 2:337–380

electronic to vibrational energy transfer, **47**:Part 2:381–418

intersystem crossing, **47**:Part 2:291–336

intramolecular energy transfer, **47**:Part 2:237–289

vibrational energy flow, **47**:Part 2:185–207

Collision energy, **10**:34, 77

reduced, **10**:77, 136–137; **12**:410

Collision frequency, **34**:27, 32, 34–36, 58, 60

Enskog theory, **34**:35

in gaseous methane, **34**:69

Poisson distribution, **34**:27

quasilattice or cell model, **34**:36

Collision-induced absorption, **33**:273

Collision-induced dissociation, **42**:30

Collision-induced emission, **42**:146

Collision-induced fragmentation, **19**:198

Collision-induced intersystem crossing,

42:235

Collision-induced light scattering, 46:1–72

Collision-induced line shape, 51:49–112

Collision integral, 36:191–194
for Brinser-Condiff theory, **31**:219
for effective potential theory, **31**:210
for ellipsoids, **31**:180–182
determined by large scale fluctuations, **38**:203
for rough spheres, **31**:169–170

Collision mechanisms, 45:196–206
C* - H$_2$, **45**:203
C$_2$H$_4^+$ - C$_2$H$_4$, **45**:200–201
H$_2^+$ - He, **45**:196–199
N$^+$ - H$_2$, **45**:201–202
N$_2$O$^+$, **45**:203

Collision number, 1:17, 18, 33, 369, 388; **5**:364ff

Collision operator, 35:142, 152; **38**:173, 182
finite, **35**:145, 189
homogeneous systems, **35**:152
infinite, **35**:146, 173, 175
inhomogeneous systems, **35**:145, 146, 173, 175, 189

Collision radius, 1:156

Collision rate, in hypersonic flow, **13**:29, 32
three-body, **13**:106

Collision term, 11:324, 329, 336
three-particle, **11**:318, 321

Collision theory, 1:119–124, 129, 133, 361, 389, 391; **5**:353ff
action space, **5**:356–358
collision number, **5**:364ff
crude, **1**:122
detailed balance, **5**:363–364, 367, 369
equilibrium distribution, **5**:379
mean first passage time, **5**:375ff

rate constant, **5**:374
steady state, **5**:379ff
rate equation, **5**:372
relaxation equations, **5**:367, 369, 370
steady state distribution, **5**:378
transition probability, **5**:354, 359ff, 363, 364, 367
translational relaxation, **5**:361–362

Collision time, 11:318, 323, 334; **13**:107; **27**:267, 275ff

Collisional activation and deactivation, 47:Part 1:257–261

Collisional alignment, 50:637

Collisional breakup, 10:199

Collisional deactivation, of excited ions, **45**:145–151
of vibrationally excited species, **47**:Part 2:12–13

Collisional energy transfer, 45:343–394
electronic, between He atoms, **21**:169, 178
between Hg and paraffins, **21**:189
see also Electronic deexcitation; Rotational deexcitation; Vibrational deexcitation

Collisional input, 51:178, 188–194
asymptotic, **51**:194–199
functional calculus, **51**:199

Collisional ionization, 42:90, 514–584; **45**:399–486
CFCl$_3$, **42**:560
Cs + CF$_3$X, **42**:548
Cs + D$_2$O, **42**:574
CsCl, **42**:579
curve crossing, **42**:517ff
deexcitation, **42**:566
differential cross section, **42**:520.564
induced by Xe* **42**:580
internal energy effects, **42**:551ff
K + Br$_2$, **42**:543
K + Cl$_2$, **42**:522

Li + Cl₂, **42**:571

Li + F₂, **42**:522

Li + I, **42**:520

Li + O₂, **42**:522

lifetime of CO_2^-, **42**:563

M + O₂, **42**:549

N₂ + CO, **42**:579

Na + Br₂, **42**:545

Na + I, **42**:520

Na + ICl, **42**:547

non-alkali, **42**:577ff

O_2^-, excited states of, **42**:573

pre-stretching, **42**:540

Th + O₃, **42**:578

total cross section, **42**:543ff

U + O₃, **42**:578

Xe + TlF, **42**:579

Collisional orientation, 50:637

Collisional output, experimental, **51**:217

filtered, **51**:208

model, **51**:217

observable features, **51**:178

Collisional processes, 25:212

Monte Carlo methods for, **13**:118

and photoselective chemistry, **47**:Part 1:58–65

vibrational and rotational, **42**:271–304

Collisional quenching of metal atoms, 50:325–394

Collisional scaling relations, 51:205

Collisional spectroscopy, 50:593

Collisional stabilization, 19:180

Colloid, aging of, **43**:238

Colloidal model of metal-ammonia solutions, 4:339

Colloidal solution, viscosity of, **6**:472

Color centers in alkali halides, 4:162, 194–198

Combinatorial analysis, 4:205–208;

49:67–68, 96

Combustion-related reactions, 47:Part 2:116–126

$O(^3P)$ + allene and alkynes, **47**:Part 2:116–119

$O(^3P)$ + carbon suboxide, **47**:Part 2:119–121

$O(^3P)$ + CF, CHF, CF₂, and CH₂, **47**:Part 2:124–126

$O(^3P)$ + CS, **47**:Part 2:123

$O(^3P)$ + CS₂, **47**:Part 2:121–123

$O(^3P)$ + OCS, **47**:Part 2:123–124

Common cage effect, 53:106ff

Communication, mathematical model of, **46**:165

Communication theory, 46:161, 190, 192

Commutation, 46:155, 188

Commutation relation, 38:80, 88

canonical, **22**:331

Weyl, **38**:80, 88, 96, 97

Commutator equation, 36:371

Commutator term, 35:145, 174

Compact operators, 25:282

Compatability relation, 38:20

Competition effects, 48:80

Competitive acceleration, 29:303

Competitive inhibition, 7:650; **29**:303

Competitive rate measurement, 41:238–242

Complementary error function, 13:286

Complete configuration interaction, 36:150, 152, 158, 166, 212, 411

Complete neglect of differential overlap, 42:67, 177

Completeness postulate, 38:53

Complex balanced current, 43:79

Complex balanced network, 43:78–85, 101

Complex biological oscillator, 55:171–176

spatial correlation, **55**:175

Complex coordinates, 25:254, 282, 293

Complex energy, 25:254, 282

Complex fluctuation function, 46:311

Complex formation, 19:82, 83, 90, 102; **44**:323

Complex graph, 43:75

Complex periodic and nonperiodic behavior, Belousov-Zhabotinskii reaction, **55**:205–217

Complex reaction networks, stability of, **43**:1–215

Complex systems, bifurcations and, **55**:177–199

adaptive chemistry, **55**:197

chiral forms, **55**:191–197

global bifurcations, **55**:183–186

local pheneomena, **55**:179–181

physicochemical examples, **55**:186–191

chaotic behavior, **55**:190–191

chirality and spatiotemporal patterns, **55**:189–190

polarity and spatial patterns, **55**:187–189

sustained oscillations, **55**:187

primary equations, **55**:178–179

Complex time, 30:118, 130

Complex vector, 43:73

Complexes, mass action, **43**:82

of defect and ion, **11**:7, 42, 43, 65–67, 71, 73

Compound nucleus theory, 40:98

Compound state resonances, 42:388, 389

Compressed gas, molecular dynamics, **44**:351

Compressibility, 4:165, 175, 182, 195, 196; **40**:353; **41**:138

adiabatic, **16**:236–239, 242–243

of fused salts, **9**:57ff

isothermal, **16**:237; **31**:37, 48, 92

of metal-ammonia solutions, **4**:308, 309, 334, 354, 355

of real fluid, **9**:54

Compressibility equation, 37:172; **41**:24

Compromise Hartree-Fock function, 2:245

Compton scattering, 53:6

Computer experiments, 17:68, 118, 120

computation time, **13**:207, 243, 244, 246, 247

Computer simulation, 34:234, 237; **35**:121, 153

of glasses, **48**:397–454

Concanavalin A, 39:85

Concave function, 51:120, 131, 141, 149

Concentration cells, 4:319

Concentration fluctuations, 24:4, 43

Concentration polyhedron, 43:18–20, 27, 207

Concentration profile, 29:118

Concentration regime, 51:25–28

Concentration variables, 33:402

Concentration vector, 43:17

Concentration zone, 29:133

Condensation, 1:295, 297; **24**:239

coefficient, **40**:139, 150

of gas, **40**:137

of solute molecules, **40**:139

Condensed media, 47:Part 2:421–679

chlorophyll *in vitro*, **47**:Part 2:579–626

coherent optical transient studies, **47**:Part 2:421–485

electron relaxation processes in liquids, **47**:Part 2:535–577

nonradiative processes in low temperature media, **47**:Part 2:521–533

proton transfer, **47**:Part 2:627–679

vibrational population relaxation in

liquids, **47**:Part 2:487–519

Condensed phase, relaxation and dephasing in, **47**:Part 1:82–90

Condenser method, for slow-ion detection, **10**:200–202

Condenser transducer method, **28**:209, 210

Conditional density operator, **46**:157, 162, 168, 173, 176, 187ff

Conditional distribution function, **34**:23

Conditional probability density, **34**:21, 52; **44**:268ff

Conductance, alametycin in membranes, **29**:358

dioleylphosphate-Millipore membrane, **29**:341

Fucus membrane, **29**:287

gramicidin in membranes, **29**:315, 316

limiting, of electrolytes, **11**:216, 225

model membranes with macrocyclic ion carriers, **29**:310, 311, 312, 358

squid axon membranes, **29**:350, 351

Conductance matching of oxidative phosphorylation, **55**:145–146

Conduction band, of disordered system, **31**:445

semiconductor properties, **53**:292–293

Conductivity, **7**:85; **17**:239, 266, 267; **38**:132, 135

anomalous, of NO_2, **12**:20

disordered structure, **52**:476

electrical, **31**:95

of metal-ammonia solutions, **4**:318–322, 365–370

ionic, of crystals, **11**:3

linear response theory, **53**:132–133

longitudinal, **27**:272

reacting systems, **16**:102, 106

related to adsorbed water, **7**:251

thermal, **1**:2, 134ff, 151; **13**:350; **16**:105, 124, 126

transverse, **27**:272

see also Thermal conductivity; Electrical conductivity

Conductivity tensor, **11**:182

Cones, complex balanced, **43**:80

conservation constraint, **43**:19

current, **43**:20–25

frame of, **43**:21

of mixing stability, **43**:66

simplicial, **43**:25

Configuration integral, **12**:289; **20**:43, 46

of ionic systems, **11**:85, 87

Configuration interaction, **1**:248; **2**:258; **14**:342ff; **23**:164; **25**:181, 184, 189, 192, 278; **26**:217; **28**:124ff; **42**:276; **45**:230–231, 236–239; **49**:277; **50**:525ff; **54**:2–3, 4

choice of configurations, **25**:189, 192

complete, **36**:150, 152, 158, 166, 212, 411

iterative natural orbital, **25**:192

in many-electron theory, **6**:318, 320, 347

using MCSCF configuration state functions, **54**:27–28

see also Hartree-Fock theory

Configuration interaction expansion, of MCSCF state, **54**:121–126

direct MCSCF , **54**:124–126

orthogonal complement states, **54**:122–124

Configuration mixing, **44**:9

Configuration space, **26**:124, 131, 136, 137, 141

Configuration space equations, **48**:127–139

Configuration state function, **26**:229; **44**:7

Configurational change, **7**:207, 210

Configurational coordinate, 24:96, 106

Configurational disorder, mechanism of melting, 6:467

Configurational energy, 16:224, 232, 235, 243, 260–261
of clathrates, 2:12

Configurational entropy, 4:218; 7:246; 16:232
excess, 16:275

Configurational free energy, 4:209, 223

Configurational internal energy, 44:402

Configurational partition function, 4:203; 6:232, 273; 16:224, 227, 231

Configurational probability density function, 44:299

Confluent hypergeometric series, 4:32, 38

Conformal ionic mixtures, 11:98ff, 107, 109, 113, 120
aqueous solutions, 11:107
pair potentials, 11:101

Conformal solution theory, 2:137
for hard sphere fluids, 6:270

Conformation, of macromolecule, 39:163, 165
of poly-(methacrylic acid), 39:163
of polyethylenimine, 39:126, 131, 135

Conformational analysis, 3:353

Conformational change, 29:325, 353
biological molecules, 33:73–152
contribution to ultrasonic absorption, 33:94, 95
cytochrome c, 29:230, 235
diffusional, 33:77, 78, 135, 136, 139
competition with nondiffusional, 33:136, 139
dioleylphosphate-Millipore membrane, 29:325, 334
elementary step of, 33:74, 90

enzyme, 39:177, 186, 187, 188, 190, 194, 197, 215

Fucus membrane, 29:288

hemoglobin, 39:86

local, 33:122, 127

local cooperative, 33:75, 115, 126, 131, 132

lyzozyme, 39:77, 79

migrational, 33:76, 131

nondiffusional, 33:77, 78, 125, 135, 136, 139

oligopeptides, 33:103

order-disorder, 33:118

polypeptides, 33:88, 94

probability of local, 33:121

proteins, 33:104–115; 39:182

rate of, 39:183, 190

regulatory, 39:195

renewal of conformations, 33:124

slow modes, 33:136–147

transitional, 33:76, 115, 131

vinyl-oligomers, 33:132

vinyl-polymers, 33:75, 115, 126–136, 144

three-bond, 33:116

Conformational mobility, in catalysis, 39:58
vibronic coupling, 39:95

Conformational partition function, 7:229

Congruence, principle of, 16:245ff

Conjugate eigenvalue problem, 20:161

Conjugate gradient method, 40:68

Conjugated organic ligands, 7:380ff

Conjugated systems, 8:78ff

Connection formula, 4:21–29

Connection procedure, 46:224

Conservation, of charges, 1:316
of flux, 36:54
of mass, 33:405
of orbital symmetry, 33:347

of probability, **1**:313

Conservation conditions, **43**:19, 46, 48

Conservation equation, **52**:429

Conservation laws, **43**:220
transport processes, **5**:262–263, 272, 273, 281, 294

Conservation matrix, **43**:19, 69

Conservative body forces, **33**:416

Conserved variable, **46**:313

Constant acceleration approximation, **12**:581, 591; **51**:68

Constant-angle modes, **27**:429, 523

Constant-field approximation, **37**:103

Constant organization, **38**:233

Constant-polarizability model, **48**:273

Constant speed approximation, **44**:416

Constants of motion, **19**:237ff, 243, 250ff, 266, 328, 347; **29**:92; **33**:180, 183, 199; **36**:27, 41; **44**:10
energy-like, **19**:255

Constitutive assumptions, **24**:96, 97, 109

Constitutive equations, **24**:106, 109

Constitutive functionals, **24**:119–120, 123–124

Constrained mixing stability, **43**:69

Constraint space, **40**:246

Constraints, change in, **38**:327
chemical systems, **38**:318, 319
transitions at constant, **38**:331
weak, **38**:321

Contact potential, effect of energy calibration, **42**:536

Contact term, Fermi, **37**:260–261

Contact transformation, **33**:176, 190

Continued fraction, **17**:106, 108, 177; **44**:272ff, 337, 345, 351, 375

Continued fraction representation, **53**:155–156

Continuity equations, **16**:105; **29**:177; **33**:406

Continuous branch, of eigenvectors, **37**:169

Continuous frequency spectrum, **36**:42

Continuous phase transition, **40**:237

Continuous-stirred tank reactor, **38**:320, 321, 334

Continuous systems, minimum entropy production, **3**:298

Continuous time random walk, **52**:369ff, 457, 460, 470, 474ff, 482, 487, 494; Random walk

Continuum, **9**:353ff, 357ff
perturbation calculations in, **9**:330
wave packets in, **9**:355, 357

Continuum approximation, **12**:300; **33**:9
in Schrödinger equation, **53**:282, 333–334
quasicontinuum, **53**:305, 311, 312
Sine-Gordon system, **53**:242, 323
soliton velocity and, **53**:284, 285
spin systems, **53**:231

Continuum eigenfunctions, *see* Singular eigenfunctions

Continuum limit, **52**:377, 381–383

Continuum mechanics, **32**:37ff, 42

Continuum model, **37**:101, 184

Continuum model polymer, **51**:24

Continuum oscillator strength, **45**:2–82

Continuum physics, variational principles in, **13**:289

Continuum theory of mesophase, **44**:385

Continuum weight function, **20**:157, 160

Convection, **19**:217; **24**:24
merging of layers, **32**:137
nonlinear instabilities, **32**:72, 74
numerical models, **32**:65ff, 67
onset of, **32**:281
oscillatory, **32**:103
penetrative, **32**:73
Prandtl number dependence, **32**:69
preferred mode of, **32**:30

produced by double diffusion, 32:135
residual, in Soret-effect cell, 13:358
steady, 32:104, 105
thermohaline, 32:281
see also Bénard convection
Convection cells, Baranowski, 29:134
Bénard, 29:133
see also Cells
Convergence of iteration procedure, 13:342
Convex body, 31:179
eccentric, 31:180, 182, 183
see also Convex molecules
Convex combination, 43:21
Convex core of molecules, 5:156ff; Convex molecules
Convex function, 40:314; 51:119–120, 142
Convex molecules, cohesive energy, 5:172
convex body, 5:148ff
parallel, 5:150
regular octahedron, 5:155
Steiner's formulas for, 5:151, 154
tetrahedron, 5:155
core, 5:156ff
anthracene, 5:157
benzene, 5:157
naphthalene, 5:157
crystal structure, 5:172
benzene, 5:178ff
cubic close-packing, 5:186ff
electrostatic interactions, 5:174
ethylene, 5:179ff
hexadecapoles, 5:188
hexagonal close-packing, 5:186
intermolecular potential, 5:172
methane, 5:172ff
multipole interactions, 5:180ff
nitrogen, 5:177ff

octupoles, 5:187
three-body interaction, 5:186
zero-point energy, 5:173
diffusion constants, 5:157ff, 161, 163
second virial coefficient, 5:164ff
multicomponent gases, 5:169
nonspherical molecules, 5:168
Convex parameters, 43:27, 29, 32, 35
Convex polyhedron, 43:19, 21
Convex thermodynamic parameters, 43:35
Convexity, 24:193, 212, 221, 222, 251; 33:439
Convolution method for Coulomb-type integrals, 13:225
Convolution theorems, 13:226, 261, 267
Cooling coefficient, 29:54
Cooling rate, for glass formation, 51:228
Cooperative defects, in crystals, 6:464
Cooperative effects, 26:121
Cooperative excitation, bond mode model, 47:Part 1:309–313
Cooperative interaction, 46:365, 367
Cooperative many-body effects, 42:310
Cooperative process, 44:263
Cooperative torsional oscillation, 44:260
Cooperativity, conformational change of polymers,
length, 33:75, 84, 95, 100, 101
local, 33:75, 115, 126, 131, 132
and nondiffusionality, 33:147
and Schottky anomaly 33:119
and sharpness of transition, 33:105, 116
slowing down by, 33:87
and two-state model, 33:115–121
dynamics of, 39:182
membranes, 29:305, 352, 353
negative, 39:192, 196
positive, 39:192, 196

yeast phosphofructokinase, 29:145
Coordinate wave function, 36:16
Coordinates, atom transfer, **30:**24, 55
    body-fixed, **49:**274, 284–285
    curvilinear, **30:**11, 33
    cylindrical, **27:**396–397, 402
    Eckart, **37:**308
    ellipsoidal, **13:**213
    general, **27:**395–401
    generalized, **32:**14
    ignorable, **13:**91
    internal, **32:**14
    laboratory, **37:**310
    molecule-fixed, **37:**310
    natural bifurcation, **30:**35
    nuclear, **13:**222
    reaction path, **30:**13
    rectangular, **27:**401, 403
    space-fixed, **49:**274, 284–285
    spherical, **13:**213; **27:**387–389, 402
Coordination number, 51:237
Coordination polymerization, 2:148, 162, 170
Copolymerization, 2:168, 171
Copper, 41:141, 182
    bulk, **49:**557–558, 571, 593
    oxygen adsorption, **49:**644–645
    partial molar heat of solution in tin, **2:**134
Copper (Cu₂), O_g^+ state, magnetic inter-action energy, **12:**91
Copper acetyl-acetonate, 7:370
Copper complexes, absorption spectra, **5:**256
Copper-gold, solid solution (CuAu₃), **2:**129
    system, **2:**123
Copper-platinum, alloy, **2:**128
    solutions, **2:**132
    superlattice, long range order of,
2:128
Copper-zinc, alloy, **2:**129
    solutions, **2:**132
Coppinger's radical, 7:319
Core field, 8:58
Core-halo structure, 26:122, 123
Core hole, 41:169
Core parameter, 12:291; **48:**201–206
Core polarization, 6:378; **14:**395
    in lithium, **6:**317
    by π-electrons, **6:**317
    potential, **6:**378
    by valence electrons, **6:**317
Core potential, 20:5
Core-valence non-penetrability approxi-mation, 16:319–320, 334–336
Coriolis coupling coefficient, 44:113
Coriolis energy, 49:183–184
Coriolis forces, 2:380
Coriolis interaction, 28:332, 341, 344, 345, 357ff, 369
Corner aproximation, 3:89
Correlated collisions, 44:355
Correlated diffusion model, 52:464–465
Correlated reorientation, 34:14, 19, 58, 82; Reorientation
Correlated wave functions, 2:207, 223, 250ff, 297, 323; **23:**184
    correlation effect in, **2:**251
    cancellation of, **2:**246
Correlation, 7:8; **19:**309; **38:**175ff, 188, 189; **41:**182
    adiabatic, **42:**32
    atomic charge, **23:**119
    attractive, propagation of, **37:**184
    clipped, **24:**19, 36
    creation of, **11:**332, 335, 372
    decay of, **15:**199ff
        harmonic oscillator models, **15:**205

independent random walk model, **15**:200

one-dimensional Ising model, **15**:212

destruction of, **11**:332, 334, 373

dynamic, **31**:179, 220–222

effect on dispersion energy, **12**:38

electric, **2**:248

electrons, **8**:3ff

    in biradical transition states, **33**:380

    contribution to intermolecular interaction, **33**:350, 354

equilibrium, **11**:182, 330, 335

fluctuations, **19**:309; **48**:344

hydrodynamical, **11**:253, 263

initial, **35**:141, 160, 173

molecular, **34**:206, 208

n-body, **40**:299

one-dimensional chain, **15**:311

pair, **11**:170, 192

potential surfaces, **42**:24, 30, 98, 140, 523

self-avoiding walk, **15**:246

static, **43**:231, 248, 254

stellar systems, **26**:114, 124–126, 129–134, 141, 142

time correlation function, **19**:269

see also Time correlation

Correlation diagram, **20**:55; **30**:204, 272, 293; **36**:170, 173; **41**:251

alkali-rare gas, **30**:438

Ar-Ar, **30**:438

Ar-K, **30**:438

Br$_2$-Cl$_2$, **30**:300

He-He, **30**:452

Ising model, using Toeplitz determinants, **15**:335

K-Hg, **30**:441

K-K, **30**:440

K-Na, **30**:441

kinetic equations, **15**:21

Ne-Ne, **30**:453

O-N$_2$, **30**:306

Correlation energy, **2**:211, 233–236, 239, 246, 254, 255, 285; **6**:317, 358; **9**:332, 350; **12**:289, 292; **14**:284; **16**:285, 326; **23**:98; **35**:136, 167, 170, 178, 184; **36**:208, 232, 239; **41**:86, 108;

approximations, **54**:294–296

neon atom, **6**:359

neon ions, **6**:359

nitrogen atom, **6**:359

nitrogen ions, **6**:359

non-closed shell system, **6**:324

numerical evaluation, **54**:258–265

Correlation error, **2**:224, 231–233, 242; **23**:183

Correlation factor, **2**:257, 300ff; **14**:301

for electrical moments, **1**:81, 83, 98, 103, 105, 116

Correlation function, **12**:553ff, 575ff, 595; **17**:239, 241, 254ff; **24**:32ff; **25**:258, 259, 266–268, 284, 285; **31**:91; **32**:91; **33**:246, 269; **34**:4, 12, 14, 16, 32, 38; **38**:202; **40**:297, 305; **41**:71, 72; **44**:263, 265, 269; **46**:113, 116, 253, 345; **48**:377–378; **53**:25, 28

amplitude, **48**:378

angular momentum orientation, **34**:8, 35, 53, 95, 97

asymptotic expression, **48**:198

atomic fluids, **53**:12–13

behavior on energy shell, **46**:116

behavior on torus, **46**:116

CD$_4$, **34**:58

CH$_4$, **34**:58

chlorine, **53**:25, 28

collective, **44**:259

current density, **44**:259

defects, **11**:32, 64, 195

dielectric, **34**:18

dipole, **34**:18

direct, **6**:181; **48**:189; **49**:407–411

directional, **17**:153

doublet, **24**:261

equation of motion for, **46**:346

extended diffusion model, **34**:27

finite volume, **40**:297, 309

fluid structure, **53**:2

Fourier transform, **12**:566, 567, 576; **53**:54–55

fourth order, **17**:156

free rotor, **34**:23

frequency matrix and, **53**:142

from memory function, **34**:99

gamma-ray diffraction, **53**:5–6

Gaussian approximation, **34**:25, 63

homonuclear diatomic fluids, **53**:22

incoherent nuclear rotational, **34**:16, 80

indirect, **6**:181

induced dipole, **51**:52

infinite volume, **40**:297, 309

infrared, **34**:17

intensity, **48**:379

isotropic small step reorientation, **34**:46

light scattering, **34**:12, 14

liquid methyl iodide, **34**:87

long time behavior of, **33**:283

longitudinal, **44**:259

model, **51**:69–71

molecular fluids, **53**:29

molecular-time-scale generalized Langevin equation, **53**:69–70

multiparticle, **44**:298, 340, 343, 400

nitrogen, **53**:24

noninteracting, **41**:116

one-sided time, **17**:84

orientational, **44**:303

Ornstein-Zernicke equation, **53**:119, 123, 124

pair, **48**:188–201, 254; **49**:407, 423; **53**:117–118

polyatomic fluids, **53**:45

quadruplet, **21**:400

rotational Raman, **34**:10

rototranslational, **44**:255–482

sine-Gordon chain, **53**:249

single particle, **44**:259, 298

solitons in spin systems, **53**:248–262

superposition approximation, **21**:401

time, **17**:65, 99, 113; **53**:128–129

translational, **34**:81

transport processes, **5**:263, 302

transverse spin density, **44**:259

two-body, **49**:407–413

van Hove, **17**:91, 191, 201

water, **53**:37, 41, 42

see also specific correlation functions

Correlation function formalism, line shapes, **15**:101

neutron scattering by liquids, **15**:129ff

spectral density, **15**:79ff

spectral function, **19**:269

Correlation integral, **31**:157, 200

Correlation length, **24**:38, 40, 47; **31**:239; **49**:313, 319, 352–353

Correlation matrix, **44**:268, 289; **51**:149

Correlation potential, **6**:344; **14**:306

Correlation properties, **52**:369

Correlation spectroscopy, **43**:261, 263, 264

Correlation theory, **9**:331ff

Correlation time, **34**:5, 32, 34, 47, 49; **40**:372, 461, 464, 465, 467; **44**:266

angular momentum, **34**:35, 59

benzene, **34**:75

CD_4, **34**:58

CH_4, **34**:58

extended diffusion model, **34**:31, 71

Gaussian approximation, **34**:25, 33, 92

isotropic small step reorientation, **34**:46

methyl iodide, **34**:86

Raman, **34**:10

random axis orientation, **34**:45

spin-rotation, **34**:7, 91

spin-spin, **34**:5

vibrational, **34**:88

Correlation wave function, 6:330

Correspondence principle, 36:42, 44, 66, 68, 86

density of states, **36**:87

in heavy-particle collisions, **36**:63–140

Heisenberg's, **36**:72

irregular spectrum, **46**:123

regular spectrum, **46**:122

strong coupling, **36**:90

Correspondence relations, 25:72, 89–91, 98, 174; **30**:81, 85

Corresponding states, 17:34, 35, 36; **30**:357, 367

principle of, **1**:230; **4**:230–232, 281–289; **9**:51, 70; **11**:118, 119, 121, 127, 132; **12**:319; **16**:224, 231ff, 248, 252, 270ff; **37**:155

for chain molecule liquids, **16**:226–235, 245, 279

for dimerization of alkali halides, **11**:95

for excess functions, **16**:228, 245, 248–252, 270–276

for mixtures, **11**:119

for normal alkanes, **16**:235–245

for spherical molecules, **16**:224–226

for surface tension, **11**:93

for vapor pressure, **11**:89, 91

single liquid model, **17**:34, 35

three liquid model, **17**:34

two liquid model, **17**:36

Corrins, 7:380, 386, 399

Cortex-thalmus, 19:348

Cortical cells, 7:269

Coster-Kronig transitions, 27:27

Couette flow, 29:134; **32**:119, 153

Coulomb correlation, 2:216, 231, 232, 253

Coulomb energy, 33:156, 209, 213, 224, 351, 356, 362, 392

of H_2, **12**:65

Coulomb exciton, 54:330–331

Coulomb field, 33:191, 220

Coulomb forces, 27:359, 435, 489, 503–504, 510, 518; **53**:65

intra-atomic, **27**:213, 218

Coulomb gas, 40:194

two-dimensional, **40**:201

Coulomb gauge, 33:155, 158–160, 177–179, 188, 190ff, 209–213, 217, 224; **41**:83

Coulomb hole, 2:216, 218, 224, 232, 305; **41**:106

Coulomb integral, 1:183, 244, 249, 256, 259; **2**:384, 389; **5**:222; **7**:25, 99

Coulomb interaction, 1:166, 167, 313, 326, 327; **5**:303; **11**:187, 195; **33**:154, 192, 210; **38**:188; **41**:84, 85, 132

collective, **11**:271

Coulomb potential, 33:160; **36**:44, 70; **41**:209 **52**:122, 154

in electron diffraction, **53**:5

for electron gas in lattice of postive charge, **6**:127–130, 143

modified, for alkali metals, **6**:129

Coulomb-type integral, 13:211, 245

convolution methods for, **13**:225

Coulomb wave, **36**:213

Coulomb's law, **4**:139

   application to ionic solvation, **6**:414, 425, 431, 443

Coulson's integral formula, **9**:159

Countertransport, **29**:303

Counting theorem, **51**:11

Coupled channel method, **19**:11; **25**:70, 149, 212, 257, 264, 279; **30**:52

Coupled-cluster approach, approximate versions, **52**:200–210

   for BeH molecule, **52**:258

   construction of, **52**:191–195

   to electronic structure, **52**:181–259

   linearized, **52**:203–205

   nondegenerate, **52**:182–234

   with one-particle perturbation terms, **52**:227–234

   quasidegenerate, **52**:234–251

Coupled differential equations, **49**:116, 156–157

Coupled electron pair approximation, **50**:535

Coupled integral equations, **49**:157–158

Coupled nonlinear oscillator systems,

   classical, **47**:Part 1:119–161

     numerical studies of, **47**:Part 1:133–147

     perturbation theory, **47**:Part 1:147–159

     properties of trajectories, **47**:Part 1:119–132

     quantum mechanical, **47**:Part 1:161–195

       compared to classical, **47**:Part 1:176–186

       ergodicity and reaction rate, **47**:Part 1:186–195

       Nordholm-Rice analysis, **47**:Part 1:162–168

       semiclassical quantization, **47**:Part 1:168–175

Coupled oscillator systems, numerical experiments, **47**:Part 1:133–147, 214–219

Coupled-pair many-electron theory, extended, **52**:205–210

   correlation energy, **52**:251–254

   Feynman-Goldstone diagrammatic construction, **52**:210–226

Coupled state method, **42**:430

Coupling, radial, **30**:423, 458

   rotational, **30**:423, 458, 468

   symmetry effects, **30**:487

Coupling factor, **39**:1, 205, 213, 214

Coupling matrix element, **30**:176, 426, 468, 482, 499, 512, 519

Covalency, **8**:15

   central field, **5**:68; **8**:53

   effect on magnetic properties, **5**:210ff

   symmetry restricted, **5**:68; **8**:53

Covalency effects in octahedral $5f^1$ complexes, **21**:353, 356

Covalent bond, **2**:258; **4**:162, 176, 178, 194; **7**:17, 203

Covalent porphyrins, **7**:74

Covariance, **44**:288

Covariant representation, **22**:337

Covering lattice, **51**:21

Creation fragment, **11**:175; **16**:11

Creation operator, **6**:90, 94; **14**:41, 95; **24**:279

Critical cluster, **40**:179, 183

Critical compressibility factor, **17**:173

Critical curve, azeotropes, **17**:4, 26

   of binary mixtures, **17**:2, 3, 4, 12ff

   liquid crystals, **17**:3

   polymerization temperature, **17**:3

   pressure maximum, **17**:4, 12, 14

   pressure minimum, **17**:14, 20, 21

   temperature maximum, **17**:4, 11, 12

   temperature minimum, **17**:4, 9, 14,

16, 18, 21, 22

Critical density, 48:295

Critical energy, 46:86, 113

Critical enhancement, 43:251, 253, 258

Critical exponents, 24:39; **37**:11, 142, 180, 184; **38**:276, 300; **49**:318, 340–341; **51**:164

  classical, **38**:300, 306

  dielectric constant, **48**:301

  nonclassical, **40**:349

Critical fluctuations, 34:302

Critical frequency, 1:78

Critical indices, 21:431, 446

Critical isotherm, 37:180

Critical mixing, 24:40

Critical nonclassical index, 49:518

Critical nucleus, 40:126, 191, 206

Critical phenomena, 24:2, 3, 8, 258; **37**:140; **51**:3

  finite size effects, **49**:3190–320

  model of Kadanoff and Swift, **24**:258

  multiple steady state systems, **49**:316–320

  plait point, **49**:442–443

  theory, **49**:358, 438, 440–444, 450

Critical point, 1:303; **17**:2, 4; **37**:180, 181; **41**:3ff, 17ff, 33ff, 40, 47, 199, 207, 208, 224; **43**:37; **46**:266; **49**:450, 560, 578; **51**:162

  carbon dioxide, **2**:97

  cell-pair correlation function, **21**:442

  cluster model, **21**:434

  critical indices, **21**:431, 446

  law of corresponding states, **21**:434

  pure substance, **2**:96

  scaling law, **21**:434

Critical point analysis, 47:Part 1:225–226

Critical pressure, 17:4

Critical ray, 43:175, 182–183

Critical Rayleigh number, 13:284

Critical region, 6:175–224

  coexistence curve, **6**:176–181

  correlation function, direct, **6**:181

    indirect, **6**:181

  dielectric behavior, **6**:224

  diffusion, **6**:222

  heat capacity, **6**:207–215

  hypernetted chain approximation, **6**:193

  light scattering, **6**:181, 194–197

    Einstein-Smoluchowski theory of, **6**:195

  Ornstein-Zernike equation, **6**:181, 192

  pressure equilibration, **6**:179

  radial distribution function, **6**:181–194

    generalized diffusion equation for, **6**:217

    numerical evaluation of, **6**:192–194

    superposition approximation, **6**:185

  sound propagation, **6**:215–222, 223

  thermal conductivity, **6**:223

  viscosity, **6**:197–199

Critical size, 40:149

Critical slowing down, 40:248; **43**:228, 253; **46**:266, 316

Critical solution temperature, 17:5, 6, 26, 27

  differentiation, **17**:26

  lower, **17**:5, 6

  upper, **17**:6

Critical temperature, 1:96, 298; **40**:138; **41**:2; **48**:295

  of air, **2**:96

  for disordering of a superlattice, **2**:129

  of ethane, **2**:100

Critical transport phenomena, model of

Resibois and De Leener, **24**:258

Critical wave number, **35**:128, 133, 182

Cross-correlation, **44**:271, 292, 309, 374, 403, 474

Cross-linking, **7**:277; **22**:88, 118–121, 123, 125; **39**:111, 157, 166

Cross relaxation, **11**:298; **35**:22

Cross section, **1**:121; **10**:35; **12**:398, 399, 414; **19**:31, 53; **20**:228, 231, 232, 234, 286; **25**:70, 71, 125, 126, 146ff, 216, 264, 268

absolute, **18**:27

absorption, **30**:48

asymptotic, **11**:337

averaging of, in non-central forces, **10**:49–50

background scattering correction to, **10**:50–53

charge transfer, **10**:197ff; **30**:458, 477, 495, 528

alkali-halogen, tabulated, **30**:509

alkali-non halogen, tabulated, **30**:522

apparent total, **10**:201

low energy, **10**:248, 250, 252–257

classical, **36**:75

collision, **19**:143, 153

differential, **10**:37–39, 83; **28**:7, 21; **51**:190, 203, 209

elastic, **30**:313, 458

electron scattering, **1ʼ**:16, 30, 39, 74

emission, **30**:422

excitation, **30**:422, 457; **36**:417

inelastic, **30**:229, 456

interference structure, **30**:234, 327, 378, 382, 476

ionization, **30**:458, 499

orientation dependence, **30**:397

oscillation of, **30**:327, 333, 344, 378, 382

reaction, **30**:7ff, 252ff

effective diffusion, **51**:203

effective viscosity, **51**:203

electron scattering, by nonpolar molecules, **36**:508

by polar molecules, **36**:513

emission, **30**:419, 431, 450

excitation transfer, **28**:280ff, 302ff, 347

fluorescence quenching, **21**:192–200

He-He scattering, **21**:175

inelastic, **30**:122

integral, **51**:188–189, 196, 202, 213

interference effects on, **12**:404, 424, 482

ion-molecule collisions, **21**:211

ionizátion, **10**:55–57, 188–190; **30**:524

krypton, **53**:16

molecular fluids, **53**:56

orientation dependence, **30**:397, 512

phenomenological, **10**:259–260

photoabsorption, **52**:288

photoionization, **52**:288

Placzek correction and, **53**:51

polar, **10**:139

pulsed neutron sources, **53**:9

quantum, **12**:481

quantum correction to, **10**:53–54

quasi-elastic scattering, **27**:56, 64, 65

Raman pressure broadening, **51**:189, 203

random phase approximation, **12**:413, 415

reaction, **13**:101; **28**:9, 11; **30**:38

relative fragmentation, **52**:341

relative reaction, **52**:297

resonances in atom-atom scattering, **12**:483

scattering, **28**:6, 7, 9

structure factor and, **53**:48

tables of, **28**:421

threshold ionization, **36**:487

total, **10**:37–39, 83, 127–131; **12**:367, 389, 400, 413, 424; **30**:315, 419, 425, 431, 458

interference structure, **30**:329, 351, 369

undulation of, **30**:17, 91, 93, 329, 351, 361, 379, 426, 436

velocity dependence, **12**:414

water, **53**:35

see also specific types of scattering

Crossed beam experiments, 10:139–141, 251, 270, 322–325; **42**:274; **50**:486, 510

quenching of excited atoms by molecules, **45**:341–398

with unstable gases, **10**:218–222

Crossing rate, as approximation to reaction rate, **13**:89

Crossing symmetry, 38:145

Crossing time, 26:110, 117, 119, 120; **35**:121, 136

Crossover phenomena, 49:461, 501–502, 527

Crude collision theory, 1:122

Crystal, band structure, **8**:29

centrosymmetric, phase determination, **16**:158–161

structure factor, **16**:139–145, 198

cleanliness, **20**:278

configurational disorder, **6**:467–469

disorder, **16**:167

electron affinity, **7**:286

etching, **20**:277

excitation band, **8**:30

ionization potential, **7**:286

isotopically mixed, **40**:373, 392

liquid, **6**:467

localized excitation, **8**:28ff

melting mechanisms for, **6**:459–480

mixed, **2**:190

trap depth, **8**:30ff, 33ff

molecular, **53**:227–287

coherent processes in, **40**:369–484

molecular rotation in, **6**:461

polishing, **20**:277

positional disorder in, **6**:463

premelting in, **6**:460

rotational disorder in, **6**:466–467

similitude rules for melting, **6**:462–463

structure factor, **16**:136

see also Molecular crystal; Plastic crystal

Crystal constants, of convex molecules, **5**:172

Crystal defects, 6:463

cooperative, **6**:464

Crystal field, 5:42, 59, 67, 251

calculation of magnetic properties, **5**:192

charge transfer transition, **8**:21ff

Kramers theorem, **5**:196

ligands, **8**:20

π bonding, **8**:22ff

σ bonding, **8**:22ff

spectra, **8**:13ff

splitting, **7**:362

see also Ligand field

Crystal growth, 4:46, 52; **21**:717; **29**:134;

analytic theory of, **40**:169

asymmetry between growth and evaporation, **40**:213

continuous fall method, **21**:767

dynamics of, **40**:157–228

microcluster models of, **40**:127–129

nucleation theory of, **40**:177–190

velocity, **40**:151

Crystal Hamiltonian, **8**:33

Crystal lattice, **53**:57; *see also* Lattice

Crystal lattice plane, evaporation of, **40**:213

Crystal melt, **40**:138, 144, 145

Crystal monochromator, **34**:162

Crystal nucleation, *see* Nucleation

Crystal nucleus, **6**:469

Crystal spectra, **8**:13ff, 27ff

Crystal structure, face-centered cubic, **4**:188, 190

Crystal surface, adsorption on, **9**:85ff

Crystal symmetry, **40**:284, 346, 354

Crystalline phase, **41**:198, 209

Crystallization, **32**:164

basal plane, **32**:199

curvature effect, **32**:188

kinetics of, **32**:184

symmetry-breaking, **22**:353

Crystallization tendency of soft sphere fluid, **48**:404

Crystallographic shear, **49**:52

Cube modes, **27**:444, 529

Cubic close-packing, of convex molecules, **5**:186ff

Cubic lattice, **52**:395, 402

Cubic potential, **53**:313

solitons, **53**:307–308

Cuboctahedron, stability of microclusters, **40**:71, 96

Cumulant, random walk, **52**:369

Cumulant expansion, **40**:13, 15, 21, 24, 25, 35; **41**:13, 22, 26

Cumulative distribution, **15**:88

Cupric dimethylglyoximate, **7**:370

Cuprite structure (Cu_2O), **49**:51–52, 92, 93–95, 105, 110

Curie point, **41**:2

Curie principle, **44**:381

Curie symmetry principle, **16**:101

Curie-Weiss law, **7**:610

Curie's law, **5**:191; **7**:167

Current, complex balanced, **43**:79, 80

equilibrium extreme, **43**:34

extreme, **43**:21, 24, 45

Hurwitz nonmiximg, **43**:125

pseudoextreme, **43**:88

stability of, **43**:63, 118–119

stability tests, **43**:125

stoichiometric matrix, **43**:114

Current cone, **43**:20–25

Current cycle, **43**:126–138, 208

critical, **43**:107, 128, 156–163

explosion, **43**:161–163

tests for, **43**:141–142, 143–144,

simple, **43**:129

Current density, **1**:316; **33**:164, 165, 188, 215; **41**:79, 85; **44**:439

longitudinal, correlation function, **44**:437

transverse, **44**:403

Current diagrams, **43**:114–119

Current polytope, **43**:20–25, 27, 207

Curve crossing, **28**:115, 119, 132, 146–151; **49**:295–300; **50**:237

adiabatic, **28**:115, 151

avoided, **28**:132; **30**:59, 347, 423, 454, 456, 469, 476

collisional ionization, **42**:517ff

conical, **30**:485

covalent-ionic, **30**:304, 306

diabatic, **28**:115, 151

mechanism, **50**:287, 480

model for reaction, **49**:300–302

pseudo, **30**:464

Cusp catastrophe, **40**:246; **43**:37

and van der Waals equation, **38**:303

Cusp conditions, **41**:96, 101, 106

Cusp singularity, **40**:245

Cusp-type effect, **18**:134
Cutoff, long-range, **35**:151, 165, 183
    short-range, **35**:136, 151, 152, 158
Cutoff frequency, **44**:327, 329
Cutoff volume, **26**:111, 136, 139, 140
Cutoff wavevector, **27**:345, 347
Cyanine dyes, **49**:34
Cyanogen, **20**:4, 12; **22**:221, 252, 291;
    36:316
    librations of, **22**:221
Cyclazocine bromide, **16**:199
Cyclazocine, **16**:199
Cycle, complex, **43**:78
    critical, **43**:128
    critical current, **43**:107
    current, **43**:126–138
    destabilizing, **43**:104
    destabilizing negative feedback,
        43:196
    destabilizing positive feedback,
        43:191
    directed graph, **43**:93
    explosions and, **43**:161–163
    feedback, **43**:105, 191, 196
    irreversible, **43**:126
    knot, **43**:127
    matrix diagram, **43**:104
    matrix feedback, **43**:86
    noncurrent knot, **43**:127
    reaction, **43**:129
    safe, **43**:104
    sign-switched, **43**:121
Cycle index, **49**:68–70
Cycle structure, **49**:68–69
Cyclic-AMP, of *d. discoideum*, **38**:383,
    384, 389, 390
    oscillations, **38**:384, 390
    pulses, **38**:367
Cyclic boundary conditions, **8**:31
Cyclic silicones, **2**:155

monomers, **2**:180
Cyclohexaglycyl, **16**:180–181
1-Cyclo-hexyl 2, 2 didecyl ethane-
    ethane system, **2**:103
Cyclobutane, **16**:177, 184, 196
Cyclopentadiene, **53**:293
Cyclophanes, electron-transfer reactions
    in, **47**:Part 2:606–611
Cyclotron frequency, **19**:142; **27**:267,
    270
Cylinder modes, **27**:396, 402
Cylindrical distribution, **44**:287
Cylindrical tubes, inflation of, **24**:139–
    145
    Hamiltonian functional, **24**:144
    inertia of, **24**:141
    Lagrangian functional, **24**:144
Cysteamine, **7**:261
Cysteine, **7**:261
Cystine, **7**:267, 268
Cystine dihydrochloride radical, **7**:550
Cytidine, infrared spectra, **7**:441ff, 455
Cytidylic acid, **7**:349
Cytisine hydrobromide, **16**:213
Cytochrome, **7**:160, 182, 377, 620;
    19:362, 365
    electron carriers, **7**:74
    electronic structure, **7**:71
Cytochrome *a*, **7**:386, 392
Cytochrome *c*, **39**:80–82, 92
    amino acid sequence, **29**:196, 197
    ancestral, **29**:218, 221
    codons, **29**:223–231
    conservative substitution, **29**:200
    constraints on evolutionary varia-
        tion, **29**:231
    covarions, **29**:226–231
    distribution of variable positions,
        29:236
    frequency of replacement, **29**:213

ion binding properties, **29**:240

radical substitution, **29**:202

replacement distance, **29**:212, 216

spatial structure, **29**:231

stereoscopic invariance and carbon diagram, **29**:204, 238, 239

structure and function, **29**:195

structure of ferrous and ferric conformations, **29**:232

tertiary structure and function, **29**:205

variability, **29**:198, 224, 226, 227, 228

Cytochrome oxidase, 7:260

Cytochrome P-450, 39:168

Cytosine, 7:52ff, 85, 97, 456

internuclear distance, 7:87

molecular diagrams, 7:94

transition moment, 7:95

Cytotoxin reaction, 38:395

D

Δ states, **52**:152

d-d transitions, 7:363, 371

D-stability, 43:63, 65

D-symmetrizability, 43:10

D-vector model, 51:15

Dalziel relationship, 7:682, 689, 698, 706, 710

Damping, **27**:380, 404–414, 419, 429–430, 438

attenuated total reflection measurement of, **27**:446–447

of coherent processes in crystals, **40**:374, 375, 378, 381, 383, 384, 389

frequency-dependent, **27**:439, 447, 503

intrinsic, **27**:522

radiative, **27**:382, 387, 438, 522

Damping factor, 4:86, 87, 154

Damping matrix, 48:89

Damping torque, 44:362

Darcy's law, 32:48

Dark electrical conductivity of aromatic compounds, 7:303ff, 304

Dark reaction, 7:5

Darwin corrections, 44:116

Darwinian system, 38:229, 233, 235

Daumas-Hérold model, 49:468, 486, 500

Davies' theorem, 38:119

Davis-Rice-Sengers theory, 37:186

Davydov splitting, 4:161; **8**:27; **23**:238; **40**:392, 393, 427, 418, 451

anthracene, **23**:311

naphthalene, **23**:329

pentacene, **23**:328

tetracene, **23**:327

DC conductivity, 52:477

de Broglie wavelength, 28:8, 65; **31**:418, 427

de Bruijn's theorem, 49:70, 97

de Haas-Van Alphen measurements, 49:558

Deactivation, collisional, of excited ions, **45**:145–151

Dead polymer, 2:173

Deamino oxytocin, 21:587

DeBoer parameter, 27:476

Debye equation for dielectrics, 1:85, 88

Debye frequency, 41:293

distribution, **42**:320

Debye heat capacity law for microclusters, 40:101, 112

Debye-Hückel theory, 4:212–214; **11**:189, 200, 202; **21**:457

modified for short-range interactions, **21**:457ff

from Ornstein-Zernicke equation,

53:121–122

Debye length, 31:94; 34:124
inverse, 11:191, 239
Debye line shape, 12:542, 543
Debye model of rotational diffusion, 44:301
Debye-Perrin model of rotational diffusion, 44:379
Debye persistence length, 41:47
Debye plateau, 44:258, 308, 391
Debye relaxation time, 44:257, 260, 274, 357
Debye scattering equation, 49:508
Debye temperature, 27:479, 482, 516; 51:249
Debye-Waller factor, 20:243, 246, 250, 289, 314; 27:63, 69; 51:244
anharmonic, 20:249
multiple scattering, 20:267, 289, 294
Decaborane, phase determination of, 16:158
Decay of excited state, 47:Part 1:13–14
exponential, 47:Part 1:13
nonexponential, 47:Part 1:14
quantum beats, 47:Part 1:14
Decay process, statistical limit, 42:254
Decay rate, 42:213, 220
Decomposition mode, 1:23
Decorated lattice-gas, 41:7
Decoupled l-dominant method, 42:286
Decoupling approximation, 36:118; 44:287, 290
Deep-chain model, 8:145, 154ff
Deep traps in mixed crystals, 8:37ff
Deexcitation, see Electronic deexcitation; Rotational deexcitation; Vibrational deexcitation
Defect diamond structure, 49:89–93
cuprite, 49:92
polyhedral molecules, 49:92
Defect scattering, 40:439, 444, 445, 475

Defects, antistructure, 11:2, 30, 40
association complexes of, 11:7, 42, 43, 65–67, 71, 73
atomic lattice, 53:310
formation energy, 11:16
Frenkel, 11:2, 3, 5, 6, 30, 41
interstitial, 11:2, 29
interstitial-antistructure, 11:3
intrinsic, 11:28, 29, 74
Schottky, 11:2, 3, 41
Schottky-antistructure, 11:2, 40
vacancy, 11:2, 29, 36, 37
Deflection function, 10:77–81, 95, 100, 114, 136–137; 12:402; 30:321, 474; 42:521
classical, 12:394
with curve crossing, 30:476, 521
Deformation frequency of complex, 1:127
Deformation polarization, 44:342
Deformation potential, 13:125, 134, 152; 44:489
Degeneracy, 4:161, 163, 174
electronic, 13:91
orbital, 8:49
slab modes, 27:471
sphere modes, 27:391, 394, 444
Degenerate kernel approximation, 20:175
Degenerate perturbation theory, 52:126, 139, 145
Degree of coupling, oxidative phosphorylation, 55:146–150
Degrees of freedom, 35:33, 104
external, 16:227–232, 234–235, 246, 278
freezing out of, 35:33
internal, 4:229; 16:224, 227, 229–231
rotational, 4:229, 246, 252, 263
separation of, 41:331

translational, **4**:234, 252, 263

vibrational, **4**:229

Delayed fluorescence of chlorophylls,
19:353, 372, 374

    membrane potential effect,
19:374

Delocalization effects, 35:152

Delocalization energy, 7:38ff, 50, 116,
150; **8**:70ff

Delocalized bond, 1:173, 176, 182, 189;
7:21ff, 87, 90, 116, 138, 146

Delocalized electrons, 2:200

Delocalized wave function, 8:29

Delta-argon-oxygen, 16:359, 381

Delta atomic orbital, 9:88, 152

Delta-function model, 12:349

Demkov coupling, 42:193

Denaturant, 46:375, 376, 385

Denaturation, 7:216

    of proteins, **46**:363–414

Dendrite, sawtooth interface, **32**:167

Dendrite soma membranes, 19:339

Dense fluids, 9:4, 49; **24**:257ff; **44**:263;
see also Fluids

Dense gas, 1:294; **11**:358

    transport properties, **3**:289

Dense liquids, dynamical processes in,
34:148

Dense random packing, 48:472

    of hard spheres, **51**:235, 238

Density, 16:237–238, 241, 245, 276

    dielectric, **1**:82

    discontinuity, **1**:216, 226

    metal-ammonia solutions, **4**:309–
311

    steady-state distribution, **13**:282,
286, 289

    time-dependent, **41**:128

    virial coefficients, **20**:41, 47

Density correlation function, 24:51, 55;
41:108

Density expansion, 38:182, 183

Density field, 48:85

Density fluctuations, 17:190; **24**:43;
46:290, 300

    spin-wave, **53**:255

Density functional theory, 41:64ff, 81,
138

    density profile, **49**:384–389, 392

        square-well, **49**:384–387

        wall-fluid, **49**:412

    distribution, **49**:396–397

    free energy, **49**:358

    Hohenberg-Kohn theorem, **54**:241–
242

    inhomogeneous fluid, **49**:405–406,
412–423

    inhomogeneous relativistic electron
systems, **54**:231–302

    Kohn-Sham theorem, **54**:242–243

Density gradient, 13:281; **46**:283, 330

Density kernel, first and second order
reduced, **26**:222, 225, 237

Density matrix, 1:232, 284, 285, 342;
2:207, 214, 227, 231, 233, 308,
309, 314, 320, 323; **5**:35, 327, 349;
25:4, 10, 11; **35**:5; **40**:386–389,
390, 469, 471; **41**:75, 78, 93;
51:116

    chemical evolution, **35**:5

    dephasing, **40**:385–391

    entropy of, **53**:353

    equilibrium grand canonical, **41**:74

    Fock-Dirac, **2**:225; **14**:240, 287;

    magnetic evolution, **35**:5

    one-particle, **41**:81

    Ornstein-Zernicke equation, **53**:119

    quantum mechanical, **51**:126–127

    reduced, **24**:213

    triplet molecule, **35**:5

    two-particle, **21**:305

    variation, **2**:319

Density of maximum displacement, **52**:418

Density of states, 13:126, 129; **17**:235; **23**:141; **31**:266, 281, 326; **40**:392, 414, 417, 420, 427, 466, 467; **42**:3; **51**:257–262, 263, 269
  alkali metals, **31**:302
  aluminum, **31**:302
  amorphous carbon, **31**:322
  bandwidth, **31**:296, 298
  beryllium, **31**:319
  bismuth, **31**:290
  complex energy approximation, **31**:285
  copper, **31**:323
  finite clusters, **31**:319
  germanium, **31**:323
  indium, **31**:303
  integrated, **31**:297, 320
  iron, **31**:323
  lithium, **31**:302
  local, **27**:213, 214
  Lorentzian approximation, **31**:285
  mercury, **31**:303, 305
  negative ion, **36**:465
  phonons, **27**:417, 474–475, 483–485
  photoemission data, **31**:303
  silicon, **31**:323
  tight-binding methods, **31**:315
  vibrational and rovibronic, **21**:145

Density operator, 38:33
  excitons in finite molecular system, **41**:321–426

Density oscillation, 41:119

Density profile, 46:330
  of liquid metals, **27**:550
  Maxwell-van der Waals fluid, **46**:331

Density response function, 41:132; **53**:115, 186–187
  solvent forces, **53**:112–113

Density wave, 35:123, 130

Deoxyadenosine, 7:349

Deoxycorticosterone, 7:411

Deoxycorticosterone acetate, 7:419
  infrared spectra, **7**:419

Deoxyguanosine, 7:349

Deoxyribonuclease, 7:461ff

Deoxyribonucleate, 7:467, 489
  infrared dichroism, **7**:467
  spectra, **7**:484ff

Deoxyribonucleic acid (DNA), 7:349, 436; **8**:177ff; **49**:34
  absorption spectra, **7**:114, 128
  band structures, **7**:142ff, 146
  chain configuration, **7**:344
  charge carriers in conduction band, **7**:148
  complementary chains of bases, **7**:217
  crystal imperfections, **7**:147
  dc conductivity, **7**:146
  denaturation, **7**:217; **22**:130, 131
    base pair strengths, **22**:140, 146
    coarse graining approximation, **22**:179
    end effects, **22**:189, 198
    four component model, **22**:190, 200
    hairpin helices, **22**:135, 136, 149, 178
    partition function, **22**:140
    stacking free energy, **22**:148, 187
    strand sliding, **22**:136, 153
    synthetic, **22**:135
  double helix, **7**:86ff, 103, 144, 148ff, 342ff
    configuration of salts, **7**:436
  duplication frequency, **7**:148, 152
  electron microscopy, **22**:185
  electronic properties, **7**:339, 346ff

electronic structure, 7:88
EPR, 7:351
flow birefringence, 33:139, 140
hydrogen bonds, 8:177
magnetic properties, 7:351
molecular structure, 7:341ff
nucleic acids, 7:460
oscillator strength, 7:128
photoreactivation, 7:586
π-electron system, 7:137
polarization, 7:148, 151, 153
protein coding, 7:135
radiation level, 7:152
replication, kinetic equations for, 15:187
secondary structure in solution, 7:436
semiconductivity, 7:146
superimposed bases, 7:128
synthesis rate, 7:153ff
transforming, 7:587, 593
tumor development, 7:148
tunnelling in, 8:178
Watson-Crick model, 8:177
Dephasing, density matrix description, 40:385–391
of coherent states, 50:112
vibrational, 40:1–48
experimental probes, 40:3–10
spectral line shape studies, 40:3–7
see also Optical dephasing; Spin dephasing
Dephasing collision, 42:275
Dephasing time, 40:417, 424, 425, 469; 42:308, 314
Depolarization, 28:328, 349–358, 370–372
Depolarized fluorescence, 17:66
Depolarized light scattering, 26:24, 30, 36, 43, 51; 31:258; 34:9, 10

benzene, 34:74
CH_3I, 34:84
classical description, 26:34
relation to binary collision models, 26:34–36
collision-induced, 34:10, 76, 88
density expansions, 26:43–49
three body structure, 26:45–48
two body effect, 26:43–45
experimental geometry, 26:3, 6, 8
fluctuating field local analysis, 26:30–33
isolated binary collision model, 26:36, 40
gases, 26:36–40
liquids, 26:40–43
lattice occupancy models, 26:50–51
molecular dynamics calculations, 26:51–55
near the gas-liquid critical point, 26:80, 81
angle dependence of cross section, 26:87, 88
double scattering, 26:78, 80
long range correlations, 26:81–87
quantum mechanical description, 26:9–11, 16–23
binary collisions, 26:17–19
higher order effects, 26:21–23
three body effects, 26:19–21
second order Raman theory, 26:49–50
vibrational Raman, 34:10
Depolarized Rayleigh scattering, 51:84
Depolarized scattering, 24:43, 44, 46, 50; 51:89–107
classical calculation, 51:95
hard sphere, 51:99–100
high-frequency wing, 51:100
Depolymerases, 7:675
Deposition rate, 40:167

Destabilizer, 43:122

Destruction operator, 6:90, 94

Destruction term, 11:175; 16:11 35:141, 146, 160, 161, 167, 190

Destructive measurement, 46:156, 161, 165, 178ff

Detailed balance, 1:122, 371, 372; 5:361–362; 12:565, 577; 15:68; 17:89; 18:84; 28:284; 36:77, 465; 38:122; 43:10, 33; 46:36; 51:53, 67, 97, 105

   in quantum theory, 13:98, 101

Determinantal equations, 7:124, 143

Deterministic path, 46:198

Detonation, 21:291, 295, 779

   shock-induced, 53:310

   single step, 3:104ff

      Arrhenius kinetics, 3:121

      equations, 3:110

      ignition temperature, 3:115

      real gases, 3:121

   see also Explosives; Hydrodynamic equations

Deuteration, 7:563

   acceptor and donor, 7:575

   of chlorophyll, 7:637

   effect in EPR, 7:563, 566

Deuterium, 1:16, 61–63

   crystal structure and intermolecular potential, 33:51ff, 67

   solid, 16:376–378

   structural and thermodynamic data, 16:346–348

Deuterium (D_2), Born-Oppenheimer approximation, 12:23

   dissociation energy, 12:22

   ground state, adiabatic approximation, 12:23

   nonadiabatic approximation, 12:23

   relativistic corrections, 12:22

Deuterium fluoride, dispersion energy of, 12:21

Deuterium substitution, 4:147

Deuteromethanes, energy transfer in, 47:Part 2:208–227

Diabatic approximation, 28:324, 329

Diabatic correlation diagram, 42:30, 32

Diabatic energies, diatomics-in-molecules method, 42:73

Diabatic function, 28:329

Diabatic potential curve, 18:138; 42:18, 566; 49:274

Diabatic potential matrix, 49:282, 287, 298, 306

Diabatic representation, 42:20; 49:275–277, 279, 283, 287, 306

Diabatic state, 23:164; 30:372, 424, 438, 464, 466; 42:190, 198

Diabatic transition, 42:16

Diagonal fragment, 11:174; 16:11

Diagrammatic perturbation theory, 48:2; 52:182

Diagrammatic techniques, 31:272, 276, 278, 284; 36:351; 52:473

   cluster diagram rules, 31:276

   momentum representation, 31:278

Diagrams, 11:168, 172, 337, 341ff

   Abrikosov, 36:224

   angular momentum or Yutsis, 14:402, 502

   anomalous, 6:93, 112, 119

   atomic, 14:475, 483

   classification of, 6:452; 11:46

   constant, 36:228

   current matrix, rules for construction, 43:118

   current, 43:114–119

   cycle, 11:49; 43:32

   directional, of cycle, 43:33

      of species, 43:32

   disconnected, 14:138

   electron gas in positive lattice, 6:118

elementary, **6**:237

extreme current, **43**:115

Feynman, **14**:219, 373, 432; **36**:224

fluid theory, **6**:235

Hill, **43**:32

Hugehholtz, **36**:224

irreducible, **6**:106; **11**:22–25, 346

ladder, **14**:156, 180; **36**:232

linked, **6**:91, 96; **11**:22, 24; **14**:134, 376; **36**:234

matrix, **43**:104

network, **43**:112

one ion, **6**:452

reducible, **6**:106; **11**:22, 24, 343

ring, **11**:199, 230, 242, 247; **14**:218; **36**:231

set-graph, **43**:107–108

semidirected, **43**:109

skeleton, **11**:344; **31**:284

stabilizer-polygon, **43**:122, 123

summation, **14**:144ff; **24**:292

in critical region, **6**:194

for Ising lattice, **6**:194

rings, **6**:118, 194

third order, **36**:233, 337

time dependent, **14**:437, 439

time independent, **14**:441

time ordered, **36**:226, 233, 242

two ions, **6**:430, 438, 442

unlinked, **6**:91, 96; **11**:22, 24

factorization of, **14**:135, 380

volume dependence of, **11**:343

Dialkylphosphate anions, spectra of, **7**:484ff

Diamagnetic anisotropy, aromatic molecules, **3**:190

London's theory, **3**:194

Diamagnetic screening constant, **23**:133; **44**:108

Diamagnetic susceptibility, **1**:276; **3**:172ff; **33**:160, 227

aromatic compounds, **12**:75

semi-free electrons, **3**:214ff

spherically symmetrical system, **2**:68

Diamond, **9**:365ff; **27**:481; **41**:159, 160

boron-treated, **9**:401ff

glide plane, **20**:10

nucleation and growth, **9**:388ff

synthesis, **9**:365ff, 378, 397; **21**:721

free energy requirement, **9**:385ff

General Electric method, **9**:365ff, 376, 383, 390, 400

high pressure apparatus, **9**:372, 376

reaction cell, **9**:378, 388ff

x-ray diffraction, **9**:371, 375, 380

Diamond structure, **21**:723

Diatom polarizability, **46**:2–72

Diatomic crystals, **27**:359, 367

Diatomic molecules, **5**:47, 48; **9**:65; **36**:20, 22; **49**:159–160, 162, 167, 179

heteronuclear, **49**:167

quantum theory of, **4**:280–300

homonuclear, **49**:167–169, 186

on plane lattice, **15**:336

vibrational relaxation of gas of, **15**:219

Diatomics-in-molecules method, **30**:139, 489; **36**:161, 164, 180; **41**:253; **42**:63–112 **49**:195, 283, 292; **50**:543, 566

Br + HI, **42**:97

F + H_2, **42**:91

H + Cl_2, **42**:92

H^+ + H_2, **42**:87, 102

H_2O, **42**:72, 81, 99

H_3^+, **42**:74

H_3, **42**:73, 74

H_4, **42**:83

K + HBr, **42**:87, 89

K + HCl, **42**:88, 90

K + HBr, **42**:89

Na + HF, **42**:87

triatomic hydrides, **42**:83

Dibenzoequinene, **16**:196

1, 2-Dibromo-1, 2-dicarbomethoxycyclobutane, **16**:177–178

Dibromobenzene, crystal structure, **5**:185

p-**Dibromobenzene ($C_6H_4{}^{79}Br_2$)**, ethylene system, **2**:102

quadrupole spectrum, **2**:195

radiation resistance of, **2**:196

1, 4-Dibromocyclohexane, **20**:12

p-**Dibromodiphenyl (($C_6H_4)_2{}^{79}Br_2$)**, quadrupole spectrum, **2**:195

radiation resistance of, **2**:196, 200

1, 4-Dibromonaphthalene, **40**:404, 453, 454, 474, 476, 477

triplet excitons of, **40**:453

Dicarboxylic acid, free radical, **7**:541ff

hyperfine coupling constants, **7**:544

p-**Dichlorobenzene**, crystal structure, **5**:185

p-**Dichlorobenzene ($C_6H_4{}^{35}Cl_2$)**, quadrupole resonance, **2**:193

quadrupole spectrum, **2**:195

radiation resistance of, **2**:196

Dichlorodifluoromethane, **55**:76

Dichloroethane, dipole moment, **2**:372

1, 2-Dichloroethane, **20**:3, 12

Dichloromethane, thermodynamic data and lattice constants, **2**:8

Dichloromonofluoromethane, thermodynamic data and lattice constants, **2**:8

Dichroism, **49**:31–32

circular, **4**:68, 70–72, 82, 83, 95, 105, 109

of crystals, **13**:193

in electric field, **13**:183

of stretched polymer films, **13**:193

Dictyostelium discoideum, **29**:152, 195; **38**:383, 384

aggregation, **38**:384

cyclic-AMP signalling system, **38**:366

differentiation, **38**:366, 384

oscillations in, **39**:30–32

Dicyanoacetylene, crystal structure, **5**:185

Dicyanoethylene, **20**:12

Dielectric behavior in critical region, **6**:224

Dielectric constant, **1**:310, 317; **4**:73, 75, 88, 139, 140; **48**:185–314; **51**:262

bulk, **12**:139

critical exponent of, **48**:301

critical properties of, **48**:295–303

dielectric anomaly of, **48**:297

high frequency, **48**:278

electrolyte solutions, **48**:303–312

high frequency, **12**:139

lattice of point dipoles, **6**:436

polar fluid, **31**:110, 114, 121, 140

frequency dependent, **31**:145–152

general statistical theory of, **31**:110–121, 130

mean spherical model for, **31**:124

Onsager model for, **31**:140

polar-nonpolarizable fluid, **48**:185–272

computer simulation, **48**:245–253

Høye and Stell formalism, **48**:185–223

hypernetted chain equation, **48**:238–244

mean spherical approximation, **48**:232–237

polar-polarizable fluid, **48**:273–302

Rosenberg and Lax expression for, 6:441

solute dependent, 48:304

solvent, 11:228

static, 4:77, 78, 98; 12:297

statistical mechanical expression for, 6:427

Van Vleck expression for, 6:441

zero-frequency, 11:196

Dielectric displacement, 37:101

Dielectric function, 17:239; 27:272ff, 357, 358; 44:390

alkali halides, 27:359, 407

complex, 27:371, 405

complex crystals, 27:428, 430

damping in, 27:361, 378, 380, 404–405, 429

diatomic crystals, 27:359, 367

electron gas, 27:361, 378

insulators, 27:280

local, 27:357, 362, 364, 379, 415

longitudinal, 27:272, 379–380, 450, 455, 457

metal, 27:361, 378

mixture, 27:444

nonlocal, 27:357–362, 379, 415, 449–459

polar semiconductor, 27:415, 420

poles in, 27:429

RPA, 27:459, 462

semiconductors, 27:279, 280

simple gases, 26:65–78

corrections to dipole-induced dipole theory, 26:70–77

density expansion, 26:66

transverse, 27:272, 379–380, 450

Dielectric loss, 1:77; 44:261, 323, 338, 505

in disordered solids, 44:391

Dielectric permittivity, 44:263, 303

Dielectric properties, 1:77

Dielectric relaxation, 1:83, 109; 17:66; 31:145–152; 34:17; 44:379

Dielectric spectroscopy, 44:256, 301

Dielectric susceptibility, 27:452, 500; 44:341

Dielectric tensor, 27:420, 425, 430, 450; 44:298

complex, 27:435

Diels-Alder reactivity indices, 21:312

Diene synthesis, 1:190, 191

Dieterici factor, 41:208

Diethylphenyl phosphates, 7:77

Diffeomorphic distortion, 40:245

Difference equation, 4:2, 3, 29; 46:253

Difference map, 16:163

Difference operator, 53:304

Differential cross section, 10:37–39, 83; 42:408, 520, 564; 49:540–541, 565; 51:86

inelastic scattering, 27:59, 60

quenching, 50:503

see also Cross section

Differential diffusion, 6:291, 292

Differential ionization energy, 5:79

Differential melting, 40:122

Differential osmotic pressure, 6:312

Differentiation, 29:126, 172, 281

molecular, 29:253

spatial, 29:253

Diffraction, 20:215; 30:131, 327

electron, see Electron diffraction

x-ray, see X-ray diffraction

see also Low energy electron diffraction (LEED)

Diffraction effect, 1:284

Diffraction pattern, 20:220

Diffraction theory of reactive scattering, 10:128–130

Diffuse surface, 27:603

second dispersion branch, 27:603

step inhomogeneity, **27**:604

truncated exponential profile, **27**:603

Diffusing systems, fluctuations in, 15:137ff

harmonically bound particle, **15**:140

particle in a box, **15**:143

plane rotor, **15**:144

spherical rotor, **15**:144

Diffusion, **4**:2, 3, 28, 29, 31–44; **6**:291ff; **19**:220, 277, 278; **37**:108; **38**:187, 190; **43**:210, 222, 240, 249; **51**:26

of atom clusters, **52**:456

of cells, **29**:177

by chromatographic column, **52**:456

competition with nondiffusional processes, **33**:136, 139

by continuous motion, **52**:375

correlated, **52**:464

coupling with chemical reaction, **39**:1–54, 229

in critical region, **6**:222

in crystals, **52**:456

differential, **6**:291, 292

dynamics of, **35**:21

effect of dimensionality, **39**:17, 34, 235, 270

elastic relaxation dissipation process, **6**:309

in enzyme membrane, **39**:234

Fick's law, **4**:2, 29, 42, 44; **19**:221; **29**:32, 344

free energy, dissipation of, **6**:309
of mixing, **6**:309

of glycolytic intermediates, **39**:34, 35

ideal gas, **6**:305

ideal solution, **51**:157–158

kinematics of, **6**:291–313

lateral diffusion constant, **39**:256, 258

lateral diffusion in membranes, **39**:255, 260, 278, 281, 282, 283

law of mass action for, **6**:303

in liposomes, **39**:236

molar friction, **6**:302, 306

multicomponent, **6**:293, 294, 298–299

Nernst-Einstein equation for, **6**:292, 299

nondiffusional processes, **33**:77, 123–125, 134–138, 141, 145

in pair space, **6**:201

phenomenological coefficients, **6**:292, 293

resistivity, **6**:293, 302

in rigid structure, **6**:309

self, **1**:10, 136, 151, 153; **6**:302–306
of oxygen in praseodymium oxide, **21**:255
see also Self-diffusion

soliton, **53**:320

studied by magnetic resonance methods, **6**:294

thermal, coefficient of, **13**:350, 357
of halides in aqueous solution, **13**:349ff

three-component, **6**:297–302, 309–311

through uniform barriers, with a source, **4**:36–42, 45

with no source, **4**:31–36, 45

two-component, **6**:295–297

volume fractions, **6**:294, 296, 297

see also Rotational diffusion

Diffusion cloud reactions, infrared laser enhanced, **47**:Part 2:43–54

experimental, **47**:Part 2:44–46

Na + SF$_6$, **47**:Part 2:46–54

Diffusion coefficient, **1**:135ff; **5**:158, 159, 263; **6**:295, 297, 310; **11**:209; **17**:67; **24**:58, 309; **29**:32, 52, 260, 274; **40**:400, 413, 437–452, 463, 472, 478; **41**:202, 204, 218; **44**:274; **48**:74; **52**:464, 473

autocorrelation function for, **48**:91

binary, **16**:104, 109, 122, 128

Brownian particle, **53**:86

coupled, **44**:299

domain wall, **53**:273

gas mixtures, **5**:161

gases, **5**:159–160

heavy molecules, **5**:157

kinks, **53**:269–270

Langevin model, **53**:107

membranes, **39**:288

polynary, **16**:104

relative, **48**:137

rigid convex bodies, **5**:160

rigid spheres, **5**:160

rotary, **1**:110

rotational, **17**:171

self, **17**:168

three-component, **6**:298–299

translational, **17**:138

truncated chain models, **53**:100

Diffusion-controlled chemical reaction, **46**:316

Diffusion eigenfunction, 48:145

construction of, **48**:175–176

Diffusion equation, 35:36, 75, 79, 80, 110; **41**:202, 203, 218; **44**:269; **46**:222; **48**:72 **52**:382, 383, 409, 471

approximate, **35**:75

bead and spring model, **33**:77, 121–125

coarse grained, **33**:78, 121, 144

Kirkwood-Riseman equation, **35**:110

Diffusion flux, 33:404, 440

boundary conditions for, **33**:429

Diffusion force, 11:219

Diffusion limit, 33:77, 123

Diffusion model, 35:13

Diffusion process, 33:77, 123–125, 134–138, 141, 145; **52**:364, 460, 470, 483

Diffusion-reaction, in membranes bearing enzymes, **39**:229–274

in structured media, **39**:229–274

Diffusion tensor, 48:80, 130–136

Diffusion velocity, 16:104

Difluoroethane, hydrate, **2**:55

water-hydrogen sulfide ternary system, **2**:55

Digital correlator, 24:3, 34

Digitalis, 16:190–191

Digitoxigenin, 16:190–192

Digonal hybrids, 7:19

Dihalide formation, 42:13

Dihydronicotine ring, 16:193

p-**Diiodobenzene,** crystal structure, **5**:185

p-**Diiodobenzene ($C_6H_4^{127}I_2$),** quadrupole spectrum, **2**:195, 197

radiation resistance of, **2**:196, 198, 200

Diisocyanates, polymerization of, **2**:163

Dilute gas, 11:357

transport properties, **3**:281

Dilute solution assumption, 35:74

Dimensionality effects, 49:461

Dimensionality scaling, 52:64–68

Dimer ions, asymmetric rare gas, **52**:284

symmetric rare gas, **52**:284

Dimer model, 41:201, 202

Dimer-phonon scattering, 40:434–435, 436, 437

Dimers, excited, **52**:287, 303, 306, 333

in molecular crystal, coherence, **40**:396, 428, 463

transitions of, **40**:411

translationally equivalent, **40**:393ff, 409, 427, 468

translationally inequivalent,

40:393, 396, 409ff, 426

photoionization studies of, acetone, **52**:345

ammonia, **52**:347

carbon dioxide, **52**:330–334

carbon disulfide, **52**:309–315

carbon monoxide **52**:301–304

ethylene, **52**:351

H_2, **52**:304

HBr, **52**:299

HCl, **52**:299

heterogeneous, **52**:266

HF, **52**:299

HI, **52**:299

N_2, **52**:301

NO, **52**:290

N_2O, **52**:330

O_2, **52**:293

OCS, **52**:324

sulfur dioxide, **52**:334

water, **52**:344

rare gas, **51**:90

see also N-mer; van der Waals dimers

Dimethylgloximate complexes, 7:369ff

spectra, 7:371ff

Dimethylgloxime radicals, 7:552

Dimethylphosphate anion, 7:484ff

Dimethylsulfoxide, 47:Part 1:641

Dimethyltetrahydrofuran, 47:Part 1:641

Dimethyltriacetylene, 20:12

m-**Dinitrobenzene**, ethylene system, 2:102

p-**Dinitrobenzene**, crystal structure, 5:185

Dinitrogen tetroxide, **16**:103, 112ff, 116–122, 125; **20**:33

Dinitrophenol, 19:368

Dioleoyllecithin membrane, 29:317

Dioleylphosphate, 29:325, 334, 341

Diophantine evaluation of integrals, **36**:152

Diophantine sums, **21**:51

Dioxetanes, chemiluminescent reactions of, 42:47

Dioxide hydrate, 2:43

Dioxide laser, 42:440

Dipalmitoyl lecithin, 41:196, 213

Dipalmitoyl phosphatidylcholine, 41:196, 224

Diphenylcyclopropene, 8:134, 136

Diphenylguinocyclopropene, 8:135ff

Diphosphoglyceric acid, 7:62

Diphosphorpyridine nucleotide, 7:67

Dipolar autocorrelation function, **44**:262, 338, 342, 360, 380

Dipolar fluid, dilute, **44**:323

polarizable, **48**:273–302

Dipole, 7:128

anisotropic overlap, **51**:81

induced, **51**:50

interference with permanent dipole, **51**:84

magnetic, **44**:388

permanent, **1**:79; **4**:165, 175

range, **51**:62, 65

resonance interaction energy, **12**:41

see also Electric dipole

Dipole absorption, dipole-induced, **44**:453

multipole-induced, **44**:446

octopole-induced, **44**:454

Dipole approximation, 38:149

Dipole-dipole energy transfer, **16**:94–95

Dipole-dipole interaction, **1**:268, 270, 311, 328; **2**:60; **12**:145, 150, 154; **44**:263, 276, 294, 343, 350

antiferromagnets, **53**:230

biological systems, **53**:287

molecular crystals, **53**:281

molecular vibrations, **53**:319

Dipole-dipole relaxation, **11**:298, 303, 309, 315

Dipole-dipole resonance, **44**:294

Dipole-dipole tensor, **6**:416, 429
   products of, **6**:429, 446

Dipole dispersion force, **51**:61, 62

Dipole effect, **51**:88

Dipole field, **1**:341

Dipole force, **1**:81

Dipole-induced dipole, **48**:276

Dipole matrix element, **51**:58, 72

Dipole model, **51**:61, 62, 79–80

Dipole moment, **1**:79, 89, 314, 342, 343; **4**:136; **12**:109, 230; **17**:239; **25**:180, 193, 204; **27**:444; **41**:154
   average, **1**:89, 93, 94
      diagrams for, **6**:437
   carbonyl compounds, **8**:136
   closed-shell diatomic, **25**:204
   electric, **37**:241–243; **44**:105, 339
   excited states, **13**:194, 196–201
   experimental, **25**:180
   heteronuclear diatomic molecule, **53**:30
   induced, **25**:9; **44**:445, 474;
      electric, **4**:74
      magnetic, **4**:74, 79
   macroscopic, **44**:298
   open-shell diatomic, **25**:205
   polyatomic, **25**:193
   transition, electric, **4**:109, 123, 157
      magnetic, **4**:109

Dipole moment operator, **25**:3; **40**:376

Dipole orientation, **44**:273, 278

Dipole potential, **37**:121, 124

Dipole-quadrupole induction energy, **12**:133

Dipole radiation, **10**:172, 178–179
   electric, **4**:163, 164
   magnetic, **4**:163

Dipole selection rules, **51**:75

Dipole strength, **4**:80–83, 92–94

Dipole sums, **23**:277
   planewise, **23**:326
   point dipole-dipole interactions, **23**:277
   retarded dipole sums, **23**:279

Dipole vector, **41**:198, 207, 225

Dipositive bond, **7**:75

Dipyrrylmethene, **7**:398, 404

Dirac bracket, **33**:176–179, 190, 191, 203, 217

Dirac equation, **18**:168; **41**:92

Dirac function, **53**:6, 83, 253, 271

Dirac theory, **1**:343; **7**:164; **41**:82
   time-dependent, **15**:75

Direct correlation function, **40**:318–321, 349, 362; **41**:45–46, 176; **46**:292; **48**:189
   atomic fluids, **53**:12
   in Ornstein-Zernicke equation, **53**:119

Direct exchange interaction, **5**:218ff, 220, 222

Direct interaction model, **30**:61

Direct transition, **49**:551, 553, 555, 560, 568, 575

Directed set-graph network, **43**:109

Directional correlation function, **17**:153

Directional force, **53**:30

Director, **46**:314

Director axes, **44**:258, 382

Dirichlet problem, **32**:39, 40, 42, 52

Discharge flow system, **42**:124

Discharges, **18**:84

Discontinuous systems, minimum entropy production, **3**:293

Discrete lattice, **52**:364

Discriminant, **44**:347

Disk-annulus system, **44**:277

Dislocation, **40**:204, 214

Disorder, 20:242, 243, 246, 276, 280, 307, 308, 331, 333
  correlated, **20:**283
  crystalline, positional, **6:**463–467
    rotational, **6:**466–467
  ion bombardment, **20:**381
  linear, **20:**281
  liquid, **20:**284
  rotational, **20:**282
  uncorrelated, **20:**284
Disordered lattice, 52:485
Disordered rate constant, 52:484–488
Disordered solid, 44:382, 391
Disordered structure, 52:468
Dispersed-crystal organic substances, 3:261
Dispersion, 1:88, 107, 113; **24:**20, 22
  optical rotatory, **4:**68, 69, 76–95, 105, 109
Dispersion energy, 12:108, 124, 125; **33:**351, 354, 356
  dipole-dipole, **12:**6
  dipole-octopole, **12:**6, 38
  dipole-quadrupole, **12:**6, 38
  dipole-quadrupole-dipole, **12:**6
  intraatomic correlation effects, **12:**38
  quadrupole-quadrupole, **12:**6, 38
  see also London dispersion force; van der Waals dispersion force
Dispersion force, 2:168; **9:**332, 336; **11:**330, 343; **42:**183
  Casimir-Polder potential, **12:**176, 177, 181, 187, 192
Dispersion function, 35:127, 129, 131
Dispersion interaction, 36:167, 168, 169
Dispersion of sound measurements of barriers of internal rotation, 2:376
Dispersion relation, 17:108; **44:**517
  bending mode, **48:**362
  electronic excitations, Al(111), **27:**194–196

  bulk excitations, **27:**56
  bulk plasmons, **27:**131
  surface excitations, **27:**57, 185
  surface plasmons, **27:**131, 191
  planar, **46:**305–306, 325, 340
  spherical, **46:**305–306
  squeezing mode, **48:**362, 383, 384
  surface modes, asymptotic behavior, **27:**368ff, 393, 404ff, 433, 456
  half-space, **27:**365–369
  nonreciprocal, **27:**422, 424, 427
  slab, metallic backing, **27:**379
    microscopic theory, **27:**492–493
  termination point, **27:**369, 373, 376–378, 416, 419, 429, 453, 518–519
Displacement, mean-square, 44:410, 417
Displacement gradient, 33:409
Displacement vector, 1:321, 322
Dissipation function, 39:329
Dissipation of energy, principle of least, 3:274, 277
Dissipation potential, 33:414, 437
Dissipative components, Langevin equation, 53:82
  molecular-time-scale generalized Langevin equation, **53:**92
  Mori theory, **53:**144, 145
  primary system motion, **53:**76–77
  velocity autocorrelation function, **53:**78–79, 80, 81
Dissipative processes, 15:11ff, 37ff
  mechanisms of, **15:**13
Dissipative structures, 19:209ff; **29:**2, 31, 64, 133, 181, 305, 343; **32:**2, 13, 61, 98, 115; **38:**264, 276, 289, 364; **39:**1, 27, 32
  aggregation and, **29:**138, 154
  in Bénard convection, **32:**130

duplication of, **29**:36
effect of fluctuations, **39**:52
localized, **29**:40
membrane excitation and, **29**:319, 335
in papain reaction, **39**:39
periodic, **29**:36, 40
quasi-periodic, **29**:40
in simple models, **39**:8, 21
steady state, **29**:35; **32**:3
in Taylor vortices, **32**:130
Dissipative terms, **35**:16
Dissociation, **1**:121; **16**:68; **41**:232, 283; **48**:82
degree of, **16**:113, 128
diatomic molecules, **13**:7
ionic, **28**:81, 396
molecular, **28**:84, 295, 300, 400, 401, 411, 454, 460
multiphoton, **41**:237
NO_2, **47**:Part 2:143–151
photosensitized, **16**:95–99
three-body, **25**:256
three-body, **13**:102
see also Multiphoton dissociation
Dissociation energy, **42**:4, 556ff; **44**:104
hexafluorides, **42**:555
many-electron molecules, **6**:316
methyl derivatives, **42**:558
N_2, **45**:265
neutron diffraction studies, **53**:12
NO, **45**:299
NO^+, **45**:335
O_2, **45**:286
O_2^-, **45**:316
Dissociation limit, **36**:45
Dissociation pressure, **2**:33
Dissociative attachment, **18**:129
Dissociative continuum, **50**:106
Dissociative recombination, in hypersonic flow, **13**:29
Distinct site, **52**:396–407, 492
Distorted atom, **12**:289
Distorted wave approximation, **18**:75; **19**:6; **28**:121, 175, 180, 182, 183; **30**:27, 54, 67, 402, 481; **42**:332, 328, 431, 471
applied to $H + H_2$, **21**:129
Distorted wave Born approximation, **36**:181, 183; **49**:193–194, 209, 225–226, 250–251, 288
Distortion dipoles, **12**:349
Distribution, angular, **16**:80–81
bimodal, **29**:82, 94
classical, **36**:75
cumulative, **15**:88
Gaussian, **29**:91, 211
Holtzmark, **15**:121
Mori, **32**:62
orbital and phase, **29**:102
Poisson, **29**:244
Distribution coefficient, **40**:219
Distribution function, **1**:203, 216, 227, 312, 319, 332, 359; **6**:424; **11**:319, 320, 321, 327, 336; **12**:263, 288; **16**:8; **26**:117, 126, 131, 133 **31**:159–160, 270; **35**:60, 66, 80; **44**:339
canonical ensemble, **53**:85, 219
chain approximation, **31**:309
classical, **4**:232–251
conditional, **26**:128; **34**:23
configurational, **4**:248, 257, 261, 277
contracted, **35**:59
defects, **11**:32, 63, 65, 72
diffusion problems, **4**:31, 40
dimensionless, **31**:271
doublet, **24**:259, 270
ellipsoids, **31**:179, 182, 205–207
equilibrium, **11**:322
formal, **16**:14

Fourier coefficients of, **11**:331
frequency, **26**:123
   bivariate normal, **26**:129
harmonic chain, **53**:75
heavy particle equilibrium, **16**:34–37
including coincidences, **31**:270
interacting reagents, **53**:182
ion pairs, **11**:43
Kramers' expression, **35**:110
Langevin equation, **53**:82ff
linear rotators, **4**:252–262
marginal, **26**:129
molecular-time-scale generalized Langevin equation matrix calculation, **53**:192–194
multivariate Gaussian, **26**:130
n-particle, **11**:318, 323, 324, 371; **40**:296, 298
nonequilibrium, virial expansion of, **11**:322
one-particle, **11**:318, 321, 322, 358
one-particle Boltzmann, **13**:303
pair, **31**:159–160, 169, 179, 205–207, 214–218 **35**:60, 67; **53**:117
   time-dependent, **12**:581
at phase boundary, **40**:291
primary system, **53**:77
quantum, **4**:232–302
radial, **1**:9, 149ff, 208ff, 226ff; **12**:271; **49**:497–498, 500, 510
reduced, **11**:326
relaxation times, **1**:86
rigid spheres, **31**:169
semiclassical, **1**:32
singlet, **24**:259; **31**:159–160, 207; **35**:60, 66
specialized, **11**:34, 66
superposition approximation, **31**:309
symmetrical tops, **4**:263–279
theory, **40**:249, 288–290

triplet, **31**:211, 214, 220
two-particle, **11**:327
unsymmetric, **12**:288
velocities, **11**:318, 325, 329, 330, 333, 334
Disulfide bond, **7**:266
Divacancies, **11**:6, 37, 38
DNA, *see* Deoxyribonucleic acid
Dodecadeltahedron, **40**:87
Dolph theorems, **40**:277ff, 328ff
Domain walls, Brownian motion, **53**:273, 274
   degenerate ground state and, **53**:289
   density, **53**:277
   mobility of, **53**:275–276
   polyacetylene, **53**:292
   propagating, **53**:274
   random walks, **53**:269, 272–273
   soliton, **53**:261
   static, **53**:264
Donnan equilibrium, **21**:685
Donor-acceptor charge transfer, **44**:396
Donor-acceptor complexes, **7**:320ff
Donor-acceptor theory, **7**:286ff, 333ff
Doob's theorem, **15**:106; **17**:103
Doolittle equation, **48**:430, 509
Doppler broadening, **17**:90; **33**:267, 279; narrowing of, **33**:271
Doppler effect, **25**:11, 16
   double resonance, **25**:54
   half width, **25**:12
   pulse trains, **25**:43
   transient absorption, **25**:22, 23
   transient emission, **25**:31
Doppler shift, **33**:258
Doppler spectroscopy, **47**:Part 2:133–152
   dissociation of NO_2, **47**:Part 2:143–151
   experimental, **47**:Part 2:143–144

multiphoton dissociation in beam, **47**:Part 2:138–143

speed distribution of NH_2, **47**:Part 2:142–143

Doppler spectrum, 50:266

Doppler width, 50:4

Dorey rule, 20:120, 121

Dose-reduction factor, 7:586

Double chain, spectrum of, **41**:402–407

Double commutant, 38:24

Double-diffusive instabilities, 32:135ff, 147

  laboratory experiments, **32**:136

  in ocean, **32**:136

  sloping boundary, **32**:140, 141

Double-dye laser technique, 49:15, 27–28

Double exchange interaction, 5:233ff

Double helix, 7:86ff, 436, 459, 493

Double hydrate, 2:52

Double ionization cross section, 36:485

Double layer, 37:107

  electric, **48**:336–338, 369–372

  free energy of interaction in, **48**:338

  pressure in, **48**:337

Double linkage, 7:20

Double McLafferty rearrangement, 19:201

Double perturbation theory, 12:154, 155, 157, 158; **37**:273, 274

  interchange theorems, **12**:155, 156, 159

Double photon emission, 10:178

Double resonance, 25:19; **50**:128, 131

  observation of, **25**:57

  steady state, **25**:19

  transient, **25**:51

Double scattering, 18:24, 25, 26

Double space, *see* Liouville space

Doublet, closure at, **33**:80

Doublet T_1 measurement, 35:24

Doublet distribution function, 24:259, 270

Doublet Kramers theorem, 7:172ff, 175

Doubly excited atom, 52:27–70

  channel coupling, **52**:28–38

  intrashell spectra, **52**:39–45

  intrashell supermultiplets, **52**:45–49

  rotor-vibrator model, **52**:49–52

Downes equation for isotope analysis, 1:43

Dressed potential curve, 42:110; **50**:238

Dressed state, 40:377

Drift term, 46:221

Drift velocity, 36:505

  electronic, Ar-N_2 mixtures, **31**:348

    hydrocarbons, **31**:355–358, 454

    liquid Ar, **31**:344, 347, 428

    liquid Ne, **31**:343

    solid Ar, **31**:348

    solid Kr, **31**:349

    solid Xe, **31**:350

dRNA, 7:219

Droplet model, 48:482, 483

Drude equations, 4:153, 154

Drude metal, 37:39

Drude sharp surface, 27:611

Drude tails, 4:95, 96, 109

Drude theory, 12:347; **49**:566; **51**:263, 264

Dual lattice, 41:2

Dumbbell model, for atomic recombination, **13**:110

  for barrier rate constant, **13**:115

Dunham expansion, 12:354; **44**:26; **49**:117, 171, 175, 178

Duroquinone radical anion, 35:26

Duschinsky rotation, 44:557; **54**:188–191

Dust, nucleation in absence of, **40**:138

Dye laser, **50**:486
  CW, **42**:362
Dye molecule, medium-sized, **49**:28
Dyes, interaction with cellulose, **21**:685, 696
Dynamic correlation, **43**:250, 258–260
  sine-Gordon chain, **53**:249
Dynamic effective charge, **27**:509–511, 513–514, 524, 527
Dynamic form factor, **17**:89
Dynamic liquid structure, **24**:2
Dynamic molecular field theory, **46**:203
Dynamic structure factor, **53**:10, 11, 251
  antiferromagnets, **53**:256
  lead germanate, **53**:276
  Toda lattice, **53**:316–317
Dynamic susceptibility, **12**:227
Dynamic viscosity, *see* Viscosity
Dynamical approximations, **51**:195
  coupled states, **51**:185
  effective potential method, **51**:214
  exponential perturbations theory, **51**:186, 214
  infinite order sudden approximation, **51**:188
Dynamical critical phenomena, **40**:197
Dynamical equations, canonical form, **24**:137
Dynamical friction in gases, **5**:159
Dynamical group, **38**:50
Dynamical matrix, **27**:470, 481, 488–489
Dynamical memory, **44**:262
Dynamical operator, **36**:30
Dynamical orientation correlation, **44**:299
Dynamical probe, **44**:386
Dynamical quantum system, **38**:51, 91
Dynamical scaling laws, **50**:589ff
Dynamical semigroup, **38**:124

Dynamical variables, **44**:263
  canonical transformation, **53**:315
  in irreversibility theory, **53**:125–126
  molecular-time-scale generalized Langevin equation, **53**:161–162
Dynamics, **38**:7, 66
  many-particle, **6**:55–80
    classical system, **6**:56
    modified Boltzmann equation, **6**:62–68
    Onsager theory, **6**:68–80
    Pauli equation, **6**:59–62
    quantum mechanical system, **6**:57
  stochastic, **32**:62
Dynamics of collisions, *see* Collision dynamics
Dyson equation, **36**:218, 220, 221, 241, 246, 262; **46**:349; **52**:473

E

e-H system, **49**:269–272
E-map, **16**:160–161, 165, 167
E. coli, *see* Escherichia coli
Earth, primitive, melting of, **55**:86–87
  temperature of, **55**:87–88
Easy-plane ferromagnet, **53**:264
Eccentric cylinders, flow of compressible fluid between, **32**:46
Eckart conditions, **2**:264, 267, 298; **44**:12
Eckart frame, **37**:308; **49**:210–216
Eckart potential, **49**:295
Eco-gas, **19**:336
Ecodynamics, **19**:325, 333
Ecological networks, **19**:326, 333
Ecology, **32**:9; **43**:15, 205
Econoise, **19**:339
Economics, **43**:15

Ecosystems, **19**:337

Edge approximation, **43**:177

Edge dislocation, **40**:214

Effective charge, **2**:274, 276

Effective collision diameter, **1**:121

Effective field distribution, **51**:253

Effective force, **11**:284

Effective Hamiltonian, **2**:226; **18**:183; **36**:391; **42**:285

Effective interaction, **52**:218

Effective Liouvillian, **44**:266

Effective local density, **9**:40

Effective magnetic moment, **7**:171, 180

Effective mass, **1**:24; **4**:232, 287; **31**:288
    cadmium, **31**:302
    charged particles, **3**:234
    electron, **13**:125; **31**:331, 374, 400, 402
    lithium, **31**:300

Effective medium theory, **31**:452; **51**:262; **52**:487

Effective potential theory, **12**:17, 18, 20; **16**:287–288, 299–300; **31**:158, 200, 203; **36**:183, 184; **42**:430; **53**:2–3, 120–121
    collision integrals, **31**:208
    nonspherical molecules, **31**:201–211
    spherical molecules, **31**:158, 200, 203
    transport coefficients, relaxation times, **31**:208, 209

Effective power function, **43**:43, 46

Effective principal quantum number, **7**:13

Effective valence, **51**:248

Efficiency coefficient, **1**:33; **46**:184

Efficiency of reaction, **1**:192

Ehrenfest equation, **51**:235

Ehrenfest model, **50**:415

Eigendistribution, **38**:159, 161, 162, 187

Eigenfunctions, **4**:211; **7**:8, 31
    expansion method, **25**:212, 213, 257,

264, 279; *see also* Variational methods, close coupling
    Hamiltonian operator, **4**:232, 236
    kinetic energy operator, **4**:233
    Liouville operator, **53**:151, 163–167
    Schrödinger equation, **4**:282
    soliton, **53**:326

Eigenpacket, **38**:29

Eigenprojection, generalized, **38**:30, 43

Eigenvalues, **4**:92, 284; **7**:8, 31
    bound-bound matrix, **25**:225ff, 270ff, 291
    energy, **4**:285
    generalized, **38**:30, 31
    Hermite equation, **4**:211
    transport equation, **4**:9–21, 25–27, 30, 53, 54

Eigenvectors, generalized, **38**:30, 31
    spin operator, **52**:177
    transport equation, **4**:9–21, 25–27, 30, 53, 54

Einstein A coefficient, **42**:195, 319

Einstein B coefficient, **50**:198

Einstein-Brillouin-Keller quantization rules, **36**:3, 6, 27, 37, 52, 56; **46**:120, 125

Einstein coefficients, **50**:407

Einstein force, **53**:92

Einstein frequency, atom-chain collision, **53**:104
    matrix, Lennard-Jones solution, **53**:215
    in molecular-time-scale generalized Langevin equation, **53**:152–153
    response function hierarchy, **53**:154

Einstein limit, **53**:88, 99

Einstein oscillator model, **53**:104

Einstein quantization, case 1, **36**:5, 22, 27, 36, 41
    case 2, **36**:5, 6

Einstein relation, **1**:151, 153; **5**:96, 273;

17:168; **31**:67; **53**:96, 273
for specific heat, **46**:392
Einstein-Smoluchowski theory of light scattering, 1:349; **6**:195
Einstein-Stokes equation, 1:136
Elastic constants, 41:159
Elastic coupling potential, 53:265–270
Elastic deformation, 41:184
Elastic dumbbell, 35:59
Frenkel, **35**:101
second-order fluid constants, **35**:101
Elastic model, excess entropy calculation from, **2**:141
solid solution, **2**:140
Elastic rhombus model, 35:50, 105
metric tensor components, **35**:50, 103
second-order fluid constants, **35**:103, 105
Elastic scattering, 18:72, 74, 80, 81; **25**:215, 259, 278, 280, 282, 285
by central forces, classical analysis of, **10**:35–47, 76–83
cross section, atom-atom, **10**:123–126
differential, **10**:76–79, 137; **12**:398, 400, 481
collisions of excited species, **10**:179, 183–187
effective, **10**:141, 166–168
low-angle, **10**:81, 105–109, 112–114
quantum, **10**:53–54, 83, 86–88, 99–100, 104–105
reactive systems, **10**:131, 141–168, 269, 362
reduced, **10**:79–82
inelastic correction to, **10**:39, 54–57, 127–131
ion-neutral reactions, **10**:269
total, **10**:39–45, 57–62, 76, 117–

120, 120–124
collisions of excited species, **10**:179–183, 186
helium-helium scattering, **10**:62–67
volume averaging of, **10**:40–42
excited species, **10**:179–187
He^+ - He, **23**:170
high-energy beams, **10**:29–73
ion-neutral reactions, **10**:248, 269
low energy, pseudopotential representation of, **16**:295–304
noncentral forces, **10**:47–50
quantum effects in, **10**:53–54, 75–134
reactive systems, **10**:135–169, 321, 329, 336–337, 362–367
see also Single channel scattering
Elasticity theory, 4:196; **32**:54; **43**:260
Electric correlation, 2:248
Electric deflection analysis, 10:9–10, 15–16, 175, 249, 326–327
reactive scattering, **10**:347–356
Electric dipole, 33:154, 203ff; *see also* Dipole
Electric dipole interaction, 25:3
selection rules, **25**:16
Electric dipole moment, 18:75; **40**:469; **44**:105, 339; *see also* Dipole moment
Electric dipole radiation, 4:163, 164
Electric dipole transition, 16:93–94; **18**:185
Electric displacement vector, 1:321
Electric field, 12:123, 125; **33**:154, 162, 191, 212ff
in chemical systems, **38**:415–450
anisotropy of translational diffusion, **38**:418, 419, 426, 427
biochemical oscillations, **38**:417, 429

circadian oscillations in cortocos-
teroid concentration, **38**:432–
433, 446–447

continuity equations, **38**:420

Debye-Hückel cut-off for patterns
in, **38**:425, 435

divergence of fluctuations, **38**:420

embryogenesis, **38**:417, 430, 433

frequency shifts in chemical oscillations, **38**:416, 427, 441

functional polarity reversal,
38:443, 445–446

glycolytic instability, **38**:428, 429

hard mode instability, **38**:428, 429

hypothalmic polarization with,
38:417, 432–433, 446–447

inversion of morphogenetic patterns with, **38**:417, 431, 443

large scale effects on self-coherent
patterns of, **38**:416, 429, 439ff

polarity asymmetry in cortical
polarization with, **38**:416, 431,
432

regenerative morphogenesis and,
38:443, 445–446

self-consistent evolution of fluctuations, **38**:421

soft mode instability, **38**:423,
425–426, 439–442

symmetry lowering chemical
instability, **38**:417, 437, 439,
440ff

trans-cellular Ca_2^+-supported self-
polarized, **38**:442–444

trans-tissue Na^+-supported self-
polarized, **38**:445–446

type I and type II patterns of self-
coherence, **38**:440–442

congruent, **10**:15–18

quenching in, **10**:175, 351

dynamic, **4**:141, 215–219

effect on multiphoton dissociation,
50:134

effect on reaction yield, **50**:397

half-space, **27**:367

Hamiltonian, **18**:178

longitudinal, **33**:159, 164, 191

slab, **27**:372, 383–385

sphere, **27**:393

static, **4**:141

transverse, **33**:164, 191

Electric field gradient, **2**:188, 189

Electric field ionization, **10**:234, 248,
268

Electric field strength, **1**:330

Electric modes, **27**:391–397

Electric multipole moment, **44**:105

of molecules, **20**:2

see also specific moments

Electric polarizability, **25**:8

Electric quadrupole interaction, nuclear,
18:199, 208

Electric quadrupole moment, **18**:65, 74,
78, 83; **33**:155, 209

carbon dioxide, **5**:176

Electrical conductance, **31**:466

Electrical conductivity, **5**:264, 267, 304

dc, **13**:173

frequency-dependent, **13**:126, 130,
144

linear response theory, **53**:132–133

metal-ammonia solutions, **4**:318–
322, 365–370

see also Conductivity

Electrical discharge, pulsed, **42**:130

Electrical fluctuations, **12**:227, 237, 243

polarization of molecule by, **12**:275

rotational motion of permanent
dipoles, **12**:237

Electrical resistivity of metallic glasses,
51:239–243

high-temperature, **51**:243–251

low-temperature, **51**:251–253

Electrochemical deposition, 40:188

Electrochemical potential, 6:146, 161; **12**:235

Electrochemical reaction mechanism, 3:13ff

carbon-halogen bond fission, **3**:14

ketone reduction, **3**:16

Electrochromism, 13:179ff, 186

Electrode reaction, 38:350

organic, bond rupture, **3**:3

chemical mechanism, **3**:5

electrochemical mechanism, **3**:13

energetic mechanism, **3**:19

processes involving more than two electrons, **3**:4

Electrodynamic coupling, 44:338

Electrodynamics, 33:153–234; **48**:357

canonical formulation, **33**:178

classical, **33**:201

covariant, **33**:154, 161

gauge-invariant, **33**:156

Hamiltonian formulation, **33**:173

macroscopic, **33**:161, 227

neoclassical, **33**:211, 212

nonrelativistic, **33**:210

quantum, **33**:154–157, 160, 201, 204, 211

semiclassical, **33**:157–159

Electrolyte solutions, 48:303–314

binary, **11**:222, 225

Debye-Hückel theory, **21**:457

dilute, **11**:246

equilbrium theory, **11**:187

heats of dilution, **21**:468

limiting conductance, **11**:216

thermodynamic properties, **11**:192

transport properties, **11**:202

with short-range interactions, **21**:457ff

Electrolytic method, 1:16

Electromagnetic field, energy density, **33**:183

Green's function for, **33**:195–197, 210

localization at surface, **27**:393–394, 409

momentum density, **33**:183

p-polarized, **27**:363, 379, 382, 409, 436, 452, 461

penetration of, **27**:364, 367, 373, 377–378, 381

s-polarized, **27**:371

Electromagnetic mixing of states, 50:195

Electromagnetic radiation, 7:260

Electron affinity, 1:250, 251, 254; **7**:46, 98, 302ff; **36**:214, 395

adiabatic, **30**:306, 513, 529

calculation by equations of motion method, **48**:1–70

methyl derivatives, **42**:515, 540ff, 558

table of, **42**:562

polyphenyls, **21**:39–47

table of, **50**:350

vertical, **30**:304, 512, 529

Electron analyzer, 45:24, 27, 30

Electron-atom interaction, 31:405

Electron-atom polarizability, 31:406

Electron-atom potential, dielectric screening, **31**:435

Electron-atom reaction rate, 31:386, 387, 388

Electron-atom scattering, 25:214, 215, 258, 267, 268, 278, 280ff

Electron attachment, 23:2, 5; **31**:338; **42**:551, 558

Electron beam method, 18:20

Electron bubbles, 41:175

Electron conductivity, 44:483

Electron configuration, 7:6
  cations, 7:375
Electron correlation, 9:321ff; 12:63, 67,
    69, 70; 25:185; 31:22; 41:67;
    50:523
  ground state He atoms, 12:69
  H_2, 12:67, 68, 69
  in-out, 12:69, 70
  left-right, 12:69
  N_2, 12:67
Electron correlation energy, 2:74
Electron density, 1:170, 176; 16:136
  in hypersonic flow, 13:29, 30
  related to ion-molecule rate constant,
    19:160
Electron density map, 16:162
  digitoxin, 16:192
  6-hydroxycrinamine, 16:191
  panamine, 16:167
  quasi-racemates, 16:171
  serotonine complex, 16:179
  thymer E, 16:184
Electron detector, 45:33–38
  coincidence method, 45:34–38
Electron diffraction, 2:372
  conformational analysis, 3:353
  fluid structure studies, 53:5
  of gases, 3:323ff
    experimental method, 3:328
    internuclear distances, 3:336
    molecular vibrations, 3:357
  internuclear distances, 3:336
  vibrational analysis, 3:357
  see also Low energy electron diffrac-
    tion (LEED)
Electron-dipole correlation, 52:15–26
  rotating polar molecules, 52:15–19
Electron donor properties of purines,
    8:169
Electron-electron correlation function,

    53:5–6, 45
Electron energy, 6:23, 32–34, 371
  at crystal surface, 9:89, 97
  distribution function, 28:427, 428,
    429–436, 459
  ion-neutral reactions, 10:258–259,
    270
Electron energy loss, 31:360, 421, 426,
    460
Electron energy transfer, 31:385
Electron exchange, 18:15, 17, 19, 80;
    25:215
Electron gas, 4:323, 338; 27:358, 361,
    550, 594; 42:187, 276, 427; 50:564
  damping effects, 27:594
  in lattice of positive charges, 6:85ff
    at zero of temperature, 6:112–113
    correlation energy, 6:124
    disordered lattice, 6:130, 135–142
      self energy of, 6:136
    exchange energy, 6:124
    ground state energy, 6:86
      multicomponent lattice, 6:132
      one-component lattice, 6:122–
        129
      randomly ordered lattice,
        6:141
    kinetic energy, 6:124
    long range forces, 6:99–103
    Madelung term, 6:110, 115, 124,
      128, 131, 136
    modified Coulomb potential,
      6:127–130, 143
    multicomponent lattice, 6:130–
      142
    one-component lattice, 6:130–
      142
    polarization energy, 6:124, 128,
      132, 135, 136
    polarization of, 6:119
    positive continuum, 6:357

superlattice, **6**:130

magnetic, **41**:102, 105

non-uniform, **14**:221

paramagnetic, **41**:104

uniform, **14**:216

Electron-hydrogen channels, 52:19–23

Electron impact, 10:257, 383

fractional abundance determination, **45**:85–92

in ion beams, **45**:92–101

ionization by, **45**:84–111

Electron impact excitation, 10:173–174, 184, 226–227; **18**:15, 17, 21, 42ff

anisotropic, **16**:82–96

Ar, **18**:21

C_2H_2, **18**:56, 57

C_6H_6, **18**:21

CO, **18**:49, 60, 71

H_2, **18**:21, 71, 72, 82, 83

H_2O, **18**:70, 71

He, **18**:21

N_2, **18**:20, 21, 60, 71, 72, 82, 83

NO, **18**:71

Electron impact ionization, 10:248–249

Electron impact spectroscopy, 45:5–7, 8, 17, 26, 29–31

Electron interaction integral, 7:99

Electron-ion recombination, 31:364

in hypersonic flow, **13**:35

Electron jump model, 10:321–322, 367, 379

Electron lenses, 45:26

Electron loss, 27:305, 310, 320

Electron loss spectroscopy, 41:485

Electron mobility, activation energy, **31**:355–357, 389

Ar, liquid, **31**:335, 346, 352–355

solid, **31**:344

binary hydrocarbon mixture, **31**:356

definition, **31**:330

field dependence, **31**:349

H_2, gas, **31**:367

liquid, **31**:335

He, gas, **31**:336–339, 367

liquid, **31**:331, 332, 335

hydrocarbon liquid, **31**:334, 335

Kr, liquid, **31**:335, 350

solid, **31**:344

liquids, **36**:514

method of measurement, **31**:368

Ne, liquid, **31**:335, 343

solid, **31**:344

nonpolar fluid, **31**:360

related to hydrocarbon properties, **31**:354

transition, **31**:337

Xe, liquid, **31**:335

solid, **31**:344

see also Drift velocity

Electron-molecule reaction rate constant, 31:392

Electron-molecule scattering, 25:278

Electron-nuclear double resonance (ENDOR), 40:400

Electron optics, 20:273, 275, 276

cathode, **20**:273

fluorescent screen, **20**:276

grids, **20**:276

Electron orbitals of metals, 7:360

Electron pair functions, 9:337

Electron pairs, 1:169; **5**:342, 349

Electron paramagnetic resonance, 2:34, 199; **7**:159, 264, 278, 330ff, 352ff, 605ff; **40**:398ff, 423, 459, 463; **41**:210, 236, 238; **46**:183;

free radicals, **7**:532, 577

metal-ammonia solutions, **4**:315, 327–332, 374–379

microclusters, **40**:58

plant systems, **7**:628

Electron promotion, 7:28

Electron radial distribution function, 34:168

Electron repulsion, 1:249, 254
cations, 7:364

Electron repulsion integral, 13:211, 223, 244
multi-center, 13:245

Electron resonance spectra, 18:149ff
BrO, 18:216, 228, 229, 231, 233
CF, 18:228
ClO, 18:216, 228, 231, 232, 233
IO, 18:216, 228, 231, 233
NO, 18:228, 233
NS, 18:228, 233
O$_2$, 18:218, 219, 220, 221, 222, 232
OH, 18:214, 228, 231, 233
SD, 18:228
SeF, 18:215, 216, 228, 233
SeH, 18:216, 228, 231, 233
SeO, 18:218, 219, 220, 221, 222
SF, 18:215, 216, 228, 233
SH, 18:228, 231, 233
SO, 18:218, 219, 220, 221, 222, 232
TeH, 18:216, 228, 233

Electron run-away problem, 20:159

Electron scattering, 27:447–449, 510–511; 45:385–391; 50:481, 503, 549
resonances, 18:17, 63, 71, 81
angular momentum barrier, 18:96
core excited, type I, 18:92, 93, 99
type II, 18:93, 94, 96
electron affinity of excited states, 18:93
Feshbach compound state, 18:92
hole-particle, 18:92
inclusion of core polarization, 18:137
induction effect, 18:96

single particle, 18:96, 99, 115, 137

Electron scavenging, 52:475

Electron-solid interaction, 27:77, 81, 88, 89, 90
adsorbed overlayer potentials, 27:112
boundary conditions, 27:91
bulk electron-ion core potential, 27:106
clean surfaces, 27:110
elastic collision model, 27:95
electron-electron interaction, 27:44, 77
electron-ion core phase shifts, 27:87
electron-ion core potential, 27:86
electron phonon interaction, 27:44
electron plasmon coupling, 27:45
Hartree-Fock approximation, 27:106
inelastic collision damping, laminar models, 27:84
multiple scattering models, 27:77
inelastic collision damping length, 27:45
inelastic collision model, 27:90, 92
ion core potentials, 27:78
lattice vibrations, 27:43, 101
local complex potential model, 27:90
optical potential, 27:45, 46, 89
overlapping atomic charge density model, 27:88, 106
phase shifts, rigid lattice, 27:98
vibrating lattice, 27:103
static potential model, 27:42, 81, 83

Electron spectroscopy, 45:3–74
CH$_4$, 45:69–72
CO, 45:54–69
CO$_2$, 45:72–73
H$_2$, 45:50–54
HF, 45:72–73
N$_2$, 45:54–69
N$_2$O, 45:72–73

NH_3, **45**:72–73
NO^+, **45**:104
noble gases, **45**:41–50
O_2, **45**:72–73
Electron spin resonance. *See* Electron paramagnetic resonance
Electron storage ring, 45:3
Electron swarms, 18:19, 71
Electron systems, inhomogeneous, spin-density-functional formalism, **41**:59–194
π-, **5**:1ff; **6**:384–387; **7**:22, 53
  absorption spectra, **5**:17ff
  bond order, **5**:23
  bonds, **7**:18ff
  characteristic function, **5**:12
    of valence band, **5**:12–13
  charge density, **5**:22
  chemical reactions, **8**:70ff, 114, 119, 124
  donors, **7**:363, 390, 395
  energies, **7**:111, 115, 124, 128
  energy spectrum, **9**:131
  Hamiltonian of, **6**:386
  Hückel theory, **5**:3
  many-electron theory, **6**:384
  orbitals, **7**:87, 117
  resonance energy, **5**:18ff
  resonance integrals, **5**:25
  Shiebe's phenomenon, **5**:4, 16
  structure matrix, **5**:9
  transitions in ethylene, **8**:5ff
  see also Relativistic electron systems
Electron transfer, 7:368
  between water and proteins, **7**:239
  polymerization and, **2**:150, 181
  theory, **7**:45
  see also Charge transfer
Electron transfer spectra, 5:76, 78, 79, 82; **8**:54, 167

  actinides, **5**:136
  calculation of, **5**:75ff
  hexahalides, **5**:125–126
  lanthanides, **5**:78, 135
  tetrahalides, **5**:82, 84
  transition metals, **5**:75ff
  see also Charge transfer spectra
Electron tunnelling, 39:244, 339, 340
Electronegativity, 20:2
  atomic terms for, **13**:72
  equalization, **13**:74
  geminal, **13**:59, 73
  Klopman's approach, **13**:72
  Moffit's approach, **13**:71
  Mulliken scale, **13**:58, 67
  Pauling arithmetic-mean rule, **13**:64
  Pauling geometric-mean rule, **13**:64
  Pauling scale, **13**:61
  relation to chemical bonding, **13**:55
    ionic, **13**:64
      Madelung-Jørgensen energy, **13**:75
      molecular orbital theory, **13**:66
      valence bond theory, **13**:58
  secondary, **13**:65
  transition metal complexes, **5**:78, 126
Electroneutrality condition, 31:92
Electronic conduction in organic molecular solids, 7:282
Electronic deexcitation, 28:277–287, 349–358, 390, 437
  atoms and atomic ions, Ca, **28**:280ff, 299ff, 348ff
  Cd, **28**:300
  Cu, **28**:300, 305
  He, **28**:411, 413
  Hg, **28**:287–294, 300, 305, 309
  K, **28**:232, 273, 280ff, 303ff, 350ff
  metallic ions, **28**:414

N, **28**:397, 410, 412

Na, **28**:280ff, 301ff, 347, 351, 364

O$^+$, **28**:409ff

O, **28**:50, 269, 388, 401ff

Rb, **28**:307ff, 348ff

molecules and molecular ions, H$_2$, **28**:240

Li$_2$, **28**:242

N$_2$$^+$, **28**:393, 395, 396, 413

N$_2$, **28**:41, 306, 387, 392, 394, 395, 413

NO, **28**:243, 412

O$_2$, **28**:243, 269, 397, 401, 402, 403

OH, **28**:249

Electronic deformation polarization, 1:77

Electronic eigenvalue problem, 49:130–131, 135, 140, 148, 160–161

Electronic—electronic energy transfer, 50:379

Electronic energy transfer, collisional, **21**:169, 178, 189

Electronic excitation, 28:14, 39, 40, 296, 392, 437

atoms and atomic ions, Ag, **28**:292

Ba$^+$, **28**:413

Bi, **28**:292

Br, **28**:39

Cd, **28**:291, 292, 293

Cr, **28**:292

Cs, **28**:288, 309, 348

Cu, **28**:292

H, **28**:411, 412, 413

He, **28**:411

Hg, **28**:63, 307

I, **28**:39

In, **28**:292

K, **28**:94, 306, 347, 398

N, **28**:397, 398, 411

Na, **28**:4, 283, 288ff, 306, 308, 347, 411, 412, 414

O, **28**:34, 39, 48, 94, 403ff

Pb, **28**:292

Rb, **28**:273, 283, 288, 309, 348

Tl, **28**:289, 292, 293

Zn, **28**:292, 293

collisional, *see* Collisions

by electron impact, **36**:416, 419, 420, 434

aniline, **36**:434

benzene, **36**:419

singlet-triplet transitions, **36**:416, 419, 420

toluene, **36**:434

molecules and molecular ions, C$_2$, **28**:42, 43

CH, **28**:44

CN, **28**:35, 40, 41, 42, 89, 414

CO$^+$, **28**:413

CO, **28**:44, 45, 189

CS, **28**:50

Hg$_2$, **28**:294, 300

N$_2$, **28**:37, 392, 394, 411, 413

NO, **28**:36, 37, 414

NO$_2$, **28**:38, 45, 46, 90

O$_2$$^+$, **28**:224, 396, 409, 413

O$_2$, **28**:45, 48, 60, 61, 401, 402, 414

OH, **28**:44, 45, 60, 413, 414

SO, **28**:50

SO$_2$, **28**:45, 46

Electronic nonadiabatic process, 49:295–309

in strong laser field, **49**:302–306

Electronic predissociation, 47:Part 1:27–28

Electronic properties, 9:322

of DNA, **7**:339ff, 347ff

Electronic quantum number, 49:131

Electronic quenching, 50:60, 325ff

Electronic radii in charge transfer states, 8:14ff

Electronic relaxation processes, 41:297–310; 47:Part 1:13–14, 29, 40–46; 47:Part 2:291–418, 529–530

collisional effects, 42:207–270; 47:Part 2:337–380

benzene, 42:260

benzophenone, 42:217, 226

biacetyl, 42:226

dephasing effects, 47:Part 2:348

Fermi resonance, 42:222, 229, 244

Hund's cases, 42:227, 242

intermediate size molecules, 47:Part 2:357–358

intersystem crossing, 42:223, 235

intramolecular dephasing, 42:219

intramolecular parameters and, 47:Part 2:363–366

intramolecular potential, 47:Part 2:368–369

long range forces, 42:225

magnetic field effects, 47:Part 2:373–376

mixed states, 42:227

naphthalene, 42:217, 260, 262

phenomenological description, 47:Part 2:341–344

photofragmentation products, 47:Part 2:372–373

primary effects, 47:Part 2:345–348

quantum beats, 42:220

radiationless transitions, 42:211

reversibility, 47:Part 2:360–362

rotational relaxation, 47:Part 2:345–346

small molecule weak-coupling case, 47:Part 2:350–357

strongly mixed states, 47:Part 2:366–368

time evolution, 47:Part 2:348–350

van der Waal complexes, 47:Part 2:370–372

vibrational relaxation, 47:Part 2:346–348, 362–363

inverse, 47:Part 1:46–51

Electronic specific heat of α-brasses, 2:131

Electronic spectra, calculation of, 6:346, 408; 9:200ff

semi-empirical theories of, 6:319, 358, 383

see also Absorption spectra; Electron transfer spectra; Spectra

Electronic states, 36:142, 159, 178

amorphous systems, 31:446

excited, 36:142, 159, 178

ground, 36:142, 178, 197, 198

Rydberg, 36:190, 197

Electronic structure, 41:64; 51:262

atmospheric systems, calculations, 45:229–231, 240–337

Born-Oppenhiemer separation, 45:231–232

configuration interaction methods, 45:230–231, 236–239

coupled-cluster approach, 52:181–262

effect of pressure on, 4:161–200

hemoproteins, 7:159

iterative natural spin orbitals method, 45:237–238

metallic surface, 41:67

N_2, 45:241–267

NO^+, 45:318–336

NO, 45:288–301

O_2, 45:265–287

O_2^-, 45:301–317

restricted Hartree-Fock method,

45:230, 234–236

self-consistent field methods, 45:234–236

of solids, 4:161, 200

variational methods, 45:232–240

see also Potential energy

Electronic transitions, 4:68, 69; 7:93; 25:150ff, 175; 53:262

ab initio approach, 37:295

crystal field approach, 37:292

d-d, 4:174, 175, 184

d-p, 4:187

f-f, 4:162, 165, 175

forbidden, 5:241ff

  aromatic compounds, 5:241

  benzene, 5:241

  Herzberg-Teller bands, 5:241

  Hückel theory, 5:247

  Jahn-Teller interaction, 5:248, 251

  orbital configuration, 5:247, 251

  oscillator strength of, 5:241, 242, 243, 252

  perturbation theory of, 5:243

  potassium titanium hexafluoride, 5:241

  symmetry, 5:251–253, 255

  transition metal complexes, 5:241

  use of Lennard-Jones approximation, 5:247

  vibronic perturbations, 5:242–243, 253

nephelauxetic effect, 37:292

semiempirical MO approach, 37:294

simpified MO approach, 37:293

singlet-singlet, 18:54, 56, 57

singlet-triplet, 18:20, 21, 43, 46, 57, 60, 67, 69

valence bond approach, 37:296

Electronic–vibrational coupling, 36:278, 286, 296

Electronic–vibrational energy transfer, 47:Part 2:381–418; 50:263, 370, 375, 475–516

curve crossing mechanisms, 47:Part 2:413–415

experimental results, 47:Part 2:395–409

experimental techniques, 47:Part 2:384–388

I_2^*, ICl^*, and NO_2^* + CO, 47:Part 2:115–116

kinetic scheme rate constants, 47:Part 2:388–395

long-range attractive forces, 47:Part 2:411–413

$Na(3^2P_j)$ + CO, 47:Part 2:113–115

$O(^1D_2)$ + CO, 47:Part 2:112–113

quantum mechanical calculation, 47:Part 2:410–411

reactions involving CO, 47:Part 2:111–116

see also Energy transfer

Electronic wave functions, 9:209, 219ff, 224

Electrons, 2:200; 7:302

bubble model, 31:330, 351, 382, 383, 437ff

density of states, 31:446; 42:216

drift velocity, 28:431, 437, 449

effective mass, 13:125

excess in hydrocarbons, 44:509

free, 1:262, 323

free energy, 31:333, 437

hydrated, 44:502

hydrocarbon glass, 31:371

Lewis pair, 13:74

in liquids, alcohols, 47:Part 2:547–550

  cluster model of solvation, 47:Part 2:550–552

  electron localization in dilute polar fluids, 47:Part 2:555–562

dynamical structures, **47**:Part 2:558–561

experimental results, **47**:Part 2:555–558

mechanisms for solvation, **47**:Part 2:561–562

laser chemistry of solvated electrons, **47**:Part 2:562–569

laser-induced electron transfer, **47**:Part 2:564–569

molecular motion, **47**:Part 2:552–555

photophysics, **47**:Part 2:564

picosecond spectroscopy, **47**:Part 2:535–577

    experimental design, **47**:Part 2:540–547

    laser saturation, **47**:Part 2:546–547

    relaxation processes, **47**:Part 2:535–577

momentum transfer cross section, **28**:430, 437

relativistic theory, **7**:164

semi-free, diamagnetism of, **3**:214, 215, 220

solvated, **44**:483–544

tunneling, **27**:323–325

Electro-osmotic flow, 29:357

Electrophilic reagent, 7:67

Electrophilic substitution, 7:44; **8**:76ff, 86ff, 100, 103, 108ff; **19**:204

Electrophoresis, 11:263

Electrophoretic effect, 11:217, 223, 241, 263

Electrostatic approximation, 27:285, 289, 295, 303

Electrostatic effects, 12:297

Electrostatic energy, 12:107, 124

Electrostatic field, effect on biological objects, **7**:155

    multipole expansion for, **44**:460

rotatory, **44**:446

Electrostatic Hellmann-Feynman theorem, 12:54, 55, 60

Electrostatic interaction, 2:391; **39**:178, 180; **44**:261

    convex molecules, **5**:174

    ion-dipole, **39**:180

    membranes, **39**:218

    metal ions, **39**:181

Electrostatic multipole interactions, 12:32

Electrostatic perturbation calculations, 12:29

Electrostatic repulsion, 7:13; **48**:336–338

Electrostatic shell model, 12:349

Electrostatic variational calculations, 12:21

Electrostriction, 4:310, 340, 347; **13**:190; **33**:77

Elementary catastrophe, 40:245

Elementary cell, 7:139

Elementary chemical reactions, 37:71, 72

Elementary excitations in liquids, 17:186

Elementary process, 33:83, 99

Ellipse metric, 20:121

Ellipsoid modes, 27:444, 507

Ellipsoidal coordinate system, 13:213

Ellipsoidal density, 53:112

Ellipsoids, bulk viscosity, **31**:183–184

    collision dynamics, **31**:180

    collision integral, **31**:180, 182

    Enskog theory, **31**:179, 223

    equilibrium pair distribution function, **31**:180, 182, 183

    excluded volume, **31**:181

    relaxation time, **31**:195, 196

    shear viscosity, **31**:183–185

    spin flux, **31**:191, 192, 194

thermal conduction, **31**:183, 186–189, 193

vortex viscosity, **31**:183, 190

Ellipsometry, **27**:547, 612

Ellipticity, **4**:71, 72, 93ff; **44**:332

Elovich equation, **7**:247

Embryos, x-rayed, **7**:640

Emissivity, **27**:438

Emulsion, **48**:332

Enantiomorphism, microcluster structures, **40**:62, 87

End-over-end rotation, **44**:261

End-to-end distance, **52**:432, 442, 445

distribution of, **52**:433, 441

Endergonic reactions, **7**:62

Endoergic reaction, **42**:85; **45**:128–144

Endothermic reactions, **29**:59

Energetic function, **33**:417, 425

Energetic reaction mechanism of organic electrode reactions, **3**:19

quantum mechanical approach, **3**:24

thermodynamic-kinetic approach, **3**:19

Energetics, bond mode, **47**:Part 1:298–302

intramolecular effects on, **47**:Part 1:306–309

of proteins, **39**:55

of reaction, **39**:58

Energy, available rate, **13**:105

theory of, **13**:96

binding, **2**:34

cohesive, **41**:138

complex, **36**:3, 53

configurational, **16**:224, 232, 235, 248

correlation, **16**:285; **41**:105, 106, 137

excess, **16**:229, 247, 274–276

exchange, **41**:137

exchange-correlation, **41**:102, 108, 167

excitation, **41**:124

first partial derivative in MCSCF calculation, **54**:145–151

force magnet system, **20**:82, 83

Madelung-Jørgensen, **13**:71, 75

stellar systems, **26**:112, 126, 143

conservation of, **26**:108, 118, 122, 131, 134

correlational, **26**:131, 132, 143

internal, **26**:115, 129, 130, 131

kinetic, **26**:110, 118ff, 130ff

potential, **26**:110, 118, 122, 131ff

total, **26**:119, 126ff

see also Stabilization energy

Energy absorption units, **7**:261

Energy analyzers, **27**:239; **42**:534

differential type, **27**:242

retarding type, **27**:240

Energy anomaly, **28**:72

Energy balance, **13**:273; **24**:96, 103; **33**:408

Energy bands, **7**:140ff

Energy-configuration criterion, **6**:5, 7

Energy defect, **10**:197, 256

Energy denominator, **9**:327

Energy-density normalization, **46**:392

Energy disposal, **42**:16; **47**:Part 1:12–13

harpooning, **10**:375–377

reactive scattering, **10**:359–360

Energy dissipation, **17**:74

principle of least, **3**:374, 277

Energy equation, **41**:24

Energy exchange, fast molecule-ion collisions, **10**:232

rotation-translation, **10**:141, 151–154

vibration-translation, **10**:141, 151–155

see also Energy transfer

Energy flux method, 37:2
Energy forms, 7:203ff
Energy function, 36:24, 27
Energy gap law, 41:288, 293, 301, 313; 42:345
Energy indices, 7:31
  pteridines, 7:59
  purine bases, 7:51
  pyrimidine, 7:54
Energy levels, 4:117; 28:386, 387
  atoms and atomic ions, C^+, 28:387
    C, 28:386
    Ca^+, 28:386
    Fe^+, 28:386
    Mg^+, 28:386
    N^+, 28:387
    N, 28:386
    Na^+, 28:386
    Na, 28:386
    O^+, 28:387
    O, 28:386
  degenerate, 4:129
  molecules and molecular ions, CH_4, 28:219
    CO_2-N_2, 28:427
    K—Rb, 28:273
    N_2^+, 28:387
    N_2, 28:386
    NO^+, 28:386, 387
    NO, 28:386
    O_2^+, 28:387
    O_2, 28:386
  perturbations of, 7:133
  semiclassical, 36:68
  see also Potential energy curves
Energy localization, 50:110
Energy loss spectrum, 30:222, 456, 457; 36:423
Energy lost per cycle, 20:96, 120–122
Energy matrix, 2:271

Energy moments of molecular system, 34:237
Energy of mixing, 16:247
Energy of protonation, 8:133, 136, 137
Energy of reorganization, 44:505
Energy partitioning among products, 42:3, 32, 155ff
Energy profiles, 4:52
Energy release, 30:287, 291, 292, 295, 304
Energy-resolved observables, 47:Part 1:14–15
Energy resonance behavior, 51:206, 216
Energy-rich bonds, 7:61ff
Energy sensitivity, 51:198
Energy sources, prebiotic synthesis, 55:89–91
Energy spectrum, discrete, 36:43
Energy storage, 47:Part 1:9–11
  excitation amplitudes, 47:Part 1:10–11
  initial condition, 47:Part 1:11
  molecular level structure, 47:Part 1:10
Energy surface, 32:195
  extended, 20:93, 95, 116, 119
Energy transfer, 7:567ff, 572; 28:359, 372; 36:178, 185; 42:354; 45:152–161; 52:492
  CD_3H, 47:Part 2:190–192
  CD_4, 47:Part 2:192–193
  CH_2D_2, 47:Part 2:193–198
  CH_3F, 47:Part 2:198–204
    highly excited, 47:Part 2:205–206
  CH_4, 47:Part 2:187–190
  coherent, molecular crystals, 40:369–485
  collisional, 28:33, 337, 343, 438, 457; 45:343–394
    electronic, 28:1, 4, 5; see also Electronic deexcitation

rotational—translational, **28**:185–193, 236–243

vibrational, **28**:428–465

vibrational—rotational, **28**:200–205, 247–256

methods of measurement, aerodynamic methods, **28**:213

condenser transducer method, **28**:209, 210

fluorescence, **28**:216

optical acoustic effect, **28**:216

optical diffraction methods, **28**:209

progressive-wave technique, **28**:211

Raman scattering, **28**:216

transport phenomena, **28**:221

vibrational—translational, **28**:175–185, 219ff

vibrational—vibrational, **28**:193–200, 218–221, 243–247

cross section, **28**:259, 280, 360, 363

dipole-dipole, **16**:94–95

E-V-R, **45**:343–346, 350–351

electronic, **36**:178

electronic excitation, **45**:152–157

electronic—vibrational, **47**:Part 2:381–418; *see also* Electronic—vibrational energy transfer

excited state, **41**:407–469

Forster, **41**:313

Hg* + CO, **45**:344

intermolecular, **41**:311

mechanism, **48**:157

methane, deuteromethanes, and methyl fluoride, **47**:Part 2:208–227

Na($3^2P_{3/2}$), **45**:368–373

Na* + CO_2, **45**:378

Na* + N_2O, **45**:378

near interfaces, **37**:1–66

rate constant, **28**:379–465

resonant, molecular crystal, **53**:277–287

rotational, **36**:178

vibration—rotation, **36**:198

vibrational, **41**:314

vibrational—translational, **50**:108, 111

see also Excitation transfer; Rotational energy transfer; Vibrational energy transfer

Energy transport, **27**:369, 412, 432; **52**:399, 468, 474

Energy trapping by molecular crystal, **53**:285–286

Energy variance functional, **36**:152

Enhancement factor, **35**:9

Enniatin B, **29**:310

Enolic form, **7**:104

Ensembles, stellar systems, **26**:112, 114, 134, 141, 142

canonical, **26**:112, 134

grand canonical, **26**:112, 134

microcanonical, **26**:113, 115, 134–142

Enskog theory, **24**:328, 330; **34**:35, 148; **37**:185, 186

'additional-to-Enskog' contribution, **31**:207

Brinser-Condiff generalization, **31**:219

contributions to effective potential theory, **31**:207

nonspherical molecules, **31**:179

rough spheres, **31**:168, 169

spherical molecules, **31**:165, 167, 179

square-well potential, **31**:158, 179, 181

Entatic state, **39**:55, 57, 58, 91, 92, 338

Enthalpy, 1:12; **17**:26, 27
  excess, **17**:26, 27
  one-dimensional plasma, **4**:209, 210, 212
Enthalpy of activation, 33:99, 100, 117ff, 129ff
Enthalpy of formation, of base pairs, **33**:100
Enthalpy of mixing of salts, 11:106
Enthalpy of reaction, 33:129, 130
Enthalpy of vaporization of salts, 11:90
Entrance reaction, 43:90, 129, 161
Entropy, 1:87, 115, 207, 228, 279; **5**:285, 286–287; **17**:27, 118; **19**:212, 252; **24**:96, 98, 103; **27**:485–487, 525; **38**:6, 91, 101, 224; **41**:90; **48**:456, 467, 468, 502
  average, **19**:253, 267
  chain, **46**:365, 368, 379, 396, 397
  clathrate, **2**:25
  combinatorial, **16**:274–276
  communal, **6**:238, 249; **48**:481, 485, 488, 505
    cluster distribution function and, **48**:482
    definition of, **48**:480
    free energy equation, **48**:471
    free volume equation, **48**:493
    heat capacity and, **48**:498, 502
    order of phase transition and, **48**:490, 492
    vanishing of, **48**:498
  configurational, **16**:232, 248; **48**:467, 473, 492, 498
  crisis of, **48**:465, 502
  crystalline, **48**:465, 498
  dissipative processes, **15**:28
  effective state, **53**:347, 353–356
  equilibrium, **53**:355
  excess, **2**:130, 131; **16**:247, 252; **19**:303; **48**:468

  production of, **19**:221, 228, 234, 245, 266
  of solution, **2**:133, 141
  generalizations of concept of, **51**:113–173
  hindered rotation and, **2**:369
  increase of, **53**:347, 362, 366, 371
  influence of internal rotation, **2**:368
  information theory, experimental resolution and, **47**:Part 1:287–288
    maximum, **47**:Part 1:250–254
  maximum, **53**:365, 369, 371, 373
  multicomponent plasma, **4**:224
  null spectrum related to conservation of, **32**:227
  one-dimensional plasma, **4**:210
  per unit mass, **33**:409
  in presence of convection, **32**:215
  residual, **16**:275; **48**:458, 469
  secular spectrum related to nonconservation of, **32**:229
  solid solutions, **2**:130
  stellar systems, **26**:108, 127, 130, 139, 140, 142; **32**:210, 212, 214
  subadditivity, **51**:137–139
  surface, **48**:483, 484
  vanishing excess of, **48**:465
    temperature, **48**:467
  vibrational, of dilute solutions, **2**:133
Entropy density, 49:413
Entropy equation, 32:215
Entropy flow, 19:213; **29**:12, 73
Entropy function, 49:396
Entropy inequality, 33:415
Entropy-like function, 51:122, 128, 130, 157
Entropy minimum, 32:211
Entropy of activation, 32:8; **33**:118, 129, 130
Entropy of adsorption, 7:239

Entropy of mixing, 2:130; **48**:467, 490, 498, 502
  abnormal, **6**:469
Entropy of reaction, **33**:129, 130
Entropy of transfer in thermal diffusion, **13**:365
Entropy production, **32**:167, 171; **38**:234, 363; **43**:95, 100
  balance, **38**:363
  critical region, **6**:199
  evolutionary feedback and, **29**:19
  excess, **29**:17, 74, 133; **43**:100
  generalized, **13**:269; **29**:337, 338
  Lyapounov functions and, **29**:14, 17
  microscopic definition of, **29**:13
  principle of minimum, **3**:293; **29**:73; **32**:21, 176, 181; **38**:137, 138
    continuous systems, **3**:298
    discontinuous systems, **3**:293
    ideal gases, **3**:307
    magnetic resonance, **3**:311
    nonlinear irreversible processes, **3**:304
    statistical theory, **3**:307
  rate, **46**:260
  steady state, **29**:17, 24
  in thermal diffusion, **13**:349, 352
Entropy production function, **33**:410
Entropy production inequality, **33**:410, 412
Entropy representation, **43**:30
Entry point, **43**:37
Enzyme action, **7**:645ff
Enzyme hydrolysis, **7**:74, 76, 665, 675
Enzyme reaction, allosteric, **38**:364, 365
Enzyme-substrate complex, **39**:184
  rate of formation, **39**:185
Enzyme systems, **7**:607; **43**:211
  magnetic properties, **7**:607ff
Enzymes, **7**:262; **29**:50, 67, 113, 270, 304, 345; **46**:378, 380
  activity loss, **7**:211, 212
  allosteric, **29**:34, 145, 278
  cleft, **39**:91, 170, 171, 172
  control by covalent modification, **39**:194
  control by environment, **39**:177–228
  control by fluctuations, **39**:222, 223
  control models, **39**:193
  covalent intermediate, **39**:186
  immobilized, **39**:36, 46
  mechanism, **39**:339
  membrane-bound, **39**:203–208
  membranes and, **29**:113, 115, 117, 119, 138
  membranes bearing, **39**:229–274
  metals in, **39**:99
  multienzyme complexes, **39**:199–202, 208, 213
  multiple polypeptide chains, **39**:178, 190–199, 208
  oscillatory, **39**:29, 31
  pH effect, **39**:6, 229–233
  polymerization-depolymerization, **39**:194, 198
  radicals, EPR spectrum, **7**:619
  reaction rate, **7**:211ff
  single polypeptide chain, **39**:178, 182–190, 208
  substrate distortion, **39**:92, 103, 186
  synthesis, **19**:345
  synzyme, **39**:109–176
  turnover number, **39**:186
  see also Allosteric; Membrane-bound enzyme; Synzyme
Epimorphosis, **29**:257
Epinephrine, **7**:518ff
  relaxation times, **7**:520ff
Epinephrine-ATP complex, **7**:524
Epitaxy, **20**:274, 310, 325, 326, 329–

333

EPR, *see* Electron paramagnetic resonance

Epstein profile, **27**:562
p-polarized reflectivity, **27**:570
s-polarized reflectivity, **27**:566, 569

Equation of conservation, **29**:33

Equation of constraint, **33**:172ff, 190, 200ff
strong, **33**:172, 175, 179
weak, **33**:172, 174, 189, 216

Equation of continuity, **29**:177; **35**:32, 56, 58
configuration space, **35**:68

Equation of state, **1**:160, 267, 281, 296, 298, 303; **11**:88; **16**:224; **17**:32, 33, 34; **40**:302
argon, **6**:277
caloric, **16**:233, 245
compressibility, **37**:172
energy, **37**:172, 175
gases, **5**:164
hard sphere fluid, **6**:229–289; **20**:39; *see also* Hard sphere fluid
high-temperature, **6**:268, 272, 275–276
liquid ^{4}He, **33**:10
mixture of hard spheres, **6**:268
one-dimensional hard sphere fluid, **6**:256
one-dimensional plasma, **4**:202, 210, 212
pressure, **37**:172
reduced, **11**:88; **16**:225, 228
solid ^{4}He, **33**:11
surface films, **6**:167–168
thermal, **16**:225, 231, 233, 245
thermodynamic, **11**:90
three-dimensional hard sphere fluid, **6**:263
two-dimensional hard sphere fluid, **6**:263
van der Waals, **17**:33
virial expansion, **17**:34

Equations of change, **35**:55
equation of continuity, **35**:57
equation of motion, **35**:58, 65
general, **35**:56
phenomenological, **13**:267

Equations of motion, **44**:278, 402
atomic lattice, **53**:297, 304, 305
center of mass of macromolecule, **35**:69
clamped chain, **53**:164
correlation function, **53**:142
degrees of freedom of macromolecule, **35**:69
effective, **44**:147
Eulerian, **44**:359
Hamiltonian, **33**:181; **53**:281, 302
heatbath modeling, **53**:204
heatbath response function, **53**:152
interacting reagents, **53**:179
iodine recombination, **53**:207–210
Liouville equation, **53**:129–131
molecular-time-scale generalized Langevin equation, **53**:116
Mori, **53**:64, 144, 147
periodic vibrations, **53**:311
recombination dynamics, **53**:106–107
soliton, **53**:270–277
spin, **53**:231
stellar systems, **26**:110, 115, 128, 129, 133, 135, 142

Equations of motion method, **48**:2–66
calculations, **48**:31–61
comparison with configuration interaction, **48**:49
electron affinity, fluorine, **48**:53
hydroxyl, **48**:53

ionization potential, boron hydride, **48**:52
hydrogen fluoride, **48**:53
neon, **48**:53
nitrogen, **48**:31
shake-up energies of nitrogen, **48**:42
derivation of equations, **48**:8–15
operators, **48**:10–22
annihilation condition, **48**:11
completeness, **48**:15, 17
linear independence, **48**:16–18
many-body, **48**:18
orthogonalization of, **48**:25
transition matrix elements, **48**:16
Equilibrium, 40:137ff
approach to, **33**:431, 434; **38**:133, 134
Brownian particle distribution function, **16**:43–44
clathrate, **2**:18
coordinate, **40**:239, 240, 241, 245
high pressure fluids, **17**:2, 12
gas-gas, **17**:2
liquid-gas, **17**:2
liquid-liquid, **17**:2, 5, 6, 12
local, **16**:124; **19**:212
local chemical, **16**:107, 122, 125, 128
mechanical, **16**:122; **33**:425, 426
metastable, **48**:506
momentum space, **35**:71
stability, **19**:228, 233
stellar systems, **26**:108, 113, 115, 124, 130ff
metastable, **26**:113, 114, 134, 141
Equilibrium average, 44:346
Equilibrium beam method, 10:207–208
Equilibrium complex, 43:33–35
Equilibrium concentration, 33:437
Equilibrium constant, 1:19, 53; **8**:65ff;

16110, 128
base pairing of oligonucleotides, **33**:100
formation of interruption in sequence, **33**:90
growth process, **33**:81
influence of internal rotation on, **2**:368
Equilibrium correlation function, 11:178, 182, 195
Equilibrium distance, 41:185
Equilibrium distribution, constrained, **40**:139, 143, 144
Equilibrium extreme current, 43:34
Equilibrium flux vector, 43:31
Equilibrium Hamiltonian, 44:339
Equilibrium hypothesis in chemical kinetics, 1:119, 124, 129, 132, 362, 363
Equilibrium mixture of spin states, 7:365
Equilibrium probability distribution, 44:280, 300
Equilibrium process, 33:427
Equilibrium properties of electrolytes, 11:159, 187
Equilibrium rate constant, 13:93; **5**:379, 382ff
Equilibrium shape, 49:116
Equilibrium state, electron distribution, **7**:147
far from, **38**:317; **38**:317–362
properties of, **33**:426
Equilibrium structure factor, 44:438
Equilibrium surface, 6:171
Equipartition theorem, 6:232; **35**:181; **48**:342, 343
generalized, **53**:171–172
Equivalence restriction, 9:358
Equivalent Hamiltonian, 14:390
Equivalent harmonic chain representa-

tion, **44**:179

Equivalent operator approach, 14:367ff, 390

Equivalent orbital transformation, 6:387
ethane, **6**:388
methane, **6**:388
water, **6**:388

Eremolactone, 16:187

Ergodic behavior, 36:40; **42**:301

Ergodic hypothesis, 31:264

Ergodic limit, 53:366

Ergodic theory, 26:112, 113, 138, 140; **38**:7
few-atom systems, **40**:95, 125

Ergodicity, 19:330; **22**:93–95, 105; **24**:156, 157; **51**:135

Erofeev equation, 15:180

Error analysis, 42:296

ESCA, 23:85ff; **27**:28
group shifts, **23**:139
multiplet splittings, **23**:148, 151, 155
Pauli charges, **23**:135
shifts, **23**:133
thermochemical model, **23**:126

Escherichia coli, **7**:587
photoprotection, **7**:587
ribosomes, **7**:462ff

ESR, *see* Electron paramagnetic resonance

Esterases, 7:74

Eta theorem, 51:116, 146–157, 160, 161
and mixing distance, **51**:147–148

Ethane, barrier of internal rotation, **2**:382, 384, 391
+ *p*-bromochlorobenzene system, **2**:101
clathrates, **2**:2
+ *p*-dichlorobenzene system, **2**:100, 101
+ hexachloroethane system, **2**:100

hydrate, **2**:32, 33, 34
thermodynamic data and lattice constants, **2**:43
infrared spectroscopy, **2**:373
+ *p*-iodochlorobenzene, **2**:101
+ 1, 3, 5-trichlorobenzene system, **2**:101

Ethyl chloride, barrier height of internal rotation, **2**:383
microwave spectroscopy of, **2**:380

Ethyl ether, + anthraquinone system, **2**:94
triple point, critical point, **2**:94

Ethyl radical, EPR spectrum, **7**:538

Ethylene, 1:50, 252, 254; **2**:102
catalytic hydrogenation of, **8**:141ff
chemisorption of, **8**:141ff
clathrate, **2**:41
clusters, **52**:350–352
crystal constants, **5**:171, 172, 179
electron energies, **6**:386
electronic transitions in, **8**:5ff
hydrate, **2**:32, 33
thermodynamic data and lattice constants, **2**:8
interatomic distances, **5**:156
+ naphthalene system, **2**:102, 103
solid, **22**:209
potential function for, **22**:261

Ethylene oxide, 2:155

Ethylenic linkage, 7:20

Ethylidene, isomerization, **42**:143

Eucarvone, 8:134, 136

Euler angles, 4:259; **13**:248; **35**:42, 47, 97; **44**:13, 447; **49**:123–125, 133, 166, 181
integrations over, **35**:47

Euler equation, 25:131; **31**:72; **33**:426, 441; **41**:86–87

Euler integral, 25:232

Euler-Lagrange equation, 13:268;

19:346; 33:170, 171, 173, 175, 188; 41:220
Euler-MacLaurin formula, 4:256, 276
Eutectic, 51:267
Eutectic line, 2:88
Eutectic point, 2:87
Evaporation, of crystal lattice planes, 40:213
    stellar, 35:154, 183
Evaporation rate, 40:140, 144
    during crystal growth, 40:167
Event, 38:57, 61, 66
    definition, 38:57
    directly perceptible, 38:59, 61, 66
    elementary, 38:57
    future, 38:66
Evolution, 29:44, 125, 256
    autocatalysis and, 29:20, 26
    chemical, 43:28
    convergent, 29:214
    criterion, 29:74, 336
        and kinetic potential, 29:79, 337
    divergent, 29:214
    ion binding of cytochrome c and, 29:240
    molecular, 29:126, 200; 43:163, 212
    parallel, 29:214
    period, 29:199, 223, 243
    prebiotic, 29:9, 44
    rate of change of cytochrome c codons and, 29:223, 229
    selection and, 29:208; 38:229, 247
        birth rate, 38:246
        death rate, 38:246
        principle of descendence, 38:247
    of species, 29:191
    stellar systems, 26:113, 128, 129, 132, 133, 134, 141–143
    through error copies, 29:44, 60
Evolution equation, 40:244

Evolution experiment, 38:233
Evolution of life, molecular level, 55:119–137
    biological complexity, 55:119–120
    by natural law, 55:128–131
    in nature, 55:131–136
    reproduction of Qβ, an RNA virus, 55:120–128
Evolution operator, 36:49
Evolutionary feedback, 29:9, 19, 44
Evolutionary stability of stellar systems, 32:207ff
Evolutionary tree, 26:162
Ewald sphere, 20:230
Exact enumeration method, 51:2, 22, 39–42
Exact self-consistent field theory, 9:340; 14:309
Excess compressibility, 11:126
Excess enthalpy, 11:126
Excess entropy, 11:126; 16:247, 252, 275
    production of, 43:100
Excess free energy, 11:126; 16:229, 247, 274–276
    of mixing, binary systems, 11:105
        Gibbs, 11:106
        reciprocal systems, 11:112
    of solution, 2:135
Excess functions, 11:118, 120, 121, 125; 16:228–229, 248–249, 253, 275–276, 278–279
    Gibbs, 16:274–277, 279
    principle of corresponding states for, 16:228, 245, 248–252, 270–276
Excess properties, 11:126
Excess specific heat, 11:126
Excess volume, 11:126
Exchange, nonlocal, energy-dependent, 41:182

Exchange amplitude, 36:418

Exchange-correlation energy, 41:102

Exchange-correlation hole, 41:101

Exchange-correlation potential function, 37:287

Exchange energy, 1:166, 167; 2:209; 4:216 33:352, 354, 356, 358; 41:86
   charge transfer states, 8:15ff
   H_2, 12:65, 66, 96
   Heitler-London, 12:45, 66
   homogeneous spin-polarized electron gas, 54:279-282
   parameterized, 35:16

Exchange equilibrium, 1:20, 51

Exchange force, 1:330; 12:339

Exchange hole, 41:101

Exchange integral, 1:244, 256; 2:384, 385; 5:220, 222, 236; 36:145, 159, 161, 162
   multiple, 36:143, 162

Exchange interaction, 35:12

Exchange limit, 40:421

Exchange narrowing, 40:399, 462, 471

Exchange of stability, 40:240

Exchange oscillation, 30:329, 479

Exchange polarization, 2:313
   excitation, 2:199

Exchange potential, $X\alpha$ method, 41:64-66

Exchange rate, 40:422

Exchange reaction, 1:125

Exchange repulsion energy, 33:325, 356, 379

Exchange scattering, 18:57, 60, 67, 70, 80, 86
   electrons, 18:15, 17, 19, 25, 26, 30, 42, 43, 46, 69
   singlet-triplet transitions, 18:20, 21, 67

Exchange terms, second order, 12:345

Exchange theory, 40:418, 424, 425

Excimers, 39:135, 140, 170
   in solids, vibrational relaxation of, 42:331

Exciplex, 35:24

Exciplex emission, 50:368

Exciplex retardation effect, 35:24

Excitability, adenylate cyclase system, 38:392
   phosphofructokinase model, 38:377

Excitation, broad-band, 40:379-383
   charge transfer, 28:394, 412
   chemical reaction, 28:3-96, 388, 412
   collinear, 36:67
   electron impact, 28:388, 394, 402ff, 427, 431, 432; 45:5-7, 8, 17, 26, 29-31
   $H^+ + H_2$, 45:159
   membrane, 29:43, 319, 343, 349
      dioleylphosphate, 29:325, 334, 341
      inactivation, 29:363
      lipid derivatives, 29:341
      sodium activation, 29:361
      squid giant axon, 29:320
      temperature dependence of, 29:322, 323
   narrow-band, 40:375-379, 394, 398
   photoselective, 47:Part 1:2-9
      biological systems, 47:Part 1:9
      coherence effects, 47:Part 1:4
      collisional effects, 47:Part 1:6-7
      high power, 47:Part 1:3
      low pressures, 47:Part 1:4-5
      matrices and mixed crystals, 47:Part 1:7-9
      solution, 47:Part 1:9
      spectral range, 47:Part 1:3
      supersonic beams, 47:Part 1:5-6
      thermal beams, 47:Part 1:5
      tunability, 47:Part 1:3

ultrashort duration, **47**:Part 1:4

rotational, **45**:157–161; *See also* Rotational excitation

simultaneous vibrational and rotational, **36**:67

vibrational, in collisional dissociation, **45**:144, 157–161

    in ion-molecule reactions, **45**:184–196

see also Vibrational excitation

vibrational and rotational, **45**:157–161

Excitation amplitude, **47**:Part 1:10–11

Excitation band, **8**:30

Excitation energy, **7**:112ff, 129, **36**:208, 395

calculation by equations of motion method, **48**:1–70

multiconfigurational time-dependent Hartree-Fock approximation, **54**:30–32

Excitation gas, **33**:2, 4, 5, 21, 22

Excitation jumping, **52**:471

Excitation transfer systems, **45**:503–505, 509–510, 539–540

cross section, **45**:573–574

elastic scattering, **45**:519–523

product angular distributions, **45**:574–579

Excited atoms, **10**:32, 171–193

charge transfer, **10**:196, 225–234

collisional quenching of, **50**:325–394

detection of beams of, **10**:174–175

elastic collisions of, **10**:179–187

lifetimes, **10**:175–179

polarizability of, **10**:20–25

production of beams of, **10**:69, 172–174

Excited ions, association reactions, **45**:127

    benzene dimer ion formation, **45**:127

charge transfer, **45**:120–123

    N^+ - Kr, **45**:121

charge transfer, **45**:128–131

collisional deactivation, **45**:147–151

collisional dissociation, **45**:135–144

    $CH_2Br_2^+$, **45**:143

    $H_2^+ + He$, **45**:143–144

    $N_2^+ + N_2$, **45**:140–141

    $NO^+ + Ar$, **45**:141–142

    reactant ion vibrational excitation, **45**:142–143

    vibrational excitation, **45**:144

electronic excitation, **45**:152–161

endoergic reactions, **45**:128–144

exoergic reactions, **45**:120–127

heavy particle transfer, **45**:123–127

    $C^+ - O_2$, **45**:124

    $CH_4^+ - C_2H_2$, **45**:124

    $C_2H_4^+ - C_2H_4$, **45**:124

    $F^+ - H_2$, **45**:123

    $N^+ - H_2$, **45**:123

    $NH_3^+ - H_2O$, **45**:126

ion impact excitation, **45**:152–157

    $N_2^+ - Ar$, **45**:155

    oscillatory structure, **45**:153–154

lifetime, **45**:86–89, 106–107

    N_2^+, **45**:106–107

nonreactive scattering, **45**:145–161

reactive scattering, **45**:120–144

rotational excitation, **45**:157–161

vibrational excitation, **45**:157–161

Excited neutrals, reactions with ions, **45**:108, 161–163; *see also* Ions, reactions with excited neutrals

Excited state potential surfaces, calculation of, **50**:517–589

Excited state proton transfer, **47**:Part 2:645–650

Excited state radical, **7**:378

Excited states, 2:263; **41**:62, 165; **45**:2–585

atmospheric systems, **45**:227–340

dipole moment of, **13**:194

dynamics of, **50**:1–642

ion-neutral reactions, **45**:83–124

molecular crystal, **53**:278–281

polarizability of, **13**:195

potential energy surfaces, **42**:169–206

radiationless transitions, **53**:363

small molecules, **50**:1–104

Exciton-exciton annihilation, **40**:439

Exciton-phonon interaction, **17**:253; **40**:433ff, 450, 452, 469ff; **41**:326, 328

quadratic, **40**:443

Exciton surface polaritons, **54**:437–455

anisotropic active medium, **54**:442–447

anthracene crystal, **54**:451–455

damped active medium, **54**:441–442

experimental excitation and detection, **54**:448–451

grating coupling, **54**:448–449

nonplanar interface method, **54**:448

particle beams, **54**:451

prism coupling, **54**:449–450

Fano and Brewster modes, **54**:441

isotropic and loss-free semi-infinite active medium, **54**:438–441

thin active medium, **54**:447–448

see also Polaritons; Surface excitons

Excitons, **17**:251, 252, 257, 258; **23**:265; **27**:463–465; **40**:369ff; **41**:171; **44**:565, 569, 581, 588, 616, 632, 635; **52**:467

anthracene, **23**:309–327

bulk, **54**:305–326

classical dipole theory, **23**:247

coherence, **40**:428, 438, 454, 457, 463, 465

collective, **23**:252

damping, **40**:432

density operator description, **41**:321–476

diffusion length, **40**:448

dilute isotopic mixed crystals, **23**:293–302

dilute mixed crystals, **23**:239–242

dimer, **41**:350–356; **44**:569, 575, 594

dipole calculations, **23**:322

double chain, **41**:402–407

dye crystals, **23**:330

electronic, **23**:230

Frenkel, **23**:230

heavily doped crystals, **23**:244

helical polymers, **23**:232, 265

helical selection rules, **44**:592, 593

interstitial impurities, **23**:266, 288

MO calculations, **23**:321

molecular crystals, **23**:230

multipole calculations, **23**:320

naphthalene and benzene, **23**:329

one-dimensional, **40**:411, 412, 414, 417, 419

ordinary chain, **41**:383–407

percolation, **40**:455

polymer, **23**:329; **44**:575, 588, 591, 599, 613

propagation of, **52**:469

pure crystals, **23**:247, 251, 261

quantized theory, **23**:281

quenching of, **52**:468, 470

random phase approximation, **44**:612, 614

scattering, **13**:125

scattering by defects, **40**:439, 444, 445

second quantized, **44**:616

selection rules, **23**:231

semi-infinite crystals, **23**:272

singlet, **40**:370, 392, 437, 440, 445ff, 456, 459, 471ff

states, **17**:230

strong coupling, **44**:571, 572, 574, 589

substitutional impurities, **23**:270, 289

tetracene and pentacene, **23**:327–329

transport of, **52**:470

traps, **40**:452

supertraps, **40**:455, 456

triplet, **40**:370, 392, 398, 411, 436ff, 445ff, 465ff, 474ff

vibrational, **23**:232

Wannier, **23**:230

wave packet, **40**:372

weak coupling, **44**:570, 572

see also Frenkel excitons; Surface excitons

Excluded site, 11:15, 20, 26

Excluded volume, 24:227, 228; **31**:166, 181

dilute polymer solution, **51**:1–44

effects of, **48**:81

Excluded volume ratio, 51:25

Exclusion principle, spin orbitals, **6**:337, 356

terms which violate, **14**:138ff, 144

Exit point, 43:37

Exit reaction, 43:90, 129

Exoergic reaction, 45:120–127

Exothermic pair potential, 53:310

Exothermic reaction, 29:59; **47**:Part2:4

Exp-six potential, 10:77, 105, 136–138, 144–147, 162, 165–168, 334–337

differential cross section for, **10**:163–164

Expansion, Barnett-Coulson, **13**:262

bipolar spherical of molecular orbitals, **13**:207

ellipsoidal, molecular orbitals, **13**:207

two-center integrals, **13**:231

Green's function, **13**:213, 236, 238

Neumann, **13**:214, 220, 231

one-center for multi-center integrals, **13**:207, 236

spherical harmonic, **13**:236

spherical wave, **13**:214

Expansion coefficient for polymers, 1:97–99, 101, 116

Expectation functional, 38:69

Experimental error, 51:180, 221

Experimental observable, 47:Part 1:13–16

coherent optical effects, **47**:Part 1:15–16

energy-resolved, **47**:Part 1:14–15

time-resolved, **47**:Part 1:13–14

Explosion, 40:138; **43**:161

stoichiometric, **43**:156–166

Explosion-extinction dynamics, 43:165

Explosives, electronic energy states, **21**:291

homogeneous granular, **21**:781

liquids, **21**:784

surface burning model, **21**:781

Exponent polytope method, 43:166–206, 210

face structure, **43**:187–193

vertices, **43**:192

Exponential approximation, 48:226

Born-distorted wave, **36**:118

Exponential decay, 4:15

Exponential departure model, 35:16, 19

Exponential gap law, 50:604

Exponential memory function, 17:169, 173, 175, 178, 180

Exponential method, 30:14

Exponential perturbation theory, **51**:186, 214

Exponential unitary operator, MCSCF optimization, **54**:6–10
multiconfigurational Hartree-Fock state, **54**:6–8
variations in total energy, **54**:8–10

Exponential wave operator, **52**:187–189

Exponomial, **43**:180

Extended coupled-pair many-electron theory, **52**:208–210

Extended diffusion model, **44**:256, 346, 353

Extended electronic state, **31**:445

Extended Hückel theory, **42**:177; **49**:75
basis of, **21**:27ff

Extended X-ray absorption fine structure (EXAFS), **49**:643; **51**:232

Extended zone scheme, **49**:551

Extensive region, **46**:219

Extensive variable, **29**:86

Extensivity ansatz, **46**:231

Extensivity parameter, **43**:225, 232, 238

Extent of reaction, **43**:19
polyhedron, **43**:25–27

Exterior derivative, **19**:350

Exterior product, **1**:356; **19**:350

External degrees of freedom, **16**:227–232, 234–235, 246, 278

External process, **44**:381

External representation, **43**:19

External species, **43**:16, 40

Extinction, **43**:156–166

Extinction coefficient, **5**:243

Extinction cross section, **27**:441, 526

Extinction theorem, **1**:310

Extrema effect, **10**:120–126

Extreme complex balanced network, **43**:80

Extreme current, **43**:21, 45
algorithm, **43**:24

Extreme network, **43**:14, 53

Eyring and Polanyi theory, **7**:78

Eyring equation, **1**:162

Eyring-Hirschfelder free volume, **1**:162

Eyring's formula of energy changes, **7**:208ff

F

F-center laser, **49**:2, 26

F-center model, **31**:472–475
absorption spectra, **31**:474, 476
excess enthalpy, **31**:472
excess entropy, **31**:472
excess partial molar free energy, **31**:473
Landé g factor, **31**:476
localized state, **31**:479
molecular orbital, **31**:475
phase diagram, **31**:473
regular solution theory, **31**:472
vibrational motion, **31**:476

f-number, *see* Oscillator strength

F polarization, **35**:22

Fabry-Perot interferometer, **24**:26, 35
finesse, **24**:26
free spectral range, **24**:26
resolving power, **24**:26
tandem, **24**:3, 27

Face-centered cubic lattice, **4**:188, 190; **9**:143; **49**:645–647

Faceting, **20**:303

Faddeev-Watson equations, **30**:11, 28, 62, 68, 167

Fading memory, **24**:108–122, 148–152

Fadley 'flower patterns', **49**:644

Falling kinks, **41**:216

Fano approach, **25**:292

Far-from-equilibrium states, **38**:317–

362

Far infrared absorption, 44:263, 275, 323, 350, 351, 397
  coefficient, **44**:284, 323
  liquid crystals, **44**:388
  pressure induced, **44**:435
Far infrared dielectric loss profile, 44:258, 267, 293, 301
  PCNB, **44**:393
  TCTMB, **44**:396
Far infrared γ resonance, 44:293, 399
Far infrared Poley absorption, 44:359
Far infrared spectroscopy, 42:258, 359
Far ultraviolet radiation, 7:584
Faraday cup, 20:276
Faraday effect, 12:119; **35**:197
Fast beam sources, 42:532
Fast exchange, 40:422
Fast flow system, 42:121
Fast particles, detection of, **10**:206–212
Fast passage, 25:50
Fast phonon-slow exciton limit, 40:444
Fast variable hypothesis, 44:426
Fatty acid, 41:196
Feedback, evolutionary, **29**:9, 19, 44
  negative, **29**:125, 303, 345
  positive, **29**:43, 125, 151
Feedback loops, **29**:49; **38**:322, 324, 332, 334, 349; **55**:247–282
  asynchronous logical description and treatment, **55**:252–266
    conditions of various sequences, **55**:259–264
    equations, **55**:255–256
    final states, **55**:264–266
    graphs and sequences of states, **55**:258–259
    involvement of time, **55**:252–253
    state tables, **55**:256–257
    variables and functions used,

55:253–255
  regulation in biological systems, **55**:248–252
    continuous and discontinuous descriptions, **55**:249–250
    homeostatic or epigenetic, **55**:248
    types of systems treated, **55**:250
  relations with other descriptions, **55**:267–278
    discrete and continuous, **55**:267–270
    methodology applied to chemical systems, **55**:272–274
    numerical and logical iterations, **55**:270–272
    synthetic (inductive) mode, **55**:274–278
Felderhof rederivation of Wertheim-Fröhlich expression, 48:279
Fermi-contact term, 37:241, 260–261
Fermi correlation, 6:325
  three-electron, **6**:356
Fermi-Dirac distribution, 46:331
Fermi golden rule, 48:43; **50**:115
Fermi hole, 2:141, 217, 218, 232, 247; **6**:353; **9**:331
Fermi interaction, 5:203
Fermi mixing, 50:127
Fermi pseudopotential, 20:149
Fermi resonance, 36:22, 45; **41**:294; **42**:222, 229, 244; **53**:319
Fermi surface, 16:378
  copper, **41**:192
Fermi systems, friction coefficient in, **16**:26–30, 46–49
  static mobility in, **16**:32–33
Fermi-Thomas approximation, 36:53, 55, 56
Fermi velocity, 27:379
Fermion gas, 6:115
Fermium, Breit and transverse interac-

tion energies, **54**:272

correlation energy, **54**:272

self-consistent field atomic calculations, **54**:265–273

self-consistent field result compared with perturbation theory, **54**:272–273

Ferric chloride (FeCl₃), 49:459, 464–466, 477, 481

Ferric dimethylglyoximate, 7:372

Ferricytochrome, 7:72

Ferricytochrome *c*, 33:108

Ferriheme, 7:190

Ferrihemoglobin, 7:169ff

Ferrihemoglobin azide, 7:172

Ferrihemoglobin complexes, 7:185

Ferrihemoglobin cyanide, 7:196

Ferrihemoglobin fluoride, 7:174, 189

Ferrihemoproteins, reactions with peroxides, **7**:612

Ferrimyoglobin complexes, 7:185

Ferrocene, 16:173

Ferrocytochrome, 7:72

Ferrodistortive material, 53:265

Ferroelectric compound, structural phase transition, **53**:265

Ferroelectricity, 7:85

Ferrohemoglobin, 7:169ff

electronic structure of iron in, **7**:177ff

Ferromagnet, 51:162

easy-plane, **53**:264

isotropic, **53**:264

kinks and, **53**:251

near critical point, **51**:171–173

optical spectra, **53**:261–262

solitary waves, **53**:237ff

spin Hamiltonian, **53**:229ff

spin vectors, **53**:231

spin waves, **53**:263

uniaxial, **53**:263–264

Ferromagnetic glass, 51:241

Ferromagnetic state, 46:314, 320

Ferromagnetic transition, 40:360

Ferromagnetism, 2:313; **20**:78, 89, 91; **34**:297; **40**:231; **51**:253

Ising model of, **40**:253–272

Ferroporphyrin, 7:72ff

Ferrous (low spin) porphyrins, 7:390

Ferroverdin, 7:378

Feshbach projection, 42:173

Feshbach resonance, 18:92; **36**:435

Feshbach theory, 18:104, 115, 116

Feshbach-type resonance, 49:200, 230–233

Feynman diagrams, 14:115, 121ff, 134ff, 139ff, 168, 172, 179ff, 373; **36**:224; **53**:126

kinetic equations, **15**:21ff

Feynman-Goldstone diagrammatic construction, of extended coupled-pair many-electron theory, **52**:205–226

Feynman identity, 36:90

Feynman path integral formalism, 28:369

polymer statistics, **22**:1, 5, 6, 32, 35, 55, 109, 110

Feynman's theorem, 20:149

Fibrinogen, 26:158

Fibrous proteins, 7:262, 263ff

Fick's diffusion equation, 4:2, 29, 42, 44

Fick's law, 13:352

Fictitious spin, 5:205

Field emission, 27:225, 226

energy distribution, **27**:225, 227

in presence of adsorbates, **27**:233

experimental aspects of, **27**:237

transfer Hamiltonian, **27**:231

Field emission distribution, 27:246

barium on tungsten, **27**:246

carbon monoxide on tungsten,

27:250
hydrogen on tungsten, **27**:248
Mo(110), **27**:247
Field emission microscope, 27:237
resolution of, **27**:238
Field gradient, 12:123, 125
Field ion emission, 27:326
Field ion microscopy, 27:28
Field momentum, 33:192
Field point formulation, 48:164–167
Field potential, 33:163, 168, 188, 216
Filters, balanced, **34**:162
Final state interactions, in photodissociation, **50**:303
Fine-grained mixing character, 51:128
Fine structure, intermediate coupling, **12**:43
J-J coupling, **12**:43
Λ-S coupling, **12**:43
resonance interactions, **12**:45
rotational energy levels, **12**:75, 91
Fine structure component, 30:433, 444; **42**:26, 198
Fine structure constant, 33:204, 208, 209, 216
Fine structure correlation, 42:97
Fine structure Hamiltonian, 18:173, 202
matrix elements, **18**:193, 202
Fine structure splitting, 42:28
alkali atoms, **28**:269
Fine structure tensor elements, 40:409
Fine structure transition, 25:158; **42**:199
Finesse of etalon, 50:12
Finite deep-channel case, 49:220
Finite difference methods, 19:9; **25**:281
Finite difference operator, 13:125
Finite-element methods, 32:40, 42, 50, 311
bicubic Hermite space, **32**:48
cubic splines, **32**:45

piecewise Hermite spaces, **32**:43
piecewise linear basis, **32**:43
spline basis, **32**:45
Finite mass case, 49:220
Finite orbital basis set, 9:330
Finite size effects, 49:313, 319, 343–346
Firing frequencies, 19:339
First-order perturbation theory, 4:129, 141, 163; *see also* Perturbation theory
First-order phase transition, 40:234, 238, 240, 250, 287, 288; **46**:370
First-order rate process, 42:11
First-order reaction, 1:36, 38
First-order wave function, 9:338
First passage time, 13:1; **52**:408, 410, 414–416, 467
distribution, **13**:12
mean, **13**:9
moments, **13**:3, 4, 5
probability, **52**:480
probability density, **13**:3
Fiske steps, 27:298
Fission model, 30:41
Five coordination, 7:370
Fixed Hessian approach, 54:67–69
numerical results, **54**:72–78
optimal use, **54**:69–72
Fixed nuclei state, 49:140
Fixed nucleus approximation, 42:200
Fixed point, 46:79
central, **46**:111
elliptic, **46**:80, 106, 109
hyperbolic, **46**:80, 106, 109
hyperbolic-with-reflection, **46**:107, 109
parabolic, **46**:105, 109
Poincaré-Birkhoff theorem, **46**:105–106
Fixed-point methods, 40:278

Fixed velocity approximation, **10**:337–339

Flame emission, **42**:12
CS$_2$, **42**:130

Flame technique, *see* Diffusion cloud reaction

Flames, **3**:59, 68, 96; **28**:4, 43, 80, 270, 301
Arrhenius kinetics, **3**:76
complex reaction chains, **3**:96
equations, **3**:88
Adams and Wilde approximation, **3**:91
Corner approximation, **3**:89
Hirschfelder approximation, **3**:93
Klein approximation, **3**:95
ignition temperature, **3**:74
unimolecular reactions, **3**:68
see also Hydrodynamic equations

Flash photolysis, **23**:43; **28**:42, 43, 48ff62, 87, 90; **42**:122, 126, 132; **50**:298, 311, 364, 369
CFBr$_3$, **42**:127
NO$_2$, **42**:129
SO$_2$, **42**:135

Flat-lattice model, **8**:145

Flavin free radicals, **7**:617

Flavoprotein, **7**:67, 617

Flip-flop phenomenon, **29**:43, 337, 346, 350

Floating Gaussian orbitals, **21**:69

Floating wave functions, **12**:53

Floquet-Bloch theorem, **31**:395

Floquet method, **53**:301

Floquet normal modes, approximate, **35**:272, 278
exact form of, **35**:285, 286
examples of, **35**:282
high frequency, **35**:297
strong field, **35**:301
theorems concerning, **35**:280–282

weak field, **35**:297, 298, 300
zero frequency limit, **35**:298

Flory-Huggins equation, **6**:203

Flow, horizontal laminar, **13**:350
rate across a surface, **13**:118
steady state, **32**:13

Flow birefringence, extinction angle, **33**:124, 126, 137–142, 145

Flowing afterglow, **19**:46, 112, 114, 117; **28**:44, 88; **45**:113–115, 427

Fluctuating cavity field, **44**:262, 263

Fluctuating polarizability model, Kirkwood and Yvon, **48**:273
van Vleck, **48**:274

Fluctuation-dissipation theorem, **17**:83, 85, 86, 104; **31**:253; **34**:271; **40**:203, 461; **41**:117; **43**:249, 261; **44**:345; **48**:84; **53**:92, 94, 177–178
inner product, **53**:167–168, 170–171
iodine recombination, **53**:208
phonon population, **53**:269
second, clamping approximation, **53**:114
Langevin equation, **53**:140ff
solitons, **53**:271

Fluctuation potential, **14**:242
many-electron theory, **6**:329, 330, 353

Fluctuation spectroscopy, **48**:376

Fluctuations, **13**:268, 305, 313; **19**:216, 219, 295, 306; **38**:297, 304, 310, 408; **39**:43, 44, 52; **40**:271, 272, 283, 291, 294, 320, 356, 361; **43**:246–264; **46**:165, 175, 401; **49**:315–316, 342–344, 353
anomalous, **46**:200
bifurcation point, **38**:306
biochemical, *in vivo*, **19**:343
canonical distribution, **19**:313
complex fluctuation function, **46**:311

concentration, **24**:4, 43

conditional probabilities, **19**:296

critical, **34**:302

current in membrane-bound gramicidin, **29**:315

density, **17**:190; **46**:326

effects of, **49**:338–343

enhancement of, **46**:206

equilibrium, **19**:297, 298

    Brownian motion, **20**:209

    continuum modes, **20**:208

    Lorentz limit, **20**:206

    power spectrum, **20**:206

    pseudo-Gaussian behavior,
20:209

    short-time approximations,
20:206

    variational solutions, **20**:208

    Wiener-Khinchin theorem,
20:206

finite, **29**:63, 66, 320

generating function, **19**:301

Hamilton-Jacobi theory of, **29**:85

hydrodynamic, **38**:304

inhomogeneous, **49**:332, 351–354

initial, **46**:206

intensity relation, **46**:200

large, **19**:305

limit cycles and, **29**:97

local, **29**:79; **46**:365, 399

Lotka-Volterra model, **29**:103

macroscopic variables, **32**:7

mean field theory, **38**:307

multiple steady states and, **29**:7, 63,
68, 91, 336

nature of, **29**:7

nonequilibrium, **19**:298

nucleation of, **38**:310

number of uninterrupted sequences
in polypeptides, **33**:79, 87, 90, 97

orientation, **24**:45, 50

origin of instabilities and, **29**:41, 66,
86

phase space description of, **32**:10

Poisson distribution, **19**:303, 308

small, **19**:303

spectrum of, **15**:75

stochastic description of, **32**:7

structural, **46**:365, 370

thermal, **15**:65ff; **32**:88

    atomic positions in solid, **13**:125,
131, 132, 153

Fluence, 50:459, 465

effect on decomposition rate, **50**:109,
148

Fluid dynamics, 44:273

coupling to reaction kinetics, **13**:21

Fluid interface, 49:357–454

Fluid mechanics, 32:42, 55

    Poiseuille-Couette flow, **32**:47

Fluid phase, 41:198

Fluid-solid transition, 40:288, 334

Fluid state molecular dynamics, 44:377

Fluids, atomic, inelasticity corrections,
53:7, 52–54

    pair correlation function, **53**:117

    sampling factor, **53**:50

    structure, **53**:12–21

dense, surface tension of, **6**:279

    thermodynamic properties of,
6:279

dense polyatomic, impulsive interactions, **31**:165, 166, 178, 180, 196,
201, 208, 226

    kinetic theory of, **31**:155–227

    moment equations, **31**:160–167

    transport, collisional, **31**:163, 164,
166, 186–189, 196, 200, 225

        diffusional, **31**:161, 163, 164,
186–189

        heat, **31**:161, 164, 173, 174,

177, 186–189, 193, 196, 209

momentum, 31:161, 171, 175, 176, 183–185, 190, 201

spin, 31:161, 164, 172, 173, 178, 191, 192, 194, 196, 209

equilibrium theory of, 53:117–124

Gibbs' surface of tension, 6:261

hard sphere, see Hard sphere fluids

hard sphere dipolar, 48:231

homonuclear diatomic, 53:3, 21–29

chlorine, 53:26–29

iodine recombination dynamics, 53:204–220

molecular-time-scale generalized Langevin equation matrix structure, 53:194–196, 198

nitrogen, 53:23–26

Lennard-Jones, 49:380, 413–415, 424–425, 437

Lennard-Jones solutions, 53:189–196

light scattering by, classical description, 26:24, 26–27

depolarized, 26:55–61

dipole-induced dipole polarizability, 26:25–27

see also Depolarized scattering

loaded sphere, 24:336

molecular, inelasticity corrections, 53:7, 54

neutron diffraction experiments, 53:29–44

pair interactions, 53:2–3

sampling factor, 53:50

structure, HCl, 53:30–33

water, 53:33–44

nonpolar, irradiated, 31:362–369

low energy electrons, 31:329–464

polar-nonpolarizable, 48:187–229

computer simulation studies of, 48:245–254

polar-polarizable, 48:273–303

polarizability, 26:55, 63, 66, 67, 69–70, 73, 74

rough sphere, 24:338

square well, 24:332

Stockmayer, 48:231, 240

structure, experimental studies, 53:1–60

see also Convex body; Ellipsoid; Hard sphere; Liquids; Loaded sphere; Polar fluids; Rough sphere

Fluorenone, 8:134, 136

Fluorescence, 19:353

cells, 28:274, 276, 298, 299

chlorophyll, 19:353

depolarization, 39:166

polymers, 33:134, 135

molecules near interfaces, 37:1–66

P870, 19:357, 358, 359

photofragments, 50:258–324

polarization, 13:193; 50:267ff, 278

quenching, 21:192; 39:136, 142, 143, 145, 165; 48:75

formation of collision complexes, 21:200

surface excitons, 54:320

vibrational, 28:216–221, 246, 247

Fluorescence spectroscopy, 39:135, 146; 47:Part 2:90–91; 49:14, 20–21, 33–34

Fluorescence up-conversion technique, 49:18

Fluorine, 41:152

atom, reaction with H_2, 41:229–268; 49:222, 232, 250, 254–257, 289

interatomic distance, 5:156

ion, electronic correlation energy, 2:240

London energy, 2:78

many-electron theory, 6:383, 384

molecule, crystal structure and inter-molecular potential, **33**:51ff, 70
intramolecular London energy, **2**:78
multiplet correlations in, **42**:95
NMR, **39**:126, 130
solid, **16**:370–374, 381
electron density contours, **16**:373
structural and thermodynamic data, **16**:346–348
see also Chemiluminescent reactions; Diatomics-in-molecules
Fluorobenzene, two-photon spectra, **54**:191–192
Fluoroethane, barrier height of internal rotation, **2**:378
Fluororacil, **7**:53
Flux, general expression for, **35**:62
mass, **35**:58, 62, 63
momentum, **35**:60, 63
one-way, **48**:83
Flux vector, equilibrium, **43**:31
nonequilibrium, **43**:31
thermodynamic, **43**:100
Fock-Dirac density matrix, **2**:225; **14**:240, 287
Fock's equation, **1**:243, 244
Fokker-Planck equation, **1**:2, 6, 79; **11**:207, 213; **13**:10, 16; **15**:66, 70, 71, 106, 137, 326; **16**:2–4, 19–23, 26–29, 34, 37–43; **17**:103; **20**:137, 195; **24**:182; **34**:254, 262, 263, 266, 277, 279, 282, 290, 302; **35**:121, 152, 178, 183; **38**:313; **43**:247; **44**:263, 299, 300; **46**:199, 335; **48**:82–83; **49**:317, 350
friction coefficient, **16**:22
linear, **34**:254, 256, 279
multivariate, **34**:279
nonlinear, **34**:263
path integral formulation, **46**:246
perfect Fermi system, **16**:46–49

polymer dynamics, **15**:328
power potentials, **20**:204
Rayleigh's solution for hard sphere gas, **20**:195, 197
solution, **46**:354
Laguerre functions, **20**:197
Lorentz hard sphere gas, **20**:200
validity of eigenvalues, **20**:199
Fokker-Planck-Kramers equation, **44**:270, 271, 360
Fokker-Planck operator, **11**:226, 244, 277; **13**:3, 5, 10; **31**:200; **46**:354
eigenvalues of, **46**:267
inhomogeneous, **11**:280
shifted, **46**:357
Fold curve, **40**:245
Fold theory, **17**:9, 26
longitudinal fold, **17**:26
transverse fold, **17**:26
Folic acid coenzymes, **7**:77
Forbidden transitions, **7**:542, 556; **10**:179; **12**:16
band width, **7**:147
charge transfer, **7**:395
electronic, **53**:262
phosphors, **7**:558
see also Electronic transitions, forbidden
Force, colloidal, **48**:330, 331
dispersion, **48**:339–340
electrostatic, **48**:332, 369
intermediate range, **12**:389, 393
inverse square-law, **26**:108, 110, 111, 116, 124
long-range, **12**:389, 393; **26**:108, 112, 114, 116, 134
mean, **48**:80, 198–199
random, **48**:84
repulsive, **48**:96
short-range, **12**:389, 393, 481

van der Waals-London, **48**:332, 373–375

interpolation formula for, **48**:374

Force correlation function, normal, **46**:295–296, 302

tangential, **46**:302

Force potential, mean, **53**:117

Ornstein-Zernicke equation, **53**:120.

Form factor, **20**:242; **53**:6, 56, 253

Formal perturbation theory, **9**:334

Formaldehyde, **36**:321; **42**:226; **47**:Part 1:6

photodissociation of, **50**:314

Formamide, **7**:461

Formic acid decomposition, **41**:502

Forster transfer, **37**:21; **41**:313

Forward scattering, *see* Low-angle scattering

Forward scattering ratio, **10**:107–108, 115

Four center reaction, **30**:300

Four centered elimination, **42**:139

Four-current density, **41**:85

Four-magnet system, in hysteresis measurements, **20**:73, 75, 81

Four-vector, **41**:83

Fourier-Bessel transform, **49**:498–499

Fourier difference map, **16**:163

Fourier E-map, **16**:160–161

hexafluorobutadiene, **16**:167–168

jamine, **16**:165

Fourier expansion, **11**:168; **35**:276, 290–292, 332

Fourier inversion, **34**:230

Fourier-Laplace transform, **53**:137, 152

Fourier law, **5**:262, 290

Fourier series, **49**:74

Fourier transform, **11**:49, 58; **34**:167; **37**:169; **49**:129, 446, 547, 566

direct, **44**:303, 458

partial, **46**:293

spherical, **20**:56

Fourier transform spectroscopy, **25**:47; **41**:486; **44**:301, 309; **50**:24

Fowler-Nordheim equation, **27**:227

Fractional moment, **52**:422, 423

Fractional parentage coefficients, **14**:507

Fractional power transfer, **28**:436–439

Fragment, creation, **11**:175

diagonal, **11**:174

Fragment-within-the-solid orbital, **49**:75–79, 90, 93–94, 104–105

Frame, of cone, **43**:21

Frame vector, **43**:22

Francium, dimer, **41**:152

Franck-Condon factor, **18**:42, 46, 48, 58, 129; **28**:396; **30**:2, 60, 82, 84, 277, 306, 490; **36**:279, 283, 293, 301; **42**:3, 211, 373, 524, 539, 568; **50**:6, 29, 78, 213, 230, 268, 283, 296, 349, 373, 498; **52**:281, 289, 290, 322

multidimensional, **50**:257, 276

Franck-Condon frequency, **44**:557, 560, 562

Franck-Condon overlap integral, **49**:222, 226, 288, 301

Franck-Condon principle, **2**:199; **10**:257, 373; **18**:124; **30**:423, 445; **35**:202; **44**:499

applied to transition metal complexes, **5**:58, 126

Franck-Condon region, **28**:144

Franck-Condon type model, **49**:200, 226–227, 260, 300–302

Franck's rule, **28**:281

Frank coefficients, **31**:237

Frank free energy, **31**:236

local stresses, **31**:256

Fréchet derivative, **24**:115; **37**:169; **40**:260, 280, 281, 323, 349

Fredholm determinant, **25**:254, 290

Free-boundary problems, **32:**38, 52
  constrained membrane, **32:**54
  elastic-torsion problem, **32:**56
Free electron, **1:**323
  molecular orbital theory, **1:**262
Free electron orbital model, **7:**559
Free electron theory, **5:**3; **7:**32, 44; **8:**5, 7; **12:**347; **31:**467
  applied to boron compounds, **21:**9
Free energy, **1:**218, 219, 279, 297; **6:**182, 186; **7:**206, 226; **12:**289; **17:**44; **48:**476, 487, 504; **51:**149, 162
  asymptotic behavior, **40:**256, 268
  charging a set of ions, **6:**420, 422, 423–425
  charging an ion, **6:**444–458
  configurational, **48:**486
  dissipation in diffusion, **6:**309
  excess, **2:**135; **6:**270; **16:**229, 247, 274–276
  excess Gibbs, **16:**274–277, 279; **17:**27
  formation of microclusters, **40:**126
  Helmholtz, **17:**32
  liquidlike cells, **48:**475, 490, 499
  local, **48:**473, 474, 475
  lower bound to, **33:**425
  of mixing, **6:**309
    excess, **11:**105
      reciprocal systems, **11:**112
    excess Gibbs, **11:**106
    excess Helmholtz, reciprocal systems, **11:**110
  one-dimensional plasma, **4:**204, 210, 212, 218, 223
  residual Gibbs, **16:**276–277, 279
  solidlike cells, **48:**490, 499
  of solvation, **6:**414
  system in contact with bath, **51:**131
Free energy density, **41:**46–49
Free energy distance, **51:**148

Free energy-like function, **51:**131, 150
Free energy surface, **6:**460; **40:**142
Free-fermion model, **41:**201
Free induction decay, **25:**29; **40:**391; **47:**Part 2:457–461; *see also* Transient absorption
Free molecule theory, **53:**37, 52
Free molecules, electrochromism of, **13:**180
Free particle dynamics, **51:**68
Free precession signal, **11:**311, 312
Free radicals, **1:**183; **7:**261; **19:**204
  in *Agrostis stolonifera*, **7:**640
  laser induced fluorescence studies of, **50:**66
  in plant material, **7:**629
  radiation-induced, **7:**621
  trajectory, **53:**109
Free rotation, **1:**94
  classical, **44:**350
  interrupted, **44:**465
  Maxwellian ensemble of molecules, **44:**474
Free rotor, **44:**275, 350, 368
Free valence, **1:**184, 185, 189, 190, 228
Free valence number, **7:**42; **8:**76, 84, 102
Free volume, **1:**155, 160, 228; **2:**26, 27, 33; **47:**493
  amorphous phase, **48:**470, 471
  distribution in liquid, **48:**473
  Eyring-Hirschfelder, **1:**162
  free exchange of, **48:**479
  liquidlike cells, **48:**475
  phase transition, **48:**493
Free volume theory, **6:**238–240; **24:**225, 226
  equation of state in, **6:**245–246, 250
  fluidity and, **48:**470, 494
  glass transition, **48:**455–526
  heat capacity and, **48:**514

internal configuration entropy and, **48**:502

limitation of, **48**:476

with structural relaxation rates, **48**:506

Freely jointed chain, 52:431–433, 443

Freezing, 20:286, 313; **34**:146

Frenet formula, 49:204

Frenkel excitons, 23:230; **54**:306–314

  calculation of exciton levels, **54**:312–313

    medium transitions, **54**:313

    strong transitions, **54**:313

    weak transitions, **54**:313

  first order energy levels, **54**:307–309

    excited states, **54**:308–309

    ground state, **54**:307–308

  transition probability, **54**:309–310

    allowed transition, **54**:310

    conservation of energy, **54**:310

    conservation of photon momentum, **54**:310

    polarization of unit cell, **54**:310

  two molecules per unit cell, **54**:311–312

Frenkel's hole theory, 48:468

Freon, 42:516

Frequency, collisional, **1**:156

  complex, **27**:380, 391, 407, 414

  molecular transit in liquids, **1**:162

  plasma, **27**:361, 378, 415, 460

  reduced, **33**:125

  relaxation, Ising chain, **33**:97

    vinyl-polymers, **33**:129, 133

  reststrahl, **27**:360

Frequency-dependent dielectric constants, 1:83

Frequency-dependent susceptibility, 12:236

Frequency difference spectrum, 27:507, 509

Frequency distribution function, 44:273

Frequency doubling, 50:17

Frequency factor, 1:113, 388, 390, 391

Frequency matrix, 46:312, 349

  adiabatic, **53**:180, 208

  equilibrium correlation function, **53**:142

  Mori equation, **53**:148

Frequency moments, 17:111

Frequency multiplication, 44:303

Frequency spectrum, 27:417, 484

  continuous, **36**:42

  microclusters, **40**:99, 102

  moments of, **27**:474–475, 483–485

  projected, **27**:506

Frequency tripling, 50:19

Fresnel coefficients, 37:27

Fresnel zone, 20:273

Friction, 11:244, 246

Friction coefficient, 1:8, 146, 150, 152, 153, 161; **6**:293, 296; **11**:203, 213, 254; **16**:22–30; **17**:169; **35**:73; **44**:293; **48**:83; **53**:95, 191

  definition, **16**:22, 26

  effective, **33**:122

  Fermi systems, **16**:26–30, 46–49

  fluid of Bosons, **16**:23–26, 44–46

  Langevin model, **53**:107

  recursive construction, **53**:218ff

  relative, **48**:158

  rotational, **17**:172

Friction force, 11:218

Friction kernel, 53:94, 175

  Gaussian model, **53**:202ff

  Lennard-Jones solutions, **53**:200ff

Friction matrix, 53:180–181, 208

Friction tensor, 1:146, 147, 154; **34**:53; **44**:301; **48**:133

Frictional acceleration, 1:141

Friedel oscillations, 27:550, 551; 41:172; 51:269

Friedrichs Hamiltonian, 38:145, 148, 149, 152, 155, 166ff

Fröhlich expression, 48:279

Fröhlich Hamiltonian, 44:488, 495, 528

Fröhlich mode, 27:505

Fröhlich's equation for the dielectric constant, 1:82, 100–102

Frontier electron method, 7:32, 44

Frontier orbital, 49:84, 94

Frontier orbital density, 8:110

Frost and Leyland method, 9:152

Frost theory, 44:459

Frozen core approximation, 26:240, 253

Frustrated total reflection, 27:590, 592

Fuchs-Kliewer mode, 27:377

hybridization with bulk modes, 27:492–495

Fuchsone, 8:134, 136

Fucus, 29:281, 285

membrane conductance, 29:287

membrane potential, 29:285

polarity, 29:283, 286

Fugacity ratio, 40:142

Fulcher equation, 48:410

β-Fumaric acid, 16:186

Functional, constitutive, 24:119–120

density, 41:64ff, 81, 138

differential equations, 24:99

Hamiltonian, 24:137

Lagrangian, 24:135

Lyapounov, 13:326, 327

maximum displacement, 52:418

mixed, 38:32

normal, 38:34

normalized positive linear, classical, 38:43

definition, 38:32

disjoint, 38:39

dispersion free, 38:43

equivalent, 38:39

factor, 38:39, 44

mixed, 38:32

normal, 38:13

pure, 38:32

quasiequilvalent, 38:39

singular, 38:35, 43

pure, 38:32

singular, 38:35

spin-density, 41:59–194

vector, 38:33

Functional inequality, 51:7, 10, 31

Functional integration, 41:219, 220

Fundamental measure, of a convex body, 5:150, 155

Fundamental solution, 35:278, 301

definition of, 35:278

Fuoss and Kirkwood factor for the dispersion curve, 1:107

Furano-*p*-cyclophane, 16:193–194

Fused salts, 9:56ff

expansivities, 9:61, 71

heats of vaporization, 9:71

as solvents for organic compounds, 21:525

Fusion cycle, 9:68

G

γ process, 44:399

g-anisotropy mixing, 35:23

g-factor, 4:327, 332; 37:265–272

g_k correction, 52:120, 130ff, 151ff, 160, 163, 176

G-Value, 7:261

GaAs, 49:545, 560, 583–587

metal overlayer, 49:618

photoexcited, 49:25

structure, **49**:99–104

surface state table, **49**:587

Galactic field, 35:121

β-Galactosidase, 29:43

Galactoside permease, 29:43

Galaxies, 26:108, 109, 115, 134; **35**:154, 181

Galerkin method, 13:318; **32**:40, 41

Galilean group, 33:162

Galilean transformation, 5:279

Gall stones, infrared spectroscopy, **7**:413

Gallium, 41:141

oriented beam of, **30**:390

Game theory, allosteric enzymes, **38**:223

cumulant, **38**:307

Gaussian white noise, **38**:87

generalized kinetic equation, **38**:174, 177, 182

generalized master equation, **38**:114

generating function, **38**:298

John von Neumann's, **38**:221

Michaelis-Menten mechanism, **38**:223

strategy, **38**:221

Gamma distribution, 19:336

Gamma-ray, 2:188, 193, 194, 197, 202; **7**:260

Gamma-ray diffraction, studies of fluid structure, **53**:3–6

Gamma-space, 2:320

GaP, 27:445

Gas, ideal, self-diffusion, **6**:305

molecular, polarizability, **46**:1–72

noninteracting soliton, **53**:256

quasiparticle, **53**:243ff

supersaturated, **40**:137

vibrational relaxation in, **47**:Part 1:367

Gas-crystal shift, 40:413

Gas diffusion, 6:305

Gas-gas equilibrium, 17:2, 8, 9, 13, 17, 24, 28

first type, **17**:10, 24

intermediate type, **17**:10, 24

meniscus, **17**:9

second type, **17**:9, 10, 16, 20, 24

surface tension, **17**:9

Gas hydrates, 2:3, 20, 22, 25, 34

mechanism of formation, **2**:4

thermodynamic properties, **2**:15

x-ray work, **2**:3

Gas-liquid coexistence, 37:162, 165

Gas-liquid coexistence curve, 44:260, 355

Gas-liquid coexistence region, 37:11, 154, 173, 190

Gas-liquid critical point, 40:350

Gas-liquid interface, 37:198

Gas-liquid transition, 37:11, 142, 172

Gas-matrix shift, 41:277

Gas mixtures, second virial coefficient, **5**:168

Gas phase collision theory, 53:65

Gas-solid collisions, 44:220

Gas-solid dynamics, surface elastic model, **44**:236

two time scale, **44**:233

Gaseous core nuclear rocket, 36:195

Gaseous discharge, 10:172–173, 178

Gaseous fraction, of liquid metals, **21**:499, 509

Gaseous phase, 41:199, 207

Gastrulation, 29:254

Gating mechanism, 29:305

Gauche **effect, 49**:99

Gauche **state**, of lipids, **41**:196, 201, 202, 204, 213ff

Gauge condition, 33:168

Gauge invariance, 44:89, 108

Gauge-invariant atomic orbital, 33:157

Gauss-Laguerre quadrature, 13:234, 260

Gauss-Seidel procedure, 54:136

Gauss' theorem, 1:313; 33:191

Gaussian approximation, 46:196; 52:378, 380, 381

Gaussian core model, 48:402

Gaussian distribution, 44:294; 46:196

Gaussian elimination technique, 2:291

Gaussian field, 24:33

Gaussian function, 49:121, 288

Gaussian intrinsic state, 49:120

Gaussian line shape, 40:385, 452, 461, 477

Gaussian-lobe functions, 26:220

Gaussian-Markov process, 15:114, 118; 17:167; 20:199

Gaussian memory function, 17:170, 173, 175, 180

Gaussian model, 40:194, 265, 271; 49:312
　　dynamic solvent response, 53:210–204

Gaussian modulation, 15:105

Gaussian noise, 53:83
　　statistical properties, 53:105

Gaussian orbital, 49:121

Gaussian overlap, 49:128, 174

Gaussian peak, 49:600

Gaussian potential, 41:6, 28, 37, 41

Gaussian probability theory, 44:197, 407

Gaussian process, 17:103; 44:268, 357

Gaussian pulse, 49:9–10

Gaussian random variable, 53:83–84

Gaussian restoring torque, 44:268

Gaussian transition probability, 17:155

Gaussian wave function, 2:276

Gaussian wave packet, 42:282

Gaussian well, 49:348

Gegen ions, 2:160

Gellation, 22:125

Geminals, 14:9

Geminate recombination, primary, 48:155
　　secondary, 48:155

Generalized actual orbital, 36:400

Generalized diffusion coefficient, 52:472

Generalized force, 44:268

Generalized functions, 36:51
　　Cauchy principal part, 32:225, 253
　　continuous modes connected with, 32:225, 253
　　Dirac distribution, 32:225, 253
　　eigenfunctions as, 32:225, 253
　　integral equation involving, 32:226

Generalized Hessian approach, 54:65–72

Generalized hydrodynamic equations, 24:2, 3, 9, 11

Generalized master equation, 52:483

Generalized mean field approximation, 48:219, 238

Generalized mean spherical approximation, 48:226, 237–238

Generalized potential, 12:228, 243, 246, 249, 268, 279

Generalized random torque, 44:266

Generalized SCF method, in many-electron theory, 6:344–346

Generalized spherical harmonics, 34:3, 9, 10, 11, 22, 38, 50

Generalized valance bond method, 50:524

Generating function, 43:250; 46:95, 100, 125; 51:19, 23

Generation, 34:251, 273

Generator coordinate method, nuclear motion, 49:115–190
　　polyatomic systems, 49:178–187
　　　inversion symmetry, 49:181–182

molecular symmetry, **49**:185–187

rotational invariance, **49**:180–181

polyatomic term formulae, **49**:193

subspace, **49**:121–122, 158

variational principle, **49**:163

wave function, **49**:123, 145, 152–153, 156, 163, 180

Generators, least-biased, **53**:362

Genetic information, **33**:100

Genetic markers, **7**:220

Geological folding, **32**:14

Geometric mean approximation, **2**:29

Geometric probability, **41**:12–15

Geometrical triangulation scheme, **49**:451

Germane, barrier height of internal rotation, **2**:391

Germanium, **27**:481; **41**:171

Gettering, **28**:279

Gibbs adsorption law, **6**:151–153

Gibbs-Bogoliubov inequality, **24**:189

Gibbs distribution, **4**:299

Gibbs dividing surface, **1**:220

Gibbs-Duhem equation, **1**:207; **2**:16; **4**:289; **6**:152; **46**:285

Gibbs ensemble theory, **19**:325, 326, 336

Gibbs entropy, **53**:355

Gibbs free energy, **2**:30; **4**:289; **6**:92

of activation, **33**:116, 117

of reaction, **33**:116, 117

state, **33**:111, 114

Gibbs function, **2**:20

excess, **16**:274–277, 279

residual, **16**:276–277, 279

Gibbs minimum principle, **41**:69

Gibbs relation, **19**:212; **43**:95

Gibbs surface, **41**:39, 40, 50; **46**:283

Gibbs surface tension, **6**:261

Gibbs theory of capillarity, **6**:145, 158

Gibbs-Wulff theorem, **40**:70, 129

Gibbsian probe, **19**:351

Gierer-Meinhardt model, **29**:260

Giesekus equation, **35**:94, 97, 111

for macromolecules, **34**:94

Ginzburg criterion, **43**:264; **49**:313, 319

Ginzburg-Landau equation, **49**:312, 325, 353–354

Ginzburg-Landau Hamiltonian, **49**:318

Ginzburg-Landau model, time-dependent, **46**:252; **49**:348

Ginzburg-Landau potential, **49**:314, 352

Glansdorff-Prigogine theory, **32**:31, 39, 61ff, 77, 78; **43**:12, 95, 101

Glass, **44**:258, 483; **48**:491

cooling rate, **51**:228

formation, **51**:271

hard-sphere, **48**:499

hydrocarbon, **31**:371

ideal, **48**:457, 467, 506

inorganic, **44**:397; **48**:513

ionic, **48**:458, 459

Lennard-Jones, **48**:399, 472

metallic, **48**:458; **51**:227–276; *see also* Metallic glass

monatomic, **48**:457

network, **48**:459, 462, 513

organic, **48**:459, 492, 500

polymeric, **48**:458, 492

primary and secondary losses, **44**:259

real system, **48**:399

repellent Gaussian, **48**:399

soft-sphere, **48**:399

tertiary process, **44**:259

viscosity, **48**:457

Glass formation, survey of computer experiments, **48**:397–454

Glass transition, computer simulation, **48**:397–454

compression rate, **48**:436

density, **48**:417–418, 436–437
diffusion coefficient, **48**:406–407
diffusivity temperature relation, **48**:410
isothermal compressibility, **48**:433
multicomponent ionic system, **48**:441
radial distribution function, **48**:436
radial fluctuation function, **48**:432–434
relaxation process, **48**:435
relaxation time, **48**:404, 412, 416, 419–421
structural characterization, **48**:435–441
transport properties, **48**:429–430
enthalpy and heat capacity, **48**:462
entropy of, **48**:443–445
first order, **48**:457, 467, 478, 491, 492, 500, 505
free volume approach, **48**:455–526
relaxation time, **48**:404
determination of, **48**:412
dielectric, **48**:412
displacement from equilibrium and, **48**:419–421
sample size effect on, **48**:421
temperature dependence, **48**:416, 419
viscosity, bulk, **48**:412
shear, **48**:412
second order, **48**:467, 478, 492, 502, 505
temperature, **44**:399; **48**:445; **51**:228
dependence on heating and cooling rate, **48**:464
lower, **48**:419
upper, **48**:419
viscosity, **48**:461

Glauber model, **46**:252
Glide plane, **20**:10
Global bifurcations, **55**:183–186
Global nitrous oxide budget, **55**:76
Global parameter scaling, **51**:212–217
complex parameters, **51**:216
nonlinear formula, **51**:179, 213, 220
use of Padé approximants, **51**:216
Globular molecule, **34**:232
Glory, **12**:369, 402, 403
Glory effect, **10**:114–117, 122; **30**:322, 328, 344, 379
with angle dependent potential, **30**:375, 402, 408
Glory extrema, **12**:389, 399, 414, 424, 482
Glory oscillation, **42**:458, 464, 490
Glory phase, **10**:115, 120, 122–124
Glory shadows, **12**:369
Glow discharge, **10**:248, 252, 268
Glucose, in enzyme membranes, **29**:118
in glycolysis, **29**:140
Glucose oxydase, **29**:119
effect of pH on, **39**:7, 231
Glutaric acid radical, **7**:546
Glycine, **16**:180
Glycine radical, **7**:546ff
Glycolate ion, **16**:186
Glycollic acid, **7**:548
Glycollic acid radical, **7**:552
Glycolysis, **19**:344; **29**:42, 138, 270
control diagram, **29**:138
diffusion of intermediates, **39**:34
intermediates in, **29**:144
oscillations in, **29**:141; **39**:30, 31; *see also* Glycolytic oscillations
phase relations in, **29**:143, 146
regulation of dissipation, **55**:169
Glycolytic oscillations, allosteric model, **38**:371

effect of random noise, **38**:368

entrainment, **38**:368

mechanism, **38**:369

molecular properties, **38**:369

phenomenological models, **38**:370

role of enzyme cooperativity, **38**:380

synchronization, **38**:380

Glycosidases, **7**:74ff

Glycylglycine, **7**:265

Glyoxal, **22**:336; **42**:224

GNS construction, **22**:327; **38**:38

Goeppert-Mayer and Sklar method, **7**:32ff

Golay pneumatic cell, **44**:324

Gold, (100), **49**:593

(111), **49**:562

(112), **49**:562

partial molar heat of solution in tin, **2**:133

Gold-copper, alloy (Au_5Cu_5, $Au_5Cu_5(S)$), calculation of thermodynamic quantities, **2**:136, 142

solid solution (CuAu), **2**:129

Gold-lead alloy (Au_5Pb_5), calculation of thermodynamic quantities, **2**:136

Gold-nickel, alloy (Au_5Ni_5), calculation of thermodynamic quantities, **2**:136, 142

solid solution (Au_5Ni_4), enthalpy of formation, **2**:144

solution, enthalpy of formation, **2**:143

short range order in, **2**:144

Gold-silver alloy (Au_5Ag_5, Au_6Ag_4), calculation of thermodynamic quantities, **2**:136, 142

Gold-thallium alloy (Au_5Tl_5), calculation of thermodynamic quantities, **2**:136

Gold-tin alloy (Au_5Sn_5), calculation of thermodynamic quantities, **2**:136

Golden mean, continued fraction representation of, **46**:110

Golden rule, **50**:373, 427

Golden-Thompson inequality, **24**:209

Goldstone boson, **46**:315, 318

Goldstone mode, **43**:254, 258; **53**:326

Goodman-Artman model, **9**:121, 127

Gordon sum rules, **44**:309, 344

Gotlieb polynomials, **1**:383, 385

Gouy balance, **7**:603ff

Gradient, of density, produced by double diffusion, **32**:147

of salinity, **32**:136, 139, 141

Gradient-based methods, second order methods and, **54**:126–127

with iterative cubic method, **54**:127–130

Gradient equation, **40**:244

Gradient theory, **49**:358, 374–389, 408–410, 452

capillary waves, **49**:448–452

computer density profile, **49**:424

correlation functions, **49**:377, 379–381, 428

exact, **49**:416

free energy theory, **49**:416

stability operator, **49**:420

thermodynamic free energy, **49**:375–376, 383

Grafting experiments, **29**:258

Grain, encroachment, **32**:200

groove, **32**:196

protruding, **32**:200

Gramicidin A, **29**:315

Gramicidin D, **19**:368

Grand canonical ensemble, **19**:336

Grand partition function, **6**:88, 109; **40**:161; **41**:81

activity expansion of, **6**:93–96

Fermi gas, **6**:97

interaction parameter expansion of,
  6:88–93
Grand potential, 41:73, 90
Graph integrals, rooted, 21:416
Graph theory, 43:16
  applied to microcluster structure,
    40:65, 93–94
Graphite, monocrystalline, magnetic
  properties, 3:222, 226
  polycrystalline, magnetic properties,
    3:222, 226
Graphite-carbon, 9:382ff
  equation of state, 9:370
Graphite-diamond, free energy difference, 9:385ff
  transformation, 9:368
Graphite intercalation compounds,
  49:459–527
  acceptor, 49:459–460, 474, 521–526
    pressure-induced transitions,
      49:525–526
    temperature-induced transitions,
      49:521–525
  classification schemes, 49:459–462
  diffraction patterns, 49:513
  distortions of carbon layers, kinks,
    49:467
    puckering, 49:469–470
    static, 49:473
  donor, 49:459, 482–521
  higher stage alkali, 49:514–515, 517–
    518
  highly oriented pyrolytic graphite,
    49:456–457, 464, 497, 502, 511
  odd-state, 49:517
  stage 1 alkali, 49:483–494, 517–518
    cesium, 49:486
    homogeneous, 49:477–479, 486
  stage 2 alkali, 49:494–525
    CsC_{24}, 49:497–503, 519–521
    KC_{24}, 49:504–512, 515–516, 519

LiC_6, 49:490–491
    MC_{24}, 49:488, 496–497
    RbC_8, 49:488
    RbC_{24}, 49:512–513
  stage 3, 49:516
    HNO_3, 49:525
    RbC_{36}, 49:517
  stage 4, HNO_3, 49:525–526
    RbC_{36}, 49:518
  transitions, ambient conditions,
    49:483–485, 499–500
    hydrostatic pressure, 49:525
    multiphase, 49:496–497, 511,
      517, 519
    order-disorder, 49:492–493, 512,
      521–523
    order-order, 49:492–493, 512,
      521–523
    pressure-induced, 49:483, 515–
      521
      KC_{24}, 49:515–516
      MC_8, 49:517–518
      RbC_{48}, 49:517
    staging phase, 49:489–490
    temperature-induced, 49:483–
      515, 521–525
Graphs, 20:44; **43**:75, 85, 91, 107
  articulation point, 20:48
  complementary, 20:54
  complex, 43:75, 76
  connected, 20:47
  disconnected, 20:46
  labeled, 20:46
  rooted, 20:47
  semidirected, 43:93
  symmetry number, 20:45
  tree, 43:75
  two center, 20:50, 52
  unlabeled, 20:46
  weight, 20:45

see also Diagrams

Grating, 27:336, 337, 440

Grating spectrographs, 7:409

Grating spectrometer, 24:25

Gravel, 21:751

Gravitational collapse, 35:122

Gravitational systems, kinetic theory of, 35:119–196

Gravity wave, 46:305

Gravothermal catastrophe, 26:115, 122, 124

Green-Kubo integrands, 38:182, 187

Green plants, photosynthesis, 47:Part 2:581–584

Green's functions, 13:5, 10, 126; 17:230, 236; 25:103, 104, 254; 27:213, 465, 517–519; 29:109; 30:115; 31:264, 265, 281, 283, 290; 32:3; 33:235, 241, 245; 36:207, 213, 214, 216, 220, 254; 37:2; 41:115, 124, 134; 42:252; 52:483

    calculation of ionization potentials, electron affinities, and excitation energies, 48:1–70

    diagrammatic approach to, 33:236, 258

    didadic, 33:250

    expansions, 13:213, 236, 238

    finite temperature, 41:73, 75

    free, 36:211, 214, 220, 264

    iterates, 13:6

    many-body, diagrammatic expansion, 48:529

        hierarchy of equations, 48:5

        one electron, 48:5, 15

        orbital basis set dependence, 48:35

        shake energies, calculation of, 48:31

        two electron, 48:5

    n-particle, 41:73, 75

    one-body, 36:213

    one-sided, 33:249

    open shell systems, 36:217

    operator, 33:250

    polymer statistics, 22:1, 49, 54, 62, 66ff, 83, 104, 109, 119, 125

    real-time, 33:237

    retarded, 33:236, 237, 241, 245

    second order perturbation theory, 31:281

    self-consistent internal propagator, 31:283, 307

    semiclassical, 36:54

    single site approximation, 31:308

    six point, 36:254

    statistical correlations in, 33:237, 263, 277

    statistical random phase approximation in, 33:257

    temperature, 33:236

    tetradic, 33:236

    two-body, 36:218

    used in calculation of ionization potentials and photoelectron spectra, 36:205–344

    zero-temperature, 41:73

Green's operator, 36:50

Griffith inequalities, 37:182

Grimm-Sommerfeld valance rule, 49:89

Groove growth, on metals and oxides, 21:669

Grotthuss proton conduction, 7:253

Ground configuration, 41:141

Ground state, 41:62, 66, 204, 218, 219, 226

    doubly degenerate, 53:251

    equilibrium, 53:333

    nondegenerate, 41:71

Ground state correlation, 36:240, 260

Ground state depletion, 16:76–78

Ground state energy, 41:70, 185

    calculation, 2:263

Ground state function, approximate, **48**:22–23

    Epstein-Nesbet perturbation theory and, **48**:23, 40

    iterative improvement of, **48**:16, 59

    multiconfigurational, **48**:24, 61

    Rayleigh-Schrödinger perturbation theory and, **48**:23, 40

    reference state of, **48**:23

    single determinant, **48**:23

Ground-state recovery, **49**:14–15

Ground state properties, **41**:160

Group, complete nuclear permutation and inversion group, **44**:14

    direct product, **32**:225

    extended molecular symmetry, **44**:11

    geometrical, **32**:225, 232

    invariants, **32**:218, 220, 223, 224, 232

    irreducible representation, **32**:218, 220, 223, 228

    Kh symmetry, **32**:226

    permutation, **32**:223, 228

    rotation, **44**:15, 375

Group invariance properties, bifurcation theory, **40**:284, 287

    physical states, **40**:303

Group orbital, **33**:370, 371, 372, 376, 390

Group theory, **9**:175, 211; **44**:10

Growth, **29**:43, 179, 183

Grüneisen constant, **16**:382

Grüneisen equation, **21**:505

Grüneisen function, **21**:503

Grüneisen theory, **27**:476

Guanidine phosphate, **7**:66

Guanine, **7**:47ff, 50ff, 85, 349

    internuclear distance, **7**:87

    molecular diagrams, **7**:94

    transition moment, **7**:95

Guanine-cytosine pairs, **7**:54ff

Guanosine, infrared spectra, **7**:449ff

Guanylic acid, **7**:349

Guest molecules, **8**:2, 30

Guest rotation, as energy acceptor, **47**:Part 2:526–528

Gyromagnetic ratio, **53**:230

H

H-function, Boltzmann, **26**:127–134, 141–143

    Gibbs, **26**:127–129, 132, 141, 142

H-quantity, **35**:168

H-theorem, **11**:333, 335; **26**:113, 126–134, 141–143; **35**:148, 153, 154; *see also* Boltzmann H-theorem

Hahn-Banach theorem, **15**:45

Hair, **7**:265

Haldane relationship, **7**:704ff

Half collision model, **50**:373

Half-line, **43**:22

Half-space modes, biaxial and uniaxial crystals, **27**:527–528

    complex crystals, **27**:428–429

    cubic crystals, **27**:362–371

    damping included, **27**:406–412

    excitons, **27**:463–465

    lattice-dynamics theory, **27**:517–518

    noncubic crystals, **27**:430–435

    nonlocal theory, **27**:452–457

    nonretarded theory, **27**:401

    polar semiconductors, **27**:414–417

        with magnetic field, **27**:420–427

    reconstruction effects, **27**:512–514

Half-wave potential, **3**:15

Half-width ratio, **4**:169–173

Halide ion, **8**:52

    absorption bands, **8**:58

Halides, Soret effect in aqueous solutions of, **13**:357ff, 361

thermal diffusion of in aqueous solution, **13**:349

Hall effect, 4:336; **44**:484

Halocarbons, atmospheric chemistry, **55**:77

Haloes, 12:364

Haloforms, 40:41–43

Halogen-containing compounds, surface tension and compressibility of, **9**:74

Halogen molecules, photon catalysed dissociation, **50**:199

Halogens, as quenchers, **50**:367

charge transfer reactions of, **10**:380–384

photodissociation of, **50**:284

solid, **22**:221, 254, 291

infrared spectra of, **22**:207

librations of, **22**:221

Hamilton-Jacobi equations, 25:102; **29**:85, 87; **36**:7, 22, 54, 68

multidimensional, **29**:88

Hamiltonian, 4:129, 141, 282; **7**:7ff, 25; **26**:108, 123, 124; **36**:214, 219, 274, 297; **38**:50, 173; **41**:69, 70, 74, 119

action variables and, **53**:315

Bethe-Salpeter's, **2**:240

continuum, **53**:235

crystal, **53**:270

Dirac, **18**:168

effective, **18**:183

nuclear spin-electron spin, **44**:31

nuclear spin-spin interaction, **44**:32

quadrupole coupling operator, **44**:32

rotational, **44**:28

spin-orbit coupling, **44**:27, 31

spin-rotation coupling, **44**:29, 32

spin-spin, **44**:28, 32

see also Effective Hamiltonian

eigenfunctions of, **4**:232, 236

electric field, **18**:178

electronic, **36**:219; **44**:5; **49**:130ff, 142, 148, 167ff, 187

non-relativistic, **44**:111

phenomenological, **44**:25

electronic-vibrational, **36**:274

external magnetic field, **18**:205, 206, 207

fine structure, **18**:173, 193, 202

function, **36**:26, 38

functional, **24**:137

interacting reagents, **53**:182

kernel, **49**:119–122, 125, 128, 173, 177, 184

magnetic hyperfine, **18**:195, 197

many-electron system, **52**:183

master equations, **53**:369

molecular crystal, **53**:278

observables, **53**:369

rotational, **18**:187, 203, 204

separated in zero, one, two electron terms, **2**:212

sine-Gordon derivation, **53**:324

spin-rotation, **18**:190, 192

spin-spin, **18**:173, 194, 202

Stark, **18**:236, 237

time-dependent, **53**:351

vibrational, **36**:274, 297; **44**:112

Zeeman, **18**:176

see also Operator

Hamiltonian methods, accuracy of, **51**:179, 218

effective, **51**:195–198

Hamiltonian system, averaged, **46**:96, 102

Barbanis, **46**:85

Chirikov, **46**:99

Henon-Heiles, **46**:81

integrable, **46**:75

nondegenerate, **46**:76

nonintegrable, **46**:78

pendulum, **46**:97

resonant, **46**:97, 99

Toda, **46**:85

Walker-Ford, **46**:93, 104, 145

Hamilton's equations, **36**:32; **53**:125, 281, 302

macromolecules, **35**:54

Hamilton's principle, **22**:55

Hammerstein equation, **40**:274ff, 299, 311ff, 328ff

Hammerstein theorems, **24**:239, 240

Hammett correlation, **33**:347, 348, 373, 381, 389, 392–394

Hammond rule, **33**:367

Hankel transform, **42**:198; **48**:235

Hanle effect, **28**:305, 307

Hansch laser, **50**:8

Hard core repulsions, **6**:344

Hard cube fluid, close-packing of, **6**:264

Hard disk fluid, **40**:338; **41**:216, 218, 219; *see also* Hard sphere fluid, two-dimensional

Hard instability, **43**:230, 255, 256

Hard mode transition, effects of inhomogeneous fluctuations, **49**:351–354

to chemical oscillations, **49**:346–354

mean field behavior, **49**:347–351

Hard rod fluid, **40**:301, 321, 322, 340, 359, 362, 262

BGY hierarchy, retaining quadruplet correlations, **21**:414

radius of convergence, **40**:301

see also Hard sphere fluid, one-dimensional

Hard sphere fluid, **9**:3, 9, 12ff, 20, 36,

53, 59; **34**:107, 110, 177; **37**:140; **38**:174; **40**:146, 318, 335, 336, 338, 359, 360;

BGY hierarchy, retaining quadruplet correlations, **21**:397

cavity formation in, **6**:255, 259, 272, 279, 282

computer simulation, **53**:64

dipoles and quadrupoles, **48**:269–271

distribution function, **21**:400, 408

equation of state, **6**:229, 253–266; **9**:19ff, 53; **21**:425

free volume theory, **6**:245–246, 250

high temperature, **6**:272, 276

argon, **6**:277

integral equation methods, **6**:235–238

mixtures, **6**:268

Monte Carlo method, **6**:250–252

power series methods, **6**:234–235

superposition approximation, **6**:235–237, 245

Tonk's, **6**:256, 257

virial coefficients, **6**:234, 235, 242ff, 263ff, 276

five terms, **6**:276

intensive properties, **6**:271

lattice theories of, **6**:238–240

cell theory, **6**:238, 239, 240

free volume theory, **6**:238

hole theory, **6**:238

linear transport properties, **20**:135ff

approximate kernels, **20**:150

degenerate kernel methods, **20**:167, 174, 190

Kuscer-Corngold theory, **20**:161, 163, 181

master equation, **20**:135, 154,

156

Monte Carlo calculations, **20**:183

Rayleigh fluid, **20**:197

Rayleigh kernel, **20**:143

Rayleigh-Ritz methods, **20**:169, 182–186

transport equations, **20**:138

Wigner-Wilkins kernel, **20**:145

WKB methods, **20**:176, 180

mixtures, **6**:268, 270, 271

equation of state, **6**:268

excess free energy, **6**:270

miscibility gap, **6**:270

Monte Carlo calculations, **6**:270

one-dimensional, **6**:274

pair correlation function, **6**:270

surface tension, **6**:268

virial coefficients, **6**:269

molecular dynamics calculations, **6**:244

numerical methods, **6**:230, 231, 263–264, 266, 268

one-dimensional, **9**:37

equation of state, **6**:256

Ornstein-Zernicke equation, **53**:123

packing, **9**:28; **40**:72, 73, 82, 87, 96

pair correlation function, **6**:233, 266, 269ff

Percus-Yevick approximation, **53**:124

pressure, **6**:234, 242, 244, 269–270, 276

scaled particle theory, **6**:230, 238, 253, 265ff; **9**:12ff

with "soft" potential, **9**:64

with surface adhesion, first-order phase transition, **21**:421

Helmholtz free energy, **21**:426

Percus-Yevick theory for, **21**:423

surface tension, **6**:255, 261, 264; **9**:23ff

two-dimensional, **9**:37

equation of state, **6**:263, 264

van der Waals loop, **6**:244, 249, 276

virial coefficients, **6**:234, 235, 265, 276; **20**:39, 49–51

fifth, **20**:51, 57; **21**:399, 405, 413

fourth, **20**:39, 51

quantum mechanical second, **20**:59

quantum mechanical third, **20**:59

second, **20**:39, 49, 50

seventh, **20**:58

sixth, **20**:51, 57

third, **20**:39, 51, 63

virial lines, **6**:277–278

Hard sphere type model, **49**:200, 216–222

light-heavy-light, **49**:216–219

light-light-light, **49**:210

square-well, **49**:218

step function, **49**:219

Hard square fluid, close-packing of, **6**:264

Harmonic approximation, **1**:20; **36**:278, 293; **49**:128, 139, 185

atomic lattice, **53**:297

lattice dynamics, **53**:294, 301–302, 307–308

molecular vibrations, **53**:317

xenon chains, **53**:316

Harmonic mode scrambling, **54**:181–188

Harmonic oscillator, **7**:204

heat bath, **40**:24

linearly perturbed, **36**:91

paradox, **42**:327

quadratically perturbed, **36**:96

two-dimensional, **52**:122, 131–132

Harmonic radiation, generation of in vacuum ultraviolet, **47**:Part 2:158–173

Harpoon potential, **10**:370

Harpooning, **10**:367–368; **50**:480

electronic structure and, **10**:379–387

mechanism of, **10**:368–379

see also Reaction mechanism

Harris variational method, **25**:214, 227ff, 243, 246, 285, 287

Hartree approximation, **2**:232, 247; **10**:6; **41**:60–64

Hartree-Fock-Brueckner SCF theory, **7**:16

Hartree-Fock energy, **14**:217, 222, 284

Hartree-Fock equations, **6**:327–330; **7**:15, 33; **14**:288; **26**:225, 227, 233; **44**:8

canonical form, **9**:329, 334

extended, **6**:324, 365

non-closed shells, **6**:323

relativistic, **6**:406

spin orbitals, **6**:328

Hartree-Fock iterative procedure, **54**:3, 63, 64

multiconfigurational, excitation energy, **54**:30–32

stability condition, **54**:32–33

time-dependent Hartree-Fock approximation, **54**:30–32

unitary transformation, **54**:6–8

Hartree-Fock magnetic properties, **2**:248

Hartree-Fock method, **2**:210, 229, 232ff, 241, 261, 296, 306, 324; **9**:322ff, 344, 348, 352, 358ff; **10**:7, 26;

12:28, 142, 157, 160; **14**:4ff; **24**:195, 197, 203, 237, 239, 245; **25**:181, 184–185, 213; **26**:224; **27**:213, 214, 223; **28**:122ff, 125, 130ff, 145; **37**:286; **41**:118, 150–151; **42**:276; **49**:631

configurational freezing, **48**:462

coupled, **12**:152, 154, 155, 157, 162; **37**:246–248

dependence on heating and cooling rates, **48**:462, 503, 509

entropy of melting and, **48**:465

helium, **2**:316

hydrogen molecule, **2**:243

isobaric cooling, **48**:462

metallic glasses, **48**:501

restricted, **45**:229, 234–236

time-dependent, **41**:121, 124, 125; **48**:15

uncoupled, **12**:152, 154, 159, 160, 162

unrestricted, **14**:326; **37**:221, 285, 286

see also Molecular orbital calculations

Hartree-Fock potential, **14**:291, 368

Hartree-Fock pseudopotential, **31**:394, 410

Hartree-Fock sea, **14**:244; **16**:286

Hartree-Fock-Roothaan approximation, **28**:122

Hartree-Fock-Slater method, **41**:63, 66; **44**:604; **49**:630

Hartree-Fock symmetry dilemma, **14**:324

Hartree potential, **31**:405

Hartree self-consistent field, **44**:489

Hartree wave functions, **2**:63, 236

Hasted-Chong correlation, **50**:347

Heat capacity, **17**:28, 31; **26**:118; **33**:418; **40**:110; **41**:19, 20

$Au_{52}Ni_{48}$ solid solution, **2**:132

influence of hindered rotation on, 2:369

critical region, 6:207–215

Debye law, 40:101, 112

glasses, 48:456, 463, 498, 500, 508

Planck-Schaefer law, 40:112

Heat conductivity, 16:105, 124, 126; 43:222

coefficient of, 16:125

effective, of a reacting system, 16:102, 106

see also Thermal conductivity

Heat flux, 1:2, 5, 6, 159; 32:14; 33:408, 440; 38:117, 127

boundary conditions for, 33:429

dense polyatomic fluids, collisional transfer, 31:164, 173, 186–189

density gradient dependence, 31:196, 209

diffusional, 31:161, 164, 173, 186–189

ellipsoids, 31:183, 186–189, 193, 209

loaded spheres, 31:196

rough spheres, 31:173–174, 177

square-well rough spheres, 31:201

thermal diffusion, 13:364

see also Thermal conductivity

Heat function for hydrates, 2:33

Heat of activation, 1:87

Heat of formation in many-electron theory, 6:316, 389

Heat of mixing, 16:252–253, 260–269, 272–276, 278–279

Heat of sublimation, 16:343

Heat of transition, 1:299

Heat of transport, 1:1–3, 6, 12

Heat of vaporization, 1:12, 13, 214; 6:281; 16:236ff, 241ff, 261, 271

salts, 11:90

Heat pipe oven, 42:5

Heatbath, coordinates, 53:171, 172

coupling constant, 53:152–153

dissipative components, 53:78–79, 80

harmonic chain, 53:74

Langevin equation, 53:81–87

molecular-time-scale generalized Langevin equation, 53:68

primary system and, 53:67

response function, 53:91–92

equivalent harmonic chain, 53:97

normalized, 53:152–153, 159–161

response matrix, 53:175

systematic and random components, 53:90, 95

Heatbath modelling, 44:147

Debye model, 44:158

effective equations of motion, 44:185

Einstein model, 44:230

local friction approximation, 44:185

Stokes law, 44:148

Heavy atom effect, 40:474; 41:304

in phase determination, 16:197–216

anomalous dispersion, 16:205–215

identical molecular units, 16:215–216

isomorphous replacement, 16:201–205

Heavy ion transfer reaction, 47:Part 1:263–264

Heavy metal ions in alkali halides, 4:162, 188–194

HeH^+, 19:43, 44, 54, 120

Heisenberg exchange interaction, 9:358

Heisenberg model, 22:349; 53:230, 263–264

reduced, 53:352–353, 368

Heisenberg model of ferromagnetism, 44:347

Heisenberg paramagnetism, **44**:275

Heisenberg representation, **27**:61; **38**:6, 50

Heisenberg uncertainty principle, **36**:176; **49**:141

Heisenberg unitary operator, **16**:20

Heitler-London energy, **12**:47

Heitler-London procedure, **49**:199

Heitler-Ma perturbation theory, **42**:326

Helical form, **7**:207

Helicity, **33**:82, 92, 95

Helicity quantum number, **42**:391

Helicity rules, **44**:582, 625

Helium, **1**:276, 282, 292, 348; **9**:67; **12**:148; **16**:112–115, 118–119; **34**:178; **41**:140, 146; **44**:446

  atom, clathrate, **2**:12

    collisional electronic energy transfer, **21**:169, 178

    electronic energies, **21**:57

  complexes of iodine and, **47**:Part 1:353–355

  compressibility, **9**:55

  electronic correlation, **2**:149, 248, 250, 258, 307, 324

  ground state energy of, **2**:252, 295, 298, 301, 302, 307, 311

  Hartree-Fock energy, **2**:281, 301

  Hopfield continuum, **52**:267

  isoelectronic series, **6**:377

  light scattering, **24**:41, 51

  liquid, **41**:175

  London coefficient for, **6**:399

  metastable, **45**:455–457

    polarizability of, **10**:21, 24

  microclusters, **40**:55

  mixtures, **17**:24

  molecule, **23**:162, 170, 179; **41**:157

    ab initio calculations, **21**:1ff, 172

    appearance potential, **23**:169

    correlation energy, **6**:356

    dispersion energy, **12**:49

    dissociation energy of $^2\Sigma_u^+$ state, **23**:167

    excited states, $A^1\Sigma_u^+$ and $^3\Sigma_u^+$, **23**:174

    $c^3\Sigma_g^+$ and $C^1\Sigma_g^+$, **23**:177

    $f^3\Pi_u$, **23**:179

    Hartree-Fock equivalent orbitals, **6**:400

    long range attractive potential, **23**:173

    mobility, **23**:179

    potential curve of $^2\Sigma_g^+$, **23**:170

    potential surface, **42**:191

    repulsive potential, **23**:170

    spectroscopic constants, **21**:4, 5

    van der Waals attraction, **6**:398, 402

  molecule (He_2^{2+}), **41**:151

  occupation numbers, **2**:280

  one electron term, **6**:361

  polarizability, magnetic susceptibility and ionization energy, **2**:71

  potential energy, **2**:71

  scattering of, **10**:180–189

  solid, **22**:272, 273

    lattice dynamics, **22**:273

  trimer-tetramer mixtures, **41**:186

  see also Noble gases; Scattering; van der Waals molecules

Helium-argon mixtures, **16**:127

  viscosity of, **13**:363

Helium + carbon dioxide system, **2**:96, 110

Helium-carbon oxide intramolecular energy, **2**:110

Helium-helium cross sections, **10**:62–67

  excited state, **10**:183–184

Helium-like ions, electronic correlation, **2**:239, 240, 300, 303, 324

  ground state energy of, **2**:299, 302,

312

Helium-neon pair, **44**:446

Helium-nitrogen dioxide-dinitrogen tetroxide mixtures, **16**:112–114

Helium seeded-beam method, **52**:272

Helium + xenon system, **2**:95, 112

Helix, double, kinetics of formation, **33**:98, 100ff

left-handed, **33**:101

nucleation of, **33**:83, 98–100, 103, 104, 108

partially paired, **33**:99

right-handed, **33**:101

α-Helix, **33**:74

Helix-coil transition, collagens, **21**:540

kinetics of, midtransition, **33**:87, 93

oligonucleotides, **33**:98–100

oligopeptides, **33**:103, 104

polynucleotides, **33**:98–100

polypeptides, **33**:80–98

poly-L-proline, **33**:101

poly-L-tetradic acid, **33**:93–98

protein unfolding and refolding, **33**:104–115

polypeptides, **15**:299

Hellmann-Feynman theorem, **2**:270; **7**:18; **24**:199; **31**:6; **41**:62, 65

generalized, **12**:52

cartesian coordinates with lab-fixed origin, **12**:60

confocal elliptical coordinates, **12**:61

electronic coordinates relative to associated nucleus, **12**:60

electrostatic, **12**:55

integral, **12**:62

Rayleigh-Schrödinger perturbation theory of, **12**:53

virial theorem, **12**:61

Helmholtz free energy, **1**:207, 332; **24**:97, 98, 129; **27**:475; **31**:28, 40,

55; **33**:418; **49**:396, 402ff, 407–414, 417, 420, 426–428, 448

binary systems, **11**:100

sticky hard spheres, **21**:426

see also Free energy

Helmholtz ponderomotive forces, **1**:338

Helmholtz theorem, **32**:55

Helmholz-Wolfsberg method, **5**:81

Hematoporphyrin, **7**:634

Heme *a* complexes, **7**:394

Heme *a* ferric complexes, **7**:392

Heme enzymes, **39**:86

Hemes, **7**:167

Hemoglobin, **7**:182, 395; **21**:570; **39**:86

adsorption isotherms, **7**:240

electronic structure, **7**:159

hydration, **7**:243

magnetic properties, **7**:159

metabolic breakdown, **7**:77

protein content, **7**:242

water adsorption, **7**:238

Hemophilus influenza, **7**:220, 593

Hemoproteins, **7**:71, 375

chemical bond, **7**:182

electronic structure, **7**:159

magnetic properties, **7**:159, 182, 607ff, 609ff

peroxide compounds, **7**:614

Hemotropin, **7**:609

Henon and Heiles Hamiltonian, **36**:38, 45, 48

Henry's law constant, **6**:282

Heparin, **21**:633, 638

Heredity, **8**:177

Hermite equation, eigenvalue of, **4**:211

Hermite polynomials, **46**:340, 350, 352

Hermitian form, **46**:246

Hermitian function, **44**:296

Hermitian model Hamiltonian, **52**:247–251

Hermitian operator, 7:7

Herpes simplex, 7:222

Hertz operator, 33:204, 205, 209

Hertz vector, 1:337, 350

Herzberg-Teller bands, 5:241

Hessian, fixed, 54:67–69
   generalized, 54:65–72

Hessian convergence characteristics, update methods, 54:88–90

Hessian matrix, 20:90, 110
   eigenvalues, 54:28–30, 156–157
      global convergence, 54:45–56
      local convergence, 54:56–59

Heteroaromatic systems, PMO method in, 8:85ff

Heteroatoms, 7:30ff

Heterocyclic bases, 7:346ff

Heterogeneous equilibria in clathrate solutions, 2:34, 35

Heterogeneous kinetics, 29:56

Heterogeneous lattice, 52:468

Heterogeneous polymerization, 2:163

Heterogeneous reaction dynamics, 41:477–516

Heterogeneous system, 38:318

Heterogenoeus nucleation, *see* Nucleation

Heteromolecules, 7:37

Heteronuclear molecule, 36:105

Heteropolar bond, 1:170

β-Hexabromocyclohexane, 20:17
   crystal structure, 5:181

Hexabromomethane, 20:28

Hexachlorobenzene, influence of radiation on, 2:198, 200
   quadrupole spectrum, 2:197

Hexachlorocyclohexane, quadrupole spectrum, 2:195, 197
   resonance lines, 2:191

β-Hexachlorocyclohexane, 20:3, 17

crystal structure, 5:181

γ-Hexachlorocyclohexane ($C_6H_6{}^{35}Cl_6$), crystal, cryoscopic measurements, 2:193

Hexachlorocyclopentadiene, dimerization of, 16:168, 170

1, 3, 7, 9-Hexachlorodibenzo-*p*-dioxin, 16:172

Hexachlorodisilane, barrier height of internal rotation, 2:382

Hexadecapole, 44:446, 453, 455

Hexadecapole-induced dipole intensity, 44:454

Hexadecapole interaction, 5:188

Hexadecapole moment, 44:450, 459

Hexadecapole term, 44:462

Hexafluoride molecules, group theory, 9:175

   infrared-active binary combination bands, 9:177

   infrared and Raman spectra, 9:172

   metal, 9:223

   nonmetal, 180ff

   nonbonding valence electrons, 9:173ff

   table of electron affinities, 42:555

   thermodynamic properties, 9:312ff

   vibrational parameters, 9:258

   vibrations of, 9:169ff

Hexafluorobutadiene, structural formula, 16:166–168

Hexagonal array, 41:198, 210, 215, 225

Hexagonal close-packed crystal, 49:649

Hexagonal close-packing, 5:186

Hexamethylenediamine, crystal structure, 5:185

Hexamethylenetetramine, 5:180, 182; 20:4, 22
   solid, 22:255
      coherent neutron scattering of, 22:208, 255

lattice dynamics of, **22**:236–237, 255

potential functions of, **22**:209, 237, 239, 255

Hexokinase reaction, 7:717

Hidden symmetry, 42:191

Hierarchical clustering, 35:121, 134

Hierarchical structure, 35:121, 122, 134

Hierarchy, BBGKY, 26:125

High density expansion, 14:232

High energy approximation, 30:332, 353; **42**:433

High pressure, effect on chemical reaction, **39**:161, 162

High pressure laser-induced reaction, 47:Part 2:10–12

High pressure techniques, 9:377

High resolution spectroscopy, 49:48–42, 116–117

High-spin, 7:169ff, 375ff

High-spin complexes, 8:51

d^5, nephelauxetic effect, **5**:69

ferric porphyrins, **7**:391

magnetic susceptibility, **5**:198

High temperature approximation, 34:118

High temperature series, 34:118

High velocity approximation for charge transfer, 10:234–238

High velocity neutral beams, elastic scattering, **10**:29–73

Highly excited vibrational state, 50:395ff

Hilbert-Schmidt projection, 53:367

Hilbert space, 33:382, 395; **38**:21, 29, 31, 41; **44**:265; **53**:347, 348, 362ff

central decomposition, **38**:41

dynamical properties, **17**:96

nonseparable, **38**:31

nonstandard, **38**:29

rigged, **38**:29, 31

structure functions as points in, **33**:382, 395

Hilbert transforms, 4:87, 89

Hilfsgas, 2:18

Hill coefficient, in allosteric model for PFK, **38**:380

in glycolytic oscillations, **38**:382

Hindered rotation, 2:33, 34

internal, **2**:382

Hirschfelder approximations, 3:93

Hirschfelder-Wigner model, 49:210, 228

History of a function, 24:106, 112

derivative, **24**:116

sets, **24**:109

smooth, **24**:117

tame, **24**:114

HKS theorem, excited state, **41**:139

finite temperature extension, **41**:67, 73

inhomogeneous liquid system, **41**:174

time-dependent, **41**:127

Hohenberg-Kohn theorem, 41:66, 71–73; **54**:241–242

Hole-burning, 41:296

in optical transitions, **47**:Part 2:434–444

Hole-particle diagrams, 14:155

Hole-particle formalism, 14:42

Hole state calculations, 23:93, 99

Hole theory of fluids, 6:238

Holes, Coulomb and exchange, **41**:95

Holtzmark distribution, 15:121

Homeostatic mechanism, 29:266, 279

Homogeneous bifurcations, nonequilibrium phase transitions and, **55**:207–209

Homogeneous broadening, related to coherent processes in crystals, **40**:379, 383–385, 473, 474

Homogeneous flow, **35**:75

Homogeneous linewidth, **41**:286

Homogeneous nucleation, **48**:401

    kinetic approach to, **40**:137–153

    see also Nucleation

Homogeneous relaxation, **47**:Part 1:557

Homogeneous system, **38**:317; **41**:117

Homonuclear molecule, **36**:106

Homopolar bond, **1**:170

Homopolymer, **2**:163, 183

Hönl-London factor, **42**:8

Hoover graph, **20**:48, 52

Hopf bifurcation, **43**:60

Hopfield continuum, **52**:267

Hopping, charge carrier, **44**:294, 391, 396

Hopping transport, **52**:475, 483, 493

Hordeum vulgare, **7**:634ff

Hormone effectors of adenylate cyclase, **39**:207

Hormone receptor, **39**:207

Horn, **7**:265

Hornbeck-Molnar process, **10**:188

Host crystals, **8**:27ff, 30ff

Hot atom reaction, **41**:250

Hot bands, **2**:374

    photoionization studies, **52**:265

Hot lattice, **2**:4, 11, 21

Hot T atom experiments, **36**:174, 194

Householder tridiagonalization, **26**:239

Høye and Stell theory, **48**:316

Hubbard model, **52**:71, 90–92, 98

Hückel approximation, **4**:105, **5**:3, 247; **6**:386; **7**:26ff, 88, 96, 108, 111, 124, 136ff, 147; **9**:122, 154, 159; **14**:297; **36**:164; **49**:50, 78; **53**:228, 287, 292

    applied to chemical reactivity, **8**:74ff, 105, 124

    carbonyl compounds, **8**:135

    extended, **21**:27–36

valency theory, **1**:179, 183, 240, 242, 248, 250, 254–256, 258, 260

Hugenholtz diagram, **36**:224

Hugoniot shock curve, **9**:369

Hulburt and Hirschfelder model, **49**:216–218, 220, 269–270

Hume-Rothery rules, **51**:269

Hund's cases, **12**:15; **28**:128, 334, 350; **30**:372, 423, 433, 468; **42**:227, 242; **44**:20, 21

    (a), **18**:155, 181

    (b), **18**:155, 156, 182

    (c), **18**:156

Hund's rules, **5**:191; **6**:318; **7**:162, 187, 193; **8**:50; **36**:217

Hurwitz determinant, **43**:53, 54

Hurwitz nonmixing network, **43**:193–203

Husimi function, **38**:84, 85

Husimi trees, **11**:377

Hutchison-King experiments, **40**:400

Hybridization, **4**:184; **7**:18; **9**:125

Hybridized nitrogen atom, **7**:91

Hydra regeneration, **29**:257

    grafting experiments and, **29**:258

    polarity in, **29**:263

    two gradient model of, **29**:259

Hydrated electron, **44**:502

Hydrates, **2**:7, 9, 21, 31, 41

    crystallization, **2**:44

Hydrocarbon-carbon monoxide systems, **17**:13, 30, 51

Hydrocarbon-methane systems, **17**:8

Hydrocarbons, **1**:34, 137, 165, 177–193, 197–199, 239ff, 250ff; **9**:63

    carcinogenic, **7**:152

    liquids, compressibility, **9**:73

        isothermal compressibility, **9**:73, 79

        surface tension, **9**:73

saturated, bond energies, 6:395

Hydrochloric acid, 34:192

Hydrochloric acid clathrates, 2:2

in hydroquinone, 2:7

Hydrodynamic approximation, 27:269, 274, 346

Hydrodynamic correlations, 11:253, 263

Hydrodynamic equations, 5:262; 32:94

flames and detonations, 3:60

boundary conditions, 3:66

steady-state, 3:65

time-dependent, 3:63

Hydrodynamic fluctuation theory, 32:97

Hydrodynamic forces, 11:253, 271

Hydrodynamic instability, 24:4

Hydrodynamic interaction, 35:33, 34, 111; 44:299

polymer dynamics, 15:325

Hydrodynamic limit, 24:12, 19

Hydrodynamic modes, 38:176, 186

Brillouin modes, 24:13

thermal diffusivity mode, 24:13

thermal relaxation mode, 24:13

viscous relaxation mode, 24:13

Hydrodynamic Navier-Stokes equation, 11:261

Hydrodynamic stability, 32:17ff

Hydrodynamic time scale, 11:323

Hydrodynamic variables, 44:297

Hydrodynamics, 40:31–34, 37, 38

generalized, 46:310

Hydrogen, 1:16, 21, 34–36, 53, 54, 58–60, 170, 171, 277; 16:127, 163; 41:140, 141, 171

adsorption, on Ni, 49:617

on Pd (111), 49:613–617

on Pt, 49:617

on Si (111), 49:617–618

on Ti (0001), 49:607–612

atom, 12:5, 146, 148; 41:138, 140;

49:358

collisional excitation, 30:454

diffusion, 35:24

random Stark modulation of, 15:120

reaction with ClO_2, 50:96

reaction with NO_2, 50:96

cubic phases, 16:378

desorption of, 41:497

diffusion of gases in, 5:161

$E \rightarrow V$ energy transfer, 47:Part 2:396, 401–402

$H + H_2$ reaction, 21:117, 120, 129; 25:140, 146

interatomic distance, 5:156

ion (H^-), 41:140, 141

in liquid argon, neutron scattering by, 15:131ff

mesospheric chemistry, 55:71–72

molecule, 12:560, 562, 568; 41:151, 152

as quencher, 50:359, 382

$B^1\Sigma_u^+$ state, adiabatic approximation, 12:24

Born-Oppenheimer approximation, 12:24

dissociation energy, 12:24

$b^3\Sigma_g^+$ state, 12:26

$b^3\Sigma_u^+$ state, 12:26

-carbon oxide intramolecular energy, 2:110

clathrates, 2:12, 20

clusters of, 52:304–306

correlated wave function, 2:300

correlation effects in, 2:250

correlation energy, 2:241, 245, 248, 258; 6:356

$E^1\Sigma_g^+$ state, 12:25

electronic energy, 2:300; 21:60

ground state, adiabatic approximation, 12:23

Born-Oppenheimer approximation, **12**:23

coulombic energy, **12**:65

dispersion energy, **12**:26, 27

dissociation energy, **12**:13, 22, 74

electron correlation in, **12**:67, 69

energy, **2**:300, 307

exchange energy, **12**:65, 92

Hellmann-Feynman theorem applied to, **12**:51, 55

hyperfine interaction, **12**:92, 97

nonadiabatic approximation, **12**:23

ortho, **12**:97

para, **12**:97

polarization correction, **12**:16

radiative corrections, **12**:74

relativistic corrections, **12**:22, 74

resonance energies, **12**:27

interaction with He, **12**:589, 595

London coefficient for, **6**:399

magnetic susceptibility of, **2**:70

metallic form, **6**:86, 125

ortho, **12**:555, 558, 567, 568, 573, 576, 587, 591, 569

ortho-para conversion, **1**:125

para, **12**:555, 573, 576

polarizability, magnetic susceptibility and ionization energy, **2**:71

potential energy, **2**:71

reaction with fluorine to form HF + H, **41**:229–268

S-P dipole resonance interaction energy, **12**:41

fine structure splitting, **12**:45

triplet state, **6**:395

wave function, **2**:243, 303

see also Photodissociation; Potential energy surfaces; Reactive scattering

molecule (D_2), **12**:572, 596

molecule (H_3), potential energy surfaces, **21**:67

molecule (H_3^+), **19**:39, 53, 59, 80, 86, 91, 97, 112, 118, 128

molecule (HD), **12**:571, 596

dipole resonance interaction energy, **12**:41

ground state, adiabatic approximation, **12**:23

Born-Oppenheimer approximation, **12**:23

dissociation energy, **12**:22

nonadiabatic approximation, **12**:23

open-shell wave function, **12**:69

relativistic corrections, **12**:22

molecule ion (H_2^+), **41**:151

negative ion (H^-), **2**:299

electronic correlation energy, **2**:249, 324

ground state energy of, **2**:302, 303, 312

reaction with oxygen, **43**:164, 166

solid, **16**:376–378; **22**:290

structural and thermodynamic data, **16**:346–348

thermodynamic data and lattice constants, **2**:8

see also Deuterium; Diatomics-in-molecules; Scattering; van der Waals molecules

Hydrogen bond, **1**:71, 138; **2**:4, 175; **7**:86ff, 111, 203, 207; **16**:163–164, 171–173, 179, 181, 186, 193; **44**:353; **49**:29–30

DNA, **8**:177

energy, **7**:232

linking γ-carboxylic groups, **33**:97

network, **31**:80

polygons, **31**:58, 79

polymers, **39**:113

polypeptides, **33**:74

proteins, **39**:178, 179, 182

solvation-desolvation, **39**:180, 181

water, **31**:17, 18, 42, 58, 793; **44**:351

Hydrogen bonded acids, deprotonation reactions of, **21**:237–242, 244

Hydrogen bonded complex, **41**:281

Hydrogen bonded solvent, **53**:65

Hydrogen bromide, clusters, **52**:299–301

E → V energy transfer, **47**:Part 2:396, 401–402

hydrate, **2**:35

Hydrogen + carbon dioxide system, **2**:110

Hydrogen + carbon monoxide system, **2**:96, 108

Hydrogen chloride, clusters, **52**:299–301

E → V energy transfer, **47**:Part 2:395–396, 401

hydrate, **2**:35

pair potential, **53**:46–47

structure, **53**:30–33

Hydrogen clathrates, **2**:2

chloride, **2**:30

+ nitrogen system, **2**:108

Hydrogen cyanide, E → V energy transfer, **47**:Part 2:400

Hydrogen-deuterium exchange, **41**:497

Hydrogen fluoride, **16**:126–127

clusters, **52**:299–301

dispersion energy, **12**:21

E → V energy transfer, **47**:Part 2:395, 400–401

formed in reaction of hydrogen and

fluorine, **41**:229–268

many-electron theory, **6**:383

Hydrogen halides, solid, **22**:207, 221

infrared spectra of, **22**:207, 252

librations of, **22**:221

Hydrogen iodide, clusters, **52**:299–301

Hydrogen-like function, **2**:274, 318

Hydrogen-like orbitals, **2**:274, 275

Hydrogen pseudocontinuum, **52**:267

Hydrogen sulfide, **36**:332

as "hilfsgas", **2**:18

hydrate, **2**:3, 42, 50

Hydrogen + xenon system, **2**:96, 112

Hydrogenated compounds, **4**:300

Hydrogenous fluid, structure of, **53**:29–44

Hydrolases, **7**:74

Hydrolytic deaminases, **7**:74

Hydrolyzable linkages, **7**:62

Hydrophobic binding in carcinogenesis, **8**:169

Hydrophobic interaction, **33**:97, 107; **39**:131, 157, 178, 181, 182; **53**:65

Hydrophobic side chain, **33**:108

Hydroquinone, cage, **2**:28

clathrate, **2**:2, 5, 6, 11, 20, 21, 29, 35, 39, 43

argon, **2**:38, 41

energies of formation, **2**:30

equilibrium pressure, **2**:30

methanol, **2**:39, 41

thermodynamic properties, **2**:5

rare gas system, **2**:35

α-**Hydroquinone**, clathrate compounds, **2**:37, 41

-gas clathrate, **2**:36

β-**Hydroquinone**, clathrate compounds, **2**:2, 6

with molecular oxygen, **2**:25

hydrogen bonds, **2**:2

lattice, **2**:29

Hydrostatic pressure, 16:106

biological effects, **21**:647

disjoining pressure, **48**:341

β-**Hydroxybutyrate dehydrogenase, 39**:203–205

6-Hydroxycrinamine, 16:190–191

Hydroxylamine, poison of photosynthesis, **7**:636

Hydroxymethylcytosine, 7:52ff

5-Hydroxytryptamine, 16:178–179

Hyleraas-Undheim-McDonald theorem, 54:25–26, 27

Hyleraas-Undheim theory, 18:102

Hyperchromicity, 7:128

Hypercycle, 38:245, 247, 249, 252, 254, 259; **43**:135, 212

chirality, **38**:259

exponential growth, **38**:248

fixed point, **38**:251

genetic recombination, **38**:259

hyperbolic curve, **38**:248

interacting, **43**:163

limit cycle, **38**:251

once=for=ever selection, **38**:250–252

sinks, **38**:251

sources, **38**:251

stability of, **38**:250

trajectory analysis, **38**:251

Hypercyclic organization, 38:243, 253, 255

compartmentalization, **38**:256

genetic code, **38**:255

genotypical, **38**:256

natural amino acids, **38**:254

phenotypical, **38**:256

proteinoids, **38**:254

replication, **38**:255

t-RNA type, **38**:254

templates, **38**:255

translation apparatus, **38**:255

Hyperfine coupling constants, 36:208

Hyperfine fields, cobalt, **41**:173

gadolinium, **41**:173

iron, **41**:173

nickel, **41**:173

Hyperfine interaction, 7:533ff, 562ff; **12**:7, 91, 92, 96; **18**:195; **35**:12; **40**:398, 400ff, 423; **44**:31, 80, 103

anisotropic, **37**:260, 261

Fermi contact term, **18**:208

Hamiltonian, **18**:175

matrix elements, **18**:200, 201, 207

Hyperfine tensor, 40:400

Hypernetted chain approximation, 6:193, 237; **40**:313; **41**:28–31, 31–36; **48**:238–243

determination of fourth and fifth virial coefficients, **6**:265

linearized, **48**:218, 238

perturbation expansion of, **48**:244–245

quadratic, **48**:223, 238

Hyperpolarizability, 12:110, 136, 137; **46**:8

Hypersonic dispersion, 24:9

Hypersonic flow, collision frequency in, **13**:29, 32

dissociative recombination in, **13**:29

intrinsic coordinates, **13**:23

ionization in, **13**:32, 35

reaction kinetics in, **13**:19ff

see also Supersonic flow

Hypersonic velocity, 24:40

Hypersound velocity, 24:21

Hypersurface of steady states, 43:160

Hypervirial constraint, 44:610

Hypervirial equivalence, 44:618

Hypervirial relations, 12:62; **44**:554, 577, 580, 600, 605ff

Hypophosphite anion, infrared and

Raman spectra, 7:478ff

Hypoxanthine, 7:47ff, 50ff

Hysteresis, 32:4, 282, 292, 308, 311; **40:**238, 241; **43:**224, 228

asymptotic, **20:**119

axon membranes, **29:**323

and bistability, **38:**329

chemical systems, **29:**6, 40, 50, 64; **39:**14, 41, 240

coordinate, **20:**93, 116

cycle, **20:**75, 76, 92, 116

dioleylphosphate-Millipore membrane, **29:**329, 335

enzyme membrane, **39:**231

genetic regulatory processes, **29:**40

influence of fluctuations, **39:**43, 44

nonlinear neural networks, **29:**40

photochemical reactions, **29:**59

virgin, **20:**118

I

IBr, *A-X* band system, **50:**24

B-X band system, **50:**25

photofragmentation of, **50:**288

potential curves, **50:**22, 40

Ice, entropy, Pauling, **31:**66

Ih, **31:**24

polymorphs, **31:**3, 25

ICl, *A* state lifetime, **50:**44

A-X band system, **50:**45

B state lifetime, **50:**45

B-X band system, **50:**46

E-A system, **50:**48

hydrolysis of, **50:**44

matrix trapped, **50:**46

potential curves, **50:**48

ICN, vacuum ultraviolet photodissociation, **50:**258

Icosahedral numbers, 40:52, 130

Icosahedron, 40:84, 129

Ideal gas, 40:140, 144; **53:**245

entropy production, **3:**307

kinetic theory of, **40:**142

Ideal localized adsorption, 2:5, 10

Ideal solution, 43:95

Identical particle scattering, 30:68, 329, 346, 361, 374, 379

IF, *B* state dissociation energy, **50:**52, 57

chemiluminescence, **50:**52

population inversion, **50:**57

potential energy curves, **50:**23

predissociation of *B* state, **50:**57

preparation of, **50:**52

Ignition temperature, detonations, **3:**115 flames, **3:**74

Image charge, 27:317, 318

Image force, 37:210, 213

Image intensifier, 42:5

Image interaction, 27:213

Image method, 37:10

Imidazole, 7:369; **39:**120

Imino forms, 7:104

Immunity, cell mediated, **38:**394

Immunochemistry of membranes, 39:275–286

Impact approximation, 42:250

Impact diameter, 1:120

Impact parameter, 1:121; **10:**38, 77; **12:**394; **25:**148

classical, **12:**403

critical, **10:**265

glory, **10:**114, 120

reduced, **10:**77, 86, 136

threshold, **10:**154, 165

Impact parameter method, 10:215, 234, 240, 255

Impact spectra, 10:123–126

Imperfection theory, singular perturba-

tion, **55**:181–183

Impulse approximation, 10:321; **13**:86; **30**:60, 66

Impulsions, average, **4**:254, 255, 266, 277

Impurity, 13:132, 134, 153; **41**:200, 206
absorption, **17**:264
atoms, **7**:255
conduction, **52**:494
effect on crystal growth rate, **40**:217, 218
energy levels, **7**:147
exciton scatterer, **40**:372, 439ff, 452, 454, 475
ion, **33**:8
segregation of during crystal growth, **40**:219

In-out correlation, 12:69

Inactivation, sodium, **29**:363, 368

Inactive medium, 27:360

Incoherent energy transport, 40:438, 452

Incoherent overlayer structure, 27:146

Incoherent scattering, 17:189

Incomplete shells, 7:161

Index function, Maslov, **36**:10, 15

Index of refraction, 4:68, 71–78, 81, 90, 91

Indicator function, 38:59

Indirect coupling theory, 40:313, 319, 336, 352

Indirect exchange interaction, 5:235ff

Indium, 41:141

INDO, 42:67

Indole group, 16:189

Induced absorption, 51:49–109
angle-dependent intermolecular potential, **44**:459
atomic-induced, **44**:446
bimolecular, **44**:464
octopole-induced, **44**:476

gas phase, **44**:445
H_2-induced, **44**:457
hexadecapole-induced, **44**:455
high microwave frequencies, **44**:445
linewidth, **44**:455
octopole-induced, **44**:474, 476
quadrupole-induced, **44**:454, 462
band shapes, **44**:445
$(CN)_2$, **44**:457
symmetric top molecules, **44**:416
translational, **44**:447, 463
see also Far infrared

Induced dipole, 44:261, 448
rotational motion, **51**:75
see also Dipole

Induced dipole-quadrupole effect, 10:119

Induced force, average, **53**:185–186

Induced polarizability, 51:88

Induced vibrational-rotational Raman spectra, 51:106–107

Induction energy, 12:107, 124
ion-molecule interaction, **12**:37
polar-nonpolar interaction, **12**:37
quadrupole-induced dipole, **12**:37

Inductive perturbations of benzenes, L_a state, **54**:209–211
L_b state, **54**:202–204

Inelastic collisions, rotational, **50**:590–642

Inelastic deflection function, 42:462

Inelastic scattering, 10:32, 39, 54–57, 65–66, 126–131, 179; **25**:255, 283; **53**:50, 101, 103
cross section, **1**:362, 363; **10**:127–131
electrons, **19**:151
energy loss in violent collisions, **10**:212–218
excited species, **10**:179, 187–190
ion-neutral collisions, **10**:199–200,

248
neutrons by liquids, **15**:129ff
rainbow, **42**:467
Inert gas, catalogue of collision cross sections, **30**:419
quenching of excited oxygen, **42**:146
quenching of metastable states, **50**:345
Inert pair effect, **49**:88
Inertia tensor, **24**:97, 103, 104, 141, 146; **44**:12, 104, 113
Inertial effects, **44**:257, 357, 360
Inertial force, **24**:97, 104
Infinite order sudden approximation, **36**:82; **42**:431; **49**:238–241; **50**:605; **51**:188, 195
j-labelled, **49**:244
Khare, Kouri, and Baer theory, **49**:240–241, 245–247, 250–251
l-labelled, **49**:244–246
matching procedure, **49**:247–250
reactive cross section, **49**:240
Infinite volume correlation function, **40**:301
Infinite wall model, **42**:383, 390
Infinite wavelength limit, **53**:314
Infinitesimal disturbances, **26**:178, 209
Influence measure, **24**:110
Information, **38**:211, 213, 215, 236, 245
colibacterium, **38**:213
Ehrenfest's urn game, **38**:217
equipartition, **38**:217
games, **38**:238; *see also* Game theory
glass bead games, **38**:216
Kullback, **32**:62
natural selection, **38**:215
quality of transfer, **38**:237, 238
bacteriophage, **38**:237
binary units, **38**:239
colibacterium, **38**:238
complementary base pairs, **38**:237
DNA polymerases, **38**:238
genotypical, **38**:239
hydrogen bonding, **38**:237
nucleotides, **38**:238
phenotypical, **38**:239
$Q\beta$-phage, **38**:238
threshold of error, **38**:239, 241
translational apparatus, **38**:237
recurrence times, **38**:213
transfer of genetic, **33**:100, 105
Information content, cooperative, **38**:253
monophosphate, **38**:253
nucleoside pyrophosphate, **38**:253
nucleoside triphosphate, **38**:253
species, **38**:236
Information entropy, **17**:160; **44**:275
Information gain, **32**:62
Information measure, **17**:118
Information theory, **15**:39, 44ff; **17**:161; **26**:127; **36**:183; **42**:292; **44**:329; **46**:154, 155, 158, 161
capacity, **46**:160, 174, 177, 191
communication theory, **46**:190
interpolative model, **17**:124
intramolecular dynamics and, **47**:Part 1:239–292
decline of entropy on improved experimental resolution, **47**:Part 1:287–288
initial state, **47**:Part 1:269–275
maximum entropy formalism, **47**:Part 1:264–269
surprisal analysis, **47**:Part 1:244–264
memory function, **17**:118, 119
Infrared absorption, **34**:17, 58; **40**:6, 39; **50**:395–474
benzene, **34**:74
CHD_3 and CH_3D, **34**:62

see also Far infrared; Multiphoton excitation

Infrared absorption coefficient, 44:301

Infrared double resonance, 23:48

Infrared laser chemistry, complex molecules, **47:**Part 1:639–659

see also Multiphoton excitation

Infrared laser enhanced chemical reaction, bimolecular, **47:**Part 2:4–13

isomerization, **47:**Part 2:13–14

Infrared laser enhanced diffusion cloud reaction, 47: Part 2:43–54

Infrared line broadening, 44:355

Infrared luminescence detection, 42:440, 453

Infrared spectroscopy, biological applications, **7:**408

medical applications, **7:**432

nucleic acids, **7:**435

sodium deoxyribonucleate, **7:**437

Inhibition, 29:260; **43:**105; **46:**377

competitive, **29:**303

Inhomogeneous broadening, 41:309; **47:**Part 1:5

related to coherent processes in crystals, **40:**379, 383ff, 391ff, 400ff, 423, 473

Inhomogeneous electron systems, spin-density-functional formalism, **41:**59–188

Inhomogeneous linewidth, 41:286

Inhomogeneous relativistic electron systems, 54:231–301; *see also* Relativistic electron systems

Inhomogeneous Schrödinger equation, 14:342

Initial condition, effect of, **48:**149–152

Initial state, 47: Part 1:269–275

activation process, **47:**Part 1:271–272

maximum entropy principle, **47:**Part 1:266–268

stationary state, **47:**Part 1:270–271

Initiators of polymerization, 2:148

Inner product, Liouville space vectors, **53:**127

molecular-time-scale generalized Langevin equation, **53:**149–150

soliton, **53:**326

Inner shell electrons, neglect of in MO calculations, **18:**277

Inner shell ionization, 36:334

Inner shell orbital, 9:356

Inosine, 7:451

infrared spectra, **7:**451

InSb, 27:414, 416–427

Insect morphogenesis bifurcations, 55:219–245

Insertion reaction, 42:118

Insolubilization, enzymes, **39:**36, 46, 229

synzymes, **39:**167

Instability, 19:234, 235; **26:**111; **32:**1, 8, 14, 87, 90; **38:**234, 317, 318; **49:**315, 322, 328, 347

amplification of signals and, **29:**346

bifurcation theory and, **19:**259

burst, **32:**30

cellular division and, **29:**18, 19

chemiacoustic, **38:**348

chemical, **19:**258, 275; **29:**34, 63; **38:**318; **43:**217–268; **49:**311–356

leading to explosion, **21:**292

convective, **32:**92

critical point, **32:**87

critical slowing down and, **29:**86

discharge valve, **29:**133

dissipative system, **29:**29, 64

exponent polytopes and, **43:**187–193

far-from-equilibrium states, **38:**318–362

fluid dynamical, **32:**6

folding, **32:**14

four-pole condenser, **29**:133
gravitational, **35**:128
Hurwitz nonmixing network, **43**:193
hysteresis, **19**:287, 289
irreversible, **20**:77, 93, 111; **29**:12, 13
Jeans, **26**:111, 112, 126; **35**:122, 185
linear description, **32**:154
local potential, **13**:267
localized, **29**:51; **38**:350
nonuniform systems, **38**:318
numerical, **26**:117, 138
one-dimensional model, **32**:153
onset of convection, **32**:94
quasi-reversible, **20**:77, 97, 112
realistic chemical networks, **43**:138–149
redox networks, **43**:149
region with opposing concentration gradients, **32**:147
role of current cycles, **43**:127
snap through, **32**:17, 73
solvable models of, **38**:143–172
subcritical, **26**:188, 205, 208
succession of, **29**:18
symmetry-breaking, **19**:221, 236, 277, 278, 282, 283, 317; **29**:2, 35, 64, 284, 291
tests for, **43**:145, 152, 195, 209
Insulators, **4**:198
Insulin, **39**:84, 335
Integrable system, **36**:44
Integral equations, **49**:119, 126, 141
diffusion eigenfunction and, **48**:175–177
polar-nonpolarizable fluid, **48**:232–245
Integral kernels, *see* Transition kernels
Integral of the motion, **26**:113, 129; **36**:38
Integral transforms, **13**:207, 232, 236

Integrals, Cauchy, **11**:180
Coulomb-type, **13**:211, 245
convolution methods for, **13**:225
Diophantine evaluation of, **36**:152
electron repulsion, **13**:211, 223, 244
ellipsoidal, **13**:214
exponential, **13**:260
four-center, **13**:247
kinetic energy, **13**:210, 218
of Legendre functions, **13**:258
linear three-center, **13**:223
multi-center, **36**:149
electron repulsion, **13**:245
one-center expansion for, **13**:236
nuclear attraction, **13**:210, 238
one-center, **13**:219, 223, 243
one-electron two-center, **13**:214
overlap, **13**:210, 238
Slater-type orbitals, **13**:205, 209
three-center, **13**:247; **36**:166
nuclear attraction, **13**:220, 243, 244
two-center, **13**:231, 243
Integration method, HEG, **37**:327, 331, 333, 335
polynomial expansion, **37**:335
Simpson's rule, **37**:327
Integrodifferential equations, **32**:46
Galerkin's method of solution, **13**:318
Intensity, allowed, **44**:558, 561
bond mode, **47**:Part 1:302–306
crystal spectra, **8**:13ff
mixed crystals, **8**:30
forbidden, **44**:558, 561, 562
ordinary and rotatory, **44**:548, 555, 561ff, 575, 588, 592, 599, 601
partial, **44**:558
Intensity correlations, **24**:31, 32
Interactant, **43**:74, 75

Interacting reagents, equations of motion for, **53**:179

equivalent chain representation for response function, **53**:178–179

low frequency parameters, **53**:180–181

molecular-time-scale generalized Langevin equation, **53**:174–178

Interaction, apolar, **39**:110, 111, 112, 131

collective Coulomb, **11**:271

collective hydrodynamical, **11**:271

electrostatic, **39**:110, 134

enzyme-membrane, **39**:217

heterotropic, **39**:192

homotropic, **39**:192

hydrophobic, **39**:131, 157, 178, 181, 182

short-range, **11**:35

Interaction energy, defect, **11**:16

Interaction parameter expansion of grand partition function, **6**:88

Interaction potential, **20**:44

exact, **20**:44

helium, **20**:64, 65

three-body, **20**:44

two-body, **20**:44

see also Potential

Interaction representation, **25**:4, 10; **35**:141

Interaction site model, **34**:236; **53**:25ff, 57

Interaction volume, effective, **13**:115

Interband transition, **51**:262, 264

Intercalation techniques, **49**:480–481

direct liquid immersion, **49**:481

electrolytic reaction, **49**:481

Hérold method, **49**:481–482

solution-reaction method, **49**:481

see also Graphite intercalation compounds

Interchange symmetry, **23**:237

Interelectronic coordinates, **25**:267

Interelectronic repulsion parameters, **5**:40, 68, 69

transition metal complexes, **5**:62–64

Interface, **32**:42, 164, 196

anisotropy, **32**:201

density profile, **40**:171

diffusive fingers, **32**:138

discontinuous, **49**:359, 389–392

multicomponent, **49**:391–392

step function, **49**:389–390

three-body connection function, **49**:390

fluid, **49**:357–454

flux across, **32**:139

influence parameter, **49**:377, 418, 423ff, 433ff, 444ff

kinetics of, **37**:110

liquid film, **48**:360

molecular fluorescence and energy transfer near, **37**:1–66

perturbation of shape, **32**:179

rounded perturbation, **32**:200

sawtooth perturbation, **32**:197

stable, **32**:166

viscoelastic properties of, **48**:365

width, **40**:195

see also Surface Phase

Interfacial tension, **49**:362–373, 415, 424

curved surface, **49**:364–365

horizontal surface, **49**:362

planar surface, **49**:362–365

Interference, **24**:46; **25**:72, 115, 129, 154, 170, 174; **36**:67, 77; **40**:381

evidenced in differential cross section, **12**:404, 424, 482

Interference function, **27**:48

multilayer diffraction, **27**:50

single layer diffraction, **27**:49

Interference pattern, 10:76; **12**:363

approximations for, **10**:98–103

glory, **10**:114–117

near rainbow angle, **10**:104–105

Interferometers, 44:321, 327

Fabry-Perot, **24**:26, 35

tandem, **24**:3, 27

Interferometry, 44:321, 325, 329

amplitude modulated, **44**:327

truncation and amplitude sampling, **44**:329

used in study of Soret effect, **13**:354

Interhalogen compounds, laser induced fluorescence; **50**:21ff

see also specific interhalogen compounds

Intermediate coupling, 5:42, 46, 61; **35**:145, 179

Intermediate normalization, 14:299

Intermediate-range forces, 12:345

Intermediate scattering function, 17:186

Intermediate-size molecule, collisional effects in electronic relaxation, **47**:Part 2:357–358

Intermode coupling, 41:293

Intermolecular degrees of freedom, 41:332–338

Intermolecular electron exchange, 35:12

Intermolecular forces, 1:5, 203, 267; **53**:21, 45, 65, 150

ab initio calculation of, **30**:163

anisotropic, long range, **30**:331, 333, 337, 390, 407

short range, **30**:393, 408–414

effective, **30**:51

three-body, **30**:365

see also Potential energy surface

Intermolecular interaction, 33:349, 350–353, 354, 355

Coulomb-type, **33**:353, 354, 358

exchange-type, **33**:354, 358

HF approximation for, **33**:350, 354, 355

perturbational treatment, **33**:349

pi-systems, **33**:362

Intermolecular light scattering, 24:44, 50, 55

Intermolecular potential, 6:400ff; **9**:30ff, 33; **10**:31–33, 76–77; **22**:207, 208; **23**:41, 65, 78, 79; **34**:117; **42**:276; **44**:262, 309, 323, 346, 445ff, 453ff, 459, 470

atom-atom, **22**:209

atom-diatom, **42**:277

charge transfer, **10**:235–236

classical, **22**:208

compressed hard sphere fluid, **9**:64

convex molecules, **5**:156, 171

dependence of collisional relaxation rates on, **47**:Part 2:368–369

ion-neutral reactions, **10**:265

nonadditivity of, **5**:173, 187

nonspherical molecules, **10**:47–49

from phase shifts, **10**:88

quadrupole-quadrupole, **22**:209

quantum calculations of, **10**:66–68

reduced, **10**:77

small distances, **10**:30, 32–33, 68–70, 82

from high-energy scattering, **10**:43, 63–68

soft, **9**:42

vibrational coordinate dependence, **42**:278

see also Interaction potential; Potential; individual potential functions

Intermolecular proton transfer, 47:Part 2:667–673

Intermultiplet transitions, 42:198

Internal conversion, 41:300

Internal degrees of freedom, 16:224,

227, 229–231

Internal energy, 1:207, 296; **7**:205; **24**:105

microclusters, **40**:110

Internal energy barrier, 1:114

Internal energy equation, 34:120

Internal energy transfer, 24:336

Internal field, 1:80, 345; **44**:262ff, 284, 341ff

correction, **44**:303, 344, 363

Internal field effect, *see* Relaxation effect

Internal force relation, 24:97, 98, 104, 105, 126

Internal libration, 44:381

Internal pressure of liquids, 1:160

Internal process, 44:381

Internal representation, 43:19

Internal rotation, 44:375

Internal species, 43:16

Internal symmetry coordinates, 5:255

Interpair excitations, 9:338

Interphase transfer, 13:274

Interstellar cloud, 36:195

Interstellar grains, 40:51, 56, 97

Interstellar molecules, 55:100–101

Interstitials, 31:33; **40**:151

Intersystem crossing, 35:2, 4; **41**:303; **42**:133, 138

anisotropy, **35**:4

collision-induced, **42**:223

dephasing, **47**:Part 2:308–311

intramolecular vibrational relaxation, **47**:Part 2:311–314

large molecules, **47**:Part 2:323–325

low pressures, **47**:Part 2:314–329

magnetic field effects, **47**:Part 2:325–329

molecular model, **47**:Part 2:300–302

pressure dependence, **47**:Part 2:329–333

review of experimental data, **47**:Part 2:294–298

small, intermediate and statistical limits, **47**:Part 2:302–308

rate, **35**:4

selection rules, **35**:4

Intraatomic correlations, 6:404

Intracellular communication, *d. discoideum,* **38**:366

Intramolecular attractions, 6:399, 400

Intramolecular dephasing, 42:219

Intramolecular dynamics, 47:Part 1:117–506

bond modes, **47**:Part 1:293–322

information theoretic approach, **47**:Part 1:239–292

onset of statistical behavior, **47**:Part 1:201–238

transfer of vibrational energy, **47**:Part 1:117–200

van der Waals molecules, **47**:Part 1:323–506

Intramolecular energy transfer, 47:Part 1:16–17, 51–56, 640; **47**:Part 2:237–289

rotational, **47**:Part 2:258–263

theory of, **47**:Part 2:241–245

vibrational, **47**:Part 2:245–258

Intramolecular isotope effect, 1:44

Intramolecular London energies, 2:74, 78

Intramolecular proton transfer, 47:Part 2:650–667

Intramolecular relaxation, 50:121, 461

Intramolecular spin-dependent interactions, 12:551, 555–557, 597

Intramolecular vibrational energy transfer, 47:Part 1:29, 117–200

coupled nonlinear oscillator systems, classical, **47**:Part 1:119–161

quantum mechanical, **47**:Part 1:161–195

Intramultiplet mixing, 28:335, 341–349, 358, 364, 372

Intrinsic coordinates for hypersonic flow, 13:23

Intrinsic defects, 11:28, 29

Intrinsic states, 49:120–128, 140–143, 163–168

nuclear, **49**:172–187

parity-projected, **49**:182

Invariance principle, 52:418

Invariant tori, 36:2, 20, 25ff, 36, 37; **46**:75

Inverse Brownian motion, 20:200, 209

Inverse dynamical transform, 53:126

Inverse electronic predissociation, 47:Part 1:28

Inverse electronic relaxation, 47:Part 1:46–51

Inverse isotope effect, 1:26

Inverse power potential, 10:32, 76, 83

low-angle scattering by, **10**:109–116

Inverse scattering transform, 53:239, 301

Inverse updates, multiconfigurational self-consistent field state, **54**:87–88

Inversion, 49:165–166

process in chemical reaction, **49**:222–230

Inversion of data, 42:294

Inversion operator, 18:223; **49**:166

Inversion procedure, 30:336, 354

Inversion symmetry, 44:298; **49**:181–182

Inverted band, 8:34

Inverted-Raman double-resonance, 47:Part 1:574–575

IO, electron resonance spectra, **18**:216, 228, 231, 233

Iodine, atom, quenching of excited, **50**:480

correlation function calculations, **53**:192, 198ff

crystal structure, **5**:184

laser, **42**:34

molecule, absorption cell, **24**:30

dimer, rotational transitions in, **42**:289

dissociation limit of B state, **50**:24

intramolecular London energy, **2**:78

laser induced fluorescence, **50**:24

photon catalysed dissociation, **50**:228

potential curves, **50**:557

quenching of, **42**:201, 203

rotational energy transfer from, **50**:614

self-quenching cross section, **50**:25

recombination in simple solvents, **53**:206–220

equations of motion, **53**:207–210

simulation results, **53**:210–22

van der Waals molecules, **47**:Part 1:353–360

argon, **47**:Part 1:358–360

helium, **47**:Part 1:353–355

neon, **47**:Part 1:355–358

see also Chemiluminescent reactions

Iodoform, ionization potential, **2**:198

quadrupole resonance of, **2**:191, 198, 201, 204

resonance line, **2**:193

sulfur crystal (CHI_3-3S_8), quadrupole resonance in, **2**:191

quadrupole spectrum, **2**:195, 196

Iodomethane hydrate, thermodynamic data and lattice constants, **2**:8

o-**Iodophenol,** quadrupole spectrum,

2:195
radiation resistance of, 2:196
p-Iodophenol, quadrupole spectrum, 2:195
radiation resistance of, 2:196, 198
Ion beams, 23:9, 12; 42:528
generation of, 30:189
inclined, 23:9, 11
internal energy, 30:191
merged, 23:13
Ion bombardment apparatus, 20:277, 281
Ion counting, 10:270
Ion cyclotron resonance, 19:65, 107, 126, 141
collision broadened, 19:157
heterodyne, 19:152, 169
pulsed, 19:147, 148
spectrometer, 19:145; 50:137
transient, 19:152
Ion-dipole interaction, 6:456
Ion ejection, 19:148
Ion energy, 19:158
Ion heating, 19:156
Ion-ion pair potential, 11:85
effective, 53:120, 121
Ion laser photolysis, 50:137
Ion-molecule collisions, 21:209
collision times, 21:221
cross sections, 21:211
trajectories, 21:221
see also Collisions
Ion-molecule half reaction, 52:298, 323
Ion-molecule interaction potentials, 19:159
Ion-molecule reactions, 45:105–106, 184–196
acetonitrile, 19:186
aliphatic alcohols, 19:185, 186, 195, 199

aliphatic ketones, 19:200
alkenes, 19:179, 181, 182, 184, 199
alkoxide ions, 19:192
argon (Ar, Ar^+), 19:161, 166, 170
B_2H_6, 19:190
2-butanol, 19:196
$C_2H_4^+$–C_2H_4, 52:350
CH_3F, 19:172
p-chloroethylbenzene, 19:198
CS_2^+–CS_2, 52:315, 320
cyclic alkanes, 19:202, 203
deuterium (D_2), 19:170
dialkyl-*N*-nitrosamines, 19:204
diatomic ions, 45:189–196
dimethyl carbonate, 19:197, 198
ethers, 19:186, 198
ethyl nitrate, 19:204
excitation of neutral reactants, 45:108–110, 112–113
flowing afterglow technique, 45:113–115
H_2^+–H_2, 52:304
HCN, 19:189
H_2O-B_2H_6 mixtures, 19:191
H_2S-B_2H_6, 19:191
H_2S-C_2H_4, 19:183
hexafluoroethane, 19:184
hydrocarbon derivatives, 19:182
hydrogen (H_2), 19:204
hydrogen molecule isotopes (H_2, D_2, HD), 19:161, 166, 170, 204
luminescent, apparatus for study, 45:118–119
diatomic reactants, He_2^+ + N_2, 45:190–193
He_2^+ + Ne_2, 45:193
mass spectrometer in study of, 45:115–117
methane, 19:169
methane mixtures, 19:170, 171, 172,

204
nitrogen (N_2), **19**:161, 166, 169
$O_2^+-O_2$, **52**:296
1-phenyltetralin, **19**:203
phosphine, **19**:204
propyne, **19**:181
protonated aldehydes, **19**:199
rare gas-hydrogen systems, **19**:204
rate constants, **19**:171
small molecules, **19**:161
styrene, **19**:202
vibrational excitation, **45**:184-196
see also Chemiluminescent reactions; Ion-neutral reactions
Ion-neutral collisions, **45**:82-224
charge transfer in, **45**:180-183
$He^+ + N_2$, **45**:190-192
$He_2^+ + N_2$, **45**:190-192
molecular systems, **45**:184-196
chemiluminescent reactions, **45**:165-196
collision mechanisms, **45**:196-202
calculations, **45**:201-206
theories, **45**:199-201
heavy particle transfer, **45**:193-196
$C^+ - H_2$, **45**:195
$C^+ - O_2$, **45**:195
$N^+ - NO$, **45**:193
$O^+ - H_2$, **45**:194
luminescence measurements, **45**:164-183
Penning ionization excitation, **45**:182
Ion-neutral reactions, high-energy, **10**:69
low-energy, electron energy variation in, **10**:258-259
isotopic effects in, **10**:262-266
repeller effects in, **10**:260-262
temperature effects in, **10**:259-

260
see also Charge transfer
Ion neutralization, **27**:23
Ion pair, **11**:43, 44, 70
Ion pair formation, **42**:24, 514
in crystal melts, **6**:466
Ion pair process, **52**:264
Ion pumping, **19**:366
photochemical, **19**:369
Ion recombination, **23**:2, 4, 5
critical radius, **23**:20, 23, 35
dissociative, **23**:2
radiative, **23**:2
Ion recombination coefficient, **23**:3ff, 19ff
Harper's, **23**:32, 33, 37
pressure dependence, **23**:19
termolecular, **23**:31
Ion scattering spectrometry, **41**:491
Ion solvation, **53**:65
Ion sources, *see* Beam sources
Ion spectroscopy, **41**:277-280; **49**:538
Ion transport across cell membranes, **21**:601
Ionic character of state, **42**:202
Ionic conductivity in crystals, **11**:3
Ionic dehydration, **19**:198
aliphatic alcohols, **19**:198
Ionic displacement, **37**:101, 110, 128
Ionic intermediate in quenching, **50**:477
Ionic mobility, **16**:3, 23
static, **16**:23, 30-33
Ionic polymerization, **2**:148, 160, 161
Ionic salts, **11**:83, 85, 87, 99
binary mixtures, **11**:99
solutions of, perturbation theory, **11**:99
Ionic solution theory, cluster theory, **34**:127
conformal, **11**:98ff

Mayer's, **34**:125

Ionic solvation, 6:413–458

Ionic structure, 19:195

$C_2H_5O^+$, **19**:194

$C_3H_6O^+$, **19**:199, 204

$C_8H_9^+$, **19**:202

Ionization, 10:55–56, 187–190, 199, 202; **38**:208

associative, **28**:394; **45**:402, 472–475

autoionization, **30**:5, 419, 448, 449, 455, 458

collisional, **42**:513–584

electron impact, **28**:395, 423, 465; **36**:485; **45**:84–111; see also Electron impact

electron transfer, **45**:475–483

He - Ba system, **45**:476–478

He - Ca system, **45**:478–480

He - Hg system, **45**:480–483

hypersonic flow, **13**:32, 35

multiple, in charge exchange, **13**:51

Penning, **28**:394, 404, 413; **30**:418, 448; **45**:401–472

cross section, **45**:426–445

elastic scattering, He - Ar systems, **45**:420–426

electron angular distribution, **45**:445–452

electron energy distribution, **45**:434–445

quantum mechanical relations, **45**:412–417

photoionization, **28**:395, 409

semiclassical approximation, **45**:417–420

slow collisions, **45**:399–486

spontaneous, **45**:401–483

threshold, **36**:487

transfer, **45**:401

see also Charge transfer; Ion-molecule reactions; Ion-neutral reactions

Ionization cross section, 48:63

intensity borrowing, **48**:44

Ionization potential, 1:250, 254; **2**:66, 67, 69, 81; **7**:46, 52, 64, 98; **10**:197, 240, 257, 379; **12**:290, 294; **19**:47, 57; **36**:206, 283, 306, 323, 330, 395

adiabatic, **36**:283, 305, 500; **52**:281

Ar_2, **52**:283

$ArCS_2$, **52**:315

benzene and benzene derivatives, **36**:489

C_2H, **52**:348

$(C_2H_4)_2$, **52**:351

calculation by equations of motion method, **48**:1–70

$(CD_3COCD_3)_2$, **52**:346

$(CH_3COCH_3)_2$, **52**:346

$(CO)_2$, **52**:302

$(CO)_3$, **52**:303

$(CO_2)_{n=1,4}$, **52**:300

$(CS_2)_2$, **52**:309

$(CS_2)_{n=3-5}$, **52**:311

CS_2, **52**:307

donors, **7**:255

first, **36**:489, 500

Green's function approach to calculation of, **36**:205–344

hydrocarbons, **6**:388

inner shell, **36**:334

many-electron theory, **6**:316

microclusters, **40**:54

molecules and clusters, **52**:263–362

$(N_2)_2$, **52**:302

$(N_2)_3$, **52**:303

$(NH_3)_2$, **52**:347

$(NO)_2$, **52**:291

$(NO)_{n=3-6}$, **52**:293

$(O_2)_{n=2-5}$, **52**:293

OCS, **52**:324, 325

(OCS)$_{n=2-3}$, **52**:325, 326

OCS—CS$_2$, **52**:325, 327

polyphenyl molecules, **21**:39–47

SO, **52**:335, 340

SO$_2$, **52**:334, 336

vertical, **36**:283, 316, 500

Xe$_2$, **52**:283

XeAr, **52**:285

XeKr, **52**:285

Ionization spectroscopy, **27**:25

Ionized gas, **23**:7, 10

dielectric constant, **23**:7

electrical conductivity, **23**:7

Ionizing radiation, **2**:197

effect on fibrous proteins, **7**:259

Ionizing reactions of metastable molecules, **19**:152

Ionophores, **19**:268, 367; **39**:304, 306, 326

distribution in membrane, **29**:352

environment of, **29**:353

field effect on, **29**:352

structure, **29**:352

as uncouplers, **19**:369

Ionosphere, **10**:254, 256

Ions, excited states, **19**:47, 62, 66, 71, 80, 111, 129

nucleation in absence of, **40**:138

reactions with excited neutrals, **45**:108–110, 161–163

Ar$^+$ + H$_2$, **45**:109

He$^+$ + N$_2$, **45**:109

Ne$^+$ + N$_2$, **45**:110

NH$_3^+$ + H$_2$, **45**:109

O$^+$ + N$_2$, **45**:109, 110

O$^-$ + O$_2$, **45**:110

O$_2^-$ + O$_2$, **45**:110

single photon dissociation, **50**:136

vibrationally excited, reactions of,

45:94–100

see also Excited ions

Iridium, (100), **49**:631

reconstructed surface, **49**:564

Iridium hexafluoride, **9**:194ff

Iron, **41**:140, 141, 161, 163

atom, energy levels of, **2**:313

microclusters, **40**:56

Iron-nickel, (Fe$_5$Ni$_5$) thermodynamic properties of, **2**:133

solid solution, **2**:125

system (γ-Fe-Ni) small deviation from ideality, **2**:133

Iron-porphyrin complexes, **7**:182

Irradiation resistance of dyed wool, **7**:266

Irreducibility condition, **11**:337ff, 358, 366, 375

Irreducible cluster integral, **37**:143

Irreducible linear operator, **18**:185, 186

Irreducible representation, **14**:96; **44**:10, 447, 449

Irreducible self-energy operator, **31**:272, 274, 277

Irreducible tensor operator, **44**:42

Irreversibility, **38**:6, 51, 145, 175, 179; **46**:154, 179ff

contraction of observables and, **53**:347

from incomplete resolution of states, **53**:341

symmetry-breaking process, **29**:12, 13

Irreversibility theory, **53**:64, 67, 124–148

generalized Langevin equations,

projection operator method, **53**:143–148

response function method, **53**:138–142

linear response theory, **53**:132–138

Liouville equation, **53**:129–131

Liouville space, **53**:126–128

molecular-time-scale generalized Langevin equation, **53**:148–172

Onsager hypothesis, **53**:82–83

perturbation theory, **53**:131132

time correlation function, **53**:128–129

time evolution of dynamical variables, **53**:125–126

velocity autocorrelation function, **53**:77–87

Irreversible cycle, 43:126

Irreversible processes, 3:33ff, 304; **53**:148–172

coordinate variable, chain representation for, **53**:163–165

equations, **53**:161–162

equipartition theorems, **53**:171–172

evolution criteria, **19**:217, 221

fluctuation-dissipation theorem, **53**:167–168, 170–171

generalized forces, **19**:214

inner product, **53**:168–171

linear systems, **3**:35

local potential, **19**:218

nonlinear, **3**:39; **19**:217

Onsager reciprocity relations, **19**:215, 223

oscillating reactions, **3**:47

phenomenological laws, **19**:214, 221

principle of least dissipation of energy, **19**:216

principle of minimum entropy production, **19**:224, 228, 316

response function, chain representation for, **53**:156–161

response function hierarchy, **53**:151–156

second law, **19**:213

statistical theorems, **53**:166–167

thermodynamics, **16**:101, 103; **19**:235; **29**:12, 30

linear, **19**:261; **29**:16, 30, 73

nonlinear, **29**:17, 31, 73, 337

quantum system weakly coupled to thermal reservoir, **38**:109–142

stability theory, **19**:228

time development of, **51**:127–130

variational methods, **3**:267

velocity autocorrelation function, **53**:150–151

see also Dissipative processes

Irreversible rate equation, 24:176, 178

Irving-Kirkwood formula, 49:366

Ising chain, 33:74

kinetics of, **33**:80–98

Ising lattice, 6:178

magnetization of, near critical point, **6**:180

three-dimensional, diagram summation techniques, **6**:194

Ising mode, 46:320

Ising model, 1:301, 303, 306; **15**:230, 249, 335; **37**:194; **40**:159, 195, 231, 265ff, 307, 327, 360; **41**:2, 3, 11; **51**:39

correlations in, by method of Toeplitz determinants, **15**:335, 343, 349

DNA denaturation, **22**:132, 138

ferromagnetism, **40**:253–272

continuum representation, **40**:255, 256

one-dimensional, applied to decay of correlations, **15**:212

polymer statistics, **22**:8, 9

two dimensions, **40**:194

Isoalloxazine, 7:617

Isobaric coefficients, 1:162

isochoric coefficients of viscosity

and, 1:137

Isobutene, polymerized, 2:157

Isochoric temperature coefficient, 1:161

Isoconjugate molecules, 8:85ff
  reactions of, 8:96ff

Isoelectronic system, 2:246

Isoenzyme, 39:175

Isoeremolactone, 16:187–188

Isokinetic relation, 33:347

Isolated molecules, radiationless transitions of, 47:Part 2:298–314
  dephasing, 47:Part 2:308–311
  molecular model, 47:Part 2:300–302
  small, intermediate and statistical limits, 47:Part 2:302–308
  vibrational relaxation, 47:Part 2:311–314

Isolating integrals, 36:41

Isomerization, 48:82
  induced by vibrational excitation, 47:Part 2:13–14
  many-electron theory, 6:316
  multiphoton induced, 50:139
  polyacetylene, 53:291

Isomerization constant, 1:53

Isomerization energy, butane, 2:75
  pentane, 2:75

Isomers, 7:416
  microclusters, 40:85

Isomorphism, 43:29

Isomorphous crystals, 16:201, 204

Isomorphous pair, centrosymmetric, 16:201, 204–205
  non-centrosymmetric, 16:202, 204

Isomorphous replacement, 16:201–205, 211, 213, 215

Isooptic systems, 17:2

Isoprene, 2:169

Isopropyl alcohol, infrared spectra, 7:494

Isopycnic systems, 17:2

Isorate-constant point, 43:37

Isotactic placement, 2:172

Isothermal compressibility, 16:237; 31:37, 48, 92; 37:159, 197

Isothermal process, 24:107

Isotope effect, 1:16, 26–29, 34, 44, 51, 63–66, 73; 19:42, 91, 97; 42:89; 51:199
  inverse, 1:26
  photodissociation, 50:280
  proton transfer, 21:244, 246
  quenching, 50:360

Isotope interaction, 39:326

Isotope separation, 23:79; 50:264, 308
  by physical thermal diffusion, 16:127

Isotopes, 4:226–230
  carbon monoxide, 4:228, 292, 294, 295
  heavy, 4:286, 300; 16:127
  methane, 4:229, 299, 300
  mixtures, 24:248
  nitric oxides, 4:229, 292, 296, 298
  nitrogen, 4:228, 292, 293

Isotopic competitive technique, 1:37, 38, 40, 41, 44, 49

Isotopic dilution, 40:14, 20, 24, 38, 40, 42, 43

Isotopic exchange constant, 1:53

Isotopic fractional technique, 1:36–38

Isotopic homogeneity, 1:48, 69

Isotopic impurity, 40:454, 474

Isotopic inhomogeneity, 1:50

Isotopic isomerization equilibrium constant, 1:53

Isotopic isomers, 1:69; 4:281

Isotopic labelling, 42:18, 145

Isotopic rate equation, 1:54

Isotopic ratio, 1:38, 48

Isotopic substitution, 4:281, 300; 30:15,

61; **53**:31

Isotopically mixed crystal, 40:373, 392

Isotropic fluid, 44:341

Isotropic hyperfine constants, 7:536

Isotropic hyperfine interaction, 7:534

Isotropic interactions, 12:564, 574, 586, 593, 597

Isotropic polarizability, induced, **51**:92

Isovaline radicals, 7:551

Iteration procedure, convergence of, **13**:342

Iterative cubic approach, 54:101–114

Itinerant librator, 44:259, 284, 394, 439, 444

planar, **44**:275

Itinerant oscillator-librator, 44:338, 351

Ito type equations, 46:234

Ivey's relation, 4:194, 195

J

J-**diffusion model, 44**:274, 285, 351, 373, 401, 429, 440, 457

j-**fragment, 4**:207, 215, 220

j-**labelling, 49**:239, 244

3*j* **symbols, 44**:41, 123

Jacobi elliptic sine function, 53:236

Jacobi polynomial, 4:266

Jahn-Teller effect, 5:209, 248, 251; **9**:171ff; **12**:9; **35**:259; **36**:159, 173, 289; **42**:193; **44**:18, 48; **53**:288, 293

Jamine, structural formula of, **16**:165–166

Jaynes prescription, 51:143–145, 159, 170

error incurred in, **51**:154–155, 156

Jeans instability, 26:111, 112, 126

Jeans length, 35:134, 136

Jeans number, 35:135

Jeans theorem, 35:124

Jeans wavenumber, 35:128

Jeffreys approximation, 10:93

Jeffreys-Born approximation, 10:93–94, 111, 117–118

Jeffries' theorem, 43:65, 105

Jellium interface, 41:171

Jepsen and Friedman equation, 48:199

Jesse effect, 45:489

Jet, supersonic free, **52**:279

Jitter amplitude function, 44:330

Joint probability distribution, 16:145–148, 156, 204, 214; **44**:284, 401

conditional, **44**:287

Jointed chain, 52:445

Jordan-Golden inequality, 24:194

Josephson effect, 27:298, 301, 324, 340

Jump frequency, 34:39, 47

Jump model, 44:374

Jump probability, in chain molecules, **33**:77, 122–124, 134

JWKB approximation, 10:89–92; **12**:359; *see also* Semiclassical approximation; WKB approximation

K

K-**meson, 38**:152

K-**region theory, 8**:165

Kadanoff block picture, 40:249, 271

Kadanoff-Wilson approach, 40:249

Kapur and Peierls methods, 25:249

Kasha's flow chart, 7:24

Kato identity, 25:219

Katriel formula, 49:228

KBr, 27:438, 445

Kekulé structures, 7:40

Kelvin ponderomotive forces, 1:330

Kennard approximation, 10:80

Keplerian pair, 26:116, 125, 126, 132

Keratin, 7:251, 254, 263, 264ff

ionizing radiation effects, 7:266

Keratin fibers, 7:266

Kernel, binary, 11:344

Kernels of integral equations, *see* Transition kernels

Kerr effect, 12:135; 44:388

for simple fluids, 26:61–65

Kerr liquids, 23:214

Ketene, 50:122, 315

Ketene formation, 42:142

α-**Keto-1, 5-tetramethyleneferrocene,** 16:173–175

Ketone reduction, 3:12

reaction mechanism, 3:16

Ketones, 4:96–98, 102, 108, 138; 44:581, 584, 585, 600

base strength of, 8:133

chemical reactivity of, 8:88

dipole moment of, 8:136

nonalternant, 8:134

octant rule, 44:627

Ketyl radical, 35:26

Kihara potential, 10:77, 105; 12:290, 295; 16:380

extrema analysis with, 10:122–123

Kinematic matrix, 36:11, 276, 294

Kinematics, 44:294

Kinetic anisotropy matrix, 35:5

Kinetic diameter, 1:120

Kinetic effects, in glasses, 48:457, 464, 465, 472, 501, 505

Kinetic energy, 41:90

classical rotational, 4:234, 252, 263

integrals, 13:210, 218

translational, 4:234

transport, 1:152

Kinetic equations, 15:21ff, 30ff; 43:18; 48:105–125

Bhatnager-Gross-Krook, 46:338

collision integral for electrons and ions, 38:206

correlation diagrams for, 15:21

Feynman diagrams for, 15:22ff

Fokker-Planck, 46:335

including inelastic processes, 38:193

integral for atoms, 38:206

Maxwell-Boltzmann distribution, 38:208

pair, 48:120–123

derivation of, 48:173–175

partially ionized plasma, 38:206, 208

singlet, 48:115–120

derivation of, 48:167–172

studied by local potential methods, 13:289

studied by variational methods, 13:302

Vlasov, 46:334

of von Neumann, 15:21

Kinetic isotope effects, 1:18, 26, 53

Kinetic parameters, 43:27

Kinetic potential, 29:74

multiple steady state model, 29:75, 337

Kinetic pressure tensor, 1:327

Kinetic sum rules, 46:332, 336

Kinetic systems, noise analysis of, 37:67–98

Kinetic theory, 38:173–192

assumptions, 35:71, 111

of chemical reactions in liquids, 48:71–182

configuration-space theory, 35:33, 34

dynamic viscosity, 35:106

using effective potential, 24:259, 263, 264, 265

delocalization effects, 24:265

doublet distribution function, 24:263

kinetic equation, **24**:264

nondissipative hydrodynamic equations, **24**:267

of fluctuations, **38**:193

of gases, **1**:131, 156, 161

variational methods, **3**:281, 286, 289

of gravitational systems, **35**:119–196

Hamilton's equations in, **35**:54

homogeneous solution of linearized singlet kinetic equation, **24**:304

of macromolecular solutions, **35**:31–118

of many-body systems, **24**:257ff

of nucleation, **40**:137–156

of plasmas, **38**:193–210

Balescu-Lenard collision integral, **38**:194

Balescu-Lenard equation, **38**:193

Boltzmann equation, **38**:193

Landau collision integral, **38**:194

Landau equation, **38**:193

nonideal, **38**:193

Progogine-Nicolas-Misguich, **24**:259, 278ff

renormalized, **48**:106

Rice-Allnatt equations, **24**:275, 276, 285

second-order fluid constants, **35**:96

stress tensor for macromolecules, **35**:87, 91, 92, 96

summary of results, **35**:108

symmetry of momentum flux, **35**:92

Kinetics, **36**:142, 153, 194

biochemical, **19**:314, 315, 317, 318, 325, 326, 327; *see also* Biochemical kinetics

of chemical lasers, **47**:Part 2:55–83

equilibrium theory, **1**:380

helix-coil transition, *see* Helix-coil transition

mass action, **43**:81

of polymer conformational changes, *see* Polymers

power law, **43**:13, 27

reaction, multi-barrier, **4**:4–8, 45, 52

reversible, on linear lattices, **15**:281, 284

see also Chemical kinetics

Kinetics studies, two-photon processes, **47**:Part 2:22–27

Kinin system, **26**:167

Kink site, **40**:220

Kink solution, of ϕ^4 model, **53**:267

Kinks, **41**:214, 216

Kinks-antikinks, **53**:237, 325ff

autocorrelation function, **53**:273–274

breather solutions, **53**:240–241

collisions with, **53**:239, 330–331

correlation functions and, **53**:250–251

diffusion of, **53**:269–270, 274ff

domain wall Brownian motion, **53**:268

at low temperature, **53**:243ff

phenomenological model, **53**:242

Kirkwood-Bauer-Magat line, **7**:425

Kirkwood correlation, **44**:343

Kirkwood coupling parameter hierarchy, **40**:290, 294, 307, 310–314, 327–339, 347

Kirkwood equation, **48**:198

Kirkwood instability criterion, **40**:334, 335, 336, 347

Kirkwood-Salsburg hierarchy, **37**:144; **40**:290, 295, 322–327

Kirkwood singlet equation, **40**:313

Kirkwood superposition approximation, **37**:11, 154, 180

Kirkwood theory for dielectrics, **1**:80, 81, 82

Klein approximation, 3:95

KMS boundary condition, 22:240; 38:120, 126

Knight shift, 4:332, 337, 383

Knot, 43:74

Knot cycle, 43:127

Knot pseudograph, 43:76

Knudsen number, 10:289, 297–300, 303; 52:270
   nozzle, 10:291
   skimmer, 10:293

Kohn-Ohmura procedure, 16:303

Kohn-Sham theorem, 41:86; 54:242–243

Kohn variational method, see Variational method, Kohn

Kolmogorov-Arnold-Moser theorem, 36:2, 37, 38, 40; 40:125; 46:74, 78

Kondo effect, 51:241, 246, 251, 253ff

Kondo Hamiltonian, 51:246

Koopmans' defect, 36:230, 238, 306, 310, 318

Koopmans' theorem, 1:250; 7:45; 9:327; 14:290; 23:93; 26:249; 28:124; 36:207, 230, 308, 450; 41:65; 48:15, 35

Koster-Slater method, 9:91, 95

Kramers diffusion equation, 1:366, 367

Kramers doublets, 7:172, 175ff

Kramers equation, 15:219, 328

Kramers-Heisenberg dispersion, 53:135

Kramers-Henneberger transformation, 44:529

Kramers-Kronig relations, 4:87–92, 94, 105; 17:75, 111, 176; 44:515
   applied to circuit analysis, 4:88
   applied to interaction of radiation with matter, 4:88
   applied to optical activity, 4:89–92
   applied to surface conductivity, 27:560

Kramers-Moyal expansion, 20:196, 197, 203; 46:198

Kramers regime, 46:250

Kramers relaxation time, 46:251

Kramers' pearl necklace, 35:101, 104

Kramers theorem, 5:196; 15:149
   second-order fluid constants, 35:101

KrAr, 52:284

Krasnosel'skii theorems, 37:168, 169; 40:280

Kronig-Penney's mode, 9:87, 151ff, 164

Krypton, 16:127; 27:476, 479; 53:44ff, 57
   clathrates, 2:20, 30, 39, 41
   clusters, 52:280
   dimer, 42:355; 46:62
   fluid structure, 53:16–21
   hydrates, 2:3, 32, 34
      polarizability, magnetic susceptibility and ionization energy, 2:70
      thermodynamic data and lattice constants, 2:8
   inelasticity correction, 53:7–9
   light scattering, 24:50
   London coefficient for, 6:399
   polarizability, 46:62
   potential energy, 2:71
   structural and thermodynamic data, 16:346–348
   see also Scattering; van der Waals molecules

Krypton laser, 49:2–3

Kubo formula, 15:75; 46:265

Kubo linear response theory, 53:132–138

Kubo relations, 17:168; 43:160

Kubo transforms, 17:78

Kubo's extensivity ansatz, 46:231

Kullback information, 32:62

Kummer functions, classical, 44:348

Kuscer-Corngold diagram, 20:164

Kuscer-Corngold theory, 20:161, 179

L

λ-cule, 9:14, 30ff

Λ doubling, 12:15; 44:30, 74, 107; 49:167, 175–176

L3 approximation, 48:237

l-dominant approximation, 42:431

l-labelling, 49:239, 233

L-matrix, 5:255

L-states, 9:139ff

l-type doubling, 44:31

Labelled elements, 7:431

Laboratory coordinate system, 10:329; see also Coordinate systems

Labotto quadrature, 13:234, 260

Labrador fox, 19:338

Ladder diagrams, 14:156, 180

Ladder double resonance, 47:Part 1:570–574

Lagrange multiplier, 26:222, 236; 41:87

Lagrange parameters, 46:167, 169, 171ff, 187, 188, 191

Lagrangian, 41:118, 119, 122, 125
    coordinate symmetries of, 13:312
    degeneracy, 33:160, 170, 171, 178, 188, 189, 216
    generalized, for local potential, 13:304, 316
    time-dependent, 41:120

Lagrangian dynamics, 44:266

Lagrangian equations, 32:13

Lagrangian functional, 24:135

Laguerre polynomials, 7:10

Lamb dip, 25:16

Lamb shift, 33:201, 202, 215, 222

Lamellar complexes, 7:326

Laminar-turbulent transition, 29:134

Landau damping, 27:274, 326; 35:130

Landau diamagnetic contribution, 4:324

Landau equation, 43:225, 243, 256, 263

Landau free energy, 31:235

Landau-Ginzburg theory, 53:266; see also Ginzburg-Landau theory

Landau-Lifschitz approximation, 10:111, 118

Landau-Lifschitz theory of fluctuating hydrodynamics, 49:315, 493

Landau-Placzek ratio, 24:15, 19

Landau-Teller approximation, 36:92

Landau theory, 40:234ff, 244, 245, 364; 49:326–335
    first order transition, 49:332–335, 491
    second order transition, 49:327–332

Landau-Zener model, 23:15, 16, 17, 18; 30:424, 432, 449, 453, 477; 42:148, 204, 237, 514, 518, 524, 545, 567; 49:274, 295; 50:239, 284, 287, 349, 480

Landau-Zener-Stueckelberg model, 28:115, 146–151, 331, 362, 369; 30:60, 469

Landé g-factor, 4:327, 332; 5:133–134, 191; 7:165

Langer theory, 49:209–210

Langevin equation, 5:158, 191; 15:7ff; 17:102; 34:270, 305; 48:82–84; 53:81–87;
    choice of ensemble, 46:116
    generalized, 46:113, 116, 316, 326; 48:106; 53:90
        chemical dynamics, 44:143–254
        Mori type, 53:95
        projection operator method, 53:143–148
        response function method, 53:138–142
    heatbath modeling, 53:98
    Monte Carlo method, 46:255

n-particle, **44**:299

nonlinear, **46**:209

rotational Brownian motion, **44**:357

solitons, **53**:270–277

see also Molecular-time-scale generalized Langevin equation

Langevin force, **13**:305; **15**:66

Langevin model, **40**:198; **53**:79, 81–87, 106

adiabatic limit, **53**:94–96

polymers in dilute solution, **15**:325ff

recombination dynamics, **53**:107, 110

vibrational energy relaxation, **53**:208, 210, 214–215

Langevin type equation, **42**:311, 346

Langmuir-Adam trough, **7**:240

Langmuir adsorption formula, **6**:170

Langmuir-Blodgett technique, **37**:12

Langmuir constants, **2**:48

Langmuir isotherm, **2**:39; **6**:167

Langmuir theorem, **2**:15

Langmuir trough, **41**:198, 200

Lanthanide complexes, **8**:55

Lanthanides, **5**:78, 133

electron transfer spectra, **5**:78, 135

interelectronic repulsion parameter, **5**:133, 134

Landé parameter, **5**:133

nephelauxetic effect, **5**:134

Laplace transform, **44**:267

heatbath response function, **53**:152

inverse, **44**:277

linear response theory, **53**:134

Laplacian operator, **7**:7

Large amplitude vibration, **52**:119

Large molecule limit, **42**:214

Large scale random fluctuations, **38**:203

Larmor frequency, **12**:557

Larmor period, **7**:510

Laser diagnostics, of reaction product energy distributions, **47**:Part 2:85–131

Laser-enhanced reaction, at high pressure, **47**:Part 2:10–12

Laser induced depopulation, **42**:440

Laser induced fluorescence, **30**:277; **41**:496; **42**:9, 274, 362, 441, 453, 483; **45**:93; **50**:257, 264ff, 298, 370

small molecules, **50**:1–105

two-photon, **50**:266; see also Two-photon excitation

Laser induced photochemistry, **41**:297

Laser induced predissociation, **50**:191ff

Laser magnetic resonance, **50**:267

Laser photodissociation spectroscopy, **45**:93

Laser pumping of Na atom beam, **45**:365

Laser saturation spectroscopy, **47**:Part 2:546–547

Laser studies of proton transfer, **47**:Part 2:643–679

Lasers, **23**:41, 43, 56; **24**:25; **28**:77, 82, 88, 89, 173, 209, 214, 218, 249ff

carbon dioxide, **42**:122, 125, 134

carbon monoxide, **23**:56, 57

chemical, **28**:52–54, 88, 89, 92

dye, **28**:82, 275, 281; **42**:117; **50**:3, 8, 14ff, 267

excimer, **50**:16ff, 257

hydrogen fluoride, **42**:138

infrared, **42**:440

iodine, **42**:34

shock waves generated by, **23**:53

tunable, **23**:43

tuning of, **50**:7

see also Chemical lasers

Lattice, **7**:502

atom-atom interaction, **53**:279–280

atomic, **38**:16, 20

longitudinal vibrations, **53**:228

mass defect, **53**:310

quantum mechanical description, **53**:314

sine-Gordon derivation, **53**:323

vibrational Hamiltonian, **53**:294, 296

Boolean, **38**:12, 14

complete, **38**:19

constant, **41**:163

continuous, **38**:16, 20

definition, **38**:19

disordered, **6**:130, 135–142

self energy of, **6**:136

equilibrium, **41**:138

ground state, **53**:262, 278

magnetoelastic coupling, **53**:264

mismatch, **20**:329, 332, 334

multicomponent, **6**:130–142

one-component, **6**:112–130

orthocomplemented, **38**:20

orthomodular, **38**:19

positive charges, *see* Electron gas

real space, **20**:227

reciprocal, **6**:114; **20**:220, 227; **53**:57

relaxation, **20**:300

rigid point dipoles, dielectric constant, **6**:436, 437

spin reversal, **53**:238

super-, **6**:130–134

Lattice defects, 46:359

as scatterers, **40**:372

Lattice displacement, polarization, **53**:265

Lattice dynamics, 22:207, 208, 217; **27**:361, 435, 465–509

ionic crystals, **27**:489–500, 528–529

librational motion, **22**:207, 217

molecular solids, **22**:206–208, 210, 216

monatomic crystals, **27**:467–475

quantum solids, **22**:208, 263, 268

small particles, **27**:483–487, 505–509

translational motion, **22**:206, 210

Lattice energy, 1:211; **41**:159

Lattice fluid equation, 49:432–436

Lattice gas, 1:301; **22**:344; **37**:183; **38**:152; **40**:159; **41**:2, 3, 4, 20, 21, 38, 42; **48**:296; **49**:492, 494, 508–510

decorated, **41**:7

two-dimensional, **6**:185, 189, 193

Lattice model, 2:123

Lattice theory, 38:19

of fluid state, **6**:238–240

cell theory, **6**:238, 240

free volume theory, **6**:238

hole theory, **6**:238

Lattice vibrations, 20:246; **24**:163, 172; **53**:228, 293–297

longitudinal, **53**:320

molecular, **53**:317–319

periodic, **53**:311–317

solitons, **53**:302–311

Toda lattice, **53**:297–302

Layer compound, 49:544–545

Layers, change with time, **32**:147

convecting, stability of, **32**:140

formation by ice block, **32**:143

formation in ocean, **32**:140, 148

merging of double-diffusion, **32**:148

production by breakdown of fingers, **32**:143

stability, **32**:148

LCAO-MO method, 7:23, 88, 116; **8**:5, 7, 59, 117; **13**:66; **28**:120, 121, 122, 129

LCAO-MO-SCF method, 28:121, 122, 125, 138, 140, 145

LDF theory of finite nuclei, 41:185

Lead, excess entropy of solution of noble metals in, **2**:133

microclusters, **40**:58

Lead germanate, 53:275

Lead oxide, red, **49**:85–88

Lead-thalium solid solution, 2:126

Lead-tin, solution, enthalpy of formation, **2**:143

system, energy of solution, **2**:143

Lead-zinc alloy (Pb$_8$Zn$_2$), calculation of thermodynamic quantities, **2**:136

Least-biased generator, 53:362

Least-biased predictor, 53:366

Least squares method, 25:248, 249, 259, 283

Lee model, 38:150

Legendre expansion, 2:294

Legendre functions, 7:10; **13**:250, 252, 258

Legendre polynomials, 1:10; **4**:253, 255; **36**:178; **44**:285

Legendre's duplication formula, 4:32

Lehman-McTague theory, 22:164

Lennard-Jones and Devonshire theory, 1:161

thermodynamic quantities of clathrates, **2**:24ff, 33, 34, 47

Lennard-Jones fluid, 34:108, 116; **37**:142; **49**:380, 413–415, 424–425, 437; **53**:64, 101

computer simulation, **34**:116

Lennard-Jones lattice, shock waves, **53**:308

Lennard-Jones matrix, 14:254

Lennard-Jones potential, 1:155, 211, 227, 230; **5**:247; **10**:77, 78–105; **12**:290, 589, 593; **16**:343–344, 380–382; **17**:128; **27**:468, 479; **30**:334, 345, 356, 381; **31**:64; **37**:145; **41**:209, 211; **44**:445; **51**:61

atomic lattice, **53**:295

extrema analysis with, **10**:120–123

Morse lattice, **53**:304

scattering of excited species, **10**:182–184

thermodynamic quantities of metallic solutions, **2**:135

Lennard-Jones solution, clamping approximation, **53**:189–191

correlation function, **53**:192–194

molecular dynamics, **53**:196

Lenses, electron, **45**:26

Lenticular cataracts, 21:661

Leray-Schauder theorem, 40:260, 271

Leurocristine methiodide dihydrate, 16:207

Level-counting technique, 10:124–125

Level manifolds, 8:35ff

Levesque-Verlet model, correlation function calculation, **53**:102, 192, 195–196, 197

Levinson's theorem, 10:124; **14**:148

Lewis electron pair, 13:71

^{6}Li-(S)-glycolate-2-d, 16:186

Libration, 36:12; **44**:351, 399

frequency, **44**:275

itinerant, **44**:259, 275, 284, 394, 439, 447

Librational phonon, 40:472

Lichtenstein-Lyapounov nonlinear operator, 40:299, 306, 322, 323

Lie algebra, 52:5–15

Lie-Koenig's Hamiltonian, 19:348

Lie-Koenig's theorem, 19:348

Liebermann-Rasmassuen model, 42:321

Lienard-Chipart theorem, 43:55–59

Liesegang rings, 29:55; **38**:350; **39**:48–51; **43**:239

Lifetime, 46:310, 313

excited state, **50**:276

finite, **38**:143

quantum state, **15**:17

radiative, **28**:271, 278, 294, 297, 300, 301, 381–385

atoms and atomic ions, C$^+$, **28**:382

C, **28**:382

K, **28**:271, 272, 295

N$^+$, **28**:382

N, **28**:382

O$^+$, **28**:382

O, **28**:382

molecules and molecular ions, CH, **28**:385

CN, **28**:325

CO$^+$, **28**:385

CO, **28**:384

N$_2^+$, **28**:383, 396

N$_2$, **28**:383, 393

NO$^+$, **28**:383, 384

NO, **28**:383

NO$_2$, **28**:384

O$_2^+$, **28**:384

O$_2$, **28**:384

Lifetime broadening, **41**:293, 306

Ligand, **4**:164, 165, 168, 176, 184

charge transfer states, **8**:13

polarizability, **7**:400

substitution reactions, **7**:376

sulfur-containing, **8**:53

Ligand field splitting, **7**:361

Ligand field theory, **5**:42, 43, 45, 48, 63, 251, 256; **7**:167ff, 187, 190, 613; **8**:47; **9**:172, 215

coefficient matrices for energy calculations, **9**:201ff

high-spin complexes, **5**:198

low-spin complexes, **5**:192, 198

magnetic susceptibility, **5**:192, 195, 197–198

octahedral complexes, **5**:43, 45, 50, 53, 62, 67, 71; **21**:353

Orgel diagrams, **5**:59

square planar complexes, **5**:54, 66

tetrahedral complexes, **5**:36, 43, 80

see also Crystal field

Ligand field transitions, **7**:363ff

Light atom anomaly, **28**:72

Light line, **27**:369, 413, 417, 428, 433

Light scattering, **1**:78, 310, 349, 354; **17**:31; **24**:1ff; **29**:175; **32**:78; **43**:261; **44**:273; **52**:441

beating, **24**:56

collision-induced, **51**:84–107

critical region, **6**:181, 194–197

cross section, **33**:169

depolarized, **24**:43, 44, 46; **51**:84–85; *see also* Depolarized scattering

Einstein-Smoluchowski theory of, **6**:195

intermolecular, **24**:44, 50, 55

line shape, **51**:49–112

measurement of, **51**:85

monatomic liquids, **24**:40

nonequilibrium fluids, **32**:87ff

Rayleigh wing, **24**:50

simple fluids, **26**:1ff

surface, **24**:55

thin films, **48**:329–396

Light source, gas-discharge, **52**:269

vacuum ultraviolet, **52**:265

Limb development, **29**:263

Limit cycle, **19**:344; **29**:5, 40, 50, 97, 146, 148, 159, 162; **34**:289, 291–293; **38**:321, 324, 330, 346; **39**:29, 31, 39; **43**:242; **49**:315, 348

noisy, **19**:345

nonuniform system, **38**:337

oscillations, in allosteric model for PFK, **38**:374

stable, **38**:324

transient, **38**:347

transitions involving, **38**:327, 328

Limit of stability, pure fluid phase,

40:297, 348

pure low density phase, **40**:361

Limiting law, **11**:227, 241, 263

Limiting static process, **33**:440

Lincoln wool, **7**:269

Lindemann melting relation, **1**:31; **51**:267

Line broadening, **12**:489; **40**:380; **50**:1113

models of, billiard ball, **33**:277, 286

Brownian particle, **33**:268, 281

nearly hard sphere, **33**:282

see also Homogeneous broadening; Inhomogeneous broadening

Line shape, **17**:109, 229; **25**:14, 49; **33**:235–294; **41**:272; **44**:549, 559, 589

absorption, **21**:319, 322

collision-induced, **51**:49–109

Doppler-Lorentz, **25**:15

electron resonance spectra, **18**:233

interference effects, **44**:565

line wings, **33**:284, 289

Lorentzian, **33**:267

Mössbauer spectra, **15**:123

motional narrowing, **15**:103, 109

NMR, **15**:109

overlapping lines, **33**:27

power saturation effects, **25**:15

pressure broadening, **25**:15

Raman scattering, **33**:238

Stark effect, **15**:120

steady state, **25**:15

stochastic theory of, **15**:101ff; **21**:339

symmetrized, **51**:105

transient effects on, **33**:267, 272

transition from resonant to non-resonant, **33**:272

translational effects on, **33**:267, 279, 290

Linear approximation, **48**:237

Linear atomic chains, **9**:126, 154, 157ff

Linear chain, **40**:395

Linear combination of atomic orbitals (LCAO) method, **6**:323; **49**:607–608

ab initio molecular orbital scheme, **37**:214

discarding muffin-tin and exchange correlation potentials, **37**:289, 290

semiempirical MO schemes, **37**:207

see also LCAO-MO method

Linear fixed Hessian approach, **54**:67–69

numerical results, **54**:72–78

optimal use, **54**:69–72

Linear lattices, reversible kinetics on, **15**:281

Linear molecule, **34**:230; **44**:355, 446, 453

Linear Newton-Raphson iterative function, **54**:11–14

Linear noise approximation, **34**:305

Linear polyene, **52**:94–109

pseudorotation geometry, **52**:94

Linear response theory, **15**:129; **17**:64, 69; **33**:241, 242; **40**:199; **44**:340, 467; **46**:265; **53**:132–138, 184–187

polarization, **33**:244

Linear systems, irreversible processes of, **3**:35, 301

Linear triatomic molecule, **9**:170; **36**:34

vibrational energy levels of, **52**:117–179

Linearization of nonlinear operator, **40**:259

Linearized coupled-cluster approach, **52**:203–205

Linearized dynamics, **43**:43–47

Linearized hypernetted-chain approxi-

mation, **48**:218, 238

Linearly forced harmonic oscillator model, 50:485, 488

Linearly independent set, 44:263

Lineweaver-Burk reciprocal plots, 7:671

Linewidth, 42:396, 414; **50**:4

collisional, **42**:357

natural, **38**:143, 146

Linkage classes, 43:76, 79

Linked cluster expansion, 9:327; **14**:113, 130ff, 372, 376ff, 386

Linked cluster theorems, 36:215, 242, 383; **52**:190–191

generalized, **52**:237–239

Liouville equation, 1:2, 3, 139, 153, 154, 163; **11**:163, 322, 323, 326, 330, 331; **15**:22; **16**:2, 7; **19**:313; **31**:162, 204; **34**:21, 52; **38**:176; **44**:263, 265, 273; **48**:89; **53**:129–131

expansion, **44**:262

pseudo, **48**:96–97

stochastic, **15**:103, 117; **34**:52

Liouville operator, 11:165; **15**:8; **17**:96; **35**:56, 57; **42**:324; **44**:265, 268; **46**:312, 332; **48**:4, 62, 88; **53**:184, 219–220

clamped chain, **53**:165

inner product, **53**:168

linear response theory, **53**:133

molecular-time-scale generalized Langevin equation, **53**:149

moments of, **46**:114, 118

Mori theory, **53**:144–145

normal modes and eigenfunctions, **53**:163–167

perturbed system, **53**:131

Liouville space, 53:126–128

double bracket notation, **33**:239, 250

invariant subspaces in, **33**:256, 258

methods in nonlinear optics, **33**:238

metric, **33**:240

operators, **33**:240, 255

rotation, **44**:265

single-molecule excitation subspace in, **33**:256, 288

Liouville-von Neumann equation, 15:22, 23; **50**:405, 423; *see also* von Neumann-Liouville equation

Liouville's theorem, 1:313; **6**:235; **19**:326, 346; **26**:113, 123, 124, 127, 130, 132

Liouvillian, 46:168

Lipid bilayer, 29:304, 309; **37**:99–138; *see also* Membrane

Lipids, densely packed, **41**:205

monolayers of, **41**:195–228

liquid crystal state, **41**:198, 208, 225

liquid expanded state, **41**:198, 207, 212, 215, 225

Lippmann-Schwinger equation, 19:6, 28; **30**:57, 67; **49**:269

Lipscomb's rule, 49:72

Liquid and plastic crystals, 44:382

Liquid crystals, 17:2; **21**:529, 531; **24**:3, 46, 49, 57; **31**:231–261; **34**:147; **46**:314

broken symmetry, **31**:238, 252, 256

cholesteric, **31**:237, 238, 244; **44**:258

critical instability, **31**:239

director, **31**:234

films of, **21**:529

Frank free energy, **31**:236

hydrodynamical fluctuations, **31**:251, 252

isotropic-to-nematic transition, **31**:239, 240

light scattering, **31**:251, 255, 256

nematic, **31**:232; **44**:258

anisotropic viscosity, **32**:156

Bénard problem in, **32**:156

Couette flow instabilities, **32**:151ff
  elasticity of, **32**:156
  focussing forces, **32**:158
  optical properties, **32**:152
  shear flow instability, **32**:158
  Taylor instability in, **32**:155
  viscosity, **32**:152
  order parameter, **31**:233, 248
  phase diagram, **31**:250
  smectic, **31**:233; **44**:258
  smectic-to-nematic transition,
    31:247
  transport coefficients, **31**:254
Liquid films, *see* Soap films
Liquid-glass transition, *see* Glass transition
Liquid metals, **13**:153; **51**:248
  cluster expansions, **31**:272
  cluster scattering, **31**:319
  coherent potential, **31**:312
  correlation function, **31**:270, 309
  density of states, **31**:286
  density profile, **27**:550
  electron states of, **31**:263–327
  electron-ion potential, **31**:267
  fluidized vacancies of, **21**:499, 509
  mercury, **27**:574
  multiple scattering, **31**:275
  pseudopotential, **31**:269, 272, 282,
    289, 304
  reflection spectroscopy, **27**:554
  structure factor, **27**:552
  structure of, **51**:234
  surface plasmon dispersion, **27**:577
  surface tension and compressibility,
    9:78
  thermodynamic properties of,
    21:501, 505
  tight-binding method, **31**:315
  transition zone, **27**:549

Liquid phase, **46**:281
Liquid-solid equilibrium, **17**:1, 3, 5, 12,
    15, 21, 22, 23
  crystallization curve, **17**:21
  undercooling, **17**:21
Liquid-solid transition, **40**:356
Liquid state properties, **41**:174–181
Liquid structure, **34**:106, 124
  attractive forces, **34**:107
  repulsive forces, **34**:106
Liquid structure factor, **40**:352
Liquid surface, **20**:284, 286, 288, 289,
    311; **48**:341
Liquid-vapor coexistence curve, **2**:87;
    40:334
Liquid-vapor equilibrium, **17**:3, 5, 22,
    23, 28
  closed solubility loops, **17**:28
  crystallization surface, **17**:3, 5, 22, 23
  immiscibility surfaces, **17**:8
  influence of salts, **17**:8, 17
  penetrable sphere models of, **41**:2–54
  pressure dependence, **17**:5
Liquid-vapor interface, **46**:279–362
  bulk-surface coupling, **46**:329, 342
  distorted bulk, **46**:298
  dynamical variables, irrelevant,
    46:310
    relevant, **46**:310, 313
  normal and tangential flow coupling,
    46:309
  undistorted bulk, pair distribution
    function, **46**:298
    singlet distribution function,
    46:298
Liquid-vapor transition, **40**:317
Liquidlike cells, **48**:481, 493, 508
  free energy of, **48**:475, 476, 490
  free volume of, **48**:475
    exchange, **48**:477, 504
Liquidlike clusters, **48**:478, 480, 493

Liquids, associated, **16**:234
  cell model of, **40**:28
  chain molecules, principle of corresponding states for, **16**:226–235
  dense, **48**:456
  dephasing of molecular vibrations in, **40**:1–48
  effect of kilobar pressures on, **44**:470
  elementary excitations in, **17**:186
  evaporation of, **40**:137
  Lennard-Jones, **48**:472, 473
  mixtures, **24**:22
  molecular, **34**:145
  monatomic, **48**:457
  neutron scattering by, **15**:129
  perturbation theory of, **34**:107
  rubidium-like, **48**:471
  supercooled, **48**:459, 462, 471, 476
  vibrational population relaxation, **47**:Part 2:487–519
    experimental studies, **47**:Part 2:50–513
    non-Markovian effects, **47**:Part 2:513–516
    theories of, **47**:Part 2:488–507
  viscosity, **48**:459
  see also Fluids
Lithium, **1**:170, 171, 172; **12**:149; **14**:203; **41**:141, 171
  atom, ground state of, **2**:313
    quenching of excited, **50**:478
  atom (^{7}Li), quadrupole moment, **23**:181
  dimer, **41**:152
    core of, **6**:377, 383
    dipole resonance interaction energy, **12**:41
    many-electron theory, **6**:383
    rotational energy transfer, **50**:615, 630

    superposition of configurations, **2**:296
  dimer (Li_2^+), potential surface, **42**:175
  ion, **6**:377; **8**:12; **9**:56
    core polarization potential, **6**:378
    electronic correlation energy, **2**:249, 254, 324
    energy of the ground state, **2**:302, 312
  ion (Li⁻), **2**:299
  isoelectronic series, **6**:377
  microclusters, **40**:54, 58
  see also Collisional ionization; Scattering
Lithium ammonium hydrogen citrate monohydrate, structure of, **16**:172
Lithium fluoride, **27**:368–370, 377, 392, 406, 438ff, 515–516; **34**:190; **41**:159, 160
Lithium glycolate monohydrate, **16**:186
Lithium hydride, **6**:383; **23**:162, 180
  dipole moment, **23**:180
  electronic correlation, **2**:324; **6**:354
  electronic energy, **23**:181
  ionization potential, **23**:181
  quenching of excited, **50**:381
  rotational energy transfer from, **50**:615, 630
  superposition of configuration, **2**:296
Littrow mount, **50**:11, 13
Liver alcohol dehydrogenase, **7**:707ff, 711
Living polymers, **2**:147, 177–179, 181, 184, 188
Living polystryene, **2**:179–182
Living seeds, **7**:623ff
Loaded sphere fluid, **24**:336; **31**:179, 196
  collisions, **31**:196
  self-diffusion coefficient, **24**:339

transport coefficients and relaxation time, **31**:196, 199

Loading parameter, 44:285

Local convergence, Hessian eigenvalues, **54**:56–59

Local density field, 41:67, 98, 99–108; **48**:85

Local density of states, 49:271

    copper, **49**:539

    nickel, **49**:593

Local energy, 2:299

Local equilibrium, 5:278ff, 291; **16**:101–102, 122, 130; **21**:476; **32**:61; **43**:221; **46**:265

Local-exchange-correlation model, 49:609

Local field, 12:269, 270

Local minimum, 40:238

Local mode, 53:318; *see also* Bond mode

Local potential, 32:38, 39, 48, 50, 58

    applied to instability problems, **13**:267

    iterative scheme, **32**:41, 50

    minimality property, **32**:39

    Prigogine's variational principle, **46**:258

    self-consistent method, **32**:41

    study of kinetic equations, **13**:289

Local-spin density, 41:138, 141

Local theory, 27:276, 278, 281, 357–359, 362, 369, 435, 438

Localizability theory, 2:258, 323

Localization energy, 1:183, 184, 187; **7**:42; **33**:381; **40**:452, 457

    correlation with stabilization energy, **33**:381ff

    integral representation, **33**:384

Localization theories, 51:247, 251, 256, 270, 271

Localized adsorption, 7:250

Localized bond, 7:21ff; **8**:117ff

Localized electronic state, 31:445

Localized excitation, mixed crystal, **8**:32

    pure crystal, **8**:28ff

Localized orbital, 14:9; **44**:611, 629

    random phase approximation, **44**:611, 631, 633

Localized state, 52:475

Localized surface state, 9:91ff

Lock-in amplifier detection, 49:16, 37–38

Logarithmic singularity, 40:231

Logic, classical propositional, **38**:15

    classical two-valued predicate, **38**:17

    definition of, **38**:13

    temporal, **38**:12

London dispersion energy, 12:38, 78, 171, 290, 335; **16**:343, 380, 382

    relativistic correction, **12**:210

London dispersion force, 1:80, 81, 267, 268; **6**:318, 398; **7**:203; **9**:332; **10**:119

    between two hydrogen atoms, **2**:60, 61, 64, 67, 68, 81, 282

    hydrocarbons, **6**:400

London equation, 36:143, 159, 162; **42**:75

London-Eyring-Polanyi method, 36:143, 160ff; **41**:251

London-Eyring-Polanyi-Sato method, 42:67, 129; **49**:195–199, 216, 222, 264

London theory of diamagnetic anisotropy, 3:190, 191

Lone pair, 5:344

    additivity of energies, **9**:355

Long-lived complex, 42:233

Long-lived variable, 46:316

Long-range interactions, 1:80, 311, 329; **11**:187, 195; **12**:146, 162, 163, 334; **26**:108, 111, 124; **42**:181, 225

H-H, **12**:162

H-He, **12**:149

He-He, **12**:147, 148, 151, 154, 155, 160

Li⁺-Li⁺, **12**:158

Li-H, **12**:150

Li-He, **12**:150

Li-Li, **12**:150

Long-range order, **1**:210; **2**:128; **37**:142; **43**:258; **46**:315; **51**:257ff

Long-time approximation, **44**:273, 277, 293

Long-time correlation, **46**:265

Long time tail, **46**:316; **48**:80

Long wavelength limit, **46**:322, 341

Longitudinal correlation function, **53**:248

Longitudinal electromagnetic field, **49**:536

Longitudinal lattice distortion wave, **49**:513

Longitudinal modes, **27**:390, 450

Longitudinal optical phonon, **27**:359, 377; **44**:488, 492; **49**:25–26

Longitudinal relaxation, **7**:503

Longitudinal relaxation time, **44**:375; *see also* T₁

Longuet-Higgins integral formula, **9**:159

Loop, **43**:75

Loop entropy model, **22**:132, 169

Lorentz condition, **4**:78

Lorentz equation, **1**:318

Lorentz field, **4**:79, 126, 435; **44**:343
    acting on molecular dipole, **13**:186, 188

Lorentz force, **33**:168, 188, 193, 201, 215

Lorentz gas, **20**:161, 200; **38**:183, 185; **52**:374

Lorentz group, **33**:162

Lorentz limit, **20**:144, 160, 201

Lorentz-local field, **31**:399, 407

Lorentz theory, **33**:161, 168

Lorentz transformation, **53**:325

Lorentzian line shape, **12**:493, 494; **40**:384, 385, 399, 421, 452, 462ff; **49**:9, 600; **51**:71, 80, 84

Lorentzian memory, **17**:175

Lorenz-Lorentz formula, **1**:339, 341, 346, 349, 353; **46**:13

Loschmidt paradox, **22**:361

Loss measurement, **44**:303, 363

Loss modulus, **35**:97

Lotka-Volterra model, **29**:10, 103, 125; **38**:324, 347, 352; **43**:236–237, 259–260
    fluctuations in, **29**:103

Low-angle scattering, **10**:78, 80–82, 96, 105–117
    energy dependence of, **10**:116–117
    glory interference effect in, **10**:114–117

Low energy electron diffraction (LEED), **20**:215ff; **27**:1, 2, 5, 29, 37, 66, 70, 83, 118, 298, 322, 475, 483, 510, 515–516, 524; **41**:479
    acoustical phonon assisted diffraction, **27**:67
    Al(100), **27**:93, 94, 97, 101, 105, 111, 120, 123, 133
    Al(111), **27**:99, 105, 111, 135, 138, 188, 192
    beam parameters, **27**:40
    computational procedures, **20**:251, 252
        Bloch functions, **20**:251, 256
        Born expansion, **20**:264, 268
        differential form, **20**:252, 255, 256
        dynamical theory, **20**:256, 262
        effective field, **20**:263
        integral form, **20**:252, 255, 260

single electron approximation, 20:264

wavematching approach, 20:256, 259, 260

conservation laws, 27:5, 6, 56

Cu(100), 27:75, 76

diffracted intensities, 27:35, 41, 70

dynamical theory, 27:73, 77, 83

  multiple scattering analysis, distorted wave, 27:1, 77

  laminar, 27:1, 83

  wave function matching analysis, 27:1, 81

elastic, Born approximation, 27:1, 47

  calculations, 27:1, 86, 118, 121, 160

impurity effects, 20:278, 281, 296, 297, 301, 303, 332

inelastic, 27:14, 37, 54, 55

  calculations, 27:2, 124, 136

  electron impact ionization, 27:14

  line broadening mechanisms, 27:2, 139

  momentum-tuned resonances, 27:127

  plasmon emission cross section, 27:15

  side band diffraction, 27:127, 129, 133

  two step, 27:124

insulators, 27:107

layer KKR method, 27:85, 92

LiF(100), 27:74, 109

liquid surfaces, 20:284, 286, 288, 289, 311

loss spectra, 27:17, 38

Ni(100), 27:107, 108

Ni(100)-c(2x2)-Na, 27:164

Ni(100)-c(2x2)-S, 27:161, 162, 164, 170, 171, 172

optical phonon assisted diffraction, 27:66

quasi-elastic, from vibrating lattice, 27:58

Rh(100)-c(2x2)-O, 27:181

see also Electron-solid interaction

Low-pass filters, 44:330

Low pressure, collisionally induced intersystem crossing at, 47:Part 2:314–329

  molecules at, 47:Part 1:4–5

Low-spin complexes, 7:370; 8:51

  magnetic susceptibility of, 5:192, 198

Low-spin ferric porphyrin, 7:393

Low-spin states, 7:169ff, 172, 368

Low temperature, photon echo decay at, 47:Part 2:473–474

Low temperature solids, nonradiative processes of small molecules in, 41:429–268

Low-velocity approximation for charge transfer, 10:239–240

Löwdin orthogonalization, 33:349

Löwdin's variational method, 9:360

Lower bound, 25:288

  on limit of stability, 37:144

  on stability of pure phase, 37:171

Lowest-order-γ-ordered approximation, 48:216, 224, 225

  soft core potentials, 48:225

Lucas modes, 27:459, 489–491, 496, 518, 529

Luminescence, apparatus, 45:118–119

  chloroplasts, 7:634

  circularly polarized, 44:595

  CrO_2Cl_2, following multiphoton excitation, 47:Part 1:661–677

  diatomic ion-molecule reactions, 45:189–196

  products of ion-neutral collisions, 45:164–183

  see also Ion-neutral collisions

Luminescent reactions, resulting from infrared multiphoton excitation, **47**:Part 1:679–711
see also Multiphoton excitation
Luxury molecule, **29**:253
Lyapounov functions, **13**:326, 327; **29**:5, 125; **32**:62; **33**:417; **43**:10, 14, 60–62, 86, 89, 98ff, 200, 208
  entropy and, **29**:14, 16
  simple, **43**:66–103
Lyapounov-Schmidt branching theory, **40**:320
Lyapounov-Schmidt method, **40**:284, 286, 357
Lyddane-Sachs-Teller relation, **27**:359
Lyman radiation, **30**:454
Lymphokines, **38**:396
L(+)-**Lysine hydrochloride dihydrate**, **16**:211–212
Lysogenic *E. coli*, **7**:588
Lysozyme, **33**:109
  active site, **39**:70
  catalytic mechanism, **39**:77, 79, 172
  conformational change, **39**:77, 79
  NMR studies of, **39**:65, 70, 79
  structure and energetics of, **39**:79, 338

M

M-**diffusion model**, **44**:256, 274, 353, 355, 373, 429
Mach disk, **10**:285, 291, 301
Mach number, **10**:277–279, 281–285, 287–289, 293–295; **52**:271
  correlation with Knudsen number, **10**:297–300, 303
  effective, **10**:282, 312
  freezing of, **10**:288–291
  terminal, **10**:289

see also Nozzle beams
Mackay icosahedra, **40**:84, 123, 130
Macro-micro correlation, **44**:298, 308, 342
Macroiteration, **54**:102
Macromeritic systems, **21**:753
Macromolecules, **7**:525; **24**:57, 58
  action of heat on, **7**:210
  binding to small molecules, **7**:523ff
  body-fixed axes, **35**:42
  conjugate momenta, **35**:152
  diffusion equation, **35**:80
  dynamic viscosity, **35**:106
  elastic rhombus model, **35**:50
  entanglement forces, **35**:70
  forces in, **35**:53
  free energy, **7**:209
  generalized coordinates, **35**:36
  Hamiltonian, **35**:52, 71
  kinetic equations, **35**:69
  kinetics of solutions of, **35**:31–118
  Liouville operator, **35**:56
  models for, **35**:39, 100, 101, 104, 108
  normal stress coefficients, **35**:100
  pairwise additivity of potential energy, **35**:53
  potential energy, **35**:52
  rigid dumbbell model, **35**:47, 104
  Rouse model, **35**:48
  second-order fluid constants, **35**:100
  secondary and tertiary structure, **7**:202
  stress tensor, **35**:87, 91, 92, 96
  swelling of, **7**:210
  viscosity and structure, **35**:100
  see also Proteins
Macrophages, **38**:401, 403
  cytotoxic activity, **38**:395
  diffusion coefficient, **38**:403
Macroscopic order, **46**:206

Macroscopic rate equation, **34**:253, 277

Madelung energy, **5**:79; **6**:110, 115ff, 124, 128ff

Madelung-Jørgensen energy, **13**:75

Magnesium, **41**:141

electronic energy of, **16**:329, 333–334, 339

quenching of ^1P, **50**:351, 365

quenching of ^3P, **50**:352

Rydberg states of, **16**:322–324

valence pair energy, **16**:336, 338–339

see also Chemiluminscent reactions

Magnesium-cadium, alloy (Mg_3Cd), **2**:129

Magnesium-cadmium, alloy ($MgCd_3$), **2**:128

Magnesium fluoride (MgF_2), **27**:435, 447

Magnesium oxide (MgO), **27**:438, 444, 505–506, 508–509

Magnesium phthalocyanide, **7**:634

Magnesium vapor studies, **47**:Part 2:167–169

Magnet, one-dimensional, **53**:229–234

Magnetic anisotropy, **7**:610

Magnetic circular dichroism, **35**:197–264; **44**:598

Magnetic deflection analysis, **10**:15, 175, 327

reactive scattering, **10**:341–343

Magnetic dipole, **33**:155, 160, 209, 228

Magnetic dipole interaction, **5**:218; **12**:85

Magnetic dipole radiation, **4**:163

Magnetic dipole transition, **18**:235

Magnetic displacement vector, **1**:322

Magnetic entropy, **2**:132

disordering, **2**:131

Magnetic field, **33**:227

congruent, **10**:15–18

effects on excited states, **47**:Part 2:373–376

fluorescence quenching, **47**:Part 2:375–376

level-anticrossing and electronic relaxation, **47**:Part 2:374–375

inhomogeneous, *see* Magnetic deflection analysis

strength, **1**:322

Magnetic field induced predissociation, **50**:198

Magnetic glasses, **51**:256–257

Magnetic heat capacity of nickel, **2**:133

Magnetic hyperfine Hamiltonian, **18**:195, 197

electron-nuclear dipole term, **18**:197

Fermi contact term, **18**:197

Magnetic hyperfine structure, **14**:21ff, 370

Magnetic induction, **33**:158, 162, 176, 217, 224–226

Magnetic intermolecular force, **12**:71

Magnetic moment, **1**:314, 315; **4**:126; **7**:164ff, 170; **12**:116, **41**:163

oscillating, **11**:312

Magnetic multipole, **5**:181

Magnetic multipole interaction, **12**:76; **33**:225

Magnetic neutron scattering, **53**:248ff

from TMMC, **53**:257

spin systems, **53**:248

Magnetic optical rotation, **35**:197

Magnetic properties, chromium, **41**:161

cobalt, **41**:161

manganese, **41**:161

nickel, **41**:161

palladium, **41**:161

Magnetic quantum numbers, **44**:447

Magnetic relaxation, benzene, **34**:74

dipolar lattice relaxation time, **34**:4

electron spin, **34**:8

gaseous methane, **34**:71

liquified methanes, **34**:63

methyl iodide, **34**:89

quadrupolar, **34**:6, 65

spin-rotation, **34**:7

spin-spin relaxation time, **34**:4

Magnetic resonance, minimum entropy production and, **3**:311

optical detection of, **40**:372, 401ff, 411, 415, 423ff, 459, 465ff; *see also* ODMR

Magnetic resonance spectroscopy, **42**:362

Magnetic shielding, **44**:89, 100

many-electron theory, **6**:316

Magnetic susceptibility, **2**:25; **3**:204ff; **5**:192, 195, 204, 210, 222; **12**:118; **27**:362, 364–365; **44**:34, 89, 99, 108; **51**:19

alkali metals, **3**:220

alkaline earths, **3**:220

atomic, **3**:204

CH_2 group, **3**:204

Curie's law, **5**:191

direct exchange interaction, **5**:218ff

double exchange interaction, **5**:233ff

exchange interaction, **5**:220, 222, 236

Fermi interaction, **5**:203

fictitious spin, **5**:205

frequency-dependent, **27**:371

high-spin compounds, **5**:198

Hund's rule, **5**:191

indirect exchange interaction, **5**:235ff

Jahn-Teller effect, **5**:209

Kramers theorem, **5**:196

Landé g factor, **5**:191

Langevin formula, **5**:191

low-spin compounds, **5**:192, 198

magnetic dipole interaction, **5**:218

measurements, **7**:183ff, 607

metal-ammonia solutions, **4**:324–

327, 371–374

metals, **3**:221

Racah parameters, **5**:201

s-d interaction, **5**:235

spin-spin interaction, **5**:202

super-hyperfine coupling, **5**:215ff

superexchange interaction, **5**:223ff, 228ff

Tellieu and Guy theory, **3**:187

temperature independent, **5**:193

tensor operators, **5**:201

transition metals, **3**:221

Van Vleck theory, **3**:181

Magnetization, **1**:316, 317, 324; **20**:97, 107; **25**:10; **33**:168; **41**:185

average, **53**:248

statistical, **20**:105

Magnetization field, classical, **33**:225, 229

quantum, **33**:226

Magnetization vector, **7**:503

Magnetoelastic coupling, **53**:264

Magnetogyric ratio, **44**:107; *see also g*-factor; Landé *g*-factor

Magnons, **53**:236, 241ff, 251ff, 262

Magnus expansion, **50**:400, 451

Magnus series, **42**:433

Magnus solution, **42**:232

Maier-Saupe theory, **41**:211

Mainstream, **32**:77, 79, 82

Majority carrier, **7**:311

Maleic anhydride, **2**:168

Malonic acid, **1**:67, 68, 72, 73

radical, **7**:541

Malthus-Verhulst equation, **34**:265, 269, 275, 299

Manganese, **41**:141, 161

Manganous phthalocyanine, **7**:402

Many-body CNDO, **49**:630–631

Many-body dynamics, **6**:55–80

Boltzmann equation, **6**:62–68
classical system, **6**:56
Onsager theory, **6**:68–80
Pauli equation, **6**:59–62
quantum mechanical system, **6**:57
Many-body effects, 46:21
Many-body force, 53:44ff
Many-body Green's function, 48:5, 15,·
31, 35, 529; *see also* Green's function
Many-body perturbation theory, 12:160
diagrammatic, **48**:2
Many-body potential, 53:3, 40, 42
Many-body problem, 22:109, 110, 117
chemical dynamics, **44**:143–254
Many-body system, 53:65
atom chain, **53**:106
harmonic chain, **53**:74–76
Mori theory, **53**:144–145
partitioning, **53**:66–68
short-time-scale, **53**:87–89
Many-body theory, 41:69, 96–98
relativistic, **19**:348
Many-electron eigenfunctions, 14:98ff
Many-electron system, correlations in,
2:304, 305, 318, 319, 323
Many-electron theory, 6:315–412;
16:286
application to He-He interaction,
12:49
Bethe-Goldstone equation, **6**:370
Brueckner theory, **6**:345
configuration-interaction, **6**:318,
320, 347
methane, **6**:347
correlation energy, **6**:317, 358–359,
360
non-closed shell, **6**:324
dynamical correlations, **6**:360
exact wave function, **6**:330–335

correlation effects, **6**:340–342,
353–357
electron clusters in, **6**:331, 337–
339
orbital orthogonality, **6**:332, 335–
337
split shells, **6**:332
Fermi correlation, **6**:325, 356
generalized SCF methods, **6**:344–
346, 363
Hartree-Fock equations, **6**:327–330
non-dynamical correlations, **6**:360
orbitals, **6**:323–324, 326
effect of correlation on, **6**:358–
363
equivalent, **6**:326
Hartree-Fock, **6**:323, 324, 365
perturbation techniques, **6**:346–349
projection operators, **6**:346, 365
π electron systems, **6**:384, 386
relativistic, **6**:406–407
semi-empirical theories, **6**:319
use of model potentials, **16**:324–329
variation-perturbation method,
6:342–344
**Many-spin problem in nuclear paramag-
netic relaxation, 11**:295, 300
Mappings, 43:25, 29, 31
area-preserving, **24**:160, 164, 174,
177, 179
Marangoni effect, 48:336
Marangoni number, 26:192, 193
Marangoni process, 29:133
Marcus equation, 49:206, 223
Marcus system, 49:201, 203–205
**Marcus theory of proton transfer,
21**:244
Marginal curve, 26:180
Marginal distribution, 44:288
Marginal stability, 26:179, 180; **38**:324;
43:228, 232, 253, 264

Markov approximation, 35:165, 168, 179; 48:90; 53:95

Markov chain, 51:28

Markov process, 5:263; 15:3, 4, 5, 67ff, 103, 224; 29:85; 35:146, 165, 174; 49:315
  lumping and, 38:297
  polymer statistics, 22:9, 30, 48, 52, 65-67, 73, 81, 90

Markovian collision operator, 38:182

Markovian generalized Boltzmann equation, 11:319

Markovian master equation, 46:399

Marriotte's law for electrons, 1:170

Martensitic transformations, 16:353, 369, 376, 379
  by deformation, 16:353
  spontaneous, 16:353

Martinez model, 29:43

Maslov index, 36:3, 7, 10, 13, 15, 24-26, 37; 46:120

Mass, closed, 33:436
  closure, 33:406
  renormalization, 33:202, 204

Mass action, 33:437

Mass action complex, 43:82

Mass action kinetics, 43:10, 82, 246

Mass action laws, 43:220

Mass analysis, fast particles, 10:208-209
  ion-neutral reactions, 10:250-252
  slow ions, 10:202-206

Mass asymmetry variable, 46:271

Mass balance, 13:273; 33:406, 435

Mass defect, 53:310

Mass density, 1:314, 315

Mass flow velocity, 52:271

Mass flux, 1:2

Mass operator, 14:231

Mass perturbations, benzenes, 54:180-191

Mass polarization, 12:6, 11, 16, 27; 44:44

Mass sensitivity, 51:198-199

Mass spectrometer, conventional, 30:186
  magnetic, 52:270
  photoionizing, 45:115-117
  quadrupole, 52:270
  scattering, 30:189
  tandem, 19:46, 85, 123; 30:186

Mass spectrometry, 42:137

Massey criterion, 10:253, 256-257, 268; 42:95, 462, 465; 50:479, 599

Massey-Mohr approximation, 10:93, 109, 117-118, 120-121

Massey parameter, 49:288

Master equation, 11:319, 333; 12:563, 595; 15:1, 5ff, 67ff, 204, 219, 221; 29:69, 86; 32:7; 34:245, 248, 249, 253, 254, 258, 259; 35:182; 38:177, 178, 298, 304; 50:437, 466; 51:128; 52:460, 468, 472; 53:344-347
  all-time, 53:349
  asymptotic, 29:86; 53:349, 365, 370-373
  chemical reactions and, 38:298
  diffusion and, 38:298
  dynamical, 53:346, 368-373
  eigenvalue properties, completeness, 20:158
    Rahman-Sundaresan theory, 20:192
    Schrödinger analogies, 20:161
    Shizuta theory, 20:163
    singular eigenfunctions, 20:159, 187
    spectral resolution, 20:165, 174
  expansion of, 15:70
  generalized, 53:342, 344, 350
  kinematical, generalized surprisal,

24:302 analysis, **53**:365–366
level population, **53**:363–365
mesurement process, **53**:362
projection scheme, **53**:356–357
quantum trajectories, **53**:360
radiationless transitions, **53**:362–363
subsystem dynamics, **53**:362
variants, **53**:366–368
Markovian, **46**:399
moment equations, **32**:8
multivariate, **34**:275
nonlinear, **32**:7
numerical solution of, **29**:71
from perturbation theory, **24**:300
practical solutions, **20**:164, 187
degenerate kernel methods, **20**:175
discretization, **20**:166
Koppel prescription, **20**:188
pseudo-eigenvalues, **20**:167
short-time expansions, **20**:168
from projection operator methods
reduced Heisenberg model and, **53**:352
singular perturbation analysis, **38**:306
integral representation of, **38**:305
stationary solution of, **38**:305
standard, **53**:342, 344
system in external field, **16**:8–14
trivial, **53**:369
vibrational relaxation of gas of diatoms, **15**:219, 221
Mataga-Nishimoto approximation, **7**:99
Material stability, criteria of, **17**:26
Matrix, diagonal diffusion, **33**:122
of internal viscosity, **33**:78, 124, 136, 145
positive definite, **43**:61, 62
positive semidefinite, **43**:61, 62
unitary, **4**:130, 246
Matrix diagram, **43**:104
Matrix elements, **9**:325ff
classical approximations, **36**:66
Hartree-Fock method, **9**:358ff
nonlinear Newton-Raphson approach, **54**:23–25
Matrix feedback, **43**:86
Matrix isolation, **41**:270; **50**:30, 45, 56
ionic microclusters, **40**:150
metal microclusters, **40**:130
Matrix stability, **43**:62–65
Matsubara's theorem, **24**:201, 202, 204, 208, 236
Matsuoka, Clememti, and Yoshimine potential for water, **53**:33ff
Maximum entropy formalism, **47**:Part 1:264–269
algorithm, **47**:Part 1:265–266
initial state, **47**:Part 1:266–268
variational approximations, **47**:Part 1:268–269
Maximum phase ("glory phase"), **10**:115, 120, 122–124
Maxwell-Boltzmann distribution, **1**:361, 363, 365, 367; **46**:333; **53**:85
shifted, **46**:356
Maxwell-Boltzmann equation, **1**:131
Maxwell construction, **37**:173; **43**:245
Maxwell distribution, **1**:310, 362; **38**:180; **44**:287, 476
Maxwell gas, **20**:158, 160, 204
Maxwell-Lorentz equation, **1**:310; **33**:160, 168, 187ff
Maxwell stress tensor, **48**:364
Maxwell's demon, **40**:143; **46**:157
Maxwell's equations, **1**:311, 317, 319–325; **12**:116; **23**:248; **27**:269, 271, 280, 363, 389, 399, 435, 450, 461; **33**:158, 161ff, 173, 179, 188, 194,

201, 215

Mayer expansion, **37**:142; **40**:301

Mayer *f* function, **20**:44, 51

Mayer graph, **20**:48, 52

Mayer ionic solution theory, **34**:125

Mayer-Montroll series, **6**:258

McDonald's theorem, **23**:93

McWheeney theory, **5**:349

Mean field behavior, **49**:320, 347–351

Mean field theory, **26**:111, 114; **31**:239; **35**:122, 134ff, 140ff, 148, 152, 157, 171; **40**:272, 273, 283, 360; **41**:17, 22ff, 40ff, 47, 51; **46**:334; **48**:219, 238, 479, 484, 492; **49**:312, 326, 338, 350

  free energy in, **48**:504

  singlet density function, **40**:312

  Temkin, **40**:168

Mean first passage time, **1**:373ff, 383, 386–388, 395; **4**:4; **5**:375ff; **52**:409

  random-walk particle, **4**:44–52

Mean force, **17**:137; **48**:80

  potential of, **48**:80, 198–199

  singlet and pair space, **1**:7, 8

Mean free path, **1**:153; **5**:290; **20**:244; **40**:438, 439, 448

  electrons, **51**:246, 270

Mean reaction field method, **48**:253–254

Mean spherical approximation, **31**:122–145; **34**:150, 151, 153, 236; **48**:216, 224, 232

  diagrammatic representation, **34**:150

  dielectric constant, **31**:124

  dipole fluid, **48**:237

  dipole-quadrupole fluid, **48**:237

  generalized, **48**:226, 237–238

  invariant expansion for, **31**:126–134

  multipolar molecules, **48**:233

  polar fluids, **31**:123–125

  polar ionic fluids, **31**:134–136

  polar mixtures, **31**:125, 126

  quadrupole fluid, **48**:237

  thermodynamics from, **31**:141–145

Mean square acceleration, **44**:344

Mean square dipole, **1**:89, 90, 93, 94, 96, 97, 99

Mean square displacement, **1**:97; **4**:42–44; **17**:134, 202

Mean square torque, **17**:138; **44**:274ff, 323, 346ff, 372, 429, 445, 465, 471

Mean trapping time, **52**:468

Mean value of observable, **36**:389

Measurement, **38**:3, 18

  of first kind, **38**:18

  of noncommuting observables, **38**:3

  theories of, **38**:3

Measurement process, quantum mechanical, **46**:153–194

  capacity, **46**:160, 174, 177, 181

  communication theory, **46**:165

  destructive, **46**:156, 161, 165, 178ff, 184

  mutual information, **46**:154, 159, 165, 175, 180

  relative, **46**:160, 173, 177

  nondestructive, **46**:169, 172, 178, 180

  repeated, **46**:186

  resolution, **46**:156, 158, 165, 172, 188

Mechanical equilibrium, **16**:122; **33**:425, 426; **49**:365, 371–372, 384–385, 394

Mechanical instability, **40**:321, 336, 357, 363

Mechanical stability, **37**:196

  criteria of, **17**:26

Mechanical theory of inhomogeneous fluid, **49**:359, 379–380

Mellin transformation, **4**:39

Melting, **6**:462–463, 478–479; **20**:286,

308, 310, 311, 312; **33**:101

mechanisms of, **6**:459–480

configurational disorder, **6**:467–469

formation of association complexes, **6**:469–474

positional disorder, **6**:463–466

rotational disorder, **6**:466–467

microclusters, **40**:122, 123

polymodal, of DNA, **33**:101

primitive earth, **55**:86–87

similitude rules for, **6**:462–463

zone, **32**:168

Melting curves, **1**:298

Melting temperature, **2**:129; **27**:479

Melts, ionic, association complexes in, **6**:474–476

conglomerate model for, **6**:476–478

ultraviolet absorption spectra of, **6**:477

polyphenyl, viscosity of, **6**:470

prefreezing in, **6**:460

quasi-crystalline, **6**:459

Membrane-bound enzyme, **39**:203–208, 214–217, 219–221

adenylate cyclase, **39**:206, 207

coupling factor, **39**:1, 205, 213, 214

coupling with transport, **39**:206, 218

electric field effect, **39**:225, 227

β-hydroxybutyrate dehydrogenase, **39**:203

phospholipid phase transition, **39**:204, 219, 220, 221

proton gradient, **39**:206

Membrane-bound protein, bacteriorhodopsin, **39**:224

Membrane channels, **39**:324

noise analysis of, **37**:67–98

Membrane potential, **19**:366; **39**:289

Fucus, **29**:285

Membrane transport, **21**:583, 601, 621

action potentials, **21**:606, 621

current-voltage relationships, **21**:607

ion channels in, **21**:601

Membranes, **19**:366

artificial, **29**:113, 310, 325

enzyme, **29**:113

lipid, **29**:309

transport processes, **39**:287–322

biological, **39**:287; **41**:200, 225

black lipid, **39**:287, 288

bulk, **39**:287

chloroplasts, **19**:366

chromatophores, **19**:367

clusters, **39**:264

collagen, **39**:229

collodion, **39**:229

components, **39**:288

composition and structure of, **29**:117, 304

conductance fluctuations, lipid bilayers, **37**:90

nerve axons, **37**:93

postsynaptic end plate, **37**:91

current-voltage relations in, **29**:293, 336, 338, 356

density and composition fluctuations, **39**:260

dioleoyllecithin, **29**:317

dioleylphosphate-Millipore, **29**:325, 335, 341

electrode, **39**:303

electron micrograph, **39**:259

excitable, **29**:43, 304, 319, 346, 349; **39**:52

fluid and solid domains, **39**:259

fluidity, **39**:227

freeze-fracture electron microscopy, **39**:259

functional organization in, **29**:119

hysteresis and, **29**:312, 323, 329, 335, 368

immunochemistry, **39**:275–286

lateral diffusion in, **39**:255, 260, 278, 281, 282, 283

lipid bilayer, **39**:264, 287, 292, 316

  admittance, **37**:111, 117

  models of transport through, **37**:99–138

liquid, **39**:288, 291, 299

memory effects in, **39**:231

models of, **29**:310; **39**:250

neutral carrier in, **39**:292

oscillations in, **39**:231

papain membrane, **39**:241

permeability, **4**:3, 6, 29; **39**:326

  variable, **29**:49

phase diagram, **39**:259, 277, 279

phospholipid, **39**:252

photoreceptor cells, **29**:343

preparation of, **29**:115

reactions at, **38**:350

role in development, **29**:269, 285

solvent polymeric, **39**:287, 288, 292

steady state conductance, **29**:312

thickness, **37**:123

Memory, **29**:35

long-time, **26**:112, 142

*n*th, **44**:276

wave, **29**:277

Memory effects, **35**:173; **44**:303

tail, **35**:159

Memory function, **17**:99, 100, 105, 108, 113; **24**:318; **31**:254; **34**:93; **42**:308; **44**:266, 274, 276, 289, 346, 357; **46**:312, 348; **51**:82

correlation function, **53**:142

delta function, **17**:175

depolarized scattering, **51**:103

derivatives, **24**:319

equation, **17**:99, 102

exponential, **17**:169, 173, 175, 178, 180; **24**:320

fading, **24**:108–122, 148–152

free rotation, **51**:83

Gaussian, **17**:170, 173, 175, 180; **24**:322

kernel, **46**:113, 114; **48**:89

  continued fraction representation of, **46**:114

  dynamic, **48**:112–113

  static, **48**:112–113

Lorentzian, **17**:175

Mori thory, **53**:144

non-Markovian, **46**:327

projection operator, **17**:94

relation to transport coefficient, **24**:320

see also History

Memory matrix, **44**:273, 290

third, **44**:289

Mercury, absorption line shape, **21**:329

Breit and transverse interaction energies, **54**:272

collisional energy transfer, **21**:189

correlation energy, **54**:272

quenching by CO, **50**:483

quenching of ^3P, **50**:166, 352, 364, 373, 382

quenching of $^3P_2$, **50**:356

self-consistent field calculation, **54**:265–273

self-consistent field theory compared with perturbation result, **54**:272–273

Mercury (Hg₂), dipole resonance interaction energy, **12**:41

Mercury hydride (HgD), $X^2\Sigma^+$ state, dispersion energy, **12**:20

Mercury hydride (HgH), $X^2\Sigma^+$ state, dispersion energy, **12**:20

Mermin model, 24:241

Mesitylene, 7:514

Mesobilene, 7:399

Mesophase, 39:276; **44**:371, 382

dynamics, **44**:258

Mesospheric chemistry of molecular hydrogen, 55:71–72

Messenger RNA, 7:219ff

Metabolic process, 29:270, 302

Metabolism, 38:224, 228, 229

nucleic acid, **38**:229

nucleoside triphosphate, **38**:229

Metal-ammonia solutions, 4:303–388

color of, **4**:305, 352

compressibility of, **4**:308, 309, 334, 354, 355

concentrated, **4**:333–337

density of, **4**:309–311

electric spin resonance, **4**:327–332, 374–379

electrical conductivity, **4**:318–322, 365–370

energetics, **4**:316–318

magnetic susceptibility, **4**:324–327, 371–374

models of, **4**:337–351

cluster, **4**:346–350, 361–363

colloidal, **4**:339

polaron cavity, **4**:342–346, 359–361

primitive cavity, **4**:339–342

unified, **4**:351

nuclear magnetic resonance, **4**:332, 333, 337, 379–383

optical properties, **4**:312–316, 357–365

photoelectric properties, **4**:316–318

solubility of metals, **4**:306, 307, 353, 354

stability of, **4**:306

thermoelectric properties, **4**:322, 323, 370, 371

vapor pressure data for, **4**:306–308, 353, 354

viscosity of, **4**:308, 309, 354, 355

volume expansion of, **4**:309–311, 355–357

x-ray scattering data for, **4**:311, 355–357

Metal chlorins, 7:380

Metal halides, photodissociation of, **50**:259, 290, 307

Metal hexafluorides, force constants, **9**:309ff

vibronic energy levels, **9**:250ff

Metal ion probe, 39:60

Metal ions, 7:182

Metal phthalocyanines, 7:380

Metal porphyrins, 7:359ff, 380

Metal protoporphyrin complexes, 7:389

Metal solutions, 2:120

Metal surface, adsorption of water, **21**:363

Metallic clusters, 40:57

Metallic complexes, electronic structure, **37**:203–304

molecular orbital calculations for, **37**:203–309

anisotropic hyperfine interaction, **37**:259, 260

dipole moment, **37**:241–243

electric polarization, **37**:249–251

Fermi-contact term, **37**:261–264

g-factor, **37**:265–272

magnetic susceptibility, **37**:251–255

nuclear quadrupole coupling constant, **37**:243

spin-spin coupling, **37**:272–283

Metallic glasses, 51:227–271

electron transport, **51**:239–253

magnetic, **51**:256–257

microcrystalline models, **51**:235
Monte Carlo studies, **51**:235
Mooij correlation, **51**:241, 243, 246, 247, 270
Mössbauer effect, **51**:232, 238, 253, 259, 266
neutron diffraction, **51**:232, 248
nuclear magnetic resonance, **51**:232, 238, 259, 266
optical properties, **51**:262–266
photoemission, **51**:257–262, 266
physical properties, **51**:230
Polk model, **51**:237, 266
radial distribution function, **51**:231
rapid cooling, **51**:229–230
soft X-ray spectroscopy, **51**:259
stability, **51**:268
structure, **51**:231–239
structure factor, **51**:232, 248
thermoelectric power, **51**:254–257
transport properties, **51**:231
Ziman theory, **51**:244–250, 254, 261, 270
Metallization concept, 49:107–109
Metalloprotein, 39:61
Metals, 27:378–379, 460
liquid, **13**:153; *see also* Liquid metals
reflection by, **23**:259, 288, 313
see also Chemiluminsecent reactions
Metastability, 29:72; **40**:143, 168, 230, 251, 306, 347
Metastable diatomic molecules, 12:18
Metastable equilibrium, 26:113, 114, 134, 141
Metastable species, *see* Excited atoms and molecules
Metastable states, 28:294ff, 300, 307, 381ff, 399, 404, 451; **34**:295; **48**:456, 457
Metastable vibrational energy states, 12:18

Methacrylonitrile, 2:155
Methane, 6:388; **34**:188
+ carbon dioxide system, **2**:97
in clathrates, **2**:30, 41
crystal constants, **5**:172
crystal symmetry, **5**:171, 173, 180
energy transfer in, **47**:Part 2:208–227
hydrate, **2**:33, 34, 47
thermodynamic data and lattice energy, **2**:8
-hydrogen sulfide-water ternary system, **2**:44
interatomic distances, **5**:156
ionization energy, **2**:71
Langmuir constants, **2**:47
London forces in, **6**:405
magnetic susceptibility, **2**:70, 71
polarizability, **2**:71
polarizability, **46**:66
-propane-water ternary system, **2**:10, 33, 46, 48
Methane thiol hydrate, 2:10
thermodynamic data and lattice constants, **2**:8
Methanol, barrier height of internal rotation, **2**:381
microwave spectroscopy, **2**:379
Methanol clathrates, 2:2, 20
in hydroquinone, **2**:7
Methene bridges, 7:376ff, 383, 397, 404
Method of images, 52:410
Method of moments, 25:232, 244–246, 248; **31**:159, 161; **35**:228; **46**:16, 17, 26
Method of steepest descent, 4:61; **25**:135
Methyl boron difluoride, 2:382
Methyl chloride, sources of, **55**:76
Methyl cyanide clathrate, 2:20
Methyl cytosine, 7:52ff
Methyl ether + iodoform system, 2:99

Methyl fluoride, energy transfer in, 47:Part 2:209–227

Methyl fluorosilane, barrier height of internal rotation, 2:383

Methyl germane, barrier height of internal rotation, 2:382, 383

microwave spectroscopy, 2:380

Methyl group, hyperconjugation, 7:90

Methyl iodide, phase relation, 40:38

Methyl ketene (CH₃CHCO), photodissociation of, 47:Part 2:104–106

Methyl methacrylate, 2:155

polymerization, 2:180

β-Methyl naphthalene, 40:456

Methyl pyridine - water systems, 17:7

Methyl radical, 7:537, 576

EPR spectra, 7:537

Methyl silane, barrier of internal rotation, 2:381, 382, 383, 391

microwave spectroscopy, 2:380

1-Methyl thymine, 16:184

Methylene, 42:226

Michaelis constants, 7:653, 670ff, 674, 721ff

Michaelis-Menten mechanism, 7:648; 26:149; 43:72, 102

Michelson interferometer, 24:27

Michelson interferometry, 44:309

two-beam, 44:321

Micro-iteration, 54:103

Micro transition probability, 46:400

Microcancer, 38:397, 400

Microclusters, argon, 40:55

astrophysical, 40:56–57

binding energy, 40:62

calcium, 40:58

catalysis and, 40:51

chemical, 40:56

cuboctahedron, 40:71, 96

dodecadeltahedron, 40:87

enantiomorphism, 40:62, 87

ESR, 40:58

free energy of formation, 40:126

frequency spectra, 40:99, 102

graph theory of structure, 40:65, 93–94

harmonic oscillator-rigid rotor model, 40:109–114

heat capacity, 40:101, 112

helium, 40:55

internal energy, 40:110

ionization energy, 40:54

iron, 40:56

isomers, 40:85

lead, 40:58

lithium, 40:54, 58

mass spectrometry of, 40:54

matrix isolation, 40:57, 130

melting of, 40:122, 123

models of crystal growth, 40:127–129

molecular dynamics methods, 40:121

Monte Carlo methods, 40:121, 123

nitrogen, 40:55

NMR, 40:58

nucleation theory, 40:125–129

photographic process, 40:58

polytetrahedral structures, 40:78

quantum mechanics, 40:129

radial distribution function, 40:85

Rayleigh scattering, 40:55

rhombicosidodecahedron, 40:81, 84, 85

rhombicuboctahedron, 40:71

rotational dynamics, 40:107–108

silver, 40:58

sodium, 40:54

sphere packing schemes, 40:72, 73, 82

stability theory, **40**:59–93

statistical morphology, **40**:85

structure, **40**:59–94

superparamagnetism, **40**:58

supersonic beams, **40**:54

thermodynamics, **40**:108–117

vibrational dynamics, **40**:99, 102

water, **40**:55, 130

see also Clusters

Microcrystals, nucleation in absence of, **40**:138

Microorganisms, motion of, **52**:488–492

protection from UV damage, **7**:592

Microscopic reversibility, **4**:6; **30**:42, 67, 418, 432; **33**:409; **40**:182; **41**:307; **42**:26, 282

Microspectrographs, **7**:409, 412

Microsymmetry, principle of, **5**:127

Microwave absorption, **44**:259, 263; *see also* Far infrared

Microwave interferometric measurements, **44**:376

Microwave klystron frequency, **44**:324

Microwave spectroscopy, **2**:368, 376, 383; **17**:66; **42**:356

Mie-Lennard-Jones potential, **16**:343

Migration, **29**:170

of impurities, **52**:470

Migration inhibition factor, **38**:396

Miller formula, **18**:101

Miller indices, **16**:136, 140, 151, 205, 216; **27**:142

MINDO, **42**:67

Mini-exciton, **40**:373

Minimal polynomial, **43**:49

Minimum energy path, **36**:154–156, 158

curvature, **36**:155, 164, 165

Minimum entropy production theorem, **43**:94, 95, 98

Minimum image method, **48**:246–249

Minimum principles, **25**:247, 248, 257–259, 283, 286, 290

Minimum uncertainty wave packet representation, **53**:314

Minimum variance method, **25**:248, 249, 283

Miscibility, neon and hydrogen, **16**:360

Miscibility gap, **6**:270

Mitochondria, **7**:238

energy conversion, degree of coupling, **55**:146–150

nonlinear energy converter, **55**:159–166

optimization of, **55**:141–167

thermodynamic buffering, **55**:150–158

Mitosis, **7**:85, 89

frequency of, **7**:152

Mixed boundary conditions, **52**:455

Mixed crystals, Bloch functions, **8**:30ff

energy acquisition and, **47**:Part 1:7–9

guest molecules, **8**:27, 30

host crystals, **8**:27, 30ff

intermolecular attractions, **8**:30

quadrupolar resonance intensity in, **2**:190

trap depth, **8**:30ff

Mixed energy release, **19**:22; **42**:85

Mixed hydrate, **2**:10, 11, 52

Mixed moment, **16**:147

Mixed states, **40**:386; **42**:227

Mixing, **24**:157, 158

asymptotically stable, **43**:67

enthalpy of, of molten alkali nitrates, **11**:106

excess free energy of, binary mixtures, **11**:105

reciprocal systems, **11**:112

excess Gibbs free energy of, **11**:106

Helmholtz free energy of, binary mixtures, **11**:100

reciprocal systems, **11**:110

molar energy of, **16**:247

stability, **43**:66–73

Mixing character, compared with ergodic theory, **51**:135

macroscopic systems, **51**:169–171

probability distribution, **51**:115, 117–135

equivalent definition, **51**:121–126

time development of irreversible processes and, **51**:127–130

Mixing distance, homomorphic functions, **51**:131

of distribution, **51**:115, 117–135

thermodynamic system in contact with bath and, **51**:130–135

Mixing homomorphic function, **51**:122, 141

Mixtures, chain molecules, **16**:245–247, 275

classical, **38**:43

nonideal, **33**:437

simple fluids, **37**:194

spherical molecules, **16**:224–226

statistical mechanics of, **11**:117ff

Mobility, **44**:507

amino acids, **39**:81, 338

domain wall, **53**:275

hapten, **39**:272

Mode coupling, **48**:91

Mode locking, **23**:190, 197, 199

BBOT, **23**:197

bleachable dyes 9860 and 9740, **23**:197

cryptocyanene, **23**:190

dimethyl POPOP, **23**:197

Gaussian shape, **23**:199

3 PF, **23**:198

polymethene dyes, **23**:190

saturable absorber, **23**:190

Mode-mode coupling, **38**:176, 186, 187; **44**:273; **46**:316, 325

Model error, **51**:180, 205

Model Hamiltonian, **52**:232

Model potentials, **19**:7; **26**:242; **50**:563

applied to beryllium, **16**:330–333

applied to magnesium, **16**:333–334

formalism, **16**:312–317

many-electron problems, **16**:324–329

non-local, **16**:309

from pseudopotential theory, **16**:317–321

two-electron equations, **16**:328–329

Model pseudopotentials, **16**:317–321

Rydberg states for beryllium and magnesium, **16**:322–324

Modes, deformation of chain, **33**:76, 125, 131, 136–147

normal, **32**:3

rotational separated from deformational, **33**:125

soft, **32**:88

Modified wave number approximation, **36**:179

Modulated beam spectroscopy, **41**:495

Moiré pattern, **32**:138

Molar area, partial, **6**:155, 159

Molar extinction coefficient, **4**:143, 313

Molar heat of clathrates, **2**:21

Molar volume of clathrates, **2**:21

Molecular alignment, **42**:468

Molecular association, **44**:323

restriction on, **44**:478

Molecular beam, **41**:249, 493; *see also* Beam

Molecular beam experiments, **28**:11–32, 60, 76ff, 87, 222ff; **37**:141

multiphoton dissociation, **50**:159

photofragmentation, **50**:259

photoionization, **52**:263–355

apparatus, **52**:276–280

time of flight spectra, **50**:307

Molecular beam scattering, elastic, **12**:358, 391

experimental arrangements, **12**:397, 421

inventory of, **12**:424ff

low-energy, **12**:389, 397, 420, 421

Molecular beam spectroscopy, **42**:310, 356, 360

Molecular beats, **23**:210

spectrum of naphthalene, **23**:212

Molecular binding energy in many-electron theory, **6**:408

Molecular chaos, **5**:290; **38**:179

Molecular complexes, charge transfer states, **8**:14ff

interaction of carcinogens with proteins, **8**:168

Molecular core field, **5**:47

Molecular correlation, **34**:206, 208

Molecular crystals

anthracene, **54**:320–325, 372–376, 381–406, 451–455

semiclassical theory, **54**:340–353

exciton surface polaritons, **54**:437–455

excitons, **54**:315–320

photon coupling, **54**:314–315

reflectivity, **54**:406–422

resonance interactions, **53**:228

resonant energy transfer, **53**:277–287

semiclassical theory, **54**:328–340

surface excitons, **54**:305–326

Molecular dynamics, **17**:120; **27**:467, 478–481, 483; **37**:146; **40**:21, 28, 36; **41**:209, 216, 226, 230, 262–264; **52**:354–355; **53**:64, 206

clamping approximation test, **53**:189, 190

classical, **41**:264–265

correlation function calculation,

53:191–201

depolarized scattering, **51**:97, 102

glasses, **48**:398–454, 471, 472

hard sphere fluid, **34**:110

liquid state densities, **53**:118

metallic glasses, **51**:232

microclusters, **40**:121

quantum mechanical, **41**:259, 264

rototranslational correlation functions, **44**:259, 267, 338, 347, 400ff, 439, 455

vibrational relaxation, **42**:310, 346

water, **31**:70

Molecular exciton, *see* Exciton

Molecular field, **46**:368, 402

Molecular incompressible volume, **1**:160

Molecular integrals, for Slater-type orbitals, **13**:205

Molecular interaction theory, **9**:91

Molecular liquid, equilibrium structure of, **34**:145

Molecular orbital calculations, **7**:191; **36**:151, 164; **49**:79–85, 107, 110

CNDO, **37**:212, 284, 285

extended Hückel, **37**:210, 211

fragment formalism, **49**:84, 93–95

Goeppert-Mayer-Sklar, **37**:210

Hartree-Fock, **37**:214

high energy orbitals, **49**:80–83

INDO, **37**:285

matrix elements, **37**:211–215, 216–220

NDDO, **37**:285, 286

nonempirical, **37**:286

one-electron properties, open or closed shell systems,

anisotropic hyperfine interactions, **37**:259, 260

electric dipole moment, **37**:241–243

electric polarizability, 37:249–251

indirect nuclear spin-spin coupling, 37:256–259

magnetic shielding, 37:251–255

magnetic susceptibility, 37:251–255

open shell systems, g-factor, 37:265–272

open shell systems, spin Hamiltonian, 37:260–261

Roothaan, 37:214

simple Hückel, 37:207–209

site preference and bond length, 49:80–83

two-electron properties, open shell systems, spin-spin dipolar interactions, 37:272–283

UHF, 37:221, 285, 286

Wolfsberg-Helmholtz, 37:284

Xα, 37:287–289

Molecular orbital configuration, N_2, 45:266

NO$^+$, 45:335

NO, 45:299

O_2, 45:286

O_2^-, 45:316

Molecular orbitals, 1:179, 183, 184, 239, 249ff, 256–258, 260; 7:19, 22, 194ff; 13:42, 52–54, 66; 30:217, 291, 451; 41:151

approximate theories, 18:249ff

energy level diagrams, 13:52

energy of, 7:104ff, 126, 143, 144

polyacetylene, 53:289

Su, Schreiffer, and Heeger model, 53:288

Molecular polarizability, 12:109

Molecular potential, 36:37

Molecular quadrupole moments, 12:137

Molecular radial distribution function, 34:173

Molecular reorientation, 12:551, 555

Molecular rotation in crystals, 6:461

Molecular scattering, elastic, 12:390, 424

Molecular spectra, 36:402

Molecular structure from infrared spectroscopy, 7:420

Molecular susceptibilities, 12:239, 251, 254, 255, 256

Molecular thickness, 8:167

Molecular wave functions, 26:247

Molecules, rigid polar, 48:188–201

surface active, 48:335

Molten salts, metals in, 31:465–480

F-center model, 31:472–475

free electron model, 31:467

thermally acitvated hopping model, 31:468

surface tension and compressibility, 9:77

Molybdenite structure (MoS_2), 49:105

Molybdenum, 41:141

Molybdenum hexafluoride, 9:197ff

Moment equations, 31:162, 163, 170–172, 182

Moment expansion, 36:285, 295

Moment of inertia, 31:168, 182; 35:45, 51; 44:269, 427

dyadic, 44:385

generalized, 35:45, 51, 105, 109

Moments, displacement, 52:368

induced absorption, 51:70

experimental, 51:54

light scattering, 51:88–89, 105

method of, 25:232, 244–246, 248; see also Method of moments

transition probability, 46:198

Momentum balance, 13:273; 33:407

Momentum conservation equation,

16:105–106

Momentum density, **46**:286

Momentum distribution function for electrons, **31**:418

Momentum equations, **32**:46

Momentum flux, **35**:60, 81

approximate expression, **35**:87

Brownian motion contribution, **35**:87

general expression, **35**:90

Giesekus equation, **35**:94, 95, 110

intermolecular contribution, **35**:64

intramolecular contribution, **35**:64

kinetic contribution, **35**:60, 63

Kramers, Kirkwood, Kataka, Fixman expressions, **35**:88, 92, 110

oscillatory shear, **35**:106

physical interpretation, **35**:65

polymer-polymer interaction, **35**:65

symmetry of, **35**:92

Momentum ratio, expansion in, **16**:16–19

Momentum representation, **36**:23

Momentum space, **25**:255

Momentum translation approximation, **33**:157

Monactin, ion transport, **29**:311

Monatomic crystals, **27**:467–487, 509–514

Monoalkylphosphate anions, spectra of, **7**:476ff

Monochromatic sound wave, **44**:296

Monochromatic spin wave, **44**:296

Monodehydroascorbic acid, EPR, **7**:636

Monodispersed polymer, **2**:177

Monodispersed system, **2**:183

Monoglycerides, infrared spectra, **7**:413

Monolayer, **7**:239, 246; **27**:298

lipid, **41**:195–228

Monomers, **40**:143

Monomethylphosphate anion, spectrum of, **7**:476

Monomolecular reactions, **10**:176, 190

Mononucleotides, **7**:468, 477ff

Monopoles, permanent, **4**:139

point, **4**:139

transition, **4**:139, 157

Monosubstituted benzenes, **54**:211–216

Monte Carlo method, **6**:250–252; **9**:2; **10**:320, 350, 359ff, 375; **17**:121; **25**:71, 116, 130, 146–149, 154, 166, 172, 175; **28**:5, 11, 32, 65ff; **30**:103, 112, 113; **31**:71; **34**:110, 234, 237; **37**:146; **40**:177, 204, 205, 212, 220, 222, 225, 338; **41**:209, 210, 226; **44**:338, 400, 402; **48**:398; **51**:2, 22, 24, 25, 33, 35, 40, 41; **53**:64, 65, 105, 106, 206

binary mixtures, **6**:270

krypton structure factor, **53**:16, 19, 20

Langevin's equation, **46**:255

liquid state density, **53**:118

metallic glasses study, **51**:235

microclusters, **40**:121, 123

one-dimensional random potential, **13**:127, 136, 152

stochastic processes, **46**:253

Mooij correlation in glasses, **51**:241, 243, 246, 247, 270

Mori continued fraction, **44**:256, 257, 271, 289, 323, 342, 345, 357, 429

Mori distribution, **32**:62

Mori equation, **51**:83; **53**:95, 147

Mori fictitious dynamics representation, **53**:165

Mori projection operator method, **53**:138, 143–148

Mori random force, **53**:94

Mori's formalism, **44**:274, 404, 467

Morphallaxis, **29**:257

Morphogenesis, 29:32, 54, 254, 282; **43**:43

Morse function, 36:301

Morse lattice, 53:302, 303, 307
  defects in, **53**:310
  shock waves and, **53**:308

Morse oscillator, 1:392
  collision of atom with, **36**:99
  period of, **13**:106

Morse potential, 10:77, 105, 382; **13**:104; **36**:70, 73
  atomic lattices, **53**:295

Morse-spline-van der Waals potential, 42:490

Mo's method, 47:Part 1:220–221

MOSFET system, 41:165, 173

Mössbauer spectra, line shape, **15**:123

Moths, 19:338

Motility, bacterial, **29**:174, 175

Motional narrowing, 15:103, 109; **40**:23ff, 35, 44, 427, 462, 468

Motor-boating, 19:342

Mott *s-d* scattering model, 51:247

Mountain peak, 24:16

Muffin-tin potential, 31:427, 435; **37**:288; **41**:161

Multiple-wave theory, 27:451, 460–465

Mulliken approximation, 18:253, 289; **36**:149, 154, 163
  extended Hückel theory, **21**:30
  SCF matrix elements, **18**:258
  SCF MO calculations, **18**:256

Mulliken-Wheland method, 7:28ff

Mulliken's scale of electronegativity, 13:58; **49**:107

Multibarrier kinetics, 4:4–8, 45, 52

Multicenter integral, 36:149
  angular momentum, **37**:268
  Coulomb exchange, **37**:217, 223, 227
  dipolar spin-spin, **37**:277

  electric field gradient, **37**:245
  electronic correlation operator, **37**:276

Multichannel quantum-defect theory, 52:288

Multichannel scattering, 25:212, 255, 283, 285, 287

Multicomponent diffusion, 6:293–295

Multicomponent gases, convex core model, **5**:169

Multiconfiguration approximation, 14:191ff, 331

Multiconfiguration Hartree-Fock theory, 52:249–251

Multiconfiguration LCAO method, 9:87, 91, 94ff, 151ff, 159, 165

Multiconfiguration self-consistent field (MCSCF) methods, 23:165; **25**:278; **26**:222, 223; **36**:153; **44**:8; **50**:530, 557

Multiconfiguration self-consistent field (MCSCF) state, 54:1–176
  characterization of, **54**:25–33
    eigenvalues of Hessian, **54**:28–30
    excitation energies, **54**:30–32
    stability condition, **54**:32–33
  cubic terms, construction of, **54**:158–163
    macro-iteration, **54**:102
    micro-iteration, **54**:103
    numerical results, **54**:107–112
  eigenvalues, full and reduced Hessian matrices, **54**:156–157
  exponential unitary operators, **54**:6–10
  generalized and fixed Hessian approaches, **54**:66–78
  gradient-based methods with iterative cubic method, **54**:127–130
  gradient-based methods with second order methods, **54**:126–127
  higher order procedures, **54**:134

large configuration interaction expansions, **54**:121–126

Newton-Raphson approach, **54**:34–66

  improved iterative methods, **54**:65–66

  initial guess of orbitals and states, **54**:62–65

  nonlinear, **54**:41–62

    global convergence, **54**:45–56

    local convergence, **54**:56–59

    matrix elements, **54**:23–25

    one-step, **54**:21–22

    step size, **54**:41–62

    two-step, **54**:22–23

  one and two-step, **54**:59–62

  small Hessian eigenvalues, **54**:40–41

  step size and sign control, **54**:35–40

Newton-Raphson iterative function, **54**:3, 4–5, 10–25

  linear, **54**:11–14

  nonlinear, **54**:14–25

partial derivative of energy, **54**:145–151

perturbative cubic approach, **54**:101–114

redundant variables, **54**:143–145

simplifications of transformations, **54**:116–121

two-electron integral transformations, **54**:164–173

update methods, **54**:78–101

  Broyden asymmetric rank-1, **54**:79–82

  Broyden family (DFP and BFGS), **54**:82–86

  Hessian, **54**:88–90

  inverse, **54**:87–88

  numerical results, **54**:90–100

variation approach, **54**:86–87

Multiconfiguration time-dependent Hartree-Fock method, 54:30–33

  excitation energies, **54**:30–32

  stability condition, **54**:32–33

Multigraph, 43:75

Multinomial distribution, 41:14

Multiphonon processes, 42:335; **47**:Part1:90–98

Multiphonon relaxation, 41:294

  Franck-Condon factors, **41**:308

Multiphonon scattering, 27:64

Multiphonon vibrational relaxation, 47:Part 2:522–526

Multiphoton absorption, *see* Multiphoton excitation; Multiphoton processes; Two-photon absorption

Multiphoton dissociation, 47:Part 1:713–733; **50**:105–190, 272, 274, 452

  acetic anhydride, **50**:122

  beam, **47**:Part 2:138–143

  CF_2, **50**:119

  CF_2CFCl, **50**:108, 123, 148

  C_2F_3Cl, **50**:176

  C_2F_5Cl, **50**:130

  CF_2HCl, **50**:110, 126

  CF_3I, **50**:121, 129

  CH_3CCl_3, **50**:156, 162

  CH_3CF_2Cl, **50**:156

  collision-free chemiluminescent reactions, **47**:Part 1:716–719

  CrO_2Cl_2, **50**:135

  cyclopropane, **50**:130

  dichlorocyclobutene, **50**:145

  ethylvinylether, **50**:119, 142, 145

  experimental methods, **47**:Part 1:714–715

  experimental studies of complex molecules, **47**:Part 1:639–659

  number of photons absorbed,

47:Part 1:647–648

real-time reaction monitors, **47**:Part 1:652–655

mechanism, **47**:Part 1:715–716

SF_6, **50**:128

SiF_4 experiments, **47**:Part 1:729–732

tetramethyldioxetane, **47**:Part 1:719–729

nature of emitting species, **47**:Part 1:726–729

time behavior, **47**:Part 1:727

uranyl compounds, **50**:141

Multiphoton excitation, **47**:Part 1:70–82; **50**:105–190, 395–474

CF_3H, **50**:447

CF_2HCl, **50**:461

CF_3I, **50**:461, 464

$C_2F_4S_2$, **50**:464

chemiluminescent reactions resulting from, **47**:Part 1:694–707

$C_2(a^3\Pi_u)$ + NO, **47**:Part 1:695–701

$C_2(X^1\Sigma_g, a^3\pi_u)$ + O_2, **47**:Part 1:701–705

CH_3NH_3, **50**:461

CrO_2Cl_2, **47**:Part 1:661–677

dephasing, **47**:Part 1:81–82

electronic luminescence resulting from, **47**:Part 1:679–711

CrO_2Cl_2, **47**:Part 1:661–677

kinetics, **47**:Part 1:687–692

mechanism, **47**:Part 1:692–694

survey of experimental results, **47**:Part 1:681–687

high-order, **47**:Part 1:47

n-level multiple resonance, **47**:Part 1:555–581

inverted-Raman, **47**:Part 1:574–575

ladder, **47**:Part 1:570–574

master equations, **47**:Part 1:567–569

open subsets, **47**:Part 1:564–567

steady-state solutions, **47**:Part 1:559–564

three-level double resonance, **47**:Part 1:569–579

N_2F_4, **50**:464

random coupling model, **47**:Part 1:78–79

reduction scheme, **47**:Part 1:77–78

SF_6, **50**:464

traditional energy level scheme, **47**:Part 1:71

vibrational quasicontinuum, **47**:Part 1:74–76

weak coupling approach, **47**:Part 1:76

Multiphoton-induced chemistry, **47**:Part 1:509–733

atom-molecule reactions, **47**:Part 1:625–638

electronic luminescence, **47**:Part 1:679–711

infrared laser chemistry, **47**:Part 1:639–659

luminescence of CrO_2Cl_2, **47**:Part 1:661–677

n-level multiple resonance, **47**:Part 1:555–581

reduced equations of motion, **47**:Part 1:509–533

SF_6 spectroscopy and coherent pulse propagation, **47**:Part 1:583–624

unimolecular dissociation, **47**:Part 1:713–733

Multiphoton ionization, **47**:Part 1:51

Multiphoton processes, molecular, **47**:Part 1:509–533

coherent driving, **47**:Part 1:540–541

correlation functions, **47**:Part 1:527–535

incoherent driving, **47**:Part 1:544–

546

intramolecular dephasing, **47**:Part 1:540–546

intramolecular dipole correlation function, **47**:Part 1:546–551

perturbative line-broadening limit, **47**:Part 1:541–544

quasicontinuum, **47**:Part 1:544

reduced equations of motion, expanded to second order, **47**:Part 1:522–527

including coherences, **47**:Part 1:535–540

reduced equations of motion for populations, **47**:Part 1:516–52

chronological ordering prescription, **47**:Part 1:518–520

partial ordering prescription, **47**:Part 1:520–522

Multiple collision expansion, 30:63

Multiple internal reflection, 44:334

Multiple isomorphous replacements, method of, **16**:203, 213–214

Multiple scattering, 34:165

Multiple short circuit tail reaction, 43:133

Multiple stationary states, 43:222–228

Multiple steady state bifurcation points, 43:37

Multiple steady states, 43:29, 35–40

Multiple trapping, 52:482

Multiplet structure, 36:264, 268

Multiplicative stochastic processes, 46:267, 271

Multipolar coupling, 44:570, 588, 614, 618

dynamic, **44**:584, 635

static, **44**:584, 588

Multipole expansion, 18:83; **33**:155, 157, 166, 220; **52**:119, 159, 162

electrostatic interaction, **48**:230

Multipole Hamiltonian, 33:155

Multipole interaction in convex molecules, 5:180ff

Multipole moments, 12:112, 114, 129; **33**:166; **44**:323, 446, 447, 451; **51**:74, 75

Multipole-multipole effects, 44:309

Multipole polarizability, 12:145, 146

Multipoles, electric, **33**:229

magnetic, **33**:229

molecular, **33**:161

see also specific multipoles

Multiquantum transitions in rotational to translational energy transfer, 50:59

Multistate random walk, 52:375, 376, 456–467

Multivariate Gaussian process, 44:267

Multivariate process, 17:102

Muon, positive, **41**:173

Murphy-Singer-Klein-McDonald model, 53:24

Muscle, infrared spectroscopy of, **7**:413

Mushroom, 7:638ff

Mutagens, 8:179; **38**:226, 228, 245

Mutations, 8:177; **38**:225

carcinogenesis, **38**:394

fluctuations and, **29**:8, 44

neutral, **29**:193, 208, 230

somatic, **38**:394

Mutual collision cross section, 10:37–39, 83, 127–131

Mutual information, 46:154, 159, 165, 175, 180

relative, **46**:160, 173, 177

Mutual neutralization, 30:495

Myoglobin, 7:160, 182, 239, 395, 615

Myosin, 7:274ff

Myristic acid, 41:196

N

N-body, *see also* Many-body

N-body calculation, **26**:116, 117, 119, 123, 134, 138

N-body density matrix, **14**:240

N-body potentials, **12**:292, 297

N-body problem, gravitational, **26**:108, 137–139, 141

N-mer, **40**:415

n-π* transitions, **6**:400

NaCl, **27**:438, 444–445, 505, 515, 517–519

NAD substrates, **7**:715, 720

NaF, **27**:445–446

Naimark's theorem, **38**:74

Nakajima-Zwanzig method, **53**:347, 356, 357, 374

Naphthalene, **1**:62, 171, 176–178, 185, 191, 246, 247, 257, 258; **20**:17; **40**:392ff, 424ff, 446ff, 457, 463ff, 473, 474; **47**:Part 1:6

  complex with sodium, **2**:153

  core parameters, **5**:157

  diffusion in air, **5**:163

  exciton bands, **23**:329

  fluorescent decay, **42**:260

  ionization potential, **7**:302

  phosphorescent yield, **42**:262

  polymerization, **2**:150

  quantum beats, **23**:212

  relaxation, **23**:208, 212

  resonance line, **2**:191

  rigid convex body constants, **5**:163

  solid, **22**:226, 239, 241, 258, 293, 299

    Raman spectrum of, **22**:207

Naphthenes, (2, 2)-*p*-cyclo-, **16**:195, 196

α-Naphthyl-4-chlorophthalamic acid, **16**:181–182

(+) α-Naphthylphenylmethylgermane, **16**:207

Narrow-band excitation, **40**:375–379, 394, 398

Native structure, **46**:364ff

Natural orbitals, **6**:366; **14**:358

Natural selection, **38**:215, 236

  population genetics, **38**:236

Natural state formalism, **49**:122

Navier-Stokes equation, **5**:262, 285; **11**:261; **16**:123; **26**:178; **44**:301; **46**:304

  viscoelastic medium, **5**:288

Near-adiabatic hypothesis, **10**:197–198

Near-degeneracy, **6**:366–368; **9**:328, 349

  atoms, **6**:366

  molecules, **6**:366

Near-resonance, in charge transfer, **10**:197–198, 254–256

Near ultraviolet radiation, **7**:586

Nearest-neighbor interactions, **4**:204; **13**:6, 7, 9; **38**:152; **44**:382

  atomic lattice, **53**:295

  heatbath, **53**:98

  molecular crystal, **53**:278

  shock waves and, **53**:310

  Toda lattice, **53**:300

  water, **53**:42, 44

Nearest-neighbor random walk, **52**:402, 410, 463, 481

Nearly free electron model, **51**:267

Negative concentrations, **19**:346

Negative energy state, **41**:84

Negative ion resonance, **36**:435

  benzene and benzene derivatives, **36**:435

Negative ions, **19**:154; **28**:81, 115, 139ff, 388, 397, 403, 411, 439, 456ff

  autodetachment, **36**:451

Negative surface ionization, **30**:494

Neglect of differential overlap approximation, **33**:356; **44**:619

correction to, **33**:356, 358, 360

semiempirical procedures based on, **33**:356ff, 378, 392, 394

two-electron interactions, **33**:357

NeH+, **19**:43, 44, 98, 120, 123

Neighbor-avoiding walk, **51**:21

Nematic liquid, **24**:49

Nematic phase, of 7CB, **44**:386

Neodymium:YAG laser, **50**:16

Neon, **14**:19, 21, 211; **34**:178; **41**:140; **49**:538

atom, in clathrates, **2**:12, 20

collision cross section, **21**:372

collisional excitation, **30**:447

complexes of iodine and, **47**:Part 1:355–358

correlation energy, **6**:359, 375

crystalline, **21**:384

dimer, **41**:157

intramolecular 6–12 potential, **2**:71

isoelectronic series, **6**:383

light scattering, **24**:40, 52

liquid, **21**:385

London attraction, **6**:399, 402

polarizability, **8**:11; **46**:46–47

polarizability, magnetic susceptibility, and ionization energy, **2**:71

second virial coefficient, **21**:379

structural and thermodynamic data, **16**:346–348

symmetry properties of, **6**:372–373

see also Scattering; van der Waals molecules

Neon-hydrogen, miscibility of, **16**:360

phase diagram of, **16**:350

Nephelauxetic effect, **8**:53, 59

Nephelauxetic series, **5**:68, 69, 134; **7**:364ff, 401; **8**:53

Neptunium, absorption line shape, 21:325

Neptunium hexafluoride, **9**:190

Nernst-Einstein diffusion equation, **6**:292, 299

Nervous nets, **19**:317

limit cycles, **19**:317

statistical mechanical, **19**:317

Nesbet's hierarchy, of Bethe-Goldstone equations, **52**:198–200

Net correlation increment, **52**:199

Net glide plane, **20**:10

Netted-chain equation, **6**:265

Network, chemical, **29**:1, 22; **43**:8, 17

entropy production, **29**:22

simple models of, **29**:32, 158, 159, 161

complex balanced, **43**:11, 78–85, 101

definition of, **43**:109

diagram, **43**:112

with diffusion, **43**:205–206

directed set-graph, **43**:109

electrical, **43**:27

equilibrium extreme, **43**:14, 53

extreme, **43**:14, 53

Hurwitz nonmixing, **43**:126, 193–203

knot-tree, **43**:87, 103, 211

mass action, **43**:82

mass action knot-tree skeleton, **43**:87–94

oxidation-reduction, **43**:149–156

pathological, **43**:81

power law, **43**:82

pseudoextreme, **43**:88

quasithermodynamic, **43**:10

realistic, **43**:138–149

reversible, **43**:78

semidirected species tree, **43**:94

skeleton, **43**:87

stability, **43**:1–215

stability classification, **43**:50, 53

stability conjectures, **43**:189

stoichiometric, **43**:103–166

thermodynamics, **43**:11

tree, **43**:11, 77, 85–86, 87

tree graphs, **43**:10

unimolecular, **43**:10, 32

zero-deficiency, **43**:11

Network model, 43:12

random, **48**:467

Neumann algebras, 38:23–25, 27, 32, 40, 46

Abelian, **38**:25, 27, 58

associated operators, **38**:23, 25

associated with quantum theories, **38**:23

atomic, **38**:46

continuous, **38**:46

definition of, **38**:24

dual, **38**:32

factor of type I, II, III, **38**:46

generated by projections, **38**:24

predual of, **38**:34

Neumann expansion, 13:214, 220, 231

Neural networks, 19:325, 327, 339, 347

Neural oscillator, 19:342

Neural tube formation, 29:254

Neuron, action potential of, **21**:606, 621

inhibitory, **19**:341

synaptic vesicles, **21**:622

Neuron gas, 19:342

Neurospora crassa, **7**:670

Neutrals, collisions with ions, excited products, **45**:163–196

interaction with excited ions, **45**:120–161

Neutron, thermalization, **20**:138

Neutron-deuteron scattering, 25:282–285

Neutron diffraction, argon, **53**:13, 14

fluid structure studies, **53**:4–5

hydrogen chloride, **53**:30–33

inelastcity corrections, **53**:7–9, 14, 16, 30, 38, 45, 48–54

magnetic, **53**:251

nitrogen, **53**:26

pulsed neutron source, **53**:9

slow, **53**:6–12

water, **53**:33–44

Neutron pair correlation function, 53:42

Neutron scattering, 17:65; **20**:136; **22**:208; **27**:505; **44**:259, 276, 285, 374; **51**:28, 39, 250

Born approximation, **20**:149

dynamical corrections, **34**:205, 220

Fermi pseudopotential, **20**:149

incoherent, **44**:308, 371

inelastic, **13**:126; **31**:87

lattice vibration measurement, **53**:294

lead germanate, **53**:276

by liquids, stochastic model for, **15**:129

magnetic, **53**:248–249, 252–253, 257

quasielastic, **34**:15, 63, 80

correlation time in liquid benzene, **34**:80

liquid CH_4 and CD_4, **34**:63

Van Hove-Glauber theory, **20**:149

Neutron structure factor, 53:42, 43

Newton diagram, 10:173, 331–332; **42**:437, 449

Newton-Raphson calculations, 2:272, 273; **54**:34–66

iterative methods, **54**:65–66

linear, **54**:11–14

nonlinear, **54**:14–25, 41–62

one and two-step, **54**:59–62

small Hessian eigenvalues, **54**:40–41

step size and sign control, **54**:34–50

see also Multiconfigurational self-consistent field state

Newton-Raphson iterative function, 54:3, 4-5, 10-25

Newton's polygon, 43:14, 178, 210

Newtonian dynamics, 33:162

Newtonian liquids, 1:148

N_2H^+, 19:38, 54, 82, 99, 128

Nickel, 41:141, 161, 163
  CO adsorption, 49:621
  H adsorption, 49:617
  N_2 adsorption, 49:627
  Na adsorption, 49:641-642
  S adsorption, 49:639-640
  surface states, 49:578
  Te adsorption, 49:641-642

Nickel complexes, absorption spectra, 5:256

Nickel (II) cyanide, trigonal bipyramid, 16:207

Nickel fluoride, 27:438

Nickel-gold, solid solution (Au_6Ni_4), enthalpy of formation, 2:144
  solution (Au-Ni), enthalpy of formation, 2:143
    short range order in, 2:144

Nickel-iron system (Fe-Ni), small deviations from ideality, 2:133

Nickel-platinum alloy, 2:128

Nicotiana tabacum, EPR, 7:629

Nicotinamide, 7:670ff; 16:192-193

Nicotinamide adenine dinucleotide (NADH), 7:586
  glycolysis, 29:139

Nicotinamide adenine dinucleotide (NADH) substrates, 7:715, 720

Nigericin, 19:368

Niobium, 41:141

Nitrates, melting point, 6:475

Nitration, 1:62, 182, 185

Nitric acid, 49:459, 522-523

Nitric oxide, 16:113-114, 127-128; 36:269
  E → V energy transfer, 47:Part 2:398, 404-405
  excited states, 47:Part 2:24
  fluorescence quenching, 47:Part 2:17-41
    by other molecules, 47:Part 2:31-33
    self-quenching, 47:Part 2:27-31
  Rydberg states, quenching mechanism, 47:Part 2:34-39
  see also NO

Nitriles, polarization of electronic transitions, 13:194

Nitrobenzene, 34:190

Nitrocellulose, flow birefringence, 33:146

Nitrogen, 16:127; 34:182
  active, 28:387
  + ammonia system, 2:98
  in clathrate, 2:30
  correlation energy, 6:359, 384
  crystal constants, 5:172
  crystal symmetry, 5:171, 177ff
  electron impact spectra, 18:20, 21, 60, 71, 72, 82, 83
  fluid structure, 53:23-26
  gas structure measurement, 53:23
  hydrate, 2:32
  inelasticity correction, 53:53-54
  interatomic distances, 5:156
  isoelectronic series, 6:381
  liquid, isotropic Raman linewidth, 40:36
  molecule, 36:301, 308; 41:152
    $C^3\Pi_u$ state, 12:21
    clathrate, 2:2
    clusters, 52:301-304
    crystal structure and intermolecular potential, 33:51ff, 67

ground state, electron correlation, **12**:67

Hartree-Fock equivalent orbitals, **6**:400

London forces, **6**:405

magnetic susceptibility of, **2**:70

Ni (110) adsorption, **49**:627–631

Pd (111) adsorption, **49**:631–632

polarizability, magnetic susceptibility and ionization energy, **2**:71

as quencher, **50**:355, 365, 371

rotational energy transfer in collisions with Ar, **50**:629

surface binding energy, table of, **49**:629

W (110) adsorption, **49**:628, 631

+ phenanthrene, **2**:99

polarizability, potential energy, **2**:71

rigid convex body constants, **5**:162–163

solid, **16**:375–376, 380; **22**:244–246

  crystal structure, **22**:247, 249

  infrared spectra, **22**:244–246, 250, 286

  intermolecular potential, **22**:244–248, 250

  lattice dynamics, **22**:225, 226, 233, 234, 244, 245–247, 248

  librations of, **22**:221, 225, 247

  Raman spectrum, **22**:244–250, 288

  translational motion, **22**:244

structural and thermodynamic data, **16**:346–348

+ xenon system, **2**:96, 112

see also Chemiluminescent reactions; Collisional ionization; Vibrational relaxation

Nitrogen-carbon monoxide, phase diagram, **16**:350, 357

solid, quadrupole-quadrupole interaction, **16**:380

Nitrogen dioxide, **16**:103, 112–114, 116–123, 125, 128–129

-dinitrogen tetroxide-argon mixtures, **16**:103, 111

-dinitrogen tetroxide-helium mixtures, **16**:112–114

Nitrogen oxide hydrate, thermodynamic data and lattice constants, **2**:8

Nitrogen tetroxide, **16**:122–123

Nitromethane, barrier height of internal rotation, **2**:382

microwave spectroscopy, **2**:379

o-**Nitroperoxybenzoic acid**, **16**:172–173

Nitrophenyl ester, acylation by, **39**:116

cleavage of, **39**:122

Nitrosyl chloride, **16**:127

Nitrous acid, isomerization of, **47**:Part 2:13–14

Nitrous oxide, E → V energy transfer, **47**:Part 2:399; *see also* N$_2$O

NO, clusters, **52**:290–293

  electron resonance spectra, **18**:228, 233

  groups, **7**:377

  as quencher, **50**:356, 371, 379, 505

  resonances, **49**:543

  see also Chemiluminescent reactions; Scattering

N$_2$O, as quencher, **50**:368

  clusters, **52**:330–334

  see also Chemiluminescent reactions; van der Waals molecules

No-retardation limit, **27**:366, 378

NO$_2$, dissociation experiments, **47**:Part 2:143–151

Noble gases, atom, polarizability, **2**:64

  excimer states, **45**:526–527

  interaction potentials, He*, **45**:567–569

metastable atoms in beams, **45**:488–580

Penning ionization, **45**:545–569

photoelectron spectroscopy, **45**:41–50

properties, **45**:491–492

scattering, Ar* + Ar, **45**:536

elastic, **45**:569–573

excitation transfer, **45**:569–579

cross section, **45**:540–541, 573–574

interfence structure, **45**:540–544

He* + He, **45**:527–536

total cross section, **45**:438–439

He* + Ne, **45**:570–571

angular distribution, **45**:574–579

Kr* + Kr, **45**:536

metastable, **45**:487–585

Ne* + Ne, **45**:536

separation of isotopes, **16**:126–127

Nodal expansion, **11**:71

Node, in diagrams, **11**:21

Noise, Gaussian white, **38**:313

Wiener process, **38**:313

Noise analysis, **39**:325

Noise burst model, **49**:6–9, 11–12

Noise forces, **20**:112

Noise power spectrum, differentiation between systems, **37**:86

eigenvalue method, **37**:79

irreversible thermodynamic approach, **37**:81

kinetic systems, **37**:67–98

master equation approach, **37**:75

matrix expression, **37**:84, 85

peaking phenomemon in, **37**:86, 87

rate constant determination from, **37**:89

relation to time correlation function, **37**:78

Noisy communication channel, **46**:156

Nonadiabatic approximation, **12**:9, 14; **49**:117, 130–139

alternative approach, **49**:156–159

correction, **49**:143

coupling terms, **49**:209, 274, 277, 283

ground state D_2, **12**:22

ground state H_2, **12**:22

ground state HD, **12**:22

Nonadiabatic collisions, **28**:321, 322, 323–325

Nonadiabatic coupling, **28**:321ff, 341, 346, 350, 361, 369; **42**:87, 89, 96, 104; **44**:38

Nonadiabatic modes, **32**:228

accumulation point of, **32**:229, 239

classification of, **32**:242

continuous branches of, **32**:230

nonconservation of entropy, **32**:229

premain sequence phase, **32**:229

secular spectrum, **32**:229, 238, 242

symmetrized secular spectrum, **32**:230

vibrational, **32**:231, 242

white dwarfs, **32**:230

Nonadiabatic region, **28**:324, 341, 342

Nonadiabatic transition, **28**:325; **30**:4, 59, 79, 418, 466

Non-alternant hydrocarbons, **7**:37

Non-anomalous scattering, **16**:205, 206, 208, 213

Non-bonded attractions, **6**:395

Non-bonding molecular orbital, **8**:80ff, 106ff

Non-central force, elastic scattering by, **10**:47–50

Non-centrosymmetric crystals, phase determination of, **16**:158–161, 202, 204, 207

Non-closed shell system, **6**:329, 363–365

Noncrossing terms, **28**:325

Noncubic crystals, **27**:430–435, 527–528

Noncurrent knot cycle, **43**:127

Nondegenerate coupled-cluster approach, **52**:182–234

Nondegenerate Rayleigh-Schrödinger perturbation theory, **52**:183–187

Nondestructive measurements, **46**:169, 178, 180

Nondiffusional processes, **33**:122–124, 134–138, 141, 145

Nondipolar liquid, **44**:445

Nondipolar molecule, **44**:261

Nondispersive correction, **44**:344

Nondispersive forces, **12**:209

Nonequilibrium diagrams, numerical studies, **24**:292
  Gaussian potential energy function, **24**:295
  Lennard-Jones potential energy function, **24**:295

Nonequilibrium fluid systems, **32**:88
  light scattering by, **32**:87ff

Nonequilibrium flux vector, **43**:31

Nonequilibrium instability, biochemical system, **38**:365

Nonequilibrium phase transition, **46**:266
  homogeneous bifurcations and, **55**:207–209

Nonequilibrium processes, **24**:24; **48**:456, 461
  anharmonic model, **21**:95–103
  local equilibrium model, **21**:476
  rates of, **21**:93ff
  Zwolinski-Eyring model, **21**:93

Nonequilibrium properties, of electrolytes, **11**:159

Nonequilibrium steady state, **32**:13

Nonequilibrium thermodynamics, chemical evolution and, **55**:43–62
  dissipative structures, **55**:46–54
  selectivity and selection, **55**:54–58
  stochastic aspects, **55**:58–60

Nonequilibrium variables, **32**:62

Nonequilibrium variational principle, **46**:257

Non-Gaussian statistics, **44**:259

Non-Gaussian variates, **44**:418

Nonhermitian model Hamiltonian, **52**:239–247

Nonhydrodynamic contribution, **48**:149

Non-ionic fluids, **9**:54, 66ff
  heats of vaporization, **9**:66

Nonlattice model, **52**:471–475

Nonlinear chemically reactive system, **38**:3–7

Nonlinear effects theory of Høye and Stell, **48**:316

Nonlinear energy converter, **55**:159–166

Nonlinear equations, **35**:288
  Floquet solutions of, **35**:290

Nonlinear equations of motion, **43**:35

Nonlinear integral equations, **37**:143, 196
  stability of solutions, **37**:143

Nonlinear irreversible processes, **3**:33ff, 39, 304

Nonlinear operator, completely continuous, **40**:281
  linearization of, **40**:259

Nonlinear operator equation, **40**:233, 260, 305

Nonlinear optics, in vacuum ultraviolet, **47**:Part 2:153–181

Nonlinear oscillator system, **24**:163

Nonlinear resonances, **24**:191

Nonlinear stochastic differential equation, **44**:357

Nonlinear time evolution equation,

51:139–143

Nonlocal theory, 27:358, 447, 449–465, 526; **41**:108–112

Nonlocalized bonds, 7:21ff

Non-Markovian approach to equilibrium, 22:361

Non-Markovian equation, 38:182, 184

Non-Markovian evolution, 11:334

Non-Markovian master equation, 11:319

Non-Markovian memory function, 46:327, 332, 348

Non-Markovian operator, 38:175, 177

Non-Markovian processes, 35:161, 164, 173, 175, 182

   relaxation, **11**:290, 301, 311, 312, 313

Non-Markovian system, 44:270

Nonmetallic elements, surface tension and compressibility, **9**:76

Non-Newtonian flow, 21:275

Non-ohmic currents, 7:313ff

Nonorthogonal transformation matrix, 49:202

Nonorthogonality, 27:220

Nonperiodic boundary condition, 48:253–254

Nonpolar attractions, 12:290

Nonpolar interactions, 12:284, 298

Nonpolar molecule, 36:508, 509

   electronic scattering cross section, **36**:508

Nonradiative energy transfer, 37:6

Nonradiative modes, 27:381, 429, 440–441

Nonradiative processes, electronic, **47**:Part 1:90–98

   in low temperature matrix, **47**:Part 2:521–533

      electronic relaxation, **47**:Part 2:529–530

   guest rotation as energy acceptor, **47**:Part 2:526–528

   host dependence, **47**:Part 2:530–531

   multiphonon vibrational relaxation, **47**:Part 2:522–526

   vibrational energy transfer, **47**:Part 2:528–529

  small molecules in low temperature solids, **41**:269–320

Nonreactive molecular process, 47:Part 1:29–40

  accessibility criterion, **47**:Part 1:30–40

  energy levels model, **47**:Part 1:32

  Hamiltonian formalism, **47**:Part 1:33

  intermediate level structure, **47**:Part 1:36–40

  level scrambling, **47**:Part 1:31

  similarity criteria, **47**:Part 1:30

  small molecule limit, **47**:Part 1:34–35

  statistical limit, **47**:Part 1:35–36

  time-resolved emission spectra, **47**:Part 1:38

Nonresonant absorption, 12:491, 539, 540

  CHF_3, **12**:544

  CH_3F, **12**:544

  NH_3, **12**:540, 544

  NO, **12**:539

  O_2, **12**:539, 544

Nonresonant intermolecular forces, first order, **12**:34

  dipole-dipole interaction, **12**:36

  fine structure effects, **12**:44

  quadrupole-quadrupole interactions, **12**:36, 43

  second order, **12**:36

  dispersion energy, **12**:37

  induction energy, **12**:37

Nonretarded interaction energy, **12**:90

Nonretarded theory, **27**:366, 393, 399, 415, 418, 421, 518

Nonrigid molecule, **49**:186

Nonspherical molecules, **10**:47–50; **12**:297; **31**:165, 179, 180, 183

    scattering of, **42**:422–512

    second virial coefficient, **5**:168

Nonunitary transformation, **38**:145, 166

Norbornyl cation, **19**:193

Nordholm-Rice analysis, **47**:Part 1:162–168

Normal alkanes, **16**:227–230, 234, 248, 263, 270, 278

    mixtures, **16**:229, 247–248, 252, 257–262, 276

    principle of corresponding states for, **16**:235–245

Normal coordinates, **36**:218, 276, 381

Normal modes, Liouville operator, **53**:163–167

Normal pressure, **1**:206, 216, 226, 227

Normal solution (Chapman-Enskog), **31**:158

Normal stress coefficients, **35**:100

Normalization, intermediate, **14**:299, 314

    trial wavefunction, **25**:249–252, 264

Normalization integral, **7**:8; **36**:30

Normalized potential, **34**:125

Notching triplet molecule, **35**:3

    rate of atom (or electron) transfer, **35**:3

Nozzle, conical, **52**:273

    sonic, **52**:273

    supersonic, **40**:54

Nozzle beams, **10**:275–317, 327

    compared with effusive beams, **10**:277–278, 282, 308–309

    mixed gases, **10**:311–315

    separation effect in, **10**:313–315

see also Beam sources

NS, electron resonance spectra, **18**:228, 233

Nuclear attraction integral, **13**:210, 238

    three-center, **13**:243, 244

Nuclear dipoles, Zeeman energy of, **7**:534

Nuclear electric quadrupole interaction, **18**:199, 208

Nuclear excited Feshbach resonance, **36**:435

Nuclear heavy ion transfer reaction, **47**:Part 1:263–264

Nuclear magnetic resonance, **2**:34, 375; **7**:499ff; **41**:210; **44**:388; **49**:523–524

    carbon-13 choline, **39**:260, 263

    carbon-13 relaxation time, **44**:375

    fluorine-19, **39**:126

    line shape, **15**:109ff

    metal-ammonia solutions, **4**:332, 333, 379, 383

    microclusters, **40**:58

    probes for nonmetalloproteins, **39**:62

    proton, **39**:128, 145, 156

    relaxation time, **42**:275; **44**:378

    spin-rotation relaxation time, **44**:391, 471

Nuclear magnetic resonance spectroscopy, **39**:55, 60, 64–66, 146

Nuclear magneton, **44**:118

Nuclear moment, **10**:26

Nuclear motion, **12**:11

    effect on charge exchange, **13**:44

Nuclear Overhauser effect, **34**:75, 90; **44**:375

Nuclear paramagnetic relaxation in solids, **11**:289ff; *see also* Relaxation

Nuclear quadrupole coupling, **44**:105

Nuclear quadrupole resonance, **2**:188

Nuclear quadrupole resonance spectroscopy, **5**:319ff, 342, 349
  asymmetry parameters, **5**:321
  density matrices, **5**:327
  net orbital populations, **5**:331
  overlap populations, **5**:331
  pair functions, **5**:344
  polarity parameters, **5**:344
  quadrupole coupling constants,
    5:322
**Nuclear spin-electronic orbit interac-
  tion, 44**:104
**Nuclear spin-electronic spin interaction,
  44**:105
Nuclear spin lattice relaxation, 42:275;
  53:248, 259, 260
**Nuclear spin-nuclear spin coupling,
  44**:107
Nuclear spin relaxation time, 17:67
Nuclear spin-rotation coupling, 44:107
Nuclear spin-spin coupling, indirect,
  37:260, 261
**Nuclear spin Zeeman interaction,
  18**:209
Nuclear symmetry, 30:65, 423
Nuclear theory, 41:135–136
Nucleation, 4:46; **32**:6, 8; **33**:98–100;
  42:354
  binary, **40**:151
  computer simulation, **48**:402, 441
    body-centered cubic nuclei,
      48:443
    characterization of, **48**:441–443
    face-centered cubic nuclei, **48**:443
    structure factor, **48**:443
    Voronoi polyhedra, **48**:443
    see also Homogeneous nuclea-
      tion
  dense system, **40**:144–147
  dilute system, **40**:138–144
  entropy driven, **33**:108

heterogeneous, **40**:138, 177
  two-dimensional, **40**:177, 181
homogeneous, **40**:137, 138
  in absence of, bubbles, **40**:138
    dust, **40**:138
    ions, **40**:138
    microcrystals, **40**:138
  kinetic approach, **40**:137–153
  length, **33**:100
  microcluster theories of, **40**:125–129
  multistep, **33**:107
  parameter, **33**:75, 83, 90, 95
  step, **33**:80, 99, 103, 104, 108, 112–
    114
  temperature, **48**:445
**Nucleation dependent sequential pro-
  cess, 33**:98–100, 103, 104, 109,
  110, 112–115
Nucleation rate, 40:139, 140, 142
  steady state, **40**:140, 144
Nucleation theory, 15:175; **40**:148–152;
  48:458
  classical, **40**:143
Nucleic acid, 7:52, 435
  biological activity loss, **7**:220
  denaturation, **7**:217
  fundamental bases, **7**:56
  infrared spectra, **7**:435, 460ff
  luminescence, **7**:347
  photoprotection of a modifiable site,
    7:594ff
  secondary structure, **7**:466
  as template for biopolymerization,
    15:185
  thermal effects on, **7**:201
  UV absorption, **7**:585
Nucleic acid derivatives, photochemical
    reactions of, stochastic model for,
    15:171
Nucleohistone, 7:462, 463
Nucleon transport, 46:271

Nucleophiles, **39**:114, 120, 146, 155, 157

Nucleophilic attack, **7**:67, 377ff

Nucleophilic displacement reactions, **19**:204

Nucleophilic substitution, **7**:44; **8**:86ff, 109

Nucleosides, infrared spectra, **7**:440ff
mixed products, **7**:452ff
prebiotic synthesis, **55**:103–105

Nucleotide bases, absorption spectra, **7**:96
molecular diagrams, **7**:94
sequence, **7**:88, 130ff
tautomers, **7**:104, 106
transition moments, **7**:95

Nucleus, crystalline, **40**:146
formed of three base pairs, **33**:98
helix, in polypeptides, **33**:75, 104
plus structure, **33**:107

Null space, **43**:22

Null spectra, **32**:226
classification of, **32**:242
connected with abundance, entropy, and momentum conservation, **32**:227, 242
effect of convection, elasticity, magnetic fields, neutrinos, photons, or rotation on, **32**:227, 241

Number density, **46**:286

Numerical integration, **17**:213; **25**:212, 266, 281ff, 292; **42**:283
Degasperis technique, **25**:292

Nusselt number, **32**:5, 18, 77, 79, 278

Nutation, **40**:391

Nylon-66, **7**:264

O

Ω **coupling**, **12**:15

Observables, actualized, **53**:343
algebra of, **38**:24
classical, **38**:29, 43, 52, 58, 66, 90; **53**:362
classification of, **37**:240, 241
contraction of, **53**:343, 347, 368
definition of, **38**:25
generating classical, **38**:67
interpretation of, **38**:25
irrelevant, **53**:345, 366
macrostates, **53**:375
nonexpansion of, **53**:368–370
truth-definite, **38**:28
value of, **38**:28

Observation, **38**:18, 40, 51
definition, **38**:18
theory of, **38**:40, 51

Occam's razor, **50**:637

Occupancy, **52**:399–407
half line, **52**:404
set, **52**:402
single point, **52**:402, 404

Occupancy of states, matrix of, **33**:102

Occupation number, geminal, **13**:73

Occupation time, **52**:493

Occupied orbitals, **9**:329

Ochkur-Rudge approximation, **18**:31, 33, 80, 81; **36**:419

OCS, clusters, **52**:326–329

Octachlorocyclobutane, **16**:177

Octafluorocyclobutane, **16**:177

Octahedral complexes, **5**:43, 45, 50, 53, 62, 67, 72
energy levels, **5**:72
$5f^1$, molecular orbital theory of, **21**:353
interelectronic repulsion parameters

for, **5**:62

Octa(methylsilsesquioxane), 20:28

Octant rule, 4:104; **44**:584, 585, 628

Octa(silsesquioxane), 20:28

Octopole, 5:187

Octopole admixture, 20:73, 100

Octopole moment, 44:452, 453, 455

Odd multiplicity, 37:170

OH, electron resonance spectra, **18**:214, 228, 231, 233

O-H bond, 1:62, 64

Ohm's law, 53:135

Old quantum theory, 36:66

Old yellow enzyme, 7:618

Oldroyd convected derivative, 35:96

Olefins, quenching, **50**:377

Oligomer, oligoadenylic acid, **33**:98, 99

   oligonucleotide, base pairing, **33**:78, 98, 99

   oligopeptide, kinetics of conformational change, **33**:103, 104

   oligouridylic acid, **33**:98, 99

   vinyl, ultrasonic absorption, **33**:129, 131, 132

Oncogenesis, 29:126

One-band approximation, 13:125

One-center expansion, 13:207, 236

One-dimensional chain, 9:94ff

One-dimensional crystal, 8:31ff, 40; **9**:111ff

   cyclic boundary conditions, **8**:31

   localized excitation, **8**:32

   transition moment, **8**:40

One-dimensional plasma, 4:201–224

One-dimensional random chain, vibrational modes of, **13**:125, 126, 146, 149, 152, 156, 161, 162

One-dimensional random potential, 13:123, 127, 128

   continuous random parameters, **13**:147

density of states, **13**:135ff

   cumulative, **13**:135ff, 138, 152

    node-counting principle, **13**:135

   differential, **13**:145

   singularities in, **13**:148

   with discrete random parameters, **13**:147

   electrical conductivity, **13**:126, 130, 144, 173

   energy gaps, **13**:153–155

   Green's functions, **13**:139ff

   limiting behavior of the probability distribution, **13**:138, 142, 147, 175

   localization of wave function, **13**:164, 167, 173

   momentum-dependent spectral density, **13**:126, 130, 139, 152, 159

   Monte Carlo methods, **13**:127, 136, 152

   randomly scattered δ-functions, **13**:132, 151, 152, 157, 158

   transfer matrix, **13**:175

   white-Gaussian noise potential, **13**:133, 152, 159, 160

One-dimensional step model, 29:176

One-dimensional walk, 52:411–413

One-electron approximation, 4:100–102; **10**:6; **9**:87; **12**:348; **52**:119, 121, 157–161

One-electron energies, 7:27

One-electron energy spectrum, 9:90

One-electron function, 2:228, 261

One-electron operator, 25:183, 201

   effect of nuclear motion on, **25**:183

   effect of (spectroscopic) perturbation on, **25**:201

   vibrationally averaged, **25**:183

One-electron properties, 25:180

One-particle excitations, 9:338

One-particle operator, 9:347

One-photon and two-photon photoselective chemistry, 47:Part 2:3–181

chemical laser kinetics, 47:Part 2:55–83

diffusion cloud reactions, infrared laser enhanced, 47:Part 2:43–54

Doppler spectroscopy, 47:Part 2:133–152

laser diagnostics, 47:Part 2:85–131

nonlinear optics and laser spectroscopy, 47:Part 2:153–181

reactant internal energy effects on chemical reaction, 47:Part 2: 3–16

two-photon excitation, 47:Part 2:17–41

One-point correlation function, 46:253

One-step process, 34:251, 258, 293

Onsager continuum, 48:210

Onsager continuum limit, 48:199

Onsager equation, 1:83, 98

Onsager flux, 33:413, 415, 437

Onsager formula, 11:250

Onsager hypothesis, 53:82–83, 138

Onsager matrix, 43:260

Onsager model, 31:139, 145

dielectric constant from, 31:140

thermodynamics from, 31:139–145

Onsager reaction field, 12:297

Onsager reciprocity relations, 1:3; 3:274; 4:6, 44; 16:104; 33:247; 38:136; 53:176

Onsager region, 32:61

Onsager relaxation theory, 11:246

Onsager theory, 53:67

many-particle dynamics, 6:68–80

Onsager's approximation, 48:198, 262

two-particle, 48:209

Onsager's principle, 43:231

Onsager's regression hypothesis, 48:357

Onset time, 46:211, 261

Opacity, 10:129–130

Opacity function, 30:48

Open kinetic system, 37:88

Open shell, 9:356ff

Open shell atoms, 14:385

Open shell molecules, 1:259; 36:213, 215, 243, 261, 264

Open systems, 19:223, 229, 244; 34:284

Belousov-Zhabotinsky reaction, 55:211–215

evolution criterion, 19:224, 233

mechanical, 19:220

Operationalism, 38:6

Operator, 4:117, 133

Balescu-Lenard, local potential for, 13:338

Boltzmann, local potential for, 13:338

continuous, 40:260

creation and annihilation, 14:41, 95

current density, 41:80, 83

density, 41:70, 80, 83, 129

destruction and creation, 41:69

effective, 14:93, 111

electric dipole, 4:135, 147

electronic spin-nuclear, 44:118

electronic spin-rotation, 44:117

electronic spin-vibration, 44:118

finite-difference, 13:125

Fokker-Planck, 11:226, 245, 277 13:3, 5, 10

inhomogeneous, 11:280

four-electron, 14:120, 124

generalized Boltzmann, 11:319, 320, 330, 353, 356, 359, 365

Hilbert-Schmidt, 53:367

inequalities, 14:320

Liouville, 11:165

mass, 14:231

momentum, 4:135

nuclear quadrupole coupling, **44**:83, 115

nuclear spin-electronic orbit, **44**:80

nuclear spin-rotation, **44**:84, 119

nuclear spin-spin, **44**:87, 119

nuclear spin-vibration, **44**:120

orbit-nuclear spin, **44**:80, 118

orbit-orbit, **44**:116

orbit-rotation, **44**:79, 118

orbit-vibration, **44**:118

orthogonal, **44**:265

positive definite, **32**:41

projection, **8**:4; **14**:320

reaction, **8**:3; **14**:300, 316, 335

resolvent, **11**:167, 283

rotation, **44**:65

secular, **14**:425

spin-orbit, **44**:53, 115

spin-other orbit, **44**:115

spin-rotation, **44**:54

spin-spin coupling, **44**:54, 116

Stark effect, **44**:115

streaming, **11**:318, 328, 359, 360, 364, 365

three-electron, **14**:110, 118

transport, **11**:320, 336, 337, 356, 364

wave, **14**:300, 316, 335

Zeeman effect, **44**:120

see also Effective Hamiltonian; Hamiltonian

Operator equation, nonlinear, **40**:259, 266

Operator-valued measure, **38**:65

Opposed resonance, of ADP and ATP, **7**:65

Optical activity, **12**:119

origin of, **55**:109

polymers, **4**:113–160

small molecules, **4**:67–112

tensor, **4**:122, 123, 126, 140

Optical coherence, **40**:381

Optical coherence effects, **47**:Part 2:425–430

free induction decay, **47**:Part 2:429–430

hole-burning, **47**:Part 2:430

nutation, **47**:Part 2:430

photon echo, **47**:Part 2:425–429

see also Coherent optical effects

Optical dephasing, **40**:385, 388–391, 472

Optical dephasing and line shape, **47**:Part 2:463–472

mixed crystal model Hamiltonian, **47**:Part 2:464–467

Redfield theory, **47**:Part 2:467–469

Optical detection of magnetic resonance (ODMR), **40**:372, 401ff, 411, 415, 417, 423ff, 455, 459, 465ff

Optical diffraction method, **28**:209

Optical electronegativity, **8**:54

Optical excitation, **41**:375–407

Optical free induction decay, **47**:Part 2:429–430

multilevel system, **47**:Part 2:481

Optical line shape, coherent effects and, **47**:Part 2:430–434

Optical model, **30**:11, 51, 67, 257, 268, 326, 333, 337

Optical modes, **27**:487

Optical nutation, **47**:Part 2:430

free induction decay and, **47**:Part 2:457–461

multilevel system, **47**:Part 2:481

Optical phonon, **40**:439, 440, 447, 472, 473, 475

Optical potential, **10**:128, 351, 362, 366–367; **44**:507, 610

Optical power density, **28**:441, 446, 447

Optical properties, metal-ammonia solutions, **4**:312–316, 358

Optical pumping, 16:71–75, 77–78; 42:440
   atom beams, 45:341ff
   distribution in excited state, 16:71–73
   polarization of emission, 16:74–75
Optical rotation, 4:76, 77, 101, 115, 116, 122–159; 7:216; 12:122; 23:302–308
   by polypeptides, 21:588, 591
   see also Optical activity; Rotatory dispersion
Optical rotatory dispersion, 44:546
Optical rotatory power, 44:546, 548, 553, 575, 579
Optical spectra, *see* Absorption spectra, optical
Optical theorem, 10:107, 121, 129–130
Optical transitions, hole-burning, 47:Part 2:434–444
Optically active molecule, photodissociation of, 50:278
Optimized cluster theory, 34:109, 132, 143
   consistency test, 34:143
Optimized random phase approximation, 48:224
Optimized valence configuration method, 26:253
Opto-acoustic detector, 50:107, 122
Opto-acoustic effect, 28:216; 50:107
Opto-acoustic methods, 42:274
Orbit, closed, 46:77, 108
   role in semiclassical quantization, 46:124
   stability properties of, 46:108
Orbital configuration, forbidden transitions, 5:247, 251
   transition metal complexes, 5:35ff, 39, 40
Orbital electronegativity, 49:107
Orbital expansion method, 9:341

Orbital population analysis, 50:362
Orbital promotion, 42:192
Orbitals, 6:323–324, 337, 387; 7:11, 167ff; 9:323; 36:152, 415
   angular momentum, 7:162
   angular momentum quantum number, 25:215, 256
   average polarization, 6:347
   benzene, 36:415
   energy, 7:178
   equipotential, 13:75
   equivalent, 6:326
   expansion formulae for, 13:262
   half-filled atomic, 13:73
   Hartree-Fock, 6:323; 7:161
   high energy, occupied, 49:80–83
   iterative natural spin, 45:237–238
   ligand, 7:168
   linear combination of atomic orbitals, 6:323
   orthogonality of, 6:332, 335–337
   π molecular, 36:415
   polyacetylene, 53:289, 290–291
   Slater-type, evaluation of integrals for, 13:205, 209
   Su, Schreiffer, and Heeger model, 53:288
Orbiting, 10:92, 105, 137, 162, 336–337, 365, 377; 12:18; 30:305, 322, 328, 339, 369
Orbiting resonances, 42:376
   enhancement of vibrational relaxation by, 47:Part 2:263–270
Orbiting-tunneling phenomenon, 10:105, 125
Order-disorder theories, 11:78, 79
   of alloys, 11:41
Order-disorder transition, 40:283, 284
Order parameter, 34:297; 38:276, 303; 40:235; 41:210, 211; 43:235; 49:312–313, 320, 338; 51:20;

53:291

Ordered configuration, 2:261; 14:305

Ordered span, 52:440

Oregonator, 43:18, 23, 35, 45, 65, 68, 74, 75, 84, 108, 115, 123, 126, 142, 145, 237

Organic electrode reaction, see Electrode reaction

Organic mixed crystals, photon echo, 47:Part 2:444–457

Organic molecular solids, electronic conduction in, 7:282

Organic molecules and polymers, prebiotic synthesis, 55:89–101

Organic peroxyacids, 16:172

Organic phosphors, 7:556ff

Organometallic compound, photodissociation of, 50:264, 306

Orgel diagram, 5:59; 8:51

Orientation, description of, 16:69–71, 73

fluctuations, 24:45, 50; 31:257

of intermediates, 16:92–98

macromolecules, 35:80

diffusion equation, 35:80

relaxation, 49:28–32

species in beams, 30:258, 389

Orientation polarization, 1:77

Orientational autocorrelation function, 44:301, 433

planar librator, 44:368

Orientational correlation function, 44:258, 382, 466

Orientational distribution function, 40:273

Orientational motions in solids, 42:333

Orientational phase transition, 40:272–283, 330

Oriented molecules, 42:447, 470ff

Origin of life, 38:242, 245

selective degeneracies, 38:242

Orlicz space, 40:277, 314

Ornstein-Uhlenbeck process, 20:209

Ornstein-Zernicke direct correlation function, 40:319; 46:334, 349

Ornstein-Zernicke equation, 6:181, 192; 31:112, 117, 126, 135; 40:352, 353, 356; 41:24, 26, 45, 174; 43:256; 46:294; 48:189; 49:499; 53:118–124

invariant expansion of, 31:126–134

site-site correlation function, 48:281

transformations of, 31:112, 117, 135

Orr-Sommerfeld equation, 32:57

Ortho hydrogen, 12:97, 138

Ortho-para hydrogen conversion, 1:125

Orthogonal complement, 14:313

Orthogonality, 9:337; 53:90, 140, 142, 326

atomic states, 53:280

Brownian motion theory, 53:83

Mori random force, 53:146

projectors, 53:355, 357

Orthogonality conditions, 9:334

Orthogonality constraints, 9:338, 341

replacement of by adding non-local pseudopotential, 16:287, 325, 334

Orthogonality relation, 4:54, 58, 59

Orthogonality theorems, 53:166–168, 170–171

Orthogonalized plane waves, 9:164

Ortholocalization, 7:44

Ortholocalization energy, 8:116

Orthonormal set, 7:14

Oscillating reactions, 3:47

Oscillation systems, conservative, 19:211

nonconservative, 19:256

Oscillations, 32:282, 297, 304; 33:435, 436; 38:320, 325, 366, 390; 43:189

adenylate cyclase, 38:390

catalytic organic reactions, 19:275

cellular, **29**:51, 148, 152, 290, 291, 292

cellular aggregation and, **29**:295

chaotic, **38**:337

chemical, **19**:249ff; **29**:36, 40, 156; **32**:98

circadian rhythm, **19**:316

conservative systems, **19**:211, 237

control processes, **19**:277

cyclic-AMP, **38**:390

D. discoideum cells, **29**:154

electric current, **29**:328

enzymatic reactions, **19**:315

enzyme membranes, **29**:121

epigenetic, **38**:366

excitation functions for ion-neutral collisions, **45**:179

glycolysis, **19**:275

limit cycles, **19**:236, 255

many-body dynamics, **53**:87–88

metabolic reactions, **19**:275

mitochondrial, **29**:52, 148

NADH, **29**:141

pattern formation and, **29**:152, 156

periodic, **38**:324; *see also* Limit cycle

phase relations of, **29**:143, 146, 148, 163

plant cells, **29**:281, 290

potential, **29**:322, 327

protein synthesis, **19**:275, 316

pyridinnucleotide and pH, **29**:142

range of glycolytic, **29**:141

reagent interaction, **53**:184, 185

relaxation, **29**:57, 159

rhythmic activity of the nervous system, **19**:315

subject to inhomogeneous noise, **29**:58

supercellular level, **19**:316

sustained, **19**:236

temporal, **43**:228–238

temporal and spatial, **43**:254–260

temporal organization, **19**:268

see also Fluctuations

Oscillator model, 12:347

Oscillator strength, 4:85, 107, 163, 175, 187, 194; **5**:241, 242, 243; **7**:94, 100, 114, 129; **10**:4–5, 14; **28**:269; **44**:555, 562, 563, 575, 577, 580, 617, 622; **50**:5

calculation of, moment method, **45**:24

CO and N_2, **45**:24

MS-Xα method, **45**:23, 24

continuum, **45**:41

ionization, **45**:40, 41

measurements of, **45**:3–74

Ar, **45**:48

CH_4, **45**:69–72

CO, **45**:54–60, 62, 66–68

H_2, **45**:50–54

H_2O, **45**:73

He, **45**:47

HF, **45**:73

N_2, **45**:54–55, 57–60

Ne, **45**:46

NH_3, **45**:40–41, 73

SF_6, **45**:73

see also Electron spectroscopy

metal-ammonia solutions, **4**:313

potassium titanium hexafluoride, **5**:252

titanium hexaquo ion, **5**:252

Oscillators, 4:78–81, 85, 89; **33**:182, 183, 187

coupled, **4**:100

harmonic, **4**:100, 286, 300

Morse, **1**:392

nonlinear, **29**:36

rotating Morse, period of, **13**:106

tandem, **43**:155

Teorell, **29**:134

Oseen interaction, 1:109; **15**:327; **48**:135

Oseen tensor, 35:21; **44**:301

Osmium hexafluoride, 9:289

vibronic coupling in, **9**:301

Osmometry, 1:78

Osmotic correction, 11:30

Osmotic pressure, 6:312; **11**:17; **22**:80, 109

Ostwald's rule, 9:396

Outer products, 14:98

Ovalbumin, 7:216

Ovalene, fluorescence excitation spectrum, **47**:Part 1:7

Oven beams, compared with nozzle beams, **10**:277–278, 282, 308–309

Overcompleteness, 27:220

nonorthogonality, **27**:220

Overdeterminacy, 16:136–139

Overhauser effect, 4:333, 384; **35**:22

Overlap atomic orbitals, 7:17ff

Overlap bonding, 16:381–382

Overlap dipole contribution, 44:457

Overlap integral, 1:244, 249; **5**:222; **7**:17, 64, 116, 121ff, 131; **13**:210, 238

Overlap interaction, 44:261

Overlap kernel, 49:119–121, 125, 128

Overlap matrix, 2:229

expansion of, **33**:350, 356, 358, 359, 379

Overlap maximum, 7:17

Overlap populations, 5:331

Overlapping lines, 51:73

Overlayers, 27:146

diffusion of impurities, **27**:145

inequivalent domains, **27**:148

nonuniform, **27**:145

out of phase domains, **27**:148

reflection registry degeneracy, **27**:149

registry degeneracy, **27**:148

rotational registry degeneracy, **27**:149

translational registry degeneracy, **27**:148

uniform, **27**:145

Overshoot, 29:345

Overstability, 26:179, 180, 207; **32**:20, 136, 277, 284, 285, 292, 296

Overtone, vibrational, **40**:18, 45

Overtone generation, 53:264

Oxalacetate, 39:152–156

Oxalic acid dihydrate, phase determination of, **16**:158

Oxidation reaction, 41:509–511

Oxidative phosphorylation, 39:104, 206, 330, 333

optimization of, **55**:141–167

degree of coupling, **55**:146–150

nonlinear energy converter, **55**:159–166

thermodynamic buffering, **55**:150–158

Oxime complexes, 7:374

Oxime oxidation, 7:377

α-**Oxygen, 16**:374–375

β-**Oxygen, 16**:369

γ-**Oxygen, 16**:369

Oxygen, 1:60, 65, 252; **14**:158ff, 174, 192; **16**:113–114, 128–129; **34**:182; **41**:140, 152, 171

α-β transformation, **16**:367–368

anti-ferromagnetic structure of, **16**:363–366

atmosphere of pure, **55**:64–67

atom, reactions of, **42**:113–168

O + alkenes, **42**:136

O + alkyl sulfides, **42**:132

O + alkynes, **42**:136, 137

O + allene, **42**:143

O + Br$_2$, **42**:119

O + CF, **42**:127

O + C$_2$F$_4$, **42**:134

O + CH, **42**:128

O + CH$_2$, **42**:135

O + C$_2$H$_2$, **42**:141

O + C$_2$H$_3$F, **42**:139

O + C$_2$O, **42**:136

O + C$_3$O$_2$, **42**:143

O + 5-chloro-1-pentene, **42**:139

O + 3-chlorocyclohexene, **42**:139

O + ClO$_2$, **42**:131

O + CN, **42**:124

O + CO, **42**:150

O + CS, **42**:123, 127

O + CS$_2$, **42**:128; **50**:96

O + CSe, **42**:124

O + CX, **42**:121

O + cyclooctene, **42**:140

O + H$_2$, **42**:100, 115ff

O + halogens, intermediates, **42**:118

O + HCl, **42**:115ff

O + NO$_2$, **42**:131

O + O$_2$, **42**:148

O + O$_3$, **42**:148

O + OCS, **42**:131

O(^1D), **42**:145ff

quenching of excited, **42**:145

autoionizing states, **30**:455

in clathrates, **2**:6, 30

delta-argon-, **16**:359, 381

electron density contours, **16**:365

in hydrate, **2**:32

ionizing radiation and, **7**:261

isotropic Raman linewidth, **40**:36

molar volume, **16**:371

molecule, **36**:264

clusters of, **52**:293–299

crystal structure and intermolecular potential, **33**:51ff, 69

electron impact spectra, **18**:71

electron resonance spectra, **18**:218ff

interatomic distance, **5**:156

O_g^+ state, long range magnetic interaction energy, **12**:91

photofragmentation, **50**:289

as quencher, **50**:357

rigid body constants, **5**:162–163

molecule clathrates, **2**:2

in primitive atmosphere, **55**:88–99

self-diffusion in praseodymium oxide, **21**:253, 255

solid, **16**:364–370

solubility of in beta nitrogen, **16**:362

structural and thermodynamic data of, **16**:346–348

superdelocalizability, **7**:67

Oxygen-containing complexes, **8**:55

Pariser-Parr-Pople parameters, **8**:135

Oxygen-fluorine system, **16**:379–381

Oxygen-nitrogen, phase diagram, **16**:350, 356, 360–361, 363

x-phase, **16**:360, 381

Oxyhemoglobin, **7**:72

Oxypurines, **7**:47

Oxytetracycline, PMR spectrum, **7**:528

Ozone, photofragmentation, **50**:262, 300

P

ϕ^4 **model**, **53**:228, 265–277, 320

conservative, **53**:265–270

Langevin approach, **53**:270–277

π **bond**, **7**:18ff

crystal field spectra, **8**:22ff

π **donor**, 7:363, 390, 395

π **electron approximation**, **14**:77; 33:358ff, 373

in many-electron theory, **6**:384

out of σ-bonds, **33**:369, 370, 376

π **electrons**, **1**:245, 246, 248, 250; **6**:384–387

Hamiltonian, **6**:386

interactions of, **4**:198

two-particle density matrix, **21**:305

see also Electron systems

π **orbital**, **1**:175, 176; **4**:93

π* **orbital**, **4**:99, 103

π-π **transitions**, **2**:201

π **pulse**, **25**:37; **40**:378, 391

π/2 **pulse**, **25**:36

p-**helicity decoupling approximation**, **42**:431

p-**polarization**, **27**:363, 370, 381–383, 409, 452, 491, 502

p-**waves**, **27**:283, 286, 294

P700, **19**:355, 356

P800, **19**:355, 356, 357

P870, **19**:355, 356

bleaching, **19**:355, 356

electron spin resonance, **19**:355

excited state lifetime, **19**:359

fluorescence, **19**:357

oxidation potential, **19**:355, 356

photochemical oxidation, **19**:357

Pacault, Lemanceau, and Joussot-Dubien apparatus, **3**:174

Packing, dense random, **48**:472

sphere, **40**:72, 73, 82

Padé approximant, **42**:488; **48**:238

Pair correlation, **14**:248, 305; *see also* Correlation

Pair correlation function, **11**:170, 192; **27**:62; **31**:75; **34**:126, 129, 205, 211; **40**:315, 325; **41**:93, 95, 100, 103, 105, 111, 116; **48**:188–201, 254; **49**:382–384, 387, 392–395, 423; **51**:18; **53**:2–3, 117–118

atomic fluids, **53**:12

cancellation of divergences in chain diagrams, **34**:129

chlorine, **53**:25, 28

diagram expansion, **34**:112, 126

fluid structure, **53**:2

fluid theory, **6**:233, 234, 266, 269–270, 275, 277, 283

hard sphere, **37**:145

heliticity, **34**:205, 209

homonuclear diatomic fluids, **53**:22

many-electron theory, **6**:349, 358, 368–372

molecular fluids, **53**:29

Pair density, **1**:11, 147

Pair distribution function, **1**:151, 220, 230; **46**:286; **48**:81; **53**:117, 192–194

quantum mechanical, **46**:289

Pair function, **5**:344

Pair integral, **33**:155, 157

Pair kinetic equations, **48**:120–123, 173–175

Pair potential, **12**:283; **48**:229; **51**:269; **53**:45ff

argon, **53**:14, 15

atomic fluids, **53**:12

atomic lattice, **53**:295, 296

bare two-body, **12**:284, 285

composition dependent, **12**:286

computer simulation of glasses, **48**:397–450

hard sphere, **48**:422–423

isobaric heat capacity, **48**:426–427

Lennard-Jones, **48**:422–423

soft sphere, **48**:422–423

thermodynamic properties, **48**:423–427

transport properties, **48**:428–430

density dependent, **12**:285, 286, 319

effective, **12**:284, 319; **31**:38

vibrationally averaged, **31**:30

exothermic, **53**:310

homonuclear diatomic fluids, **53**:21

ionic, scaling of, **11**:85

krypton, **53**:16–18, 19, 20

molecular crystals, **53**:278

Morse lattice, **53**:304

power-law, **53**:317

water, **31**:33ff

Paired electron wave function, 9:336

Pairing symmetry, 52:3

doubly excited atom, **52**:56

molecule, **52**:71, 81–94

Palladium, 41:141, 161, 172

(100), **49**:634

(111), H adsorption, **49**:581–582, 613–617

Palladium hexafluoride, 9:241

Panamine, structural formula, **16**:166–167

Papain, membrane, **39**:241

oscillator, **39**:36, 239

***Para* hydrogen, 12**:97, 138

Parachor, 9:80

Parallel body, 5:150

Paralocalization energy, 7:43; **8**:114

Paramagnetic relaxation, nuclear, **11**:289; *see also* Relaxation

Paramagnetic resonance, *see* Electron paramagnetic resonance

Paramagnetic species of biological interest, 7:602

Paramagnetic state, 46:315

Paramagnetic susceptibility, 5:191ff; *see also* Magnetic susceptibility

Paramagnetism, 4:315

Pauli spin, **4**:324

temperature independent, **7**:167

Parameterization of energy levels, 14:369

Parameters, convex, **43**:27

convex thermodynamic, **43**:35

kinetic, **43**:27

redundant convex, **43**:32

thermodynamic, **43**:27

Parametric oscillator, 50:128

Pariser-Parr model, 7:35ff

Pariser-Parr-Pople model, 52:70

carbonyl bonds, **8**:135

unitary structure of, **52**:71–74

Pariser-Parr-Pople parameters, carbon, **8**:135

oxygen, **8**:135

Parity, eigenfunction, **49**:167

electronic, **49**:167–169

invariant, **49**:165, 181–182

nuclear, **49**:167–169, 175

Parmenter-Seaver correlation, 42:225

Partial bond order, 7:41

Partial correlation function, 53:29, 41

Partial orthogonalization, 6:333, 360

Partial structure, phase determination, **16**:161, 190–191, 196–201

Partial structure factor, HCl, **53**:32

metallic glass, **51**:232

Partial valence, Thiele's theory of, **8**:76, 78

Partial volume factor in diffusion, 6:293

Partial waves, 10:84; **25**:215, 216, 255, 256

multichannel scattering, **25**:255

single channel scattering, **25**:215, 216

Partially absorbing trap, 52:409

Particle in a box, 4:194, 196

Particle scaling, 9:12ff, 13, 45

packing problem, 9:49

radial distribution function, 9:40

thermodynamics, 9:45

Partition function, 1:33, 54, 179–181, 183ff, 219, 279, 300, 388; 3:134; 4:202ff, 215, 230ff, 245, 261, 277–279, 286; 7:206, 226, 229ff; 20:40, 41; 25:166;

   asymptotic behavior of, 40:326, 327

   canonical, 20:40

   classical, 13:93

   configurational, 4:289; 16:224, 227, 231

   grand canonical, 20:40–41

   kinetic, 4:289

   liquid water, 21:492

   quasigrand, 22:170, 178

   sine-Gordon system, 53:242, 244

   system constrained to surface, 13:95

Partitioning theory, 8:3; 48:24–26

   Born expansion, 48:25

   primary space for ionization potential, 48:26–29

   unshifted ω-dependent denominators method, 48:25

Parvalbumin, 39:83

Pascal's law, 1:206, 220

Passage time, first, 13:1ff, 10

Past, 38:93, 95, 99

Patey approximation, 48:218, 223, 238

Path integral, 25:102, 150–152, 154, 164; 29:87; 36:67; 41:219, 220; 52:445, 447

   Fokker-Planck equation, 46:246

   series expansion for, 15:355ff

Pattern formation, 29:152, 156, 254

Pattern recognition, 38:51

Patterson function, 16:132, 134, 138–139, 197, 213

Patterson vector maps, 16:186, 201

Pauli equation, 4:3; 6:59–62; 11:300, 308, 313; 15:26; 38:115, 122; 50:418

Pauli exclusion force, 52:312

Pauli matrix, 33:185

Pauli principle, 7:8, 11, 186, 557; 31:411; 42:77; 44:11, 609

Pauli spin paramagnetism, 4:324

Pauli's hole-equivalency theorem, 5:44

Pauli's master equation, 43:246

Pauling charges, 23:135

Pauling scale of electronegativity, 13:61

Pauling-Slater principle, 7:17

Pauling-Wheland method, 7:30ff

Pausing time density with infinite mean, 52:388–391

Pearson diagram, 49:59–65, 107, 110

Peierls distortion, 49:76

Peierls inequality, 24:193

Peierls instability in polyacetylene, 53:288

Pekar polaron, classical, 44:493

Pendulum, 36:13

Penetrable sphere models, 41:2–54

   gas-liquid surface, 41:38–51

   heat capacity, 41:19, 20

   orthobaric diameter, 41:19, 20–21

   primitive forms, 41:4–5

   primitive mixtures, 41:31–38

   transcribed forms, 41:4–5

Penetration depth, 20:274, 290, 333; 27:556

Penetration effect, 7:13

Penetration theory, 13:281

Peng-Robinson equation, 49:428–430, 432–434, 436

Penicillin, relaxation rates, 7:514, 525

   albumin concentration and, 7:514

   spectra, 7:524

Penicillin-protein system, 7:527

Penning ionization, 10:187–189; 28:394,

404, 413; **30**:418, 448; **42**:195; **45**:105–106, 401–472, 489, 496–497, 506–509

associative, **45**:452–454, 472–475

atomic targets, **45**:460–463

cross section, **45**:426–445, 564–567

elastic scattering, **45**:420–426, 545–557

electron angular distribution,
He(2^3S) - Ar, **45**:433, 447–449, 564–567
He(2^3S) - CO, **45**:448
He(2^3S) - Hg, **45**:433
He(2^1S) - Hg, **45**:432
Ne* - Kr, **45**:430, 432

electron energy distribution, **45**:434–445
He(2^3D) - Ar, **45**:438
He(2^3S) - Ar, **45**:437
He(2^3S) - H, **45**:437, 439–441

He* + Ar, **45**:545–573

He* + D$_2$, **45**:554

He* + Li, **45**:555

ion angular distribution, **45**:557–564
He* + Ar, **45**:557–559
He* + CH$_4$, **45**:562
He* + H$_2$, **45**:560
He* + O$_2$, **45**:560
Ne* + Ar, **45**:563

molecular targets, **45**:463–472

optical model, **45**:502–503

potentials, He(2^3S) - H, **45**:464

quantum mechanical relations, **45**:412–417

Pentacene, 49:21–24

absorption spectrum, **47**:Part 1:8

exciton bands, **23**:327

fluorescence excitation spectrum, **47**:Part 1:6

Pentadecanoic acid, 41:207

Peophytine, 7:634

Peptidases, 7:74

Peptides, 49:34
chains, 7:262
prebiotic synthesis, **55**:101–103
water adsorption on, 7:242

Percolation, 40:455, 456, 457, 459, 463; **52**:484

Percolation model for electronic motion, 31:336, 447

Percolation theory, 44:509; **48**:477, 493
cluster distribution function, **48**:478, 482, 487, 507
continuum, **48**:482
critical exponents, **48**:478, 482
environmental, **48**:477, 481
mean field approximation and, **48**:479
probability distribution, **48**:482, 487, 500
threshold, **48**:477, 479, 499

Percolation transition, 48:445

Percus-Yevick approximation, 9:3, 27; **53**:14, 58, 124, 210
argon, **53**:13, 15
hard spheres, **21**:397, 423
matrix Gaussian model, **53**:202
pair theory of atomic fluids, **53**:12, 13
sticky hard spheres, **21**:421

Percus-Yevick direct correlation function, 40:320, 352

Percus-Yevick equation, 22:76; **34**:110, 133; **37**:142, 147; **40**:319, 338, 352, 353; **41**:28–31, 174, 175; **49**:424–425, 499, 520

Perinaphthenone, 8:134, 136

Periodic array of traps, 52:409

Periodic boundary conditions, 7:138; **17**:123; **27**:472; **52**:375
atomic lattice, **53**:295, 297
Hamiltonian, **53**:311

Periodic defects, **52**:461

Periodic model sequences, **7**:88

Periodically connected lattice, **52**:462

Peritoneal fat, infrared spectroscopy, **7**:414

Permeability, concentration dependence of, **29**:50

  for H + H_2 reaction, **21**:129

  kinetics, **29**:302

  membrane, **4**:3, 6, 29

Permutation symmetry, **14**:94ff; **25**:258

Peroxidase, **39**:168, 175

Peroxyacids, **16**:172

Peroxydase, **7**:160, 182, 615

Perrin hypothesis, **50**:408

Perrin's equation, **44**:357, 360

Persistence length, **52**:445

Persistent complexes, **10**:264–265

Persistent random walk, **52**:375–377

Perturbation, parabolic type, **32**:189

  sinusoidal, **33**:89

  small, **32**:167

    of equilibrium, **33**:102

  stepwise, **33**:89

  unstable, **32**:13

Perturbation expansion, **14**:391; **46**:210

  quantal, **36**:30, 34

Perturbation theory, **7**:9; **8**:69ff, 74ff; **9**:238ff, 334; **10**:2–3, 6–7, 14, 85, 175; **12**:179, 188; **14**:460; **19**:6; **23**:183; **24**:207, 231; **25**:213, 267; **37**:145, 186, 306, 313, 315, 317, 320; **40**:260; **42**:180; **47**:Part 1:147–159; **49**:117, 131–139, 142, 145, 156–159, 170; **53**:14, 45, 126, 131–132, 278, 279, 317

  asymptotic, **18**:107

  Born-Oppenheimer, **49**:117, 131–139, 142, 145, 156–159, 170

  Brillouin, **8**:4; **12**:46

  Brillouin-Wigner, **14**:394; **36**:348

  contact transformation, **37**:317

  continuum, **9**:330

  convergence of, **35**:314, 338

  degenerate Rayleigh-Schrödinger, **35**:275, 276, 338

  Dirac's, **1**:343

  Dirac's time-dependent, **15**:75

  double perturbation, **37**:273, 274

  energy denominators, **9**:354

  first-order, **4**:129, 141, 145, 163

  first-order time-dependent in collision dynamics, **6**:37–40

  H_2, **12**:45, 46

  infinite order, **8**:4

  intermolecular forces, intermediate separations, **12**:45

    large separations, **12**:31, 32

    very small separations, **12**:51

  ionic solutions, **11**:99

  liquids, **34**:107

  many-electron theory, **6**:346–349

  multipole expansion, **12**:31

  nondegenerate Rayleigh-Schrödinger, **35**:275, 309; **52**:183–187

  Nth order equation, **37**:314

  partitioning, **35**:275, 276

  Rayleigh-Schrödinger, **14**:304, 342, 384, 393; **36**:355

  Schrödinger, **8**:4; **49**:131–132, 134, 137, 160

  second order, **4**:145; **49**:79–80

    error bounds for, **15**:91

    quantal, **36**:33

  singular, **35**:275, 276, 277

  steady-state, **35**:316, 317

  Sternheimer techniques, **12**:45

  time-dependent Rayleigh-Schrödinger, **35**:273

  two-state quantum system, **35**:265–350

united atom expansion, **12**:51

see also specific perturbation theories

Perturbational MO method, 8:76ff, 115

Perturbative cubic approach, 54:101–114

Perturbed benzenes, two-photon spectroscopy, **54**:117–230; *see also* Benzenes

Perturbed stationary-state method, 42:331

Perturbing orbitals, 9:353

pH jump technique, 33:108, 109; **46**:398, 410

Phase coexistence, 37:171

Phase coherence, 40:374, 385, 390

Phase determination, 16:150–158, 161

   heavy atom techniques, **16**:197–216

   probability measures, **16**:154–155, 159–160

   symbolic addition procedure, **16**:159, 161–197

   tangent formula, **16**:152–153, 160–161, 187–191, 196, 198–199, 204

Phase diagram, 16:349–350, 363; **43**:242

   argon-carbon monoxide, **16**:350, 352, 354, 356, 359

   argon-fluorine, **16**:350, 356, 359

   argon-nitrogen, **16**:350–351, 354–355

   argon-oxygen, **16**:350, 356, 358–359

   neon-hydrogen, **16**:350

   nitrogen-carbon monoxide, **16**:350, 357

   oxygen-nitrogen, **16**:350, 356, 360–361

Phase diffusion, 29:55

Phase distribution function, 53:184, 185

Phase function, 36:30

Phase grating, 49:17

Phase increment at caustics, 36:14

Phase integral, 30:100

Phase memory, decay rate, **40**:467

Phase mixing, 35:121

Phase modulation, 44:324, 327, 328

   interferogram, **44**:331

   spectra, **44**:333

Phase shift, 10:84–86, 91–93, 95, 100; **12**:402, 403; **20**:61–62; **25**:70, 210ff, 255, 270ff, 283, 286–288, 291; **30**:88, 100, 330, 347, 472

   glycolytic oscillations, **38**:376

   Kohn, **25**:273, 275, 291

   low-angle scattering, **10**:106, 110

      random phase approximation, **10**:105–109, 117, 120, 127

   maximum or "glory", **10**:115, 120, 122–124

   near rainbow angle, **10**:103

   reduced, **10**:78, 85

   soliton collisions, **53**:239

   see also Quantum scattering

Phase shift technique, 50:369

Phase slip, 34:293

Phase solition, 53:292

Phase space, 19:328, 331; **25**:88, 166, 172, 173; **26**:124, 126, 138; **30**:16, 38, 66; **42**:149

   density of points, **13**:88

   division into cells, **26**:126

   Γ-space, **26**:123, 124, 133

   generalized velocity of a point in, **13**:88

   μ-space, **26**:113, 124–126, 133

   of stellar evolution, **32**:216

      continuity equation in, **32**:220

      decomposition into direct term, **32**:220, 248

      definition of, **32**:216

      norm in, **32**:216, 221, 249, 251

      nullspaces, **32**:223

uncertainty in, **32**:240

volume in, **13**:88, 90

Phase space coordinates, inner product definition, **53**:169

Phase space trajectory, **36**:3, 10

Phase space variable, angular velocity, **44**:270

coupled angular and linear velocity, **44**:270

orientation, **44**:270

Phase transition, **1**:294; **6**:229, 246, 268; **20**:282, 297, 301; **24**:3; **29**:66, 325; **32**:6, 8, 88, 163; **34**:297, 306; **37**:143, 172; **38**:394

continuous, **40**:237

cooperativity in, **33**:100

discontinuous, **40**:238

evaporation, **20**:307

freezing, **20**:286, 313

Landau theory of, **40**:234

Maxwell construction, **29**:72

melting, **20**:286, 308, 310, 311, 312

nonlinear problems in the theory of, **40**:229–364

order-order, **33**:101

orientational, **40**:272–283, 330

phospholipids, **39**:204, 219, 220, 221

polymorphic, **40**:283–287

second order, **40**:234, 237, 239, 250ff; **53**:266

steady state system, first order, **49**:332–335, 345–346

frequencies, **49**:353

hard mode, **49**:346–354

to nonuniform steady state, **49**:320–346

second order, **49**:316, 327–332, 335, 350

symmetry-breaking, **49**:320, 329, 354

structural, **53**:265

van der Waals theory of, **29**:66

Yang-Lee theory, **37**:144

see also Critical region; Glass transition; Helix-coil transition

Phase velocity, **24**:20

Phase volume, **26**:115, 123, 124, 135–137, 139–142

Phenanthrene, **40**:473

triplet state, **7**:569

Phenazine, **40**:394, 398, 406, 417, 434ff, 457ff

Phenomenological coefficients, **1**:2; **6**:293

Phenomenological equations, **1**:2, 8; **13**:267; **15**:66

Phenomenological variables, **44**:262, 263

Phenylacetylene, **54**:220–224

α-**Phenylcinnamoylchloride**, **16**:181

Phillips-Kleinman pseudopotential, **16**:289–294

Phillips-van Vechten scheme, **49**:63–64, 108

Phonon-assisted vibrational relaxation, **42**:348

Phonon induced echo relaxation, **47**:Part 2:474–475

Phonon sideband, **41**:274, 309

Phonons, **8**:19; **27**:490; **33**:19, 20, 22; **40**:400, 419ff, 429, 452, 467ff; **53**:243–244, 316

acoustic, **40**:439, 440, 447, 472; **44**:488, 492

bulk, **27**:381, 391, 456, 471–474, 490, 491, 510

collisions with solitons, **53**:245, 268–270, 287, 328

density, **33**:20

excitation spectrum, **33**:19

frequency, **44**:488

harmonic, **53**:314, 317

lifetime in ionic crystals, **8**:19

mean free path, **33**:22
optical, **40**:440, 441, 447, 472ff
as periodic vibrations, **53**:311
scattering, **40**:454, 476
spectrum, **42**:315
surface, **27**:473; *see also* Surface modes
temperature dependence of, **53**:275–276
translational, **40**:472
Phosphate, **7**:467
absorption spectra, **7**:467ff
high-energy bond, **7**:70
ion, **7**:64
Phosphenol pyruvate, **7**:66
Phosphite anions, vibrational spectra, **7**:469ff, 472ff
Phospho-diester link, **7**:219
Phosphofructokinase, **29**:42; **38**:366, 368; **39**:30, 198
activity, **29**:144
allosteric model, **29**:34, 146; **38**:371; **39**:30, 31
cooperativity, **29**:145
role in glycolytic oscillations, **38**:368
Sel'kov model of, **29**:146
Phosphoglycerate kinase, **39**:85
Phospholipid, **41**:196
phase transitions, **39**:204, 219, 220, 221
Phosphomonoester group, **7**:468
Phosphors, alkali halide, **4**:188–194
Phosphorus, **9**:368
superdelocalizability, **7**:67
Phosphorylation, **19**:366
chemiosmotic hypothesis, **19**:366
coupling to electron transport, **19**:366
uncouplers, **19**:366
Photoacoustic effect, **50**:263, 308

Photoaggregation, **41**:209
Photobiology, **49**:34
Photobleaching of electrons in hydrocarbon glasses, **31**:371, 382
Photochemical *cis*-effect, **7**:403
Photochemical electron acceptor, **19**:357, 362
oxidation potential, **19**:358, 362
Photochemical electron transfer, **19**:360, 364
cyclic and noncyclic, **19**:363
Photochemical ion pumping, **19**:369
Photochemical oxido-reduction, **19**:354
stabilization of products, **19**:360, 362
Photochemical process, triplet state, **7**:576
Photochemical reaction, **1**:54
dissociation, **2**:196
nucleic acid derivatives, **15**:171
rate of, **21**:155
Photochemistry, **40**:372
molecular crystals, **53**:277–278
Photoconductivity, **7**:290
of aromatics, **7**:310, 316ff
impurity effects, **7**:318ff
pre-exponential factor, **7**:308ff
pressure dependence, **7**:297ff
pulsed, **7**:290ff
temperature dependence, **7**:296ff
Photoconductor, **34**:260, 261
Photodetachment, **25**:279; **42**:542, 579
ion cyclotron resonance, **19**:193
SH, **19**:193
Photodichroism, **13**:193
Photodiffusion, **41**:310
Photodissociation, **30**:85, 304; **36**:187, 188; **47**:Part 2:91–111; **48**:149
CF$_3$CN, **50**:258
CH$_3$COI, **50**:281
CH$_2$I$_2$, **50**:260

CH$_3$I, **50**:259, 309

CN compounds, **50**:258, 265, 275, 292, 297

Fe(CO)$_5$, **50**:263, 309

H$_2$, **50**:284

HCCBr, **50**:263

HCN, **50**:258

HN, **50**:266

mercury halides, **50**:273, 282, 306

N$_2$O, **50**:259

NO containing compounds, **50**:304

O$_2$, **50**:289

O$_3$, **50**:262, 300

OCS, **50**:258

OCSe, **50**:258

OH containing compounds, **50**:298

PH$_3$, **50**:259, 312

Rb$_2$, **50**:268

single photon, **47**:Part 2:92–106

   HN$_3$, **47**:Part 2:92–106

   ICN, **47**:Part 2:92–96

   methylketene, **47**:Part 2:104–106

TlI, **50**:259

UF$_6$, **50**:260, 308

VUV multiphoton, C$_2$H$_2$, **47**:Part 2:107–111

   CH$_3$Br, **47**:Part 2:106–107

see also Photofragmentation

Photodissociation product, **46**:356

Photoejection, **33**:17

Photoelectric properties of metal-ammonia solutions, **4**:316–318, 334, 338, 352, 357–358

Photoelectron-photoion coincidence, **52**:355

Photoelectron spectroscopy, **36**:205–340; **41**:489; **42**:537; **45**:7, 29–33, 41–44, 104

angle-resolved, **41**:483

ultraviolet, **41**:490

x-ray, **41**:490

Photoemission, **10**:177; **27**:244, 326, 338, 343

ESCA, **27**:245

x-ray, **27**:343

Photoemission probes of surfaces, angle-integrated, **49**:534, 583

angle-resolved, **49**:533–656

angle-resolved core level, **49**:639

azimuthal pattern, **49**:634–635, 641–643

chalcogen adsorption, **49**:605

core level, **49**:594, 634, 638–645

   azimuthal dependence, **49**:641, 643

dangling bond surface state band, **49**:589–591

diffraction, **49**:642–643

final state, free electron, **49**:557

   plane wave, **49**:547–549

   resonances, **49**:543–544, 625

full width at half maximum, **49**:567, 592–593

low energy electron diffraction, **49**:565, 583–588, 633–634

one-step model, **49**:565

three-step model, **49**:564–565

Photoexcitation, **49**:549–551

Photofragmentation, triatomic, **50**:275ff

   impulsive model, **50**:281

   vibrationally excited molecules, **50**:268

Photofragments, Doppler spectroscopy of, **47**:133–152

   multiphoton dissociation in beam, **47**:Part 2:138–143

   one photon dissociation of NO$_2$, **47**:Part 2:143–151

   electronic relaxation of, **47**:Part 2:372–373

Photographic process, **40**:51, 58, 372

Photoionization, **10**:189–190, 248, 252, 257, 260, 270; **19**:59, 62, 64, 111, 118; **28**:395, 404; **31**:376; **45**:7, 41–73, 101–105, 115–117; **49**:535–570

of clusters and molecules, **52**:263–362

gas-phase, **49**:540

H_2, **45**:103

mass spectrometer, **45**:115–117

one-dimensional ordered system, **49**:535–544

periodic potential, **49**:549–551

three-dimensional ordered system, **49**:544, 551–565

two-dimensional ordered system, **49**:544–549

see also Ionization; Photoelectron spectroscopy

Photoionization cross section, **52**:288

NH_3 and NH_4^+, **45**:126

Photoionization mass spectrometry, **52**:267–270, 352

Photolysis, **2**:200; **41**:236

Photomicrography, **29**:174

Photon catalysed dissociation, **50**:191–254

Photon coupling, **54**:314–315

Photon echo, **40**:391; **47**:Part 2:425–429

multilevel system, **47**:Part 2:472–481 **47**:Part 2:479–481

anomalous stimulated, **47**:Part 2:477–479

low temperature, **47**:Part 2:473–474

optical free induction decay, **47**:Part 2:481

optical nutation, **47**:Part 2:481

$^3P_0-^3H_4$ transition of Pr^{3+}/LaF_3,

phonon induced relaxation, **47**:Part 2:474–475

stimulated, **47**:Part 2:475–477

TPM in TPA, **47**:Part 2:473

organic mixed crystals, **47**:Part 2:444–457

Photon field, incident, **49**:565–566

longitudinal, **49**:566

screened, **49**:565

spatially varying, **49**:565–570

Photon simulation, **45**:5, 7

Photons, **27**:294, 328

coupling with surface plasmons, **27**:294, 304, 328, 339

coupling with surface polaritons, **27**:340

s- and p- polarized, **27**:332

virtual, **33**:204, 205, 213, 221, 222

Photons as catalysts, **50**:191–254

criteria for, **50**:243

semi-classical theory, **50**:235

Photoprotection, **7**:584, 586ff

Colpidium colpoda, **7**:594

E. coli B, action spectrum, **7**:590, 596

from far ultraviolet, **7**:584

killing in *E. coli*, **7**:589

microorganisms from UV damage, **7**:592

Pseudomonas acruginosa, **7**:591, 594

Streptomyces griseus conida, **7**:591

subcellular systems, **7**:593

temperature independence of, **7**:590, 595

x-ray effects, **7**:585

Photoreactivation, **7**:589

Photoreception, **29**:343

Photoselection, **41**:284

Photoselective chemistry, **47**:Part 1:1–734

biophysics and, **47**:Part 1:98–102

coherent optical effects, **47**:Part 1:65–69

collision effects, **47**:Part 2:185–418

collisional processes, **47**:Part 1:58–65

condensed media, **47**:Part 2:419–679

electronic relaxation, **47**:Part 1:40–46

inverse, **47**:Part 1:46–51

energy acquisition, **47**:Part 1:2–9

energy disposal, **47**:Part 1:12–13

energy storage, **47**:Part 1:9–11

experimental observables, **47**:Part 1:13–16

intramolecular energy flow, **47**:Part 1:51–56

multiphonon processes, **47**:Part 1:90–98

multiphoton excitation, **47**:Part 1:70–82

nonreactive molecular processes, **47**:Part 1:29–40

one-photon and two-photon processes, **47**:Part 2:3–181

predissociation, **47**:Part 1:19–28

unimolecular reaction, **47**:Part 1:56–58

van der Waals molecules, **47**:Part 1:343–360

Photosensitization, **50**:326, 359, 382

Photosensitized dissociation, **16**:95–99

Photosynthesis, **7**:379; **40**:372

EPR investigations of, **7**:628ff

main sequences, **7**:636

prebiotic origin, **55**:113–118

primary processes in, **47**:Part 2:581–586

Photosynthetic bacteria, **19**:354

Photosynthetic materials, EPR studies, **7**:632

Photosynthetic reaction centers, *see* Reaction centers

Phthalocyanine, **47**:Part 1:6

Phthalocyanines, **7**:370, 383, 402; **16**:198

metal substituents, **7**:387

Phycomyces, **7**:638ff

melanin content, **7**:639

Phylogenetic trees, **29**:210, 216, 224

Physarum polycephalum, mitotic cycle, **38**:378

Physical thermodiffusion, isotope separation by, **16**:127

Physisorption, **49**:636–638

Picosecond laser pulses, generation of, **23**:213

light shutters, **23**:214, 220

CS_2 birefringence, **23**:214

Kerr liquids, **23**:214

measurement of, **23**:193, 194, 196

dibenzanthracene—benzene, **23**:194

dye solution, **23**:196

three photon method, **23**:196

two photon method, **23**:193

Picosecond spectroscopy, **23**:189ff; **40**:2, 3, 7, 37, 44; **49**:1–46

argon laser, **49**:2–4

coherence coupling phenomenon, **49**:16–17

coherence spike, **49**:12–13, 16–17

coherent anti-Stokes Raman scattering experiment, **49**:15, 27–28

dephasing, **49**:21, 24, 26–28

electrons in liquids, **47**:Part 2:535–577

emission techniques, **49**:17–18

energy relaxation, **49**:21, 25

Enskog diffusion constant, **49**:30

fluorescence decay, **49**:20–21, 33–34

fluorescence up-conversion technique, **49**:18

isotropic motion, **49**:30

krypton laser, **49**:2–3

longitudinal optical phonons, **49**:25–26

Lorentzian spectral profile, **49**:9

noise burst model, **49**:6–9, 11–12

orientation relaxation, **49**:28–32

phase grating, **49**:17

pulse duration and structure, **49**:5–11

pulse shortening mechanism, **49**:4

pulse-to-pulse reproducibility, **49**:11–14

pump-probe technique, **49**:14, 16–18, 25, 27

pyrazine vapor, **49**:33–34

synchronously pumped laser, **49**:2–10, 14–15, 19–21, 26, 34–42

vibrational studies, **49**:21–28

    liquids, **49**:26–28

    mixed crystals, **49**:21–25

    semiconductors, **49**:25–26

Piezoelectricity, 7:85

Pigmentary epithelium, 29:343

Pigments, photoreduction, 7:635

Pine's wave function, 2:306

Pinocytosis, 21:637

Piperidine, photoreduction with ascorbic acid, 7:635

pK_a of carbonyl compounds, 8:136

Planar itinerant libration, 44:276ff, 287, 289

Planar itinerant oscillator, 44:259

Planar libration, damped, 44:275

Planar XY model, 40:201

Planck relation, 36:43

Planck-Schaefer heat capacity law, 40:101–102

Planck's constant, 1:19, 179, 272

Plane wave approximation, 48:43, 63–69; **49**:560

    density of states, 48:64

Plane wave spectrum method, 37:45

Plane waves, final state, **49**:547–549

    free-electron, **49**:557

    nonlocal, 27:450

Plants, EPR investigations, 7:628

    irradiation of, 7:640ff

    non-photosynthetic material, 7:638ff

    viruses of, photoreactivation of, 7:586ff

Plasma, 18:84; 27:266; 35:122, 126, 132, 139, 181

    bulk oscillations, 27:238

    degenerate, 27:267, 274, 281

    gaseous, 27:266

    kinetic theory of, 38:193–210

    laser-excited, 41:186

    multicomponent, 4:220

    nondegenerate, 27:267, 274

    one-dimensional, 4:201–224

    solid-state, 27:266

    surface oscillations, 27:286

    turbulent, 38:176, 187, 188

    wavelength, 27:278

Plasma-dynamic approximation, 11:230, 272

Plasma frequency, 2:306; 27:266, 274, 361, 378, 414, ; 37:39

Plasma model, 2:207, 304, 306, 318, 319, 323

Plasma radiation, 27:33

Plasmon, bulk, 27:45; *see also* Surface plasmons

Plasmon frequency, **49**:566

Plastic crystals, 16:357; 44:258, 391, 471, 474

    benzene, 44:393

    nondipolar, 44:470

    significant structure theory of, 21:517

Platelets, 26:157

Platinum-cobalt alloy, enthalpy of formation, 2:144

Platinum hexafluoride, 9:195ff

Plethysm, 14:106

Plutonium hexafluoride, 9:190

PMO method, 8:69ff

steric effects in, **8**:66ff

PMR, *see* Proton magnetic resonance

Pneumococcus, 7:220

PO, quenching of B state, **50**:87

radiative lifetime of B state, **50**:86

spectroscopy of, **50**:84

Pockel's cell, 50:107

Poincaré-Bertrand theorem, 20:190

Poincaré-Birkhoff fixed point theorem, 46:105–106

Poincaré-Keller method, 40:253

Poincaré recurrence, 19:329, 330, 335; **53**:343

Poincaré surface of section, 24:164; **47**:Part 1:205–206

Poincaré theorem, 17:117

Point-defect interactions in solids, 11:1ff

Point molecule, 9:47

Point of no return, 4:44–48

Poiseuille-Couette flow, 32:47

Poisoning of ionization detectors, 10:324, 341

Poisson-Boltzmann equation, 4:212–214; **48**:337

Poisson brackets, 19:349; **33**:159, 168, 172ff, 189ff, 200, 216–218; **44**:467; **53**:126, 219, 220

Poisson distribution, 32:8; **34**:287; **36**:285; **41**:14

Poisson equation, 4:202–204; **11**:190, 219; **35**:123

Poisson summation formula, 36:51, 56

Polanyi rule, 33:347, 381, 382

Polar fluids, 31:103–154; **44**:483

correlation function, **31**:117, 135

dielectric constant, **31**:110, 122, 130

dielectric relaxation, **31**:145–152

mean spherical model, **31**:125, 126, 141–145

microscopic structure, **31**:108, 115, 130

Onsager model, **31**:139–145

rigid, **31**:110, 119, 130

thermodynamic properties, **31**:138–145

Polar molecules, 12:297

electron scattering cross sections for, **36**:513

surface tension and compressibility, **9**:75ff

Polaritons, 23:254, 257; **27**:369, 404; **40**:370

dispersion curves, **54**:334–338

anisotropic media, **54**:338–340

infinite molecular crystals, **23**:285

quantized theory, **23**:272, 285

quantum theory, **54**:353–372

diagonalization of Hamiltonian, **54**:367–370

energy dependence, **54**:370–373

Hamiltonian, **54**:360–367

semi-infinite theory, **23**:272

semiclassical theory, **54**:328–340

application to anthracene, **54**:346–353

Coulomb excitons, **54**:330–331

longitudinal, **54**:330–331

nonlongitudinal, **54**:331

dispersion curve, **54**:334–338

anisotropic media, **54**:338–340

damping and local field, **54**:337–338

retarded interactions, **54**:331–334

approximations, **54**:333–334

dielectric tensor, **54**:331–333

linear medium, **54**:334

local field, **54**:333

no spatial dispersion, **54**:334

no Umklapp effect, **54**:334

rigid lattice approximation, **54**:334

surface excitons and, **54**:303–468

bulk and surface excitons, **54**:305–326

crystalline anthracene, **54**:381–406

exciton surface polaritons, **54**:437–455

experimental methods, **54**:372–381

infinite crystals, **54**:327–372

reflectivity, **54**:406–422

see also Exciton surface polaritons; Surface polariton

Polarity, hydra, **29**:263

insect epidermis, **29**:265

membrane, **29**:269

membrane potential and, **29**:285, 288

plant cell, **29**:281, 282

Polarity parameter, **5**:344

Polarizability, **1**:79, 272ff, 311, 337, 342ff, 358; **4**:75–88, 123–125, 157, 218; **9**:348; **25**:8; **41**:149, 157; **44**:35, 102, 447, 449, 451; **52**:334

angle-averaged, **19**:157

anion binding and, **39**:335, 336

anisotropy, **13**:195, 202; **46**:6

atomic, **10**:1–27

alkali atoms, **10**:8–20

anisotropy of, **10**:5–6, 21

average, **10**:4

beam measurements of, **10**:9–26

metastable $^3P_2$ argon, **10**:20–25

tensor, **10**:3–4, 20–21, 25

calculations, *ab initio*, **46**:10

density functional formalism, **46**:11

electron gas approximation, **46**:12

carbon, **2**:75

diatomic, **46**:2, 4

models of, **46**:44

dipole-induced dipole model, **46**:6

dynamic, **46**:7, 18, 43

electric, **37**:260, 261

excited states, **12**:128; **13**:195

field fluctuations, **46**:5

helium, **46**:42

hydrogen molecule, **2**:65, 75

induced, **51**:88

anisotropic, **51**:92

interference with permanent polarizability, **51**:107–109

invariants, **46**:5

ionization potential data and, **2**:70

matrix elements, **51**:89–90

model, **46**:20

molecular, **10**:26; **12**:109; **42**:118

many-electron theory, **6**:316

multipole, **12**:145, 146

range of, **46**:19

rigid rod, **21**:232

static, **46**:7, 18, 43

Polarizability approximation, **4**:151

Polarizability distortion in simple fluids, **26**:34

Mazur theory, **26**:34–35

Polarizability tensor, **4**:108, 449; **12**:112, 135; **34**:9, 10, 11, 12, 14, 84; **44**:108, 449

Polarizable molecule, **48**:272

Polarization, **1**:77, 81ff, 316, 324ff, 336ff; **25**:5; **27**:360, 487; **33**:161ff, 183, 185; **35**:4, 140, 152, 155, 182; **44**:341, 342

after π, $\pi/2$ pulse train, **25**:42

after $\pi/2$, π pulse train, **25**:39

asymptotic, **35**:20

averaged dynamical, **38**:199

binding, **8**:169

electronic transitions, anthraquinone derivatives, **13**:193

  nitriles, **13**:194

  tropones, **13**:193

ellipse, **33**:351, 356, 362, 392

emission, **30**:85, 304

fluorescence, **13**:193

lattice displacement and, **53**:265

ligand by cation, **7**:361

light, **33**:183, 187

macroscopic, **25**:5

measurements of, **7**:329

neutron beams, **16**:91–92

phase, **25**:16

of radiation emitted by optically pumped atoms and molecules, **16**:74–75, 79–80

state, **33**:184, 197, 221

target in electron scattering, **18**:72, 74

transient absorption, **25**:22, 23

transient emission, **25**:30

velocity dependence, **25**:12

Polarization diagrams, **1**:261

Polarization effects, **4**:103

Polarization energy, **4**:383; **7**:286; **42**:174

Polarization field, **33**:161ff, 209, 215ff

  electric, **33**:155ff, 213ff

  magnetic, **33**:155ff, 165, 215

Polarization force, **10**:255, 263, 268; **44**:497

Polarization matrix, **40**:387

Polarization potential, **9**:332; **16**:296–299; **28**:145, 335, 336, 347

Polarization resonance, **49**:41–42

Polarization scattering potential, **36**:509

Polarization spectroscopy, **49**:39, 42

Polarized fluorescence, **50**:267, 278, 298, 509

Polarized light, **35**:5

Polarized light scattering, by simple fluids, **26**:4, 15

  classical description, **26**:27, 28, 34

  Einstein relations, **26**:16

  lineshape, **26**:15

  quantum mechanical description, **26**:9, 15, 16

Polarized orbitals, **25**:267

Polarizing interferometer, **44**:327, 331

Polaron, **40**:432, 433, 442ff; **44**:487, 495, 496

  Pekar, **44**:493

  small radius, **44**:502

Polaron cavity model of metal-ammonia solutions, **4**:342–346, 359–361

POLCI calculations, **50**:527, 560, 565

Pole strength, **36**:211, 221, 222

Poley absorption, **44**:259, 276, 308, 309, 350, 355, 396, 399

Polio virus, **7**:222

Polk method for binary alloy glass, **51**:237, 266

Pollution of atmosphere, **55**:80–84

Poly-(acrylic acid), **39**:163

Poly-DL-glutamic acid, **33**:95

Poly-L-glutamic acid, **21**:593; **33**:93, 94

Poly-L-proline, **33**:101, 104

Poly-(methylacrylic acid), **39**:163

Polya theorem, **15**:231

Polyacetylene, kink diffusion in, **53**:277

  solitons in, **53**:287–293

Polyadenylic acid, **7**:493

  infrared spectra, **7**:456

Polyamide, **2**:181

Polyatomic bimolecular reaction, **47**:Part 2:7–10

Polyatomic fluids, **53**:45

Polyatomic molecules, **12**:597; **36**:20,

45; **47**:Part 2:185–289; **52**:345–352

collision induced vibrational energy transfer, **47**:Part 2:237–289

multiphoton excitation of, **50**:452

scattering from, **42**:492

vibrational energy flow, **47**:Part 2:185–235

Polyatomics-in-molecules, 42:109

Polybutadiene, 2:170, 181

Polybutyl methacrylate, 1:99–101, 104–106, 114

Polycrystalline solid, 32:196

Polycytidylic acid, infrared spectra, **7**:455

Polydispersed system, 2:183

Polydispersity effects, 51:26

Polyenes, valence shell correlation in, **52**:70–109

Polyethylenimine, binding properties, **39**:112, 131

catalysis of decarboxylation by, **39**:146

catalysis of ester cleavage by, **39**:120

conformation, **39**:126, 135

deacylation by, **39**:116

see also Synzymes

Polyfunctional polymer, 2:178

Polygonized spiral, 40:206

Polygons, 43:121, 122

Polyhedra, concentration of, **43**:27

extent of reaction, **43**:25–27

nearest neighbor, **31**:45

properties of, **43**:27

representations, **43**:19

Polymer adsorption, statistical mechanics of, **51**:1–44

Polymer chain, configurations of, **52**:365, 430ff, 463

Polymer chain-adsorption at surface, 52:449–456

Polymer clusters, structure of, **42**:354

Polymer dynamics, bead and spring model, **33**:76, 121–125

dilute solution, **15**:325

stochastic model, **33**:77, 121–124

Polymer films, dichroism of, **13**:193

Polymer physics, 52:430–456

Polymer solutions, 21:531; **51**:5–28

statistical mechanics of, **51**:1–44

Polymerization, 19:103; **46**:378, 379, 382

butadiene, **2**:163

ionic, **2**:148, 160, 161

low temperature, **39**:244, 245, 246

prebiotic, **55**:101–105

nucleosides and polynucleotides, **55**:103–105

peptide synthesis, **55**:101–103

propagation of, **2**:158

shock-induced, **53**:310

solid acetaldehyde, **2**:163

styrene, **2**:150, 172, 175

trans, **2**:170

vinyl monomers, **2**:154

vinylidene, **2**:164

vinylidene dicyamide, **2**:155

Polymers, 1:78, 89–96, 99–106, 113–115; **4**:113–160

closure, at doublets, **33**:80

conditions, **33**:80

at quadruplets, **33**:82

at triplets, **33**:82, 83

flexible, **4**:150, 158

interactions, between triplets, **33**:100

codon-anticodon, **33**:100

hydrodynamic, **33**:125, 137, 139

hydrophobic, **33**:97, 107

nearest neighbor, **33**:80, 83

jump probability, **33**:77, 122–124, 134

kinetic coefficient, double helix formation, **33**:100
helix-coil transition, **33**:83
sequential processes, **33**:103, 106, 110, 112–114
kinetics, fast step in unfolding and refolding, **33**:105, 109, 112, 113
intermediates, base-pairing, **33**:99
concentration of, **33**:110
denaturation, **33**:106, 108–115
metastable, **33**:108
sequential processes, **33**:101, 103, 104, 108–115
unfolding and refolding, **33**:105, 108–115
models of, bead and spring, **33**:76, 121–125, 137–139, 144, 145
dumbbell, **33**:78
elastic sphere, **33**:78, 125, 138
four state, **33**:109, 113
sequential, **33**:98–100, 103, 104, 106–115
three-state, **33**:106, 108, 131
two-state, **33**:80, 105, 115–121
two-state, non-Arrhenius, **33**:115–121, 129, 132
magnetic analogy, **51**:14–20, 38
melts, **51**:26
monodispersed, **2**:177, 183
pH-jump technique, **33**:108, 109
physical boundary present, **52**:449
polydispersed, **2**:183
polyfunctional, **2**:178
restricted environments, **51**:28–44
sequences, disordered, **33**:88, 90, 95, 97
fluctuations of number of uninterrupted, **33**:92, 95
ordered, **33**:88
of states, **33**:99, 101, 103, 104, 108–114
in slits and tubes, **51**:28–35
star-shaped, **2**:183
statistics of,
cross links, **22**:88, 118–121, 123, 125
diffusion, **22**:1, 12, 28, 32, 39ff, 55, 66, 72, 104, 109
end to end vector distribution, **22**:10–12, 20, 26, 36, 39, 47, 61, 62
entanglements, **22**:94, 95, 100, 105, 124
excluded volume problem, **22**:1, 4, 6, 76, 77, 119
Feynman path integrals, **22**:1, 5, 6, 32, 35, 55, 109, 110
Fokker-Planck equation, **22**:43, 45, 50
free energy, **22**:68, 89, 96
Gaussian random fields, **22**:70, 83, 84, 110–112, 120, 124
Green's functions, **22**:1, 49, 54, 62ff, 70, 83, 104, 119
internal constraints, **22**:2, 6, 94–97, 100ff
Ising model, **22**:8, 9
Lagrangian, **22**:34, 50, 55, 86, 110, 116, 117
Markov process, **22**:9, 30, 48, 52, 65–67, 73, 81, 90
mean square end to end distance, **22**:17, 60
microcanonical ensemble, **22**:88, 92, 95, 97, 99
partition function, **22**:2, 12, 13, 26, 34, 41, 56, 89, 96, 123
persistence lengths, **22**:21, 22
random walks, **22**:13, 14, 23, 115, 169
rotational isomeric model, **22**:20, 37

second quantization, **22**:109, 110

self-avoiding walk, **22**:64

self-consistent field, **22**:1, 61, 67–70, 73

self energy, **22**:75

short-range interactions, **22**:8, 10, 22

Smoluchowski-Chapman-Kolomogorov equation, **22**:52, 81

θ point, **22**:6, 9, 53, 80

torsional energy, **22**:48

stereoregular, **2**:165

synthetic, having enzyme-like activity, **39**:109–179

temperature jump, **33**:99, 107, 109

units of, **33**:88

viscosity, internal, **33**:77, 78, 138–146

matrix, **33**:78, 124, 136, 145

intrinsic, **33**:125

shear, **33**:131–144

of solvent, **33**:123, 124, 128, 133–147

volume change, conformation change, **33**:87, 97

formation of interruption, **33**:91

loss of disordered sequence, **33**:90, 97

see also Chain; Macromolecules

Polymethylmethacrylate, **2**:180

flow birefringence, **33**:141

ultrasonic spectra, **33**:127

Polymorphic phase transition, **40**:283–287

Polynomial potential, **36**:38

Polynomials, diagrammatic, **43**:119–125

Polynucleotides, **7**:136ff, 140, 144

chains, **7**:86ff

emission characteristics, **7**:347ff

prebiotic synthesis, **55**:103–105

Polypeptides, **7**:135; **21**:584; **33**:74, 79, 94, 101

chains, **7**:203, 220, 239

Polyphenyl molecules, ionization potentials and electron affinities, **21**:37ff

Polyribonucleotides, **7**:135, 454

infrared spectra, **7**:454ff

Polysaccharides, **21**:638

Polystyrene, **2**:172

flow birefringence, **33**:139–141

ultrasonic spectra, **33**:127–129, 133–135, 139–143

Polystyril carbanions, **2**:154

Polysulfones, **33**:140

Polytetrahedral structure, **40**:78

Polytope, concentration, **43**:27

current, **43**:20–25

Polyuridylic acid, **7**:459, 514

infrared spectra, **7**:459ff

relaxation rates, **7**:514

Polyvinylpyridine, ultrasonic spectra, **33**:127ff

Polyvinylpyrrolidone, **39**:111, 113

Ponderomotive forces, **1**:311, 325, 330, 331

Pople SCF treatment, **8**:74ff, 105, 124ff

POPOP as quencher, **50**:366

Population analysis, Coulson and Longuet-Higgins, **37**:225

Mulliken, **37**:226

Population indices, **37**:227, 228

Population inversion, **28**:427; **30**:28; **41**:311

Population-time curve, **19**:337

Pores, **41**:205, 206

Porous dissipative solid, **32**:14

Porphyrin, **7**:72, 380, 383

spectra, **7**:381

Porphyrin complexes, photochemistry, **7**:404ff

Positional disorder in crystals, 6:463
Positional information, 29:253, 278, 295
Positive definite matrix, 43:61
  test for, 43:62
Positive operator-valued measure, 38:65
Positive orphant, 43:17
Positive semidefinite matrix, 43:61
  test for, 43:62
Positron annhilation, 51:246
Positron scattering, 18:86
Potassium, 41:141
  beams, 10:139–168
  dimer, 41:152
    dipole resonance interaction energy, 12:41
    ion, core polarization potential, 6:379
  see also Collisional ionization; Diatomics-in-molecules; Scattering
Potassium bromide, 27:438, 445
  Soret effect, 13:361, 362, 366
Potassium hypophosphite, spectra, 7:479
Potassium iodide, Soret effect, 13:366
Potassium titanium hexafluoride, charge density, 5:251
  electronic energies, 5:251
  forbidden transitions in, 5:241, 251–252
  oscillator strength, 5:252
  symmetry operations, 5:255
  vibrations of, 5:244, 251–252
  vibronic perturbations, 5:251, 252
Potential, 12:289; 49:317–318
  Brazovski, 49:338
  Brazovski generalized, 49:325
  Buckingham Exp, 12:394
  central forces, 51:72, 73, 76
  centrifugal, 28:8
  deformation, 13:125, 134, 152

dipole-dipole, 28:337ff, 358, 360, 363, 366, 372
  effective, 12:323; 16:299–300; 41:91
    self-consistent one-body, 41:86
  electron-electron, 28:118, 119, 128
  electron-nuclear, 28:118, 119
  exchange correlation, 37:287
  external, 41:68–69
    static nonlocal, 41:69
    static spin-dependent, 41:69
    time-dependent, 41:128
    velocity dependent, 41:69
  finite temperature, 41:69
  hard sphere, 51:64
  Hartree, 41:115
  He-He repulsive, 12:393
  induced absorption, 51:57, 61
  induced light scattering, 51:91
  inner, 20:237, 238, 239, 241, 293
  intramolecular, 31:10
  ion-induced dipole, 28:130
  Kihara, 12:396
  kinetic, 49:317
  Lennard-Jones, 11:118; 12:396; 28:17, 181, 182, 189, 195, 223, 228ff
  local, 13:268, 300ff; 41:68
    applied to instability problems, 13:267
    used in study of kinetic equations, 13:289
  local external scalar, 41:72
  Mataga-Nishimoto, 37:219
  Morse, 12:396; 13:104; 28:68, 69, 177ff, 228ff
  muffin-tin, 37:288
  Mulliken-Rudenberg, 37:217, 223
  nonlinear, 41:134
  nonlocal, 41:81
  nonlocal external, 41:80–81

normalized, **34**:125
nuclear-nuclear, **28**:118, 119, 128
one-body, **41**:67, 86
one-dimensional random, **13**:123ff
polarization, **16**:296–299
renormalized, **34**:109, 125, 133
   cancellation of, **34**:136
   definition of, **34**:133
   divergence of, **34**:136
   rescreened, **34**:125
   smaller at high density, **34**:137
scalar, **41**:68, 72
scalar external, **41**:71–73
scattering, **25**:215, 259, 278, 280, 282, 285
spin-dependent, **41**:69, 76
spinor external, **41**:75–78
static, **41**:68, 69
static external four-vector, **41**:84
thermodynamic, **13**:300
   in study of stability problems, **13**:325
time-dependent, **25**:214
van der Waals, **28**:335, 336, 345, 350, 358
variation, **16**:293
velocity-dependent external, **41**:78–80
see also Pair potential
Potential curve, **41**:272
   adiabatic, **18**:138
   diabatic, **18**:138
   maxima, **23**:177
   near dissociation limit, **41**:282
Potential distribution theorem, **41**:39, 41, 54
Potential energy, **31**:11
   atom-chain, **53**:102
   atomic lattice, **53**:296
   Coulomb interaction, **1**:166

intermolecular, **4**:229, 252, 263, 280ff, 295ff
Lennard-Jones solution, **53**:192ff
multi-component plasma, **4**:220
nonadditivity, **1**:267, 270–272, 288, 290, 294; **31**:23, 42
   three molecule interaction, **31**:54
one-dimensional plasma, **4**:202, 204
solvent-reagent, **53**:176
transport, **1**:152
Potential energy barrier, **1**:181; **52**:301, 304, 354
   crossing of, **33**:122–124, 144, 147
   to internal rotation, **2**:368, 374
      measurements of, **2**:369
   to reaction, **4**:10–29, 42, 48
   vinyl-polymers, **33**:75
Potential energy surface, **1**:20, 31, 53ff, 125, 127; **3**:139; **28**:11, 44, 67, 71–74, 95, 116–146, 321ff; **30**:8, 137ff; **36**:65, 142–173; **41**:230, 251–255; **42**:1–584; **44**:104; **45**:201–205; **49**:140, 193, 198, 295; **50**:282ff, 517–588
   ab initio, **30**:139, 140, 371, **37**:321, 322, 328ff; **50**:300, 361, 476, 490, 517–588
   adiabatic, **23**:14
   AlH, **50**:551
   alkali—alkali, **28**:338, 339, 340, 369
   alkali halides, **50**:240
   alkali + halogen atom, **42**:32
   alkali—noble gas, **28**:336, 369
   anisotropic, **30**:229, 332, 337, 375, 380, 393, 407, 410
   anisotropy parameters, **42**:454, 457, 459, 466, 488, 494
   Ar_2F, **50**:576
   ArH_2^+, **30**:165
   atom-diatom systems, **42**:424
   atoms-in-molecules, **30**:158

Au_2, **50**:557

avoided crossing, **28**:150

B^+ and rare gases, **50**:565

BeF_2, **30**:145

BeH, **50**:549

BeH_2, **30**:165

BH_2, **30**:145, 162

Br - Br, **30**:254

Br_2H, **30**:165

$C^+ - H_2$, **45**:203

C_2, **28**:42; **50**:553

C_3, **30**:150; **50**:573

CaF_2, **30**:145, 146

CCl_3 - Cl, **30**:254

Cd_2, **50**:556

CF_2, **28**:138

CH_2, **30**:145, 150; **50**:567

$CH_2 + H_2$, **30**:146, 169

CH_2^+, **30**:145

$CH_3 + H$, **30**:169

$CH_3 - I$, **30**:254

CH_3NC, **30**:146

CH_5^+, **30**:146

$Cl + H_2$, **30**:146

$Cl^- + CH_3Br$, **30**:169

Cl_2H, **30**:165

Cl_2S_2, **30**:162

ClO, **50**:560

closed shell interaction, **30**:162

CN, **28**:41

$CN^- + CH_3F$, **30**:146

CO_2, **30**:145, 150

collisional effects on electronic relaxation processes, **42**:207–270

collisional ionization, **42**:513–584

configuration interaction, **30**:148

Cs - Hg, **30**:358

diabatic, **23**:14

diatomic molecules, **28**:130–134

diatomics-in-molecules, **30**:163; **42**:63–112

Diels-Alder reaction, **33**:360, 373

Dunham's form, **37**:320, 345, 347

electronic chemiluminescence, **42**:1–62

empirical, **30**:170, 334, 345, 351, 356, 381

excited state, **42**:169–206

experimental, **37**:342, 365

$F + CH_3F$, **30**:146

$F + H_2$, **30**:150, 167; **50**:569

$F + H_2O$, **30**:146

F_2H, **30**:165

fitting, **30**:334, 355

force constants, **37**:320, 321

geminal method, **30**:156

generalized valence bond, **30**:147, 154

H_2, **50**:547

H_3, **30**:165, 166, 174; **50**:547

H_3^+, **30**:150, 165, 174; **45**:203

H_4, **30**:165, 166

H_n^+, **30**:165

$H + Cl_2 - HCl + Cl$, **28**:67

$H + F_2$, **30**:150

$H + H_2$, **30**:150

$H_2 + H_2$, **30**:150, 166

H_2 - He, **30**:391

$H^- + HF$, **30**:146

harpooning, **10**:375–377

Hartree-Fock, **30**:141, 143

H_2Br, **30**:165

HCHO, **30**:162

HCl, **50**:551

H_2Cl, **30**:165

HCN, **50**:571

HCO, **50**:571

$He + H_2^+$, **30**:146

$He + H_2$, **30**:145, 150

HeH$^+$ + CH, **30**:150
He - He, **30**:363, 366, 373
HeH$^+$ + H$_2$, **30**:150
He$_2^+$ - N$_2$, **45**:205
HeBeH$_2$, **30**:146
HeH$_2^+$, **30**:165
H$_2$F, **30**:165, 174
HFF, **50**:572
HgCl - Cl, **30**:254
HgCl$_2$, **50**:575
H$_2$I, **30**:165
HNO$^+$, **50**:571
HNO, **30**:145
H$_2$NO, **30**:145
HNSi, **50**:573
HO$_2$, **30**:150
H$_2$O, **30**:145, 146; **50**:569
HOF, **50**:572
H$_2$S$^+$, **50**:570
H$_2$S$_2$, **30**:162
I - Br, **30**:511
I$_2$H, **30**:165
interhalogen, **50**:21
iterative natural orbital, **30**:174
K$_2^+$, **30**:169
K$_2$, **30**:169
K - Br, **30**:306
K + Br$_2$, **30**:174
K$_2$ + Br$_2$, **30**:264
K + CH$_3$I—KI + CH$_3$, **28**:67
K + Cl$_2$, **30**:490, 507
K - Hg, **30**:358
K + NaCl, **30**:168
KCl$_2$, **30**:169
K$_2$Cl, **30**:168
KRb, **28**:367
KrF$_2$, **30**:150
Kr$_2$F, **50**:569
LEPS, **30**:22, 171; **37**:375

Li$_2^+$, **50**:552
Li$_3$, **30**:146, 162, 165; **50**:573
Li$_3^+$, **30**:165
Li$_4$, **30**:165
Li + F$_2$, **30**:176, 307, 489, 507
Li$^+$ + H$_2$, **30**:145
Li + HF, **30**:146
Li - Hg, **30**:358
Li + LiF, **30**:176
Li$_2$Cl, **30**:168
Li$_2$F, **30**:168
LiH$_2^+$, **30**:165
LiH$_2$, **30**:162, 165
Li$_2$H$^+$, **30**:165
LiKCl, **30**:168
LiN, **30**:165
LiNaCl, **30**:168
LiO$_2$, **30**:145
maximally paired HF, **30**:147
MgF$_2$, **30**:145
MgH$^+$, **50**:550
microclusters, **40**:59
N - H, **30**:150
N - NO$^+$, **30**:206, 207
Na$_2^+$, **30**:169·
Na$_2$, **30**:169
Na + Cs$_2$, **30**:274
Na - Hg, **30**:358
Na$^+$ + H$_2$O, **30**:146
Na + K$_2$, **30**:274
Na + Rb$_2$, **30**:274
NaBr + KCl, **30**:172
NaCl, **30**:162
Na$_2$Cl, **30**:168
NaK$^+$, **30**:169
NaK, **30**:169
NaKCl, **30**:168
negative ions, **28**:143
NH, **30**:150

NH_2^+, **50**:568
NH_3, **30**:150
NO, **28**:36
N_2O and N_2O^+, **28**:152–159
noble gas molecules, **28**:88, 89
noble gas molecules, **28**:88, 89
$O^+ + N_2$, **30**:150
O_2, **30**:150
O_3, **30**:150; **50**:573
O_3^-, **30**:150
oxygen atom reactions, **42**:113–168
perturbation methods, **30**:152
PH_2, **50**:570
polyatomic molecules, **28**:134–142
Porter-Karplus, **30**:25, 54
pseudopotential, **30**:139, 160, 162
radical addition, **33**:363, 365
reactive systems, **10**:147–148, 360
Rittner, **30**:168, 172
scattering of nonspherical molecules, **42**:421–512
SCF, **30**:143
sensitivity of the cross section to, **36**:119
separated pairs, **30**:157
SPF form, **37**:371, 375
spin optimized SCF, **30**:146
Taylor series, **37**:361
Ti_4^+ - F, **30**:162
TiF_3^+, **30**:162
truncation, **37**:320
van der Waals molecules, **42**:353–420
vibrational and rotational collision processes, **42**:371–404
vibrational relaxation of small molecules in dense media, **42**:305–352
Watson's term, **37**:313, 338, 352
Zn_2, **50**:556

see also Diatomics-in-molecules; Electronic structure; Energy levels
Potential gradient, electric, **13**:354
Potential of mean force, **44**:300; **48**:80, 198–199; **53**:117, 120, 183
between ions, **6**:420–422
infinite dilution, **6**:421
non-polar solvent, **6**:425–435
between three ions, **6**:435, 444
between two ions, **6**:435–444
Bellemans'equations for, **6**:435
defects, **11**:32, 33, 47, 63
Potential parameters, atom-atom, **12**:466, 467
H—rare gas, **30**:371
Hg—alkali metal, **30**:357
rare gas—rare gas, **30**:361, 368
atom—diatom, **12**:467, 469
atom—molecule, $H-H_2$, **30**:379
rare gas—H_2, N_2, O_2, NO, CO, CH_4, CO_2, **30**:378
gases, **16**:111
H_2—N_2, O_2, CO, CH_4, NO, CO_2, NH_3, SF_6, **30**:382
ion—molecule, **30**:236, 239, 513
molecule—molecule, H_2—H_2, **30**:379
orbital radii, **12**:478, 479
from scattering data, **12**:390, 425, 460ff, 477–480
Potential well, **44**:258, 280
curvature at minimum, **12**:418
Potentiometric studies of adsorbents, **3**:258
Potentiometry used in study of Soret effect, **13**:355
Potts model, **41**:6, 36
Powder electrodes, compound, **3**:246
construction of, **3**:239, 247
potential of, **3**:251
used to study contact substances, **3**:253

Power absorption coefficient, 44:260, 274, 301, 447

Power absorption plateau, 44:357

Power broadening, 50:216, 406

Power function, 51:147–150

Power spectrum, 17:85, 120; **44**:410

Poynting effect, 2:106

Poynting vector, 27:369, 412, 432
complex, **37**:6

PR enzyme, 7:587, 596

Prandtl number, 32:18, 78, 82

Praseodymium oxide, self-diffusion of oxygen in, **21**:255

Prebiotic synthesis of organic molecules and polymers, 55:85–107
amino acids, **55**:91–94
carbonaceous chondrites, **55**:99–100
energy sources, **55**:89–91
interstellar molecules, **55**:100–101
mildly reducing and nonreducing atmospheres, **55**:94–96
polymerization process, **55**:101–105
purine and pyrimidine, **55**:96–98
sugars, **55**:98–99

Precession signal, free, **11**:311, 312

Predator, 19:333

Predator-prey model, 29:129
immune response against cancer, **38**:396

Predictor, least-biased, **53**:366

Predictor-corrector method, 44:402

Predissociation, 2:199; **12**:354; **18**:133; **28**:35, 36, 37, 43, 325; **42**:2, 372, 387, 388; **50**:27, 39, 47, 191–254, 295; **52**:464
collisional, **50**:229
electronic, **47**:Part 1:27–28; **52**:304, 332
heterogeneous, **50**:30, 38, 44, 65
homogeneous, **50**:44
inverse electronic, **47**:Part 1:28

inverse predissociation, **28**:35
natural, **50**:231
photoselective chemistry and, **47**:Part 1:19–28
rotational, **47**:Part 1:28; **52**:332
solids, **41**:285
Stark induced, **50**:229
vibrational, **47**:Part 1:24–26, 453–464; **52**:304, 306, 332, 341

Preequilibrium, 33:100, 107, 118

Prefreezing, 6:460

Premelting, 6:460

Prepattern, 29:43, 256

Prepoisoning, 10:324

Pressure, 1:148, 159, 204, 216, 224ff, 333
effect on chemical reaction, **39**:161, 162
hydrostatic, **16**:106
measurement, **10**:61
negative, **17**:7
normal, **1**:206, 216, 226, 227
reduced, **16**:225, 228, 233, 240
tangential, **1**:206, 216, 221, 223, 226
virial coefficients, **20**:41, 42, 47

Pressure broadening, adiabatic, **12**:489, 490
Anderson's theory, **12**:493ff, 509ff
diabatic, **12**:490
impact theory, **12**:489, 491, 492
intermolecular potentials, **12**:497, 505, 522, 524, 526
line widths, BrCN, **12**:525
CH_3F, **12**:527
CHF_3, **12**:527
H_2O, **12**:527
HCl, **12**:523
NH_3, **12**:529–531, 533, 535, 536
N_2O, **12**:525, 526
O_2, **12**:526

OCS, **12**:516, 522, 535

PF$_3$, **12**:527

SO$_3$, **12**:528

temperature dependence, **12**:512, 516, 528

nonadiabatic, **12**:490

phase shift, **12**:490

semiclassical theory, **12**:513

Pressure equation, 34:120; **41**:24

Pressure equation of state, 37:172, 175

Pressure flow, 29:337

Pressure gradient in cylindrical containers, 21:737

Pressure-induced vibrational absorption, 36:195

Pressure tensor, 1:148, 158, 327ff; **31**:161; **49**:359–374, 410, 417, 418–420

bulk viscosity, **31**:173, 175, 196, 201

collisional transfer part, **31**:171

diffusional part, **31**:161

ellipsoids, **31**:183

loaded spheres, **31**:196

molecular theory, **49**:365–371

Rice-Allnatt rough spheres, **31**:201

rough spheres, **31**:171, 173, 175, 176

shear viscosity, **31**:173, 175, 196, 201, 208, 224

vortex viscosity, **31**:173, 176, 183, 196, 201

Pressure tuning of laser, 50:19

Pressure-volume-temperature data, 16:236–237, 241, 245

Pre-stretching in collisional ionization, 42:540

Pretransitional phenomena, 32:93

Prey, 19:333, 341

Prigogine-Glansdorff principle, 38:234

Prigogine-Herman theory, 32:9

Prigogine-Lefever reaction, 38:352, 354

Prigogine-Nicolis-Misguich-Davis theory, 31:203

nonspherical molecules, **31**:201–211

spherical molecules, **31**:158, 167

Prigogine-Nicolas-Misguich kinetic theory, 24:259, 278, 280, 281, 283, 285

doublet correlation function, **24**:281

Prigogine's theory, 11:329, 356

Primary bifurcation, 55:178–179

Primary cone, 49:562–563

Primary ions, 10:225–228, 248–249

Primary photochemical process, 7:631

Primary system, 53:66–67

dissipative-stochastic character of motion in, **53**:76–77

phase space distribution, **53**:85

Primary trap, 23:243

Primitive atmosphere, composition of, **55**:89

oxygen in, **55**:88–89

Primitive cavity model of metal-ammonia solutions, 4:339–342

Primitive earth, melting of, **55**:86–87

temperature of, **55**:87–88

Principal moments of inertia, 44:357, 360, 449

Principal polarizability, 2:80

Principal quantum number, 7:10

effective, **52**:312

Principle of averaging, 42:342

Principle of least action, 33:155

Principle of microsymmetry, 5:127

Prior distribution, 50:377

Prism spectrographs, 7:409

Probability, absorption, **52**:409, 415

conditional, **26**:128, 129, 141

dissociating base pair, **33**:99

first return, **52**:453

forming base pair, **33**:99

jump, **33**:77, 122–124, 134

particle crossing potential barrier, **33**:122, 123

return to origin, **52**:466

sequence, **33**:79

Probability current density, 44:299

Probability density, 12:288; **44**:259, 268, 279, 405

Probability density wave, 49:513

Probability distribution, 40:235

    calculation of, **53**:77

    centrosymmetric case, **16**:139–145

    harmonic chain, **53**:75

    Langevin theory, **53**:82ff

    linear response theory, **53**:133

    Liouville equation, **53**:129–131

    non-centrosymmetric case, **16**:145–150, 155

    vibrational energy, **53**:215

    vibrational energy relaxation, **53**:211

Probability function, 38:298, 299

    correlation function, **38**:110, 111

    moments of, **38**:299

Probability measure, 16:154–155, 159–160; **38**:34, 69

    σ-additive, **38**:34

Probability operator measure, 38:65, 80, 85

Probability theory, 38:34, 57

Probe force field, 44:340

Product inhibition, determination of enzyme action, **7**:645

Product orientation, 42:18

Product state distribution, 41:247–249, 257, 258

    multiphoton dissociation, **50**:151

    vibrational, **42**:132; **50**:347

Progress zone, 29:263

Progressive wave technique, 28:211

Projected configuration, 2:311

Projected Hartree-Fock scheme, 14:328

Projection, Hilbert-Schmidt, **53**:367

    linearly independent, **2**:293

    Nakajima-Zwanzig, **53**:344, 347

Projection lattice, 38:21

    Boolean, **38**:25, 26, 39

Projection matrix, 53:90

Projection operator, 2:217, 228, 285ff; **8**:4; **14**:320; **17**:230, 250, 259, 265; **18**:100, 111; **19**:27; **25**:258, 259, 288; **36**:50, 246, 248, 254, 259; **38**:177; **40**:24; **43**:250; **44**:266, 297; **46**:312, 346; **49**:125

    bulk phase, **46**:328

    for non-closed shells, **6**:365

    surface, **46**:298, 322

Projection operator formalism, 2:293; **46**:312; **53**:138, 143–148

Projection postulate, 38:2, 3, 5, 6, 13, 63

Projector, one-dimensional, **53**:355

Promotion, 43:105

Propagator, 25:97, 100ff, 121, 151, 152, 160; **36**:50; **44**:265, 266; **48**:5; **53**:344, 349, 352

    electronic, **25**:151, 161, 165, 168

Propane, hydrate, **2**:10, 33, 43, 46, 47

    thermodynamic data and lattice constants, **2**:8

    + iodoform system, **2**:99

    Langmuir constant, **2**:47

    -water-hydrogen sulfide ternary system, **2**:53

Propeller action, 44:428

Propensity rules, 50:631

Propionic acid, 16:181

Proposition, 38:10, 12ff, 28, 32, 40

    atomic, **38**:10, 16

    classical, **38**:17, 40

    compatible, **38**:13

    definitions of, **38**:10, 13

    generalized, **38**:32

    incompatible empirical, **38**:17

state as, **38**:15
temporal, **38**:11, 12
timeless, **38**:11, 12
truth definite, **38**:28
Propositional system, 38:12, 13, 39
Boolean, **38**:12, 39
compatible, **38**:13
interpretation of, **38**:13
structure of, **38**:12
temporal, **38**:12.13
timeless, **38**:12
Propylene, barrier height of internal rotation, **2**:383
-propane-water ternary system, **2**:44
Propynal, 50:131
Prosthetic group, 7:160
Protein-chloranil complex, 7:254
Proteins, 41:200, 226; **49**:34
active site, **39**:55-108
anion binding, **39**:84
β pleated sheets, **39**:90
binding sites, **39**:77
biological activity, **33**:105
clefts, **39**:77
coding, **7**:135
collagen, **21**:539
conductivity of wet, **7**:253
conformation, **21**:539, 559, 567, 583, 591
conformational changes of, **33**:104-115
cross-linked, **39**:91
crystal, adsorption on, **7**:249
crystallization, **39**:69, 84
denaturation, **7**:216; **46**:363-414
denatured structure, **33**:105-114
dry, **7**:243
energy-transducing, **39**:102
fibrous, ionizing radiation effect on, **7**:259

flexibility, **39**:78, 79, 95
fluctuations of side chains, **39**:73-75, 83, 91
fold, **39**:56, 69, 70, 338
globular, **7**:262; **39**:69, 90
helices in, **39**:90
hydrogen-bonding, **39**:178, 182
hydrophobic regions, **39**:91, 92
incorrectly folded form, **33**:109
interior, **39**:73-75
metal binding to, **39**:86, 87, 88
metals in, **39**:66, 88, 99, 105, 338
motion in, **39**:57, 96, 338, 341
native structure, **33**:105-114
noncovalent interactions, **39**:178, 182, 186
pancreatic, **33**:106
partition function, **7**:229
radiation effect on, **7**:262
random coil, **39**:88-90
refolding of, **33**:80, 98, 104-115
regaining of biological activity, **33**:113
rigid, **39**:78, 79
secondary bonds, **46**:365ff, 379, 384, 394
statistical thermodynamics of, **46**:363-414
structural transition, **46**:365ff, 370, 375, 378ff, 400ff
structure, **46**:365ff
structure and energetics, **39**:55
structure in solution, **39**:73-79
surface, **39**:75-77
swelling, **7**:228
synthesis, kinetic equations for, **15**:191
synthesis rate, **7**:153ff
thermal denaturation, **46**:365, 369, 377, 386, 391, 397
thermal effects on, **7**:201

thermal expansion, 7:224
unfolding of, 33:80, 98, 104–115
x-ray diffraction studies, 39:69–71
see also Macromolecules
Protoanemonin, 16:177
Protohaem complexes, 7:393
Proton affinity, 19:40, 59ff, 191, 193
ammonia, 19:61, 67; 52:347
carbon dioxide, 19:68
deuterium, 19:168
ethyl nitrate, 19:204
HBr, 52:299
HCl, 52:299
HF, 52:299
hydrogen, 19:59, 168
hydrogen sulfide, 19:65, 68, 191
methane, 19:68
water, 19:40, 60, 64, 191; 52:345
Proton magnetic resonance (PMR), 7:511ff
relaxation rates, 7:514, 517ff
spectra of epinephrine, 7:519ff
Proton transfer, 39:184, 186; **47:**Part 2:627–679; **53:**66
isotope effects, 21:244, 246
laser studies of, 47:Part 2:643–679
excited state, 47:Part 2:645–650
generation of pH jump, 47:Part 2:674–676
intermolecular, 47:Part 2:557–573
intramolecular, 47:Part 2:540–557
solvent cage effect, 47:Part 2:673–674
Marcus theory, 21:244
models for, 47:Part 2:639–641
rhodopsin, 47:Part 2:631–636
primary processes, 47:Part 2:632–636

tunnelling, 47:Part 2:637–639
Protonated adenosine, 7:446, 459
Protonated aniline, 7:443
Protonated cytidine, 7:442
Protonated pyridine, 7:443
Protonation, 8:133, 136
Protonmotive force, 19:366
Protons, energy, 39:103
gradient, 39:206
jellium and, 41:172
NMR, 39:128, 145, 156
Protoporphyrin, 7:72
Proximate encounter, 35:143
Pseudoantiresonance, 25:213, 214, 233ff, 270ff
Pseudoatom, 1:183
Pseudocharge density, 49:584
Pseudocontainer, 9:49
Pseudocrossing, 28:338, 339, 340, 366
selection rules, 28:145, 151, 338
Pseudodimer, 16:171
Pseudograph, 43:75
interactant, 43:75
knot, 43:76
species, 43:75
Pseudomolecule, 42:311, 339, 344
Pseudonatural orbitals method, 36:151, 152
Pseudophoton, 45:5, 6
Pseudopotential, 16:283–341, 378; **20:**241, 260; **26:**240; **31:**269, 272, 282, 289, 304; **35:**17; **42:**173; **49:**50; **50:**536, 559
absence of bound states, 16:302
energy-dependent, 31:289, 297
generalized Phillips-Kleinman, 16:289–294
Heine-Abarenkov, 31:290
Hermitian, 16:293
local, 16:317–318

mercury, **31**:304
model, **16**:317–321
nonlocal, **16**:287, 290, 321; **31**:289
oscillation effect, **16**:318
representation of electron-atom scattering, **16**:295–304
representation of Rydberg states, **16**:304–307
screened, **27**:550, 551
structural map, **49**:65, 74
valence electrons, **16**:325–327
weak-scattering liquid metal, **31**:282
see also Electron-solid interaction; Model potential
Pseudoreaction, 43:17
Pseudoresonance, 25:213, 233, 238ff, 270, 276, 288, 291
Pseudorocking curve, 27:36, 38
Pseudorotation symmetry, of linear polyene, **52**:104
Pseudoselectivity, 50:118, 143, 174
Pseudostate, 25:252
Pseudostationary state, 38:326, 330
transitions involving, **38**:332
Pseudosurface modes, 27:467, 472, 497, 518; *see also* Surface modes
Pseudoyohambine, 7:414
Pteridines, 7:55, 59
Puckered sheets, 49:99–106
$A_3S_3H_6$, **49**:102
cadmium halide, **49**:104
$CdCl^+$, **49**:105
GaSe, **49**:99–101, 104
6^3 graphitelike, **49**:100
Pulse echoes, 25:39, 42, 47
Pulse soliton, 53:254, 263
Pulse velocity in Toda lattice equations, 53:300
Pulsed neutron source, 51:234; **53**:38
Pulsed nozzle beam source, 42:445

Pump parameter, 43:226, 228
Pump-probe technique, 49:14, 16–18, 25–27; **50**:363
Pure spin state, 2:227, 230, 307
Pure state, 40:386, 388, 389
Purine, 7:47ff, 51, 97
electron donor properties, **8**:169ff
prebiotic synthesis, **55**:96–98
removal from RNA by heat, **7**:219
Purine-pyrimidine pairs, 7:131; **8**:163
Purpurins, 7:398
Pyncometer, radioactive counting, **7**:224
Pyrazine, 42:226
vapor, **49**:33–34
Pyrene attachment to polyethylenimine, 39:135
Pyridine, 36:329; **47**:Part1:641
energy indices of bases, **7**:54
nucleotide enzymes, **7**:67
Pyridine deuterochloride, 7:444
Pyridine hydrochloride, 7:444
Pyridoxal phosphate enzymes, 7:77
Pyridoxine hydrochloride, 16:186
Pyrimidine, 7:52ff, 97, 385ff
prebiotic synthesis, **55**:96–98
Pyrollidine ring, 16:190
Pyrrole rings, 7:385
Pyruvate dehydrogenase complex, 39:201

Q

QL algorithm, 26:239
Quadratic functional, 32:39, 41, 52, 55
Quadratic hypernetted-chain approximation, 48:223, 238
Quadrature, Gauss-Laguerre, **13**:234, 260
Lobatto, **13**:234, 260

Quadrature fields, **19**:155

Quadruple point, **41**:33, 34, 35

Quadrupole and overlap induction, **44**:457

Quadrupole coupling constant, **5**:322

   nuclear, **37**:243, 246, 260, 261

Quadrupole effects, **2**:25

Quadrupole Hamiltonian, **18**:175

Quadrupole interaction, in carbon dioxide crystal, **5**:178

   nuclear electronic, **18**:199, 208

Quadrupole mass filter, **42**:532

Quadrupole moment, **2**:188; **10**:5–6; **12**:109, 516, 522ff, 535, 537; **16**:380

   heteronuclear diatomic molecules, **53**:30

   molecular, **20**:3, 4; **44**:459

   quantum effect, **12**:138

Quadrupole moment tensor, cartesian, **44**:450

Quadrupole-quadrupole interaction, **12**:577, 581, 592; **16**:380; **17**:128

Quadrupole radiation, **10**:178–179

Quadrupole resonance absorption, **2**:190

Quadrupole satellite, **40**:406, 407

Quadrupole shielding factor, **10**:2

Quadrupole tensor, **40**:404

Quadrupole term, dipole-dipole, **53**:19

Quadrupole transition, **18**:63, 65

Quality factor, charge exchange, **13**:50

   classical resonance, **13**:41

Quantal perturbation series, **36**:30, 34

Quantization, **36**:143

   Einstein-Brillouin-Keller, **36**:52

   semi-classical, **25**:71, 72, 167, 171, 172, 174

Quantization conditions, **36**:17, 26

Quantization rule, Schwarzchild-Epstein, **36**:4

   Sommerfeld-Wilson, **36**:4

Quantized rotation, **44**:353, 371

Quantized vorticity, **33**:6, 7, 18

Quantum beats, **40**:381, 383; **42**:220

   naphthalene, **23**:212

Quantum biochemistry, **7**:3, 44ff

Quantum Brownian motion, **16**:27–29

Quantum chemistry, **33**:157

   crystal surfaces, **9**:85ff

Quantum corrections, classical scattering theory, **10**:53–54

   cluster integrals, **1**:283, 287, 291, 292

   partition functions, **1**:22, 25, 67

Quantum defect, **16**:304, 323; **52**:288

   doubly excited state, **52**:38

Quantum dynamical semigroup, **38**:125

Quantum effects in scattering, **12**:399

Quantum efficiency, **28**:426

Quantum electrodynamics, **12**:167, 176, 195

Quantum electron gas, **6**:86

Quantum electronics, **38**:3

Quantum ergodicity, **46**:121, 139

Quantum field theory, **14**:36

Quantum interference phenomena, **12**:363, 405

Quantum logics, **38**:5

   Hilbert-space model of, **38**:21

Quantum master equation, **46**:269

Quantum mechanical coupled nonlinear oscillator system, **47**:Part 1:161–195

Quantum mechanical distribution function. *See* Distribution function, quantum mechanical

Quantum mechanical measurement process, **46**:153–194

Quantum mechanical trajectory method, **50**:422

Quantum mechanics, **38**:1–108

   algebraic, **38**:5, 21

   applied to DNA, **7**:84

completeness, **38**:52

generalized, **38**:4, 12

Hilbert space model of, **38**:20

lattice-theoretical formulation, **38**:5

limitations of traditional, **38**:3

microclusters, **40**:129

qualities, **38**:10, 28

reduced description, **38**:46

traditional, **38**:12, 46

Quantum numbers, magnetic, **44**:447

Quantum oscillations, 12:424

Quantum scattering, 10:75–134

Quantum systems, finite,

entropy of effective states, **53**:353–356

irreversibility, **53**:343–344

master equation, **53**:344–347, 348–352

dynamical, **53**:368–373

kinematical, **53**:357–362

reduced Heisenberg picture, **53**:352–353

Quantum temperature, 16:30, 35, 52–56

Quantum theory, 32:63

de Boer's, **4**:230, 231

Quantum time, 15:15

Quantum tunnelling, 19:23

Quantum yield, 39:141, 164; **41**:283; **50**:262

Quartz, 27:435

Quasiadiabatic potential energy surface, 42:20

Quasibound levels, 25:235, 238, 267, 284, 288, 291; **42**:387

Quasichemical approximation, 2:122, 123, 128, 137; **4**:202

point defects, **11**:5, 31, 40, 77

Quasicontinuum approximation, 50:107, 134; **53**:305, 311, 312

Quasicrystalline melts, 6:459

Quasidegenerate coupled-cluster

approximation, 52:234–251

Quasielastic scattering cross section, 27:56, 64, 65

Quasiequilibrium condition, 16:2

Quasifree electron, 31:405

effective mass, **31**:405

kinetic theory, **31**:416

Quasigrand partition function, 22:170, 178

Quasiharmonic and relaxation oscillations, Belousov-Zhabotinskii reaction, **55**:209–210

Quasiharmonic interactions, 27:475–476

Quasiharmonic Poley absorption, 44:294

Quasilattice hypothesis, 16:226, 228

Quasilinear approximation, 43:249, 252, 256

Quasilinearized extended coupled-pair many-electron theory, 52:218

Quasilocal algebra, 22:344

Quasinormalized structure factor, 16:150–151, 154, 200, 205

Quasiparticle, 24:197, 198, 202, 204, 207, 208, 235, 237; **36**:223, 244; **53**:284

Quasiparticle gas, 53:243, 244, 247

correlation function, **53**:249

Quasipotential, 16:295

Quasiracemates, 16:170–171

Quasiresonance interaction, 40:395

Quasiresonant charge exchange, 13:48

Quasiresonant Poley absorption, 44:396

Quasispecies, 38:234

Quasispherical approximation, 12:291, 295

Quasispin, 52:3, 99, 100, 108

pairing, **52**:81–89

Quasistatic process, 6:150

Quasithermodynamic network, 43:10

Quasithermodynamics, **1**:12

Quench lamp, **45**:514–515

Quenching, **7**:170; **36**:188, 190, 197; **45**:342–394

absolute cross section, **50**:332

alkali atoms, **45**:346–353

beam detection by, **10**:175

by CO, **45**:369–373

collisional, **30**:375, 380, 418, 432, 442, 490

crossed beam experiments with Na(3^2P), **45**:358–368

by D_2, **45**:369–373

gyroscopic, **10**:351

by H_2, **45**:369–373

inverse, **28**:306; **45**:391

of K*, **45**:392–393

of metal atoms, **50**:325–394

models of, **45**:251–258

Na* + C_2H_4, **45**:379

Na* + CO, **45**:350

Na* + CO_2, **45**:378

Na* + N_2, **45**:352–358

Na* + NO_2, **45**:378

Na* + O_2, **45**:377

of scattering pattern, **42**:460, 482, 483, 486, 490, 492

see also Electronic deexcitation; Rotational deexcitation; Vibrational deexcitation

Quenching kinetics, **47**:Part 2:24–26

Quenching mechanisms, **47**:Part 2:26–27

NO Rydberg states, **47**:Part 2:34–39

by other molecules, **47**:Part 2:31–33

Quirk-Ruppert-Maybee theorem, **43**:64–65, 85, 104, 105–107

R

R-matrix method, **25**:250, 251, 279, 283, 292; *see also* Variational methods, derivative matrix

Rabi approximation, **35**:326

Rabi field, **42**:440

spectroscopy, **42**:361

Rabi flopping frequency, **50**:124, 150, 202, 208, 224, 406

Rabi frequency, **40**:377

Rabies virus, **7**:222

Racah parameters, **4**:175, 176; **5**:40, 70, 201; **7**:364ff; **44**:42, 123

Radial coupling, **42**:21, 26, 90

Radial distribution function, **1**:9, 149ff, 156, 208, 226ff; **9**:11, 16, 40; **12**:583; **20**:52, 242, 289; **34**:112, 168, 173; **46**:289; **49**:497–499, 520

critical region, **6**:181–194

density correlations, **6**:181

diagrammatic methods, **6**:192–194

experimental determination of, **6**:194–197

generalized diffusion equation for, **6**:217

superposition approximation, **6**:185

hard rods, **21**:414

hard spheres, **21**:397, 423

metallic glass, **51**:231

microclusters, **40**:85

Radial wave function, **10**:84–85, 88, 89–91

Radiant heat transfer, **36**:195

Radiation, **8**:179; **43**:17

absorption of, **12**:119

angular distribution, **28**:271

diffusion, **28**:279

electric dipole, **4**:163, 164
electromagnetic, **7**:203
magnetic dipole, **4**:163
mutation induced by, **7**:85
natural, **7**:146ff, 152
polarization, **28**:271
quadrupole, **4**:163
stimulated emission of, **12**:119
trapping, **28**:271, 272, 282, 284, 301
Radiation chemistry, **2**:191
Radiation resistance, **2**:194, 199; **7**:271
Radiation theory, **36**:71
Radiationless transitions, **23**:198, 199;
    38:152
electronic, **42**:65, 207, 211, 332
master equation, **53**:362–363
Radiative capture, **10**:264
Radiative decay, **37**:3; **42**:194
lifetime, **42**:22, 113
Radiative lifetime, **50**:2; *see also* Life-
    time, radiative
Radiative recombination, **30**:276
two-body, **42**:13
Radical attack, **7**:44
Radical complex mechanism, **48**:157
Radical-pair theory, **35**:10
Radical scavengers, **7**:272
Radical substitution, **8**:86ff
Radius of gyration, **52**:434ff
Radius ratio rules, **49**:53–58
Rahman free-molecule theory, **53**:37
Raich-Gillis model, **53**:24
Raich-Mills model, **53**:24
Railway-car soliton, **53**:308
Rainbow angle, **10**:78–79, 137–138;
    12:402; **28**:8, 17
Rainbow effect, **30**:323, 324, 327, 333,
    344
double, **30**:376
Rainbow scattering, **10**:78ff, 86, 103ff,

137ff, 144, 147, 157, 161, 377;
    12:364, 399, 402, 409, 481;
    42:422, 459, 521
maximum, **12**:409
primary, **12**:410, 424
supernumerary, **10**:105, 137; **12**:409,
    410, 424
Raman gain technique, **49**:35–37
Raman scattering, **17**:66; **28**:216; **50**:231
time-resolved, **50**:128
see also Stimulated Raman scatter-
    ing
Raman spectroscopy, **7**:469ff, 476ff,
    478, 484ff; **34**:9, 74; **49**:34–38
anisotropic, **40**:6
benzene, **34**:74
CH_4, **34**:59
collision-induced, **46**:32
isotropic, **40**:5, 19, 36
liquid CH_3I, **34**:85
molecular solids, **22**:207, 277, 285
second order, **24**:51, 55
vibrational, **34**:9
Ramsauer effect, **31**:345; **42**:185
Ramsauer-Townsend effect, **44**:508
Random chain, vibrational modes of,
    13:125, 126, 146, 152, 156, 161,
    162
Random coil, **4**:150; **7**:207
helix-coil transition, **7**:231
power of confusion, **7**:229
Random coil protein, **39**:88
chromagranin A, **39**:89
histones, **39**:89
scavenger for metals, **39**:88
Random distribution, **12**:302; **38**:202
intensity of random source, **38**:204
moments of random source, **38**:204
Random flight, **1**:90, 97, 101; **52**:434
chains, **52**:433–440

Random force, **44**:262, 296; **46**:206; **53**:105

Brownian motion theory, **53**:82

common cage effect, **53**:115

fictitious dynamics, **53**:165

heatbath, **53**:90, 91, 95

Langevin equation, **53**:83–84

Mori, **53**:94, 146

primary system motion, **53**:76, 77

solvent, **53**:78, 111, 112

Random hopping, from discrete sites, **44**:384

Random-mixing approximation for disordered lattice, **6**:130, 135

Random-packed hard sphere structure, **51**:266

Random phase approximation (RPA), **10**:105–109, 117, 120, 127; **26**:251; **27**:273, 275, 346; **36**:248, 259; **41**:124, 125; **42**:325; **44**:601, 608, 622

coupled 2ph RPA, **36**:257

exciton, **44**:612, 614

localized orbital, **44**:611, 631, 633

uncoupled 2ph RPA, **36**:255

Random torque, **44**:357

Gaussian, **44**:268

Random variable, **29**:86

bounded, **38**:58

classical observable, **38**:24

definition, **38**:57

expected value, **38**:58

realization of, **38**:57

variance, **38**:58

Random walk, **1**:368, 373; **4**:1–66; **15**:230; **34**:293; **41**:202, 217, 219, 226; **51**:2, 28, 37, 39, 43; **52**:363–506

bacteria, **52**:488

boundary conditions for, **52**:407–430

continuous time, **52**:369ff, 380, 457, 460, 474ff

nonseparable, **52**:381, 476

relation to master equation, **52**:372–375

domain walls, **53**:269, 272–273

lattice, **52**:366

with longer-range transitions, **52**:412

maximum displacement, **52**:416–427

migration, **40**:400

model for decay of correlations, **15**:199

persistent, **52**:375–377, 489

separable, **52**:370, 374

span-constrained, **52**:424

transport, **40**:372, 438, 455, 478

use of Toeplitz determinants, **15**:334

see also Self-avoiding walk

Randomly connected lattice, **52**:462

Randomness of finite sequence, **38**:53

Rankine balance, **7**:605

Ranunculin, **16**:177

Raoult's law, **4**:307

Rare earths, optical spectra, **4**:162–174

Rare gas, expansivity, **9**:60, 65

heat of fusion, **9**:69

solids, **22**:208, 274

lattice dynamics, **22**:208

Rare gas excimer, **50**:547

Rare gas excimer laser, **42**:34

Rare gas mixture, **17**:24

collision-induced translational absorption, **51**:56–71

Rare gas oxide, potential curves, **50**:563

Rashba effect, **23**:240–241

Rate, of activation, **1**:388

of deactivation, **1**:33

of denaturation, **33**:105

of flow of phase fluid across surface,

13:91
of non-equilibrium processes, **21**:93; *see also* Non-equilibrium processes
of reaction, **43**:17
of refolding, **33**:106
of renaturation, **33**:105
of shear tensor, **1**:148
of strain tensor, **1**:148; **33**:408
of unfolding, **33**:106
Rate coefficient, 48:73
correlation function expression for, **48**:88–94
diffusive, **48**:81
one-way flux, **48**:105
time dependent, **48**:75
Rate constant, 1:19, 127, 128, 166, 179, 181ff, 388, 390; **8**:67ff; **13**:92; **19**:65, 109, 344;
Arrhenius equation for, **1**:17, 65, 120, 388
barrier, **13**:112, 117
bimolecular, **19**:157
chemiluminescent reactions, **42**:9
collision theory, **5**:374, 379
differential, **13**:100
energy dependence of, **19**:157, 161
equilibrium, **13**:93
high-pressure, **13**:94
intrinsic, **48**:78
ion-molecule reactions, **19**:171
organic electrode reactions, **3**:29
statistical, **13**:98
two-state model, dissociation, **33**:99
recombination, **33**:99, 100
sequential process, **33**:103
Rate determining step, 33:100
Rate equation method, 46:262
Rate equations, 33:78, 79, 84, 85, 101, 102

collision theory, **5**:372
irreversible, **24**:176, 178
vibrotational population, **47**:Part 2:57–59
Rate kernel, 48:75
analysis of, **48**:139–149
dynamic process contributing to, **48**:139–143
kinetic theory expression for, **48**:125–127
nonhydrodynamic states contribution to, **48**:147–149
from pair theory, **48**:138–139
projection onto diffusion modes, **48**:143–146
Smoluchowski equation result for, **48**:162–164
structure of, **48**:101–105
t-space expression for, **48**:79
Rate law, 48:73
generalization of, **48**:75
Rational fractions, 25:254
Rayleigh-Brillouin spectrum, 24:19, 38, 41
Rayleigh-Fokker-Planck equation, 20:137, 195, 197; *see also* Fokker-Planck equation
Rayleigh gas, 20:161, 197
Rayleigh kernel, 20:143
Rayleigh law, 20:120, 121
Rayleigh limit, 20:144
Rayleigh number, 13:283, 284; **24**:24; **26**:178, 190, 191; **32**:5, 18, 77, 78, 95, 110, 131, 278
critical, **26**:178, 180ff, 188ff, 193ff, 202ff
high, **26**:209
supercritical, **26**:209
Rayleigh peak, 24:23, 25, 39, 40
Rayleigh piston, 20:135
Rayleigh regime, 20:138

Rayleigh-Ritz method, 7:9; **20**:156, 208; **32**:40, 41; **41**:62, 69, 72, 129

Rayleigh scattering, **34**:13; **44**:355; **50**:194, 231

depolarized, **44**:301, 303, 401

from microclusters, **40**:55

Rayleigh-Schrödinger perturbation theory, **12**:29; **36**:355

application to long-range interactions, **12**:31

generalized Hellmann-Feynman theorem, **12**:53

nondegenerate diagrammatic, **52**:183–187

Rayleigh wave, **27**:357, 471, 496–497

Rayleigh wing, depolarized, **44**:374

Reactance matrix, **25**:225, 249, 257ff, 264, 268, 283, 286

derivative matrix, **25**:249

trial T-matrices, **25**:283

Reactant complex, **43**:74

Reactant isolation, **15**:166

Reaction, chemical, *see* Chemical reaction

complex vector, **43**:73

cycle, **43**:129

electrochemical, *see* Electrochemical reaction

electrode, *see* Electrode reaction

elementary, stochastic models for, **15**:157

enthalpy of, **33**:100, 106, 107, 129, 130

entrance, **43**:90, 129

entropy of, **33**:107, 129, 130

exit, **43**:90, 129

extent of, **43**:26

external, **43**:87

forward, **43**:88

initial bimolecular step, **33**:101, 103

intermediate, **33**:101

internal, **43**:87

kinetic energy operator, **19**:15

matrix, **41**:285

networks, stability of, **43**:1–216

nonadiabatic, **41**:251, 263

proton transfer, **33**:94

reverse, **43**:88

sequential, **33**:88, 98, 101–115

short circuit, **43**:133

simplex, **43**:19

tangent, **43**:129

three-step, **33**:109

two-step, **33**:107

unimolecular, **33**:85, 101

velocity, **43**:17

see also Biochemical kinetics

Reaction barrier, **42**:85

Reaction center, model of, **33**:390, 392

photochemical, electron spin resonance, **19**:357

electron transfer in, **19**:360, 361

fluorescence, **19**:358

Reaction complex, **10**:190, 261, 263, 266–267

persistent, **10**:264–265

Reaction coordinate, 1:28, 29; **7**:209; **19**:11; **48**:86; **49**:203–205

broken path model, **49**:205

curvilinear coordinate system, **49**:205

Johnson's modification, **49**:205

Marcus equation, **49**:204, 223

Reaction diameter, 1:120

Reaction-diffusion equations, **29**:33, 52, 56, 162, 272; **38**:265; **43**:240

pattern formation and, **38**:266

Reaction dynamics, atom-chain collisions, **53**:101–106

clamping approximation and common cage effect, **53**:111–116

correlation functions, **53**:188–206

harmonic chain, **53**:74–76

heatbath modeling, **53**:98–101

heterogeneous, **41**:477–516

iodine recombination in simple solvents, **53**:206–220

irreversible process theory, **53**:124–128

Langevin equations, projection operator method, **53**:143–148

    response function method, **53**:138–142

Langevin model, **53**:81–87

molecular-time-scale generalized Langevin equation representation, irreversible dynamics, **53**:148–172

primary system motion, **53**:76–77

recombination, **53**:106–111

short-time-scale many-body dynamics, **53**:87–89

velocity autocorrelation function, **53**:77–81

Reaction field, **1**:80, 95; **12**:227–229, 233ff, 270; **13**:187; **44**:262, 263

cavity radius, **12**:236, 268, 269

complex susceptibility, **12**:238

continuous medium, **12**:227, 233

continuum model, **12**:228, 235, 239, 260ff, 279

dipole moment, **12**:234

fluctuating moment, **12**:228, 237

higher-order moments, **12**:234

nonpolarizable dipole, **12**:229

polarizable charge distribution, **12**:233

reaction potential, **12**:231, 233, 234, 263, 279

response function, **12**:229, 238, 261

spherical cavity, **12**:229

static moment, **12**:227, 228

static reaction, **12**:267

Reaction kinetics, **10**:128; **33**:415

coupling to fluid dynamics, **13**:21, 25

hypersonic flow, **13**:19

Reaction mechanism, collision complex, **30**:37, 66, 253, 265, 297, 302

electron jump, **30**:60, 67, 249, 257, 260, 263, 301, 507

four center, **30**:260, 262, 300

impulsive, **30**:60, 66

rebound, **30**:248, 252

stripping, **30**:60, 64, 201, 248, 250, 275, 284

Reaction operator, **8**:3; **14**:300, 316, 335; **36**:350

finite systems, **6**:345

Reaction probability, **13**:98; **30**:12, 34, 61, 94, 121

Reaction product energy distribution, **47**:Part 2:85–131

CO laser-probing technique, **47**:Part 2:86–90

combustion-related reactions, **47**:Part 2:116–126

E → V energy transfer involving CO, **47**:Part 2:111–116

experimental details, **47**:Part 2:86–91

laser-induced fluorescence measurements, **47**:Part 2:90–91

photodissociation reactions, **47**:Part 2:91–111

Reaction rate, **3**:139, 151; **33**:423

absolute, **1**:30, 124, 362, 391; **13**:94; **33**:117

    variational theory of, **13**:95

catalytic, **21**:791

flow analogy, **13**:89

non-equilibrium effects, **6**:55

statistical theory of, **13**:98

termolecular, **30**:41, 48

tunnelling corrections to, **3**:131ff

upper limit to, **13**:89

variational theory of, **13**:85, 89, 95

see also Absolute rate theory; Activated complex theory; Rate; Transition state theory

Reaction threshold, 30:94

Reactive scattering, 10:128–131, 319–393; **25**:257, 259, 268

angular distribution in, **10**:328–340, 360–361

apparatus for, **10**:322–328

$Ar + H_2$, **49**:292–295

cross section, **10**:128–131

differential, **10**:362–366

multichannel, **10**:128

ion-molecule reactions, **10**:259–262, 267, 269

total, **10**:129, 157–165

harpooning, **10**:374–375

reduced, **10**:157

e-H system, **49**:270–271

electric deflection analysis of, **10**:347–356

energy disposal in, **10**:359–360

$F + H_2$, **49**:289

$H + H_2$, **49**:289–292

Hulbert-Hirschfelder model, **49**:269–270

$I + H_2$, **49**:231–232

magnetic deflection analysis of, **10**:341–343

mechanisms of, **10**:321–322, 361–367

optical potential model, **10**:128, 351, 362, 366–367

quantum-mechanical approximate treatments, **49**:191–310

velocity analysis of products, **10**:343–347

see also Chemical reactions

Reactive systems,

one-dimensional models, **49**:200, 206–216

exchange reactions, **49**:212

single curve model, **49**:209–216

one-dimensional to three-dimensional transformation technique, **49**:239, 257–261

information theoretic approach, **49**:257–258

rotation-decoupled-vibration model, **49**:258–261

quantum-mechanical approximate treatments, **49**:191–310

Reactivity index, 33:346, 347, 368, 381, 382

Reactivity number, 8:84

Reactor theory, 20:137

Read-in-read-out technique, 49:15

Real-angle modes, 27:383

Real fluids, heats of vaporization, **9**:62

isothermal compressibility, **9**:53

statistical thermodynamic theory, **9**:28ff, 53

thermal expansivities, **9**:58

Rearrangement operator, 30:57

Rearrangement reactions, 16:192–197

Rebound mechanism, 10:321, 329, 358, 361–367

Reciprocal lattice, 6:114; **27**:5, 152, 154

Reciprocal lattice vector, 20:220, 230, 253; **27**:448

Reciprocal relations, 17:79, 83

Reciprocal systems, conformal ionic solution theory, **11**:109

excess free energy of mixing, **11**:110, 112

Reciprocity relations, 16:104; **33**:414

Onsager-Casmir, **33**:414, 415

Recombination, 34:273; **35**:20; **36**:187,

196; **42**:1; **53**:106–111

atom-atom and atom-molecule, **28**:33, 34, 35, 46

atomic, **13**:110

dissociative, **13**:29

electron, **13**:112

electron-ion, **13**:35; **28**:390, 397, 404, 409, 458

geminate, **48**:155

ion-ion, **28**:459

radiative, **28**:36, 37, 38, 87

rate-determining step, **33**:100

three-body, **13**:102; **28**:33, 34, 35, 43, 413, 414

see also Ion recombination

Recombination coefficients, 23:3, 4, 5, 19ff, 28, 31

bimolecular, **23**:10, 11, 13

Recombination energy, 10:252–253

Reconstruction, 27:468, 480, 509–517, 524

thin films, **27**:514

Recontinuation rate, 35:20

Recoupling coefficient, 44:41

Recrossing corrections, 13:118, 119

Rectangle function, 44:329

Recurrence, 36:49

Recurrence frequency, 46:111

Recurrence relations, Bessel functions, **13**:255

Clebsch-Gordan coefficients, **13**:253

expansion of Slater-type orbitals, **13**:263

Fourier transforms, **13**:261

Legendre functions, **13**:252

Recursion formula, 4:54, 57

Recursive cubic approach, 54:101–114

Redich-Kwong equation, 37:172

Reduced dimensionality, 43:261, 263

Reduced equations of motion for multi-

photon excitation, 47:Part 1:509–553

Reduced Hessian matrix, eigenvalues, **54**:156–157

Redundant variables, multiconfigurational self-consistent field state, **54**:143–145

Ree-Eyring equation, 21:273

Reencounters, 35:14

random, **35**:19

Reference frame, body-fixed, **49**:162–166, 169, 181

Eckart, **37**:308; **49**:183

space-fixed, **49**:130

see also Coordinates

Reference state, 9:323, 353

Reflectance, 27:290, 293, 330, 332, 436–440, 462, 501

attenuated total, **27**:339, 445–449, 519–523, 525–526

minimum, **27**:409, 439, 519, 521–523, 525

Reflecting boundary condition for random walk, 52:407, 428–430

Reflection, surface excitons, **54**:319–320

total internal, **27**:445

Reflection high energy electron diffraction (RHEED), 27:29

Reflection spectroscopy, 23:257, 312; **27**:554

anthracene, **23**:312, 317

dye crystals, **23**:331–333

Reflectivity, 27:559

optical, **51**:262, 265

Refraction, 1:335

Refractive index, 1:82, 98, 101, 310, 335, 339–341, 350–352; **4**:126, 128, 152; **12**:119, 121, 122, 135, 136; **17**:239; **44**:303

Regeneration, hydra, **29**:257

wave-broom model, **29**:269

Regge-Watson-Sommerfeld transformation, 30:347

Region of dominance, 43:186

Regression hypothesis, 43:250; 48:1116; 53:82–83

Regular region of phase space, 36:36, 38, 41, 42

Regular solution, 1:1, 12, 13; 17:27

Regular spectrum, 36:2, 36, 42, 43, 53

Regulation, 29:35, 42

biological systems, 55:248–252

continuous and discontinuous descriptions, 55:249–250

feedback loops, 55:250–252

homeostatic or epigenetic, 55:248

types of systems treated, 55:250

epimorphic, 29:257

morphallactic, 29:257

pattern formation, 29:255, 257, 278

Relative coordinates, 12:10

diatomics, 12:10, 11

polyatomics, 12:10

Relativistic corrections, 44:115

Relativistic dispersion energies, 12:78

Relativistic effects, 41:81; 44:52; 50:540, 557, 563

many-electron theory, 6:406–407

Relativistic electron systems, electron gas, 54:243–255

inhomogeneous, Breit interaction approximation, 54:296–297

correlation energy, 54:294–296

Coulomb and transverse energies, 54:297–298

density functional formalism, 54:232–298

exchange-correlation energy, 54:255–258

formal theory, 54:240–258

numerical calculations, 54:258–273

correlation energy, 54:258–265

mercury and fermium, 54:265–273

dielectric constant expressions, 54:290–294

renormalization, 54:239–240

spin-density theory, 54:273–288

effective exchange potentials, 54:282–288

exchange energy, homogeneous spin-polarized electron gas, 54:279–282

spin-density-functional formalism, 54:273–279

spin-density-functional formalism, 54:238–239

survey of earlier work, 54:234–237

Relativistic interaction energies, nonretarded, 12:90

retarded, 12:86

Relativistic intermolecular forces, 12:71, 75

Breit-Pauli Hamiltonian, multipole expansion of, 12:75

H_2, radiative corrections, 12:74

at large separations, 12:75

magnetic dipole interaction, 12:85

retarded dispersion energy, 12:78

Relaxation, 23:106; 26:111; 27:474, 483; 35:121, 125; 41:236

atomic, 23:106

chemical, 33:85

cross, 11:298

differential, 23:107

effect (in conductance), 11:217, 223, 230, 246

effects of impurities, 41:291

enhanced, 35:171

excited state, 16:68–70; 49:17

extra-atomic, **23**:107
final state, **16**:68–70
ions, **41**:280
large molecules, **41**:271
molecular, **23**:107, 189ff
Mountain peak, **24**:16
nuclear paramagnetic, dipole-dipole
    temperature, **11**:298, 303, 315
    many-spin problem, **11**:295, 300
    solids, **11**:289ff, 296, 300
    stochastic approach, **11**:291
orientation, **49**:17
oscillatory, **35**:154, 164
polyatomic molecules, **41**:293–297
role of rotation, **41**:288, 302
rotational, **23**:46–49
spin-spin, **11**:305
studied by picosecond spectroscopy,
    23:189ff
temperature dependence, **41**:288,
    290
trajectory models, **41**:289
via real intermediate states, **41**:297–
    302
vibrational, **23**:41ff
    small molecules in dense media,
    42:305–352
violent, **35**:121, 125, 164, 181
Zeeman temperature, **11**:298, 303
see also Electronic deexcitation;
    Rotational deexcitation; Vibra-
    tional deexcitation
Relaxation energy, 49:623–624
Relaxation equations in collision theory,
    5:367, 369–370
Relaxation frequency, ultrasonic
    oligomers, **33**:129, 131, 132
    sensitivity to chain structure, **33**:127
    vinyl-polymers, **33**:127, 129, 133
Relaxation matrix, 37:75; **43**:248
Relaxation measurements, 7:505ff, 514,

515ff
Relaxation methods, 12:377
Relaxation processes, 24:14, 20, 41, 43,
    46
Relaxation property, 24:113, 128
Relaxation theory, Onsager, **11**:246
Relaxation time, 1:78, 84–88, 107, 109,
    112, 113, 115; **4**:327–330; **7**:500;
    11:323; **17**:92; **28**:184, 185;
    31:157; **35**:136, 165, 179, 181,
    182; **46**:265, 336; **51**:67, 257
    angular momentum, **34**:63
    associate amplitude, **33**:86, 87, 110,
    119
    bead and spring model, **33**:77, 125,
    137
    chemical, **48**:74
    cluster in a melt, **6**:476
    correlations of one-dimensional
    chain, **15**:316
    density dependent, **46**:336
    diffusing systems, **15**:145ff
    dipole-dipole, **11**:309
    elastic sphere, **33**:125, 142
    glass transition, **48**:404, 412, 416,
    419–421
    helix-coil transition of oligonucleo-
    tides, **33**:99, 104
    Ising chain, **33**:84, 86, 87, 90
    local, **33**:128–136
    longest, Ising lattice, **33**:90
    sequential process, **33**:104, 110
    three-state model, **33**:107
    longitudinal *see* T_1
    macroscopic, **44**:263
    mean, of Ising chain, **33**:87
    mixed crystals, **2**:191
    from nuclear magnetic resonance
    and fluorescence depolarization,
    33:135
    momentum, **31**:170, 195, 208, 224

nuclear spin, **17**:67
orientational, **31**:157, 196
rotational, **33**:134
single, **17**:167
spin, **17**:171; **31**:172, 195
spin flux, **31**:172, 224
structural, **48**:506
temperature dependence, **33**:117, 128
thermal, **31**:171, 195
TMMC, **53**:259
transverse, *see* T_2
two-state model, **33**:115–120
Renaturation, **46**:364
Renin-angiotension system, **26**:166
Renner effects, **9**:170ff; **12**:9
Renner-Teller effect, **44**:18, 27, 48
linear triatomic molecules, **52**:117–179
Renner-Teller Hamiltonian, **52**:124–128, 167
Renner-Teller parameter, **52**:119, 137, 143, 169
Renormalization group, **37**:184; **40**:196, 203, 233, 249, 271; **41**:207; **49**:312, 318, 352; **51**:25, 39, 42; **52**:493
Renormalization theory, **33**:201, 202, 209, 210; **38**:185, 189
Renormalized cluster theory, **34**:139
Renormalized interaction, **36**:271
Renormalized potential, *see* Potential, renormalized
Reorganization, **36**:238, 330
Reorganization energy, **44**:492
Reorientation, correlated, **34**:14, 19, 58, 82
models of, anisotropic small step, **34**:48, 80, 91
collision frequency, **34**:48
correlation function, **34**:51

cyclic Markov chain, **34**:37
diffusion, **34**:96
eight well, **34**:44
extended diffusion, **34**:26, 47, 58, 96
fields of arbitrary symmetry, **34**:44
free rotor, **34**:21, 59, 96
inertial, **34**:20, 59
internal rotation, **34**:44
isotropic small step, **34**:30, 45, 96
jump model, **34**:37
large step, **34**:47
one-dimensional rotational diffusion, **34**:49
random axis, **34**:38, 44
rotational diffusion, **34**:51
rotational Langevin equation, **34**:53, 71
single axis, **34**:39, 40
stochastic, **34**:37
Repeated measurements, **46**:186
Repeated randomness assumption, **15**:68
Repeller effects in ion-molecule reactions, **10**:260–262
Replication, **29**:126
Representability problems, **24**:213, 216
Representations, B* algebra, **38**:37
canonical cyclical, **38**:38
cyclic, **38**:37
disjoint, **38**:38
factor, **38**:38
GNS, **38**:38
irreducible, **38**:37
quasiequivalent, **38**:38
spectral, **38**:154, 155
subrepresentations, **38**:38
unitary equivalent, **38**:38
vector, **38**:49

Repression, **29**:40, 50

Repulsion envelopes of "globular" organic molecules, 6:465

Repulsion integrals, 1:183, 244, 250

Repulsive force, 10:30, 32–33, 68–70, 82; **48**:96

anisotropy of, **42**:423

electrostatic, **48**:336–338

screening, **34**:109, 125

Reserpine, 16:188–189

Reservoir state, 42:15

Residual excess entropy, 16:275

Residual Gibbs free energy, 16:276, 279

Resistivity, electrical, of metallic glass, 51:239–243

Resolution of quantum mechanical measurement, 46:156, 158, 165, 172, 188

Resolvent, 14:304

Resolvent operator, 11:167, 283

Resonance, 7:116; **25**:213, 233ff, 254, 267ff, 282, 288, 291; **35**:298, 299, 303, 318, 326; **36**:41; **42**:182; **46**:92, 93, 100

compound state, **30**:21, 30, 66

Feshbach, **30**:104

half width, **46**:98, 103

isolated, **46**:93

overlapping, **46**:93

shape, **30**:21, 30, 66, 99, 330, 369

Resonance energy, 5:18ff; **12**:127; **33**:352, 356, 362, 375

Resonance fluorescence, 28:51, 77, 270–273

Resonance integral, 1:183, 244, 250; **5**:25, 28; **7**:124

forbidden transitions, **5**:247

Su, Schreiffer, and Heeger model, **53**:289

Resonance interaction, 12:40, 41, 171, 199, 202

as dipole-dipole transfer, **53**:281

energy transfer

molecular crystal, **53**:277–287

polyacetylene, **53**:287–293

molecular crystal, **53**:228

molecular vibrations, **53**:319

weak, **53**:284–286

Resonance line, emission, 30:437, 438

polarization, **30**:425, 431, 437

Resonance operator, 44:265, 294 ┆

Resonance radiation, 28:269–316

detectors, **28**:276, 277

trapping, **28**:271, 272, 282, 284, 301

Resonance structures, 2:189

Resonance transfer, 7:571

Resonant cavity, 18:244

Resonant charge exchange in atomic collisions, 13:41

Resonant delta function, 44:450

Resonant reabsorption, 24:29

Resonant state, 42:195

Resonant transfer, vibrational, 40:13–15, 18, 34, 39, 43, 45

Resonator cavity, chloroplasts in, 7:630

Respiration, cellular, 29:148

oscillation in, **29**:148, 149

Respiratory coenzymes, 7:67

Response function, 12:236; **33**:242; **36**:52, 248, 249; **41**:114; **44**:149, 340; **53**:138–142

adiabatic frequency, **44**:156

chain representation, **53**:156–161, 178–179

continued-fraction representation, **44**:175

damping kernel, **44**:163

Einstein frequency, **44**:155

equations of motion, **53**:204

generalized Langevin equation, **53**:138–142

heatbath, **53**:152–153

hierarchy of, **53**:215

linear, **41**:113, 116, 134

linear response theory, **53**:132–138, 184–187

local density of states, **44**:189

molecular-time-scale generalized Langevin equation, **53**:93, 154–156, 176–178

normal modes, **44**:188

retarded density, **41**:178

spectral density, **44**:154

static linear, **41**:112–118

Resting potential, 29:363

Resting state, 29:346

Restoring force constant, 44:405

Restoring torque, 44:278

Restricted Hartree-Fock method, 36:219, 265; **45**:229, 234–236

Restricted walk, 15:230, 234

Reststrahl frequency, 27:360

Retardation, 27:336, 425, 435, 488, 503

Retardation effect, 10:119; **12**:139, 170; **27**:282, 292, 295, 303, 381

covariant calculations, **12**:182, 195

dipole resonance energy, **12**:88

dipole-dipole interaction, **12**:80, 82, 175

dispersion energy, **12**:8, 78

fine structure splittings, **12**:89

magnetic dipole interactions, **12**:85

noncovariant calculations, **12**:187

nondispersive, long-range limit, **12**:209

nonresonant interactions, **12**:85

off-resonance energy, **12**:176

^1S - ^1P, **12**:86

^2S - ^2P, **12**:89

with spin, **12**:221

Retarded motion expansion, 35:96, 111

Retina, 29:343

Return to origin, 52:365, 391–393, 396, 492

Reversibility, 26:113; **43**:78; **46**:185

collision-induced electronic relaxation, **47**:Part 2:360–362

principle of microscopic, **4**:6

Reversible electronic relaxation, vibrational relaxation by, **47**:Part 2:362–363

Reversible kinetics on linear lattices, 15:281

Reynolds equation, 32:48

Reynolds number, 10:285, 300; **32**:47, 89

Rhenium hexafluoride, 9:276ff

Rheological properties, solutions of macromolecules, **35**:31–118

dynamic viscosity, **35**:106, 111

second order fluid constansts, **35**:96, 100, 111

viscometric functions, **35**:96

Rhodamine B, 7:266; **19**:372

Rhodamine 6G, vibrational relaxation, **23**:215

Rhodium, 41:141

Rhodium hexafluoride, 9:198

Rhodofying carbonyl group, 7:386

***Rhodopseudomonas spheroides*, 19**:355, 370

***Rhodopseudomonas viridis*, 19**:355, 373

Rhodopsin, 29:344

primary processes in, **47**:Part 2:362–363

proton transfer, **47**:Part 2:631–636

***Rhodospirillum rubrum*, 19**:355, 363

Rhombicosidodecahedron, 40:81, 84, 85

Rhombicuboctahedron, 40:71

Ribitol dehydrogenase, 7:712ff

Ribonuclease, 7:246, 462

Ribonuclease A, 33:106, 109, 111–113;

39:187–190

Ribonucleates, infrared spectra of, 7:492ff

Ribonucleic acid (RNA), 7:84, 436
magnetic properties, 7:351
nucleic acids, 7:460
photoreactivation, 7:586
scission, 7:219
secondary structure, 7:436
synthesis rate, 7:153
trinucleotides, 7:135, 136

Ribose, infrared spectra, 7:494

Ribose pyrophosphate adenosine, 16:192

Ribosomes, 7:219

Ricatti equations, 35:289

Rice-Allnatt theory, 24:285; 31:158, 199
doublet distribution function, 24:275
doublet kinetic equation, 24:275
generalized, 24:289
rough spheres, 31:199–201
singlet kinetic equation, 24:275
spherical molecules, 31:158, 167
transport coefficients, 24:276

Riemann surface, disjoint, 17:53
function with essential singularity, 17:54
function with logarithmic singularity, 17:54

Rigid band theory, 51:263

Rigid convex body constants, determination of, 5:163

Rigid dipole reorientation, 44:396

Rigid dumbbell model, 35:47, 59, 88
effective free, 35:59
metric tensor components, 35:47
momentum flux, 35:88
second-order fluid constants, 35:101

Rigid ion model, 27:488, 491, 496, 528

Rigid lattice model of solvent structure,
6:436, 452

Rigid rod, 22:22, 47, 61

Rigid rotor, 36:179, 182; 49:183, 185
linear, 4:230, 248, 250, 252, 261

Rigid shell model, 42:461

Rigid sphere model, 1:154, 163

Rigid sphere potential, 10:102
low-angle scattering by, 10:106–109, 113

Rigid symmetric top, 4:230, 248, 250, 256, 263, 278, 299

Ring chelates, chemical reactivity, 7:402ff

Ring closure in polymers, 51:10

Ring collision event, 48:107

Ring diagrams, 11:199, 230, 242, 247; 41:105, 106

Ring ligands, 7:380

Ripples, 41:201

Ripplons, 24:55, 57; 46:319

RISM model, 34:236

Rittner potential, 10:370–371; 50:547

Ritz-Galerkin method, 32:42

Ritz method, 32:40, 42

Ritz's variational principle, 49:118

RKR inversion, 42:355

RKR potential, 50:7, 41, 58

RNA, see Ribonucleic acid

Rock salt structure, 49:54, 67
break up of, 49:95–99
GeP, 49:96
$Hg_3S_2Cl_2$, 49:97
$La_2Be_2O_5$, 49:97

Roginsky-Zeldovich equation, 7:247ff

Roginsky-Zeldovich kinetics, 7:255

Rolls, convective, 26:181, 183, 187–191, 194, 205, 209
circular, concentric, 26:194–197, 204, 206–208
counter-rolls, 26:207

Roof-top reflector, **44**:332

Root-mean-square angular velocity, **44**:260, 309, 355, 372, 399

Root-mean-square error, **49**:613

Root-mean-square torque, **44**:280, 350

Roothaan's equations, **7**:33; **21**:34, 35

Rosenfeld formulation, **4**:73–87

Rosenthal effect, **30**:436

Rotary diffusion constant, **1**:110

Rotary-inversion axis, **20**:11

Rotating principle axes of inertia, **44**:360

Rotating system, **35**:132, 155, 166, 181

Rotating wave approximation, **38**:149; **40**:376; **50**:401

Rotation, **7**:203; **25**:254, 282; **36**:17, 22

Rotation correlation function, **44**:301

Rotation-decoupled-vibration model, **49**:258–261

Rotation diagrams, **27**:38

Rotation group, **49**:123

Rotation-translation coupling, **42**:21, 26; **44**:267, 284, 439, 440

Rotation-vibration energy states, **44**:447

Rotational absorption cross section, **44**:308

Rotational autocorrelation function, **44**:292

   velocity, **44**:377, 391

Rotational average, **49**:177

Rotational barrier, **13**:105

   maximum, radius of, **13**:106

Rotational Brownian motion, **44**:267

Rotational collision number, **34**:61

Rotational collision processes, **42**:271–304

Rotational constants, **37**:319, 320

Rotational coordinate transforms, **13**:211, 241, 248

Rotational deexcitation, **28**:4, 173, 180ff, 185–193, 222, 236–243, 256ff

C_2H_4, **28**:256

CH_4, **28**:237, 256

Cl_2, **28**:186, 237

CO, **28**:236, 237

CO_2, **28**:237, 257

DCl, **28**:243

DF, **28**:243

H_2, **28**:186, 188, 189, 192, 193, 222, 237–242

H_2O, **28**:256

H_2S, **28**:256

HCl, **28**:237, 242, 243

HCN, **28**:256

HF, **28**:243

Li_2, **28**:242

N_2, **28**:188, 189, 237

NaCl, **28**:4

NH_3, **28**:256

NO, **28**:186, 237

O_2, **28**:188, 189, 237

OCS, **28**:257

Rotational diffusion, **24**:43; **31**:86, 257, 259; **44**:274, 391

   anisotropic, **44**:379

   Brownian, **44**:276

   Debye model, **44**:301

   inertialess, **44**:257

   small step, **44**:380

Rotational diffusion constant, **1**:110; **17**:171; **34**:50, 52, 55, 78; **44**:357

   hydrodynamic theory for, **34**:55, 81, 92

   inertial effects, **34**:52, 78, 92

Rotational diffusion equation, **34**:50, 54

Rotational disorder in crystals, **6**:466–467

Rotational dynamics, **44**:258, 277, 478

Rotational energy, **7**:205

   kinetic, **44**:344

   promoting reaction, **42**:86

Rotational energy levels, 18:209

Rotational energy transfer, 42:422; 47:Part 2:258–263; 50:589ff

Rotational equilibrium assumption, 47:Part 2:65–72

Rotational excitation, 10:126; 18:15, 70; 25:118; 28:13, 39, 47–62, 81, 95, 296; 30:227, 258, 377

   Cl_2, 28:186

   CO_2, 28:223

   CsI, 28:28

   DBr, 28:83

   DCl, 28:83

   DI, 28:83

   H_2, 28:186, 188, 189, 192, 183, 203, 204, 252

   HBr, 28:83

   HCl, 28:59, 83

   KI, 28:28

   N_2, 28:188, 189

   NaCl, 28:4

   NO, 28:186

   O_2^+, 28:224, 225

   O_2, 28:186, 224, 225

Rotational friction coefficient, 17:172

Rotational friction tensor, 34:56; 44:357

Rotational Hamiltonian, 18:187, 203

   matrix elements, 18:204

Rotational hole filling, 50:111, 140

Rotational invariance, 48:230; 49:161–165

Rotational isomers, 7:416ff; 33:75, 77, 126, 129, 147; 52:365, 443

   structure determination, 7:485

Rotational line shape in induced absorption, 51:80

Rotational magnetic moment, 12:134

Rotational nonequilibrium, 47:Part 2:70–72

Rotational partition function, 2:24

Rotational perturbation, 41:298

Rotational polarization analysis, 30:258

Rotational predissociation of diatomics, 47:Part 1:20–21

Rotational rainbow, 51:194, 211

Rotational relaxation, 42:15, 243, 444

Rotational state distribution after multiphoton dissociation, 50:164

Rotational states, distribution of, 47:Part 1:261–263

Rotational strength, 4:80–83, 92–111

   polymer, 4:128–158

   reduced, 4:96

Rotational temperature, 31:160, 164, 170, 186–189; 42:26, 444; 52:265, 272, 307

Rotational transitions, 2:377, 379, 380; 30:331, 332, 337

   collision-induced, 42:199, 271ff; 50:589–642

Rotatory dispersion, 4:68, 73–82, 86, 92, 95, 105, 109, 128, 147, 153–155

   infrared, 4:147

Rotatory strength, 44:555, 562, 575, 577ff, 595, 617, 622

Roton, 33:7, 19, 20, 22, 42–45

   creation by charges, 33:7

   density, 33:20

   effective Hamiltonian, 33:45

   excitation spectrum, 33:19

   kinetic restrictions, 33:42, 43

   localization around k_0, 33:19, 43–45

   mean free path, 33:22

   pairs, 24:51

   spectrum, 24:52

Roton-phonon crossover, 33:20, 32

Rotor, asymmetric, 44:11, 351

   linear, 31:160, 163, 166, 180, 182, 205, 223

   rigid linear, 4:230, 248, 250, 252, 261

   spherical top, 31:160, 171

see also Free rotor

Rototranslational correlation function,
44:255–482
  aluminum oxide, 44:396
  ammonia, 44:351
  argon (liquid), 44:404, 410, 417
  benzene, 44:259, 261, 351, 353, 388
  benzonitrile, 44:348
  broad-band spectroscopy, 44:320
  bromobenzene, 44:400
  carbon tetrabromide, plastic crystals,
    44:474
  carbon tetrachloride, 44:261, 309,
    351, 353, 386, 388, 471
  trans-carotene, 44:396
  $CBrF_3$, 44:384
  $CClF_3$, 44:384
  CD_4, 44:445
  CH_2CF_2, 44:348, 350, 351
  $(CH_3)_3CCl$, 44:391
  CH_2Cl_2, 44:257, 303, 309, 342, 371,
    385ff, 398
    in glassy decalin, 44:398
  CH_2Cl_2 (liquid), neutron scattering,
    44:373
  CH_3Cl, 44:348, 350
  CHF_3, 44:384, 385
  2-chloro-2-nitropropane, 44:363
  chlorobenzene, under kilobars of
    external pressure, 44:399
  chloroform, 44:333
  cholesteric mesophase, 44:385
  cholesteryl lineate, 44:385, 388
  cholesteryl oleyl carbonate, 44:388,
    390
  $(CN)_2$, 44:445, 457, 458, 459, 465
  CO, 44:404, 410
  CO_2, 44:445, 458
  CS_2, under kilobars of external pres-
    sure, 44:399
  CS_2 (liquid), 44:470, 473

4-cyano-4'-*n*-heptyl biphenyl (7CB),
    44:386, 388
  cyclohexane, 44:351, 353
  cyclooctene, 44:382
  decalin, 44:258, 388, 398
  p-dimethoxybenzene, 44:375, 376
  dimethyl carbonate, 44:375
  doped silica, 44:396
  ethane, 44:350
  ethylene (compressed gas), 44:465
  fluorobenzene, 44:259
  furan, 44:393
  halobenzenes, 44:367, 400
  hexasubstituted benzenes, 44:392,
    393
  HF, 44:263
  hydrogen halides, 44:351
  hydrogen, 44:446
  methane, 44:445
  methoxy and methyl-group internal
    rotations, 44:375
  p-methoxybenzylidene-*p*'-*n*-
    butylaniline (MBBA), 44:382, 384
  methyl carbon relaxation time,
    44:379
  methyl halides, 44:355
  N_2, 44:261, 267, 446, 464, 465
    simulations of, 44:407
  naphthalenes, 44:399
  NH_3, 44:263
  nitrous oxide, 44:334, 445
  O_2 (compressed gas), 44:453
  O_4 formation, 44:457
  pentachloronitrobenzene (PCNB),
    44:392, 393
  poly(ethyleneterphthalate), 44:321
  propyne, 44:384
  selenium (amorphous), 44:396
  SF_6, 44:263, 446
    mixtures, 44:351

o-terphenyl, **44**:398

tetrachloroethylene, **44**:376

tetrahydrofuran in decalin, **44**:400

Th^6(Pa^3) monoclinic phase, **44**:475

vanadium phosphate glass, **44**:396

water, **44**:353

Rough annuli, 44:370, 401

Rough-hard sphere, 44:477

Rough sphere fluid, 24:338; **31**:168;
44:285, 374, 401

bulk viscosity, **31**:171, 172, 175, 201

collision integrals, **31**:169

collisions, **31**:168

Enskog theory, **31**:169

relaxation times, **31**:170

Rice-Allnatt theory, **31**:199–201

self-diffusion coefficient, **24**:339

shear viscosity, **31**:170–172, 175, 201

spin flux, **31**:178

thermal conductivity, **31**:174, 177, 201

vortex viscosity, **31**:171, 176, 201

Roughening transition, 40:190, 191, 194ff

Rouse matrix, 35:49

Rouse model, 35:47, 48, 102

metric tensor components, **35**:47

momentum flux, **35**:65

second-order fluid constants, **35**:102

Routh-Hurwitz theorem, 43:54–55

Rovibronic relaxation, 50:26

Rowin's theory, 7:76

RRK model, 50:281, 297

RRKM theory, 21:145; **30**:45, 46, 215, 298, 300, 303; **47**:Part 2:313; **50**:113, 147, 181, 313, 364, 450

lifetime, **42**:127, 141, 159, 223

Rubber, 1:89

elasticity, **22**:5, 84, 85, 88, 94, 100, 126

Rubidium, 41:141

Rubidium (Rb$_2$), 41:152

dipole resonance interaction energy, **12**:4

Rubidium bromide, 27:438

Rubidium chloride, 27:438

Ruby, energy levels, **5**:129, 130

optical spectrum, **5**:57, 58

Rudenberg approximation, 18:266, 290

Ruelle bound, 40:301

Ruelle theorem, 40:297, 325, 361

Rule of geometric mean, 5:28ff

Runga-Kutta-Gill method, 17:123, 215

Runge-Lenz vector, 52:58, 66, 68

helium, **52**:34

linear Stark effect, **52**:11

SO(4) symmetry of hydrogen atom, **52**:9

Russell-Saunders coupling, 4:163; **5**:39, 41, 42, 46; **7**:197; **28**:116, 117, 118; **42**:26

Ruthenium, 41:141

Ruthenium hexafluoride, 9:302ff

Rydberg-Klein-Rees method, 12:351

Rydberg-Klein-Rees potential, 12:21

Rydberg series, 42:13, 175, 579; **52**:288

carbon dioxide, **52**:330

carbon disulfide, **52**:307, 309

N$_2$O, **52**:333

NO, **52**:288

Rydberg states, 16:285; **41**:275; **50**:138, 257, 268, 303, 568

beryllium and magnesium, **16**:322–324

calculation of energies, **16**:304–307

nitrogen and benzene, **16**:312–316

orbitals, **9**:357

Rytov splitting, 44:296

S

σ bond, **7**:18

in crystal field spectra, **8**:22ff

σ donor, **7**:363

σ-field, **38**:57

σ orbital, **1**:173, 176

s-d interaction, **5**:235

s-fragment, **4**:221

S-matrix, **10**:126–128; **25**:71, 102ff, 120ff, 139, 146, 150, 174; **30**:38, 40, 93, 99, 347, 481; **49**:240–247, 250

body-fixed coordinates, **49**:240, 242–246

classical, **30**:20, 59, 78, 86, 90, 93, 94; **36**:65; **42**:280, 281, 433

classical trajectory, **33**:276

forbidden processes, **30**:35, 114, 117, 118

L-space, **33**:255

space-fixed coordinates, **49**:240–243, 246

s-p-d hybridization, **51**:259

s-polarization, **27**:371, 381, 409, 491, 502

S - T_{-1} mixing, **35**:11, 13, 24

S - T_0 mixing, **35**:11, 16

Sachs-Teller effective mass, **53**:52

Saddle point, **36**:145, 149, 153, 155, 159–164

nonsymmetric, **36**:142, 143, 155

Sagittal plane, **27**:470, 491

Salsburg hierarchy, **40**:294

Salt domes, **32**:15

Salt fingers, **32**:137

effect of horizontal gradients, **32**:143

stability of, **32**:143

Salts, enthalpy of vaporization, **11**:90

fused, **6**:283

isothermal compressibilty, **6**:284–285

solvents for organic compounds, **21**:525

interaction with cellulose, **21**:689, 691

molten, compressibility, **6**:273

surface tension, **6**:273, 283

Sampling experiments, with nozzle beams, **10**:304–306

Sampling factor, **53**:11, 50

Sancheria, **7**:634

Sand, **21**:751

Sandwich cells, **7**:311ff

Satellite line, **49**:537

Sato method, **36**:161, 162

Sato parameter, **42**:75, 99

Saturated ligand, **7**:362

Saturated spin system, **7**:502

Saturated vapor, **1**:204

Saturation, **29**:303

Sayvetz conditions, **44**:12

Scalar potential, **33**:163, 173, 188, 216

Scalar product, **1**:356; **46**:332, 351

Scaled particle theory, **9**:1ff, 13; **49**:413

critical, **49**:439

hard sphere fluids, **6**:230, 238, 253, 266, 272ff

fourth and fifth virial coefficients, **6**:265

isothermal compressibility, **6**:284

mixtures, **6**:271

see also Particle scaling

Scaling, **36**:155, 157, 163, 165; **43**:41

configurational integral for ionic systems, **11**:87

ionic pair potential, **11**:85

Scaling arguments, **51**:25–28, 33, 35, 43; **52**:472

Scaling conditions, **35**:126, 132

Scaling hypothesis, **37**:182; **38**:311;

51:164; **52**:486

Scaling laws, **24**:38; **49**:358, 439, 443

Scaling limit, **46**:215

Scaling parameter, **6**:361

Scaling procedure, **2**:219, 269

Scaling property, **46**:208

Scaling regime, **46**:219
  existence of, **46**:224

Scaling solution, **46**:235

Scaling theory, **46**:211, 215, 231

Scaling transformation, infinitesimal, **46**:287

Scandium, **41**:141

Scattering, **20**:239; **25**:69, 70, 116, 120, 123–125, 132, 143, 175; **31**:159, 162, 227, 423; **36**:65, 169, 197, 198

  boundary conditions, **25**:216–218, 220, 254, 268

  Bragg, **20**:232

  charge transfer, **30**:463

  coherent, **17**:189

  of composite particles,
    trial functions, **25**:212, 257–260, 266–279, 282
      algebraic close coupling, **25**:267
      close coupling, **25**:257
      correlation method, **25**:258, 282
      Harris and Michels, **25**:266, 267
      including closed channels, **25**:260

  depolarized Rayleigh, **31**:162

  diffuse, **27**:277

  dipole-dipole interactions, **12**:482

  double, **18**:24, 25, 26

  elastic, **20**:229; **25**:69, 71, 72; **36**:169, 174, 177, 188, 194, 195, 197; **45**:521–523

He* + Ne, **45**:570–571

  electron-atom, pseudopotential representation of, **16**:285, 295–304

  electronically inelastic, **21**:169, 189; **36**:176, 186, 188

  of electrons in fluids, **31**:423

  excited state, **45**:510–521

  He* - He, **45**:527–536

  identical particle, **30**:68, 329, 346, 361, 371, 379

  incoherent, **17**:189

  inelastic, **20**:243, 244, 253, 264, 275, 288; **25**:70–72, 116, 132; **30**:78, 223; **36**:65

    neutron, **13**:126

  ion-molecule, Ar⁺-D₂, **30**:199, 224
    C⁺-D₂, **30**:209
    C₂H₂⁺-C₂H₄, **30**:212
    C₂H₄⁺-C₂H₄, **30**:212
    D⁺-HD, **30**:195
    H⁺-Ar, **30**:234, 237
    H⁺-CO, **30**:220
    H⁺-H₂, **30**:194, 224
    H₂⁺-H₂, **30**:210, 225
    H⁺-N₂, **30**:238
    I₂⁺-C₂H₄, **30**:215
    Kr⁺-D₂, **30**:202
    Li⁺-H₂, **30**:226
    N⁺-H₂, **30**:203
    N₂⁺-H₂, **30**:211
    N⁺-O₂, **30**:203
    O⁺-D₂, **30**:62, 208, 240
    O⁺-HD, **30**:241
    O⁺-N₂, **30**:204

  kinematic, **20**:228, 234, 254, 280

  of light, by thin films, **48**:329–396
    collision-induced, **46**:1–72
      cross section, **46**:31, 32
      depolarized, **46**:24
      intensity, **46**:32

photon counting, **46**:32
polarized, **46**:24, 38
virial expansion, **46**:2
low-angle, classical, **12**:389, 408, 481
quantum effects, **12**:394
multiple, **20**:228, 233, 250, 254, 262, 267, 288
neutron, by liquids, **15**:131
neutron-deuteron, **25**:282–285
nonrelativistic, **20**:253
nonspherical molecules, **42**:421–512
anisotropy parameters, H_2 + He, **42**:476
Ar + cesium halides, **42**:480
Ar + CO_2, **42**:461
Ar + CsF, **42**:448
Ar + LiF, **42**:448, 481
Ar + LiH, **42**:453, 498
Ar + N_2, **42**:431, 453, 472, 486
H_2 + rare gases, **42**:448, 473
HCl + rare gases, **42**:453
HD, rotational transitions in, **42**:474
He + CO, **42**:434, 500
He + CO_2, **42**:490, 491
He + H_2, **42**:281, 430, 496
He + HCN, **42**:427, 431, 498
He + Na_2, **42**:430, 469, 500
Hg + I_2, **42**:485
K + CO, **42**:485
K + CO_2, **42**:452, 488, 502
K + COS, **42**:488
K + CS_2, **42**:488
K + N_2, **42**:485
K + NH_3, **42**:492
Kr + N_2, **42**:461, 485
Li^+ + H_2, **42**:461
Na_2 + Ne, **42**:453
Na + tetrahedral molecules, **42**:492

Ne + D_2, **42**:475
Ne + Na_2, **42**:484
Newton diagram, **42**:437, 449
NO, **42**:483
NO + rare gases, **42**:448, 458, 483, 500
potential surface, Ar + HCl, **42**:427
Ar + HF, **42**:427
Ar + N_2, **42**:427
TlF + rare gases, **42**:447, 448, 498
reactive, **25**:70–72; **36**:65
alkali atom, superthermal, **30**:256
thermal, **30**:49, 250
alkali dimer, alkali atom, **30**:272
+ halogen atom, **30**:269
+ halogen molecule, **30**:260
+ hydrogen atom, **30**:270
+ polyhalides, **30**:265
alkaline earth atoms, **30**:274
halogen atom—halogen molecule, **30**:283
halogen atom—hydrogen, **30**:294
hydrogen atom—hydrogen molecule, **30**:296
hydrogen atom—unsaturated hydrocarbon, **30**:296
methyl radical—halogen molecule, **30**:289
oxygen atom—halogen molecule, **30**:292
related to coherent processes in crystals, **40**:420, 426, 428, 452
rotationally inelastic, **36**:169, 176, 178, 184, 188, 194ff
shadow, **10**:106–107, 128–130
specular, **27**:277
thermal diffuse, **20**:249, 289
total, **36**:170, 174
vibrationally inelastic, **36**:176, 184, 188, 194, 198

see also Potential energy surface; specific types of scattering

Scattering amplitude, **10**:83–84, 94–99; **20**:239; **25**:70, 116, 216; **36**:416
glory, **10**:114–115
low-angle, **10**:107–111
near rainbow angle, **10**:103

Scattering apparatus, **45**:515–516

Scattering calculations, **25**:212, 225, 249, 255ff, 279, 281

Scattering cross section, **20**:228, 231, 232, 234, 286; **27**:441 **34**:212, 217; **40**:444
Born approximation, **34**:212, 213, 218
collision dynamics, **6**:40–42
dynamic corrections, **34**:224
see also Cross section

Scattering data, bibliography of, **12**:389, 424
inversion of, **12**:390, 418; **30**:336, 354

Scattering factor, **16**:137, 205, 207–210

Scattering geometry, **10**:35–37, 44–47, 64, 253

Scattering intensity, **24**:6; **45**:364–365

Scattering length, **16**:303; **42**:184

Scattering matrix, **25**:225

Scattering parameter, **25**:215
stationary property, **25**:221, 224, 230, 251, 280

Scattering potential, **36**:65
theory of, **45**:496–510

Scattering probability, **25**:268

Scattering theory, **14**:25ff; **52**:495
asymptotic constants and, **53**:370
role of sensitivity analysis, **51**:177–225
semiclassical, **12**:360, 403

Scattering time, **40**:449

Scattering transformation, inverse, **53**:301

Scattering vector, **20**:240; **44**:285

Scattering volume, **10**:36, 45–46

Scheffers and Stark experiment, **10**:2, 13

Schiebe's phenomenon, **5**:4, 15

Schiff base, **39**:119, 135, 145, 152, 156

Schiff-Landau-Lifshitz approximation, **10**:118–119, 121–123

Schlieren techniques, **32**:139

Schlögl model, **43**:37, 225, 245; **49**:316

Schmidt procedure, **2**:283; **40**:253

Schrödinger equation, **2**:212; **4**:253, 264, 282, 283; **7**:7, 12, 187; **9**:86; **41**:86, 145; **44**:4, 5; **49**:73, 196, 200ff, 220ff, 238, 247–249, 263, 276ff, 296; **53**:280–283
cubic, **53**:283–287
Heisenberg picture and, **53**:352
hydrogen-like atoms, **7**:10
lattice electronic state, **53**:280–281
nonlinear, **53**:320, 321, 332–334
time-dependent, **6**:53–55; **41**:130, 131
time-independent, **6**:47–53
distorted waves, **6**:49–51
partial waves, **6**:47–49

Schrödinger perturbation theory, **8**:4

Schrödinger representation, **38**:6, 11, 50

Schrödinger wave function, **9**:323

Schwartz-Christoffel formula, **49**:205

Schwartz inequality, **31**:40; **46**:315, 321, 343

Schwarzchild distribution, **35**:121

Schwarzchild-Epstein quantization, **36**:4

Schwinger transformation, **33**:204, 206, 207, 214

Scission, **7**:219, 224

Sclera, **7**:270

Screened pseudopotential, **27**:550, 551

Screening, 7:13; 10:6, 9; 26:108, 111, 112; 27:266; 51:27
of Coulomb potential, 34:124
of repulsive force, 34:109, 125
in van der Waals lattices, 16:380
Screening length, 14:233; 27:267, 274; 51:26
Debye, 34:124
Screw axis, 20:10
Screw dislocation, 40:205
Sea urchin, development of, 29:257
Second harmonic generation, 26:23, 24
Second law of thermodynamics, 24:106–108, 122, 123; 46:154, 163
Second moment condition, 31:92
Second moment of rotational spectra, 34:89
Second order fluid constants, 35:96, 100
Second order phase transition, 2:119; see also Phase transition
Second quantization, 9:324ff; 22:109, 110
Second virial coefficient, 2:107, 112; 30:359, 365
convex molecules, 5:164ff
Secondary bifurcation, 55:180
Secondary electron emission, 10:174, 201; 27:17
Secondary ion mass spectrometry, 41:491
Secondary trap, 23:243
Sector, 38:44
Sector rule, 44:584ff, 627
Secular determinant, 7:92, 100, 124
Secular equation, 2:270, 271; 7:93
discrete variational method, 37:289
MO model, 37:232
multiple-scattering wave method, 37:289
solution of, 37:289, 323
Secular operator, 14:425

Secular phase, 35:154, 185
Secular term, 35:320, 321
Seeded beams, 10:312; 30:430; 42:444, 480, 531; 50:489; 52:272
Seeds, 7:623
irradiated by x-rays, 7:641
SeF, electron resonance spectra, 18:215, 216, 228, 233
SeH, electron resonance spectra, 18:216, 228, 231, 233
Seigbahn-Lin-Truhlar-Horovitz *ab initio* surface, 49:198–199
Sel'kov model, 29:146
Selection, 38:224, 225, 228, 233, 236
Darwinian system, 38:225, 233, 235
see also Evolution
Selection principle, 38:235
Selection rules, 25:71, 118, 120; 28:116, 135, 145–151, 186, 338–340, 355, 359; 30:58; 36:44; 52:284
optical, 30:221
photon catalysed dissociation, 50:194, 206
spin, 30:221
Selective advantage, 38:247, 248
once=for=ever selection, 38:249
Selectivity, cation, 39:292, 298
complex formation, 39:309
enantiomer, 39:304, 306
ion, 39:287, 297, 335
partition reaction, 39:309
potentiometric, 39:289
Selenium, adsorption, 49:639–640
Selenium (Se$_2$), quadrupole-quadrupole interaction, 12:21
Selenium hexafluoride, 9:184
Self-avoiding walks, 22:64; 51:6–14
on lattices, 15:229ff, 235ff
length, 15:239
polygonal closures, 15:237
radius of gyration, 15:245

with nearest neighbor interactions, 15:261ff, 268

Self-consistency conditions, Brueckner, 2:319

Self-consistent electron pair method, 50:536

Self-consistent equations, 1:253, 256, 260

Self-consistent field configuration interaction method, 50:300

Self-consistent field (SCF) methods, 1:242, 243, 244; 5:303; 4:141; 7:14; 9:330, 348, 356; 25:181; 28:124, 125, 131; 35:123, 138, 140; 42:426; 48:35; 50:520ff; 51:3, 26, 43, 44

basis set, 25:183, 184

exact, 14:309

Hamiltonian, 23:165

LCAO-MO method, 7:102

molecular orbital (SCF-MO), 7:33; 8:133

in chemical reactivity, 8:105

wavefunction, 23:165

$X\alpha$ scattered wave method, 26:258

see also Hartree-Fock; LCAO-MO-SCF methods

Self-consistent linearization, 46:203

Self-consistent Ornstein-Zernicke approximation, 48:226, 238

Self-consistent phonon method, 40:104, 116, 117

Self-consistent potential, 41:63

Self-diffusion, 6:302; 24:274, 277, 309, 313, 321ff, 333, 339, 392; 37:186

Self-diffusion coefficient, 1:10, 136, 151, 153; 17:168

Self-energy, 31:272, 275, 279; 33:156, 220, 224; 36:207, 208, 220, 224, 230, 243, 306; 41:81, 114, 124

kernel, 36:245, 262

lattice of positive charges, 6:128

Self-field, 33:199, 201, 202

Self-free energy effect, 53:251

Self-frequency matrix, 33:236, 259, 262, 270, 278

binary-collision, 33:265

classical trajectory approximation, 33:278

diagonal elements of, 33:266

off-diagonal elements of, 33:270

proper, 33:259, 262

Self-gravitating gas, 26:109, 110, 115, 116, 126, 132, 139; 32:208–209

isothermal, 32:209

Jeans criterion for, 32:246

λ point in, 32:209

phase transition, 32:209

specific heat singularity, 32:209

stability of, 32:245

statistical mechanics of, 32:208

thermodynamics of, 32:208, 209

Self-induced transparency, 53:227

Self-interaction, 33:201–204, 209, 210

Self-organization, 29:2; 38:211–262

catastrophe game, 38:218

Darwinian phase, 38:258

Ehrenfest's urn game, 38:217

equipartition, 38:217

glass bead games, 38:216

macromolecule, 29:25, 44

microstate, 38:221

nucleic acids, 38:258

Onsager's relations, 38:218

proteinoids, 38:258

proteins, 38:258

random walk, 38:219

t-RNA type, 38:258

translation, 38:258

Self-polarizability in chemical reactivity, 8:109ff

Self-quenching, 50:25

NO, **47**:Part 2:27–31

Self-replicating cycle, 29:60

Self-replicative units, phenotypical, **38**:242

Selvedge, 27:145

Semi-free electron, diamagnetism of, **3**:214, 215, 220

Semi-grand partition function, 46:375

Semi-infinite chain, 9:105

Semi-infinite crystal, 9:87, 147

reflectivity, **54**:406–411

see also Half-space

Semi-invariants, Thiele, **11**:20

Semiclassical approximation, 10:54, 76, 86, 88–131, 137, 167–168, 362; **28**:323–328, 332, 341; **30**:77, 78, 320, 341, 353; **42**:231

periodic vibrations, **53**:315–316

Toda lattice, **53**:301

Semiclassical distribution function, 1:32

Semiclassical dynamics, 42:280

Semiclassical energy spectrum, 36:24

Semiclassical equivalence relationship, 10:95, 97

Semiclassical expansion, 36:3

Green's functions, **36**:54

Semiclassical mechanics, 36:13, 25

many degrees of freedom, **36**:25

Semiclassical perturbation series, 36:34

Semiclassical phase space wave function, 36:16

Semiclassical quantization, 36:1

of coupled oscillator systems, **47**:Part 1:168–175

Semiclassical scattering theory, 19:29

Semiclassical SO(4) symmetry, 52:25–26

Semiclassical theory, bound states, **36**:1–62

of polariton, **54**:328–340

Semiclassical transformation, 36:14

Semiclassical wave function, 36:18, 22

Semiconducting diamond, 9:399ff

Semiconductor, 34:260; **44**:294

antibonding band, **53**:292–293

nonpolar, **27**:361, 481–483, 510–514

polar, **27**:360, 414–428

in magnetic field, **27**:420–428

surface state, **49**:583–591, 617–619

Semiempirical atomic orbital, 7:13, 35

Semiempirical calculation, 12:333; **18**:285

Semiempirical potential, 12:293

Semigroup, 53:366

Semirigid complex, 42:369, 415

Semistability, theorems on, **43**:125–126

Semistable linear system, 43:49

Seniority group, 52:71, 83

Sensitivity analysis, 42:297

feature sensitivity analysis, **51**:206–212

role in quantum scattering theory, **51**:177–225

Sensitivity and selection, bifurcation diagram, **55**:57

nonequilibrium systems, **55**:54–58

Sensitivity coefficients, elementary, **51**:178, 180–199, 220

evaluation of, **51**:180–187

feature, **51**:179

higher order, **51**:216, 220

orbital, **51**:223

valuation, **51**:199

Sensitized fluorescence, 28:270–273, 274

SeO, electron resonance spectra, **18**:218, 219, 220, 221, 222

Separability condition, 52:189

Separability theorem, 36:373

Separable systems, 36:36, 37

Separated electron pairs, 5:342, 349

Separation, by thermal diffusion, 16:103, 122–123
chemical, 16:122–123
Separation effect in nozzle beams, 10:313–315
Separation factor, 16:115–121, 129
Separation of electronic and nuclear motion, 49:130–131
Sequential processes, conformational changes treated as, 33:80, 98, 101–115
double helix formation, 33:98, 101
melting of DNA, 33:101
nucleation-dependent, 33:98–100, 103, 104, 109ff
Series analysis, 51:2, 33, 35, 39, 40
Series expansion, 11:356
Serotonin, 16:178–179
Serum albumin, 39:110, 111, 112
Set-graph diagrams, 43:107–108
SF, electron resonance spectra, 18:215, 216, 228, 233
SH, electron resonance spectra, 18:228, 231, 233
Shadow, 12:367
glorified, 12:402
Shadow scattering, 10:106–107, 128–130
Shadowgraph techniques, 32:139
Shake up lines, 36:212, 222
Shallow traps in mixed crystals, 8:30ff, 33ff, 36
Shape distribution, 52:438
Shape factor analysis, 44:363
Shavitt-Stevens-Meri-Karplus potential, 49:216
Shear-horizontal modes, 27:470, 474
Shear rate, 44:438
Shear resistance, 4:196
Shear-vertical modes, 27:472–473
Shear viscosity, 1:148, 150–152, 158,

160; 5:297; 24:272, 277, 284, 286, 291, 310, 329, 334; 37:154
Shear waves, 24:50; 31:259
dynamic, 24:44
Shell model, 27:496, 499, 515, 518
Shielding, 1:156; 4:162
Shock fronts in explosions, 21:295, 786
Shock tube, 10:269, 309–310; 28:44; 41:238
Shock wave, 17:1
in atomic lattice, 53:308–310
laser generated, 23:53
Shockley's surface states, 9:99, 106, 113, 117ff, 130, 136ff, 138ff, 150, 156
Goodwin-Artmann's model, 9:121ff
graphite, 9:158
interaction model, 9:126ff
molecular orbital, 9:138
Short-range forces, 1:81; 10:30, 32–33, 68–70, 82; 12:338; 27:435, 489, 510; 42:184
Short-range order, 1:210; 51:257ff
Short-time approximation, 44:273
Short-time-scale many-body dynamics, 53:87–89
Shpolskii crystals, 17:231
Shuler's rule, 30:58
Sigma-two formula, 16:188, 199
Sign matrix set, 43:63
Sign stability, 43:15, 16, 63, 85–86, 118–119, 203–205
stability classifications, 43:64
Sign-switched cycle, 43:121
Significant structure theory, 21:483, 487
liquid water, 21:489
plastic crystals, 21:517
Silica-stishovite transformation, 9:368
Silicon, 1:89; 27:481–483, 509–510, 527; 41:171
(111), Cl adsorption, 49:618

H adsorption, **49**:617

metal overlayer, **49**:618

surface state, **49**:589

Silicon dioxide, **34**:186

Silicon tetrachloride, **34**:186

interatomic distances, **5**:156

Silicon tetrafluoride, **20**:4, 22

crystal structure, **5**:180, 183

interatomic distances, **5**:156

Silicon tetraiodide, **20**:22

Silk, **7**:240, 264ff

Silver, **27**:440, 525–526; **41**:141; **53**:37, 38, 44

microclusters, **40**:58

partial molar heat of solutions in tin, **2**:133

surface, **27**:476, 479, 486

Silver-cadmium alloy (α-Ag$_6$Cd$_4$), calculation of thermodynamic quantities, **2**:136

Silver-copper, energy of solution, **2**:142

Silver-gold, excess entropy, **2**:132, 136

excess free energy, **2**:136

Silver-lead alloy (Ag$_5$Pb$_5$), calculation of thermodynamic quantities, **2**:136

Silver-zinc, alloy (Ag$_5$Zn$_5$), **2**:129

alloy, calculation of thermodynamic quantities, **2**:136

solution, **2**:132

Similitude rules, **6**:462

Simple current cycle theorem, **43**:129–133

Simple fluids, definition of, **26**:2

light scattering by, **26**:1ff

Simplex equilibrium, **43**:33–35

Simplex reaction, **43**:19

Simplicial decomposition, **43**:24

Simply-connected double resonance, **47**:Part 1:558

Sinanoğlu exact pair equation, **16**:325–327

Sine-Gordon equation, **53**:320, 321

breather solution, **53**:240–241, 263, 264, 331–332

correlation function, **53**:248–262

derivation of, **53**:325–328

dynamics of, **53**:235–241

spin systems and, **53**:229–234

statistical mechanics, **53**:241–247

thermodynamic properties and, **53**:248

Sine-Gordon soliton, Bäcklund transformation, **53**:328–332

polyacetylene, **53**:292–293

stability of, **53**:325–328

Single bases, electronic system, **7**:90ff

MO energies, **7**:92

mutation and tautomerism, **7**:103ff

spectroscopic properties, **7**:90

Single-channel scattering, **25**:212–215, 255, 259, 278, 280ff

Hulthén, **25**:213

Single collisions, method of, **10**:208–209

Single-defect systems, **11**:36

Single-dipole approximation, **6**:456

Single particle phase space density, **46**:286

fluctuations, **46**:326

Single photon counting, **49**:19–21, 34

Single quantum exchange theory, **50**:116

Single-scattering approximation, **49**:643–644

Single-substrate single-product reactions, **7**:648ff

Single superchain approximation, **48**:238, 288

Single-surface adsorption problem, **52**:456

Single transition approximation, **44**:604, 622

Single trapping site approximation,

52:470

Singlet density distribution, **46**:326

Singlet distribution function, **24**:259

Singlet exciton, **40**:370, 392, 437ff, 456, 459, 471ff

Singlet kinetic equation, **48**:115–120, 167–172

Singlet-singlet transition, **7**:96; **18**:54, 56, 57

Singlet state, **7**:287, 331ff; **41**:145

Singlet-triplet coupling, **42**:230
magnetic field dependence, **42**:239

Singlet-triplet transition, **7**:96; **18**:20, 21, 43, 46, 57, 60, 67, 69

Singular decomposition theory, **43**:39

Singular eigenfunctions, **20**:159, 187
Koppel prescription, **20**:172
Rahman-Sundaresan prescription, **20**:192

Singular matrix, **25**:229, 230, 235, 239, 264, 288
algebraic, **25**:264, 276

Singular motion, **36**:13

Singular perturbation, imperfect theory, **55**:181–183

Singularity, **25**:235; **40**:243, 244, 362
of stable differential mapping, **40**:364
unfolding of, **40**:244

Sink Smoluchowski equation, **48**:129, 136–139

Sink term, **48**:81

Sinusoidal modulation, **44**:327, 331

SiS, $X^1\Sigma^+$ state, **25**:195

Site-binding of counterions, **33**:94, 95

Site preference, **49**:83

Site-site correlation function, Ornstein-Zernicke equation for, **48**:281

Site-site model, **53**:24

Site-site potential, chlorine, **53**:25ff

Six-vertex model, **40**:196

Size parameter, **10**:136–137, 144

in reactive systems, **10**:154–157, 162, 168

Skeleton, connected and disconnected, **14**:57
degenerate, **14**:83

Skimmer, **10**:277–278, 292–304; **52**:277; *see also* Nozzle beams

Skin, **7**:270

Slab, dielectric, **27**:378–380
reconstruction in, **27**:514

Slab modes, complex crystals, **27**:428
cubic crystals, **27**:371–387
damping included, **27**:413
experimental studies, **27**:436–441
lattice dynamics theory, **27**:467–483, 488–505, 528
nonlocal theory, **27**:455–459
polar semiconductors, **27**:417
reconstruction effects, **27**:509, 514
unretarded, **27**:403–404

Slater coordinate, **1**:66, 67, 70, 71

Slater determinant, **2**:210, 224, 230ff, 261, 293, 307, 309, 316; **9**:323; **26**:218; **41**:63; **44**:7

Slater exchange potential, **23**:95

Slater exponents, **49**:107

Slater-Kirkwood formula, **2**:67

Slater orbitals, **1**:254; **4**:102

Slater sum, **1**:280, 285; **4**:232; **22**:34

Slater-type orbitals, **7**:13; **26**:220; **44**:9; **49**:121

Slater's theory, **42**:322

Slip boundary conditions, **44**:286; **49**:29–31

Slip coefficient, **44**:285

Slow exchange, **40**:399, 422

Slow ions, detection of, **10**:200–206

Slow neutron diffraction, **53**:6–12

Slow phonon-fast exciton limit, **40**:444

Slowing down, **43**:264

Slowing down density, **20**:178

Small molecule weak-coupling case, collisional effects in electronic relaxation, **47**:Part 2:350–357

dephasing and rotational relaxation, **47**:Part 2:353–354

pure dephasing, **47**:Part 2:352–353

Small norm solution, **40**:275, 280, 363

Smallness parameter, **46**:198

Smoluchowski-Chapman-Kolomogorov equation, **22**:52, 81

Smoluchowski diffusion operator, **35**:20

Smoluchowski equation, **15**:137, 139, 140, 149, 328; **48**:80–81, 130

sink, **48**:128, 136–139

Smooth sphere fluid, **44**:374

Smoothness postulate, **51**:24

Snap-out reactions, **28**:86

Snell's law, **27**:290

SnO, A-X band system, **50**:69, 71

laser induced fluorescence studies, **50**:69

natural lifetime of A state, **50**:74

preparation, **50**:70

SnO$_2$, **27**:444

SO, **25**:198

A-X band system, **50**:87

electron resonance spectra, **18**:218ff

potential energy curves, **50**:88, 560

radiative lifetime of A state, **50**:91

Soap film, **48**:330, 335–338

black spots on, **48**:330

color interference, **48**:330

contact angles, **48**:333

diffraction of light by, **48**:347–350

diffusion in, **48**:365–366

equilbrium film, **48**:341

equipartition of energy, **48**:334

evaporation, **48**:373, 374

fluctuations, **48**:356–367

instability regions, **48**:355

growing fluctuations, **48**:375

scattering of light by, **48**:334–346

on substrate, **48**:367

surface fluctuations in, **48**:352–354

thickness profile, **48**:333

Sodium, **41**:137, 141, 168, 169, 171

$3s$-$4d$ transition, **49**:39–41

atom, **41**:138

crossed beam quenching experiments, **45**:358–393

quenching, by diatomics, **45**:368–377

by polyatomics, **45**:379–380

by triatomics, **45**:377–379

polarization studies, **45**:380–393

see also Alkali atoms

collisional excitation, **30**:421

dimer, **6**:370; **40**:35; **41**:152

dipole resonance interaction energy, **12**:41

quenching of, **42**:201

rotational energy transfer from, **50**:616

excited state potential surface, **42**:176, 192, 201

generation of excited atoms, **50**:346

ion, core polarization potential, **6**:379

microclusters, **40**:54

quenching, **50**:479, 487

of ^2P, **50**:370, 371

see also Collisional ionization; Diatomics-in-molecules; Scattering; van der Waals molecules

Sodium bromate, **16**:207

Sodium bromide, Soret effect, **13**:361, 362

Sodium chlorate, **16**:207

irradiated with neutrons, **2**:202

quadrupole spectrum of, **2**:196, 203

radiation resistance, **2**:196

Sodium chloride, **27**:438, 444–445, 505, 515, 517–519; **41**:150

Soret effect, **13**:361

Sodium fluoride, **27**:445–446

Sodium hypophosphite-*d*, spectra of, **7**:480

Sodium ribonucleate, infrared spectra of, **7**:463ff

Sodium sulphate, **2**:92

Soft instability, **43**:230, 242, 255, 256, 258

Soft modes, **32**:88; **53**:275

Soft potential, **9**:13, 20

Soft sphere fluid, crystallization, **48**:404

hard sphere diameter, **34**:114

temperature, **34**:114

Soft X-ray spectroscopy, **51**:259

Sojourn time, **52**:375, 457, 459, 460

Solar radiation, absorption of, **55**:64

Solid-gas boundary curves, **2**:87

Solid-liquid boundary curves, **2**:87

Solid-liquid transition, **37**:142

Solid-on-solid model, **40**:159–164

Solid-solid equilibrium, **17**:1

Solid state, electron affinities, **7**:298, 303

ionization potentials, **7**:298ff

metastable, **44**:420

structure, **49**:47–114

Solidification, dynamics of, **32**:165

Solidlike cells, **48**:475, 490, 499, 504

Solids, amorphous and disordered, **44**:396

nonradiative processes of small molecules in, **41**:269–320

nucleation from melt, **40**:146–147

polycrystalline, **32**:196

porous dissipative, **32**:14

proton and muon motion, **41**:165

Solitary wave solution, ϕ^4 model, **53**:267

Soliton-phonon gas, **53**:243ff

Soliton-soliton interaction, **53**:292

Solitons, **44**:496; **53**:225–340

amplitude, **53**:308

biological systems, **53**:287

Brownian motion of, **53**:228

collisions, with phonons, **53**:245, 268–270, 287, 328

with solitons, **53**:328, 330–331

compressional, **53**:306–307

damping, **53**:270–271

density, **53**:245, 250–251

diffusing, **53**:277, 320

dispersion interaction, **53**:284

domain walls, **53**:261

energy, **53**:243ff, 283

free energy, **53**:247

gas of, **53**:256

kink diffusion, **53**:277

lattice vibrations, atomic lattice, **53**:302–311

molecular, **53**:317–319

periodic, **53**:311–317

Toda lattice, **53**:297–302

mass, **53**:259, 284

mobile, **53**:292

molecular crystals, **53**:277–287

momentum, **53**:308

nonlinear Schrödinger equation, **53**:332–334

phase, **53**:292

ϕ^4 model of diffusion, **53**:265–277

polyacetylene, **53**:287–293

propagation of, **53**:264

pulse, **53**:254, 263

sine-Gordon, **53**:229–261, 292–293, 325–332

spin, **53**:262–264

correlation functions, **53**:248–

262

spin-vector profile, **53**:238

stability of, **53**:237, 268

topological, **53**:320

trapped, **53**:286–287

velocity, **53**:283, 284

wave train, **53**:314

Solubility, gas in liquid, **17**:4

in compressed gas, **2**:92

metals in metal-ammonia solutions, **4**:306, 307, 353, 354

parameter, **17**:32

quartz, **2**:99

solid, **17**:3

solid in liquid, **2**:86

Solute atom, isolated, test of clamping approximation, **53**:189–191

Solute-atom limit, isolated, **53**:187–188

Solute molecules, chemical potential, **9**:13

connection formula, **9**:16

fluctuations, **9**:44

λ-cule, **9**:14

pair correlation with solvent, **9**:13

partial pressure, **9**:48

Solutions, **24**:22, 57, 59

Bjerrum theory of, **11**:7, 43

cell model, **11**:117, 118

conformal, theory of, **11**:120

Debye-Hückel theory of, **11**:7

ideal, **43**:95

formed by point defects, **11**:3, 5

ion-pair theory of, **11**:44

ionic, **48**:303–314

Mayer theory of, **11**:8, 46, 56

McMillan-Mayer theory of, **11**:17, 30

theory of, **24**:230

Solvated electrons, **44**:483–544

optical spectroscopy, **44**:486, 514

reactivity, **44**:512

see also Metal-ammonia solutions

Solvation, **41**:274

energy, **52**:293

free energy, **6**:414

layers, **12**:302

Solvation shell, **53**:89

heatbath modeling, **53**:98

recombination dynamics, **53**:106, 110

Solvay conferences, astrophysical problems, **55**:27–30

in chemistry, (1922–1978), **55**:37–42

proceedings of 17th (1980), **55**:1–282

in physics, **55**:7–35

search of last constituents of matter, **55**:9–23

solid-state and statistical mechanics, **55**:23–27

Solvent, hydrogen bonded, **53**:65

Solvent cage effect, in proton transfer, **47**:Part 2:673–674

Solvent effects, **1**:62; **12**:284; **49**:34

Solvent forces, **53**:179

common cage effect, **53**:111–116

Lennard-Jones solutions, **53**:193, 194

systematic and random, **53**:78

see also Interacting reagents

Solvent response, **53**:201–204

Solvent shift, **40**:429, 431

Solvent structure, **12**:306

Solvolysis, **1**:63, 64

Sommerfeld-Wilson quantization rule, **36**:4

Soret band, **7**:382, 393

Soret coefficient, **1**:1, 3; **13**:352ff

Soret effect, aqueous solutions of halides, **13**:357

experimental methods for study of, **13**:351, 355

Sound, propagation in the critical region, **6**:215–222, 223
  velocity, **53**:283, 292, 300, 303, 309
    in liquids, **21**:483
Sound wave, **46**:313
Source strength, **31**:163, 164, 166
Space groups, **20**:10
Span distribution, **52**:417, 422, 439
Span end-to-end distance, **52**:426
Spans, **52**:416–427, 439
  of one-dimensional chain, **52**:417
  of polymer chain, **52**:440
  of star-branched polymer chain, **52**:440
Spark gap generation of picosecond pulses, **23**:193
Spatial correlation, **38**:174
  principle of, **55**:175
Spatial dispersion, **27**:449–465; **44**:488, 495; *see also* Nonlocal theory
Spatial structure, time-independent, **38**:349
Spatial symmetry breaking, **46**:323
Spatiotemporal patterns, chirality and, **55**:189–190
Special points method, **49**:74–75
Species, external, **43**:16, 40
  internal, **43**:16
  pseudograph, **43**:75
Specific activity, **1**:40
Specific heat, **26**:139, 140; **28**:205; **48**:295
  metallic glass, **51**:269
  negative, **26**:115, 122, 139
  non-interacting electrons, **13**:126
  rotational, **28**:206
  translational, **28**:206
  vibrational, **28**:205, 206
Specific rotation, **4**:72
Spectator model, **10**:377–379

Spectator-stripping model, **30**:60; *see also* Reaction mechanism
Spectra, band systems, **7**:95
  conjugated chains, **7**:398ff
  electronic, band contours, **20**:353, 356
    diffuse transitions, **20**:380
    dipole moments of excited states, **20**:382
    emission spectra, **20**:385
    geometries in excited states, **20**:376, 387
    molecular crystals, **20**:370, 381, 382
    polarizations, **20**:371, 387
    polyatomic molecules, **20**:341
    rotational structure, **20**:346
    table of analyzed rotational band contours in
      cyclic molecules, **20**:358, 369
  see also Electronic Spectra
Spectral density, **12**:251; **36**:51; **51**:53
  error bounds for, **15**:79, 84ff
  of fluctuations, **12**:245; **38**:189
    electric field, **38**:198
  heatbath response function, **53**:153
  moments of, **15**:80
  response function hierarchy, **53**:154
  velocity autocorrelation function, **53**:151
  see also Line shape
Spectral function, bismuth, **31**:295
Spectral line shape, **42**:275, 294; *see also* Line shape
Spectral moments, **44**:344, 346, 559, 560, 574, 577
Spectral operator, **36**:50
Spectral representation, **36**:220, 245
Spectral resolution, **44**:329; **49**:155
Spectral theory, **38**:26

connection with stability, **32**:219, 220, 221, 228, 230

continuous modes in, **32**:225, 226, 228

modes, **32**:222, 226, 228

partition of, **32**:218, 247

symmetrized, **32**:221, 230

Spectrochemical series, **5**:65; **7**:363ff, 371; **8**:52

Spectrometer, ion-cyclotron resonance, **45**:117–118

Spectrophone, **23**:43, 60

Spectroscopic constants, **44**:24; **49**:155, 178

N₂, **45**:267

NO⁺, **45**:336

NO, **45**:300

O₂, **45**:287

O₂⁻, **45**:317

Spectroscopy, broad-band, **44**:320

classical dielectric, **44**:256

Doppler, of photofragments, **47**:Part 2:133–152

fluorescence, **52**:264

multiphoton, **47**:Part 2:17–19

nonlinear, **47**:Part 2:153–181

photoabsorption, **52**:264, 324

photoelectron, **52**:264, 290, 307, 324

picosecond, **47**:Part 2:535–577

time of flight, **47**:Part 2:134

see also specific spectroscopies

Spectrum, irregular, **36**:2, 36, 42–45

regular, **36**:2, 36, 42, 43, 53

Spectrum analyzer, **24**:36

Specular scattering, **27**:380, 452, 459

Speed of sound in the critical region, **6**:219, 220

Sphere modes, **27**:359–360

complex crystals, **27**:428

cubic crystals, **27**:387–396

damping included, **27**:413

nonlocal theory, **27**:526

optical studies, **27**:441–445

polar semiconductors, **27**:418

unretarded, **27**:402

Sphere packing models, **12**:317

Spherical cavity, **4**:311, 339; **9**:6, 41

Spherical cutoff method, **48**:246

Spherical harmonics, **13**:236; **14**:496

generalized, **34**:3, 9, 10, 11, 22, 38, 50

Spherical ion, **9**:55

Spherical model, **40**:268, 271

Spherical molecules, **37**:175

principle of corresponding states for, **16**:224–226

Spherical sample, **48**:200

Spherical shells, **24**:145–148

Hamiltonian functional, **24**:148

inertia of, **24**:146

Lagrangian functional, **24**:147–148

Spherical symmetry, **52**:442

Spherical top, **4**:279; **31**:160, 171

Spider, dangling, **24**:100–102, 138, 139

Spin, **1**:172, 240, 243, 259; **7**:11, 365, 501; **31**:161, 182; *see also* High-spin; Low-spin

Spin angular momentum, **7**:162

Spin angular velocity, **44**:286

Spin basis function, **52**:146, 175, 177–179

Spin conservation rule, **42**:20, 133, 149

Spin correlation, **41**:109; **53**:261–262

Spin correlation length, inverse, **53**:256

Spin density, **7**:537ff; **41**:76, 81, 85, 139; **44**:403

Spin-density-functional formalism, **41**:59–194

relativistic electron systems, **54**:238–239

Spin-density theory, relativistic electron

systems, **54**:273–288

Spin-dependent interactions, 12:551, 555–557, 597

Spin dephasing, 40:385, 386–388, 472

Spin diffusion constant, 12:596

Spin echo, 12:552, 596; **40**:424, 426

Spin-exchange scattering, alkali atoms, **12**:97

  inelastic, **18**:25

Spin flux, 31:161, 164, 171–173, 191–192, 194, 196, 209

  collisional transfer part, **31**:164

  density dependence, **31**:196, 209

  diffusional part, **31**:161, 164

  ellipsoids, **31**:191, 192, 194

  rough sphere, **31**:171, 173

Spin-forbidden transitions, 7:396; **18**:19, 21; **42**:65, 144, 224

Spin-free Hamiltonian, 18:172

Spin Hamiltonian, 5:204; **7**:554; **37**:260, 261

  atomic lattice, **53**:294, 296

  continuum approximation, **53**:248, 292

  excited lattice, **53**:279

  ferromagnets and antiferromagnets, **53**:230ff

  isotropic Heisenberg chain, **53**:263

  polyacetylene, **53**:288

  uniaxial ferromagnet, **53**:263–264

Spin-lattice relaxation, 7:502; **42**:354; **53**:259, 260

Spin-lattice relaxation time, 35:24; *see also* T_1

Spin-orbit coupling, 30:176; **40**:401, 411ff, 423, 428; **42**:4, 21, 26, 94, 99, 117, 145; **44**:106, 115; **49**:637–638; **52**:144–149, 174–176, 307

Spin-orbit coupling constant, atomic, **37**:267

molecular, **37**:266, 268

Spin orbital, 2:258, 277, 279; **6**:328; **7**:162; **14**:287; **26**:219

Spin orientation in ion-neutral reactions, 10:268

Spin-pairing, 7:365

Spin-pairing energy, 5:76, 78; **8**:50

  hexa- and tetrahalides, **5**:82

Spin polarization, 9:356, 359; **35**:6

  effect of polarized light, **35**:7

  effect of triplet quencher, **35**:9

  of doublet radical, **35**:6

  of duroquinone, **35**:9

Spin polarization orbitals, 9:164

Spin relaxation time, 17:171; **31**:172, 195

Spin-rephasing, 35:12

Spin-rotation interaction, 18:190, 192; **35**:23; **44**:107, 119

  matrix elements, **18**:192

Spin-rotation tensor, 34:7, 63

  methane, **34**:63

  methyl iodide, **34**:90

Spin saturation, 7:502

Spin selective reaction, 35:17

Spin-sorting process, 35:11

Spin-spin correlation function, 44:347

Spin-spin dipolar integrals, 37:280

Spin-spin interaction, 5:202; **12**:7, 96, 97; **18**:173; **37**:260, 261; **44**:107, 116

  H_2, **12**:96

  matrix elements, **18**:194, 202

Spin-spin relaxation, Caspers theory of, **11**:305

Spin-spin relaxation time, *see* T_2

Spin states, 7:365, 376

Spin susceptibility, 41:185

Spin system, 22:344

  polyacetylene, **53**:292

sine-Gordon equation, **53**:229–234

solitons in, correlation function, **53**:248–262

Spin temperature, **11**:296

Spin velocity, **31**:158, 171, 172, 187, 207

linear rotor, **31**:182

rough spheres, **31**:171

Spin wave, transverse, **53**:263

Spin wave dispersion of nickel, **41**:164

Spinodal line, **17**:58

Spinor, Dirac, **41**:82

Pauli, **41**:76

Spinor-density matrix, **41**:76

Spiral growth, **40**:204

Split shells, **6**:332

Splitting, Bethe, **4**:161

Davydoff, **4**:161

Spontaneous emission, **35**:269; **40**:381

Einstein theory of, **15**:12ff, 17

Spontaneous fluctuations, **43**:246

Spontaneous mutation, **7**:85

Spreading pressure, **7**:245

Sputter source, **42**:533, 577, 578

Square planar complex, **5**:54, 66

Square-well fluid, **24**:332; **31**:158; **37**:139–202; **40**:304, 316, 340, 348; **49**:386–387

equation of state and transport coefficients, **24**:333

nonspherical molecules, **31**:181–196, 210

spherical molecules, **31**:158, 168, 179, 210

Square-well potential, **1**:295; **9**:33, 50

calculation of thermodynamic quantities of clathrates, **2**:33

Squared structure, **16**:200–201, 205

Squid, giant axon of, **29**:319

sRNA, **7**:219

ST2 potential, **31**:73

Stability, **26**:113, 133, 134, 143; **35**:125, 127

asymptotic, **19**:224; **29**:4

bicyclic olefins and ketones, **19**:193

bifurcating solutions, **40**:303

bifurcation theory and, **19**:159

characteristic equation, **38**:268

characteristic exponents, **19**:241

chemical waves, **38**:346

classifications, **43**:53

of networks, **43**:50

sign stability, **43**:64

complex reaction networks, **43**:1–216

constrained mixing, **43**:69

critical point, **46**:89

D, **43**:63, 65

definitions of, **43**:48

diagram, **32**:110; **38**:327; **43**:9

dioleylphosphate-Millipore membrane, **29**:334

domain of, **43**:69, 71

of equilibrium, **29**:15

exponent polytopes and, **43**:187–193

exponential asymptotic, **32**:217

fluid, **40**:356

global, **29**:125, 128; **46**:90, 145

Hurwitz criteria, **19**:241; **43**:53, 54

hydrodynamic, **32**:43, 57

hydromagnetic, **32**:43, 57

in respect to small parameters, **29**:7

infinitesimal, **19**:233, 244

linear, **26**:178; **29**:5; **32**:3; **38**:268; **49**:314, 328–332, 347

local, **46**:90, 145

Lyapounov, **19**:225, 227; **29**:4

Lyapounov functions, **19**:226; **43**:60–62

marginal, **19**:241, 254, 305; **38**:270, 272

nonoscillatory, **19**:280

matrix, **43**:63

metastability, **32**:10

mixing, **43**:66–73, 101

mixing asymptotic, **43**:67, 78, 82

nonlinear, **26**:200, 203, 208

nonlinear equation, **37**:143

normal mode, **19**:240, 244, 249, 259, 266, 288

orbital, **19**:225, 243; **29**:4

oscillatory, **19**:261, 279

passage from unstable to stable state, **46**:195–278

polynomial, **43**:59, 119–125

  dominant terms, **43**:171

  number of terms, **43**:167

  off-diagonal terms, **43**:194

qualitative, **43**:8–9, 15

secular equation, **19**:241

sign, **43**:15, 16, 63

singular points, **19**:239, 241

space-clamped membranes, **29**:368

squid giant axon, **29**:320, 349

stationary state, **38**:322

strict, **32**:217

structural, **29**:8, 9; **38**:265, 267

successive primary branches, **38**:282

tests for, **43**:61, 69, 71, 72, 91ff, 101, 126, 193, 206

thermodynamic, **29**:14, 30, 74; **38**:264

total, **43**:63, 65

variational equations, **19**:225, 239

weak, **32**:217

see also Instability; Oscillation

Stability analysis, **29**:5, 34, 53, 345; **37**:176, 184

  application of local potential methods, **13**:267ff, 324ff

  flow systems, **13**:267ff

kinetic equations, **13**:324ff

Stability condition, multiconfigurational Hartree-Fock state, **54**:32–33

Stability criterion, **20**:90, 110, 111; **32**:168, 182

Stability of form, **32**:165

Stability theory, **19**:224

  microclusters, **40**:59–93

Stabilization energy, correlation with localization energy, **33**:381ff

  integral representation, **33**:384, 388

Stabilization energy method, **33**:347, 348, 362, 366

  failure for biradical transition states, **33**:380

Stable equilibrium structure, **49**:415

Stable law, **52**:386, 393, 420, 427

Stable mapping, singularity of, **40**:242

Stable periodic trajectory, **36**:39, 41, 52

Stable regular potential, **40**:297

Stable wave function, **12**:53

Stacking fault, **16**:344–345; **49**:488, 505–506

  energy of argon, **16**:346

Stacking sequence, **49**:495, 505–506

Stacking transition, **49**:516–517

Staging, **49**:460–467

  disorder, **49**:464

  impure, **49**:469

  phenomenologial model, **49**:463, 472–474, 478

  see also Graphite intercalation compounds

Staphylococcal nuclease, refolding, **33**:108

Star-branched molecule, **52**:440

Star graphs, **11**:345, 346, 350, 351

Stark dielectric constant, **12**:297

Stark effect, **1**:261; **2**:377, 378; **10**:351; **12**:134; **18**:105, 236, 237; **30**:389,

395; **44**:34, 89, 101
line shape, **15**:120
linear, **30**:397
matrix elements, **18**:236, 237
negative, **30**:396
quadratic, **30**:396
rotational spectra, **13**:180
see also Electric deflection analysis
Stark field mixing, 50:195
Stark modulation, 18:241
line shape, **18**:242
Stark shift, 42:412, 439; **50**:212
Stark switching, 25:34
Stars, binary, **26**:109, 116, 137, 140
dynamical formation of, **26**:116, 132ff, 143
formation of, **32**:209, 210
Keplerian pairs, **26**:116, 125, 126
minimum entropy in main sequence, **32**:209, 210
nonequilibrium thermodynamics applied to, **32**:209, 210
stellar evolution equations, **32**:211
Eulerian form, **32**:211, 214, 216
Fréchet derivatives of, **32**:216
Lagrangian form, **32**:215
thermodynamic instability of, **32**:209, 210
see also Stellar systems
State area principle, 20:102
State function, 7:6
State-path-sum, 41:261
State preparation, 41:495
State-to-state reaction constant, 42:274
State-to-state reaction dynamics, 50:91, 326
States, 38:11, 15, 18, 27, 32, 36, 44
atomic, **38**:27
definition of, **38**:11
ergodic, **22**:339; **38**:34

extremal G-invariant, **22**:339
extremal KMD, **22**:342
generalized, **38**:32
individual system, **38**:11
invariant, **22**:336
KMS, **22**:340
L-admissible, **38**:15
n-clustering, **22**:337
observable, **22**:324
partial weak mixing, **22**:357
physically pure, **38**:44
primary, **38**:18
as propositions, **38**:15
R-equivalent, **38**:44
stationary, **38**:317, 321, 332
continua of, **38**:326
existence of two, **38**:329; *see also* Bistabiity
interaction with, **38**:332
pseudo, **38**:326
stability of, **38**:322
transitions involving, **38**:328
uniformly clustering, **22**:342
vector, **38**:27
weak mixing, **22**:357
Static approximation, 34:215
distinct scattering, **34**:216, 217
self-scattering, **34**:215
Static correlation, 43:231, 248, 254
Static dielectric constant, 1:77, 78, 98
Static dipole field, 33:198
Static equilibrium, 27:468–469, 474–475
Static mobility of ions, 16:23
Boson fluid, **16**:30–31
Fermi fluid, **16**:32–33
Static process as limit process, 33:440
Static selection of hydrogen halides, 42:482
Static stability, 33:436

Stationary action principle, **41**:131

Stationary distribution, **43**:248; **46**:267

Stationary phase approximation, **10**:94–97, 103–104, 114–115; **25**:80–82, 87, 106, 135, 143, 168, 169, 174; **30**:32, 58, 81, 82, 114, 117, 321, 474; **36**:14, 58; **42**:236

periodic vibrations, **53**:315

Stationary process, **44**:357

Stationary property of scattering parameters, **25**:221, 224, 230, 251, 280

Stationary state, **7**:6; **43**:224

eigenfunction, **9**:327ff

Statistical fluctuations, **12**:228, 236, 237, 260

density, **12**:236

Statistical limit, **23**:201

Statistical mechanical ensemble theory, **49**:396

Statistical mechanical variational principles, **24**:188, 189, 193

Gibbs-Bogoliubov inequality, **24**:189

Jordan-Golden inequality, **24**:194

Peierls inequality, **24**:193

Statistical mechanics of polymer solutions and polymer adsorption, **51**:1–44

Statistical morphology, **40**:85

Statistical power gap law, **50**:603

Statistical thermodynamics of fluids, **9**:1ff

Statistics, Bose-Einstein, **41**:70

Fermi-Dirac, **41**:69–70

Steady flow reactor, **38**:320, 324

continuous stirred tank reactor, **38**:320, 321

Steady state, **19**:215, 234, 291; **29**:31; **33**:101, 104, 110

concentration, **43**:32

convex parameters for, **43**:27–33

flow, **32**:13

hypersurfaces of, **43**:160

linear density distribution, **13**:282

metastability and, **29**:72

multiple, **19**:291, 318; **29**:40, 50, 59, 63, 94, 345; **43**:35–40

Bierman's model, **19**:287

excitable membranes and, **19**:319

Monod-Jacob type models, **19**:291

thermodynamic properties, **19**:294

protein refolding and unfolding, **33**:110–112

quasi, **33**:110, 111

rate constant, **5**:370ff, 382ff

transitions between, **29**:340

transport processes, **5**:267

Steady state compliance, **35**:97

Steam distillation, extraction, **17**:15

Steam + quartz system, **2**:99

Steepest descent method, **25**:135; **30**:115; **38**:306; **40**:267, 326, 327, 361

Steiner's formulas, **5**:151, 154

Stellar systems, **26**:108ff, 121, 124ff, 131, 134, 142

angular momentum, **26**:118, 122, 129, 135

binary collisions, **26**:111, 123, 133

boundary conditions, **26**:119, 121, 122, 142

clusters, **26**:109ff, 119–122, 133ff, 140–141, 142, 143

galaxies, **26**:108, 138

correlations, **26**:114, 124–126, 129–134, 141, 142

distribution function, **26**:117, 126, 128, 129, 131, 133

energy, conservation of, **26**:108, 118, 122, 131, 134

correlational, **26**:131, 132, 143

internal, **26**:115, 129, 130, 131

kinetic, **26**:110, 118, 123, 126, 130, 133, 138

potential, **26**:110, 122, 126, 131ff, 141–143

total, **26**:119, 126ff, 141–143

ensemble, canonical, **26**:112, 134

grand canonical, **26**:112, 134

microcanonical, **26**:113, 115, 134–142

ensemble averages, **26**:117, 131, 138, 140

entropy, **26**:108, 127, 130, 139, 140, 142

equation of motion, **26**:110, 115, 128, 129, 133, 135, 142

equilibrium, **26**:108, 113ff, 124, 130ff, 140–143

metastable, **26**:113, 114, 134, 141

evolution, **26**:113, 128, 129, 132, 133, 134, 141–143

fluctuations, **26**:121

numerical experiments, **26**:115, 125, 133, 134, 142, 143

thermodynamics, **26**:112, 115–118, 130

time averages, **26**:110, 137, 138, 140

time-scales, dynamical, **26**:110, 123, 124, 133

evolution, **26**:113

secular-relaxation, **26**:111, 113, 123, 124, 133, 134, 141

see also Binary stars

Step free energy, 40:179

Step function behavior, 52:288

Stephan's constant, 1:214, 215, 229

Stepped surface, 41:500

Steps of infinite variance, 52:384–388

Stereoconfiguration, molecular, **16**:163–164, 173, 177

Stereographic drawings, 16:177–178

Stereoregular polymer, 2:165

Stereospecificity, 2:166, 169

Steric factor, 1:18, 121; **42**:20

Steric force, 41:198, 209, 211, 213, 214, 215, 220

Steric hindrance, 1:262; **2**:376, 391

Steric repulsion, 2:75, 389, 390

Steric stabilization, 51:30, 44

Stern-Gerlach experiment, 38:2; **46**:192

Stern-Gerlach magnet, 30:395

Stern-Volmer plot, 21:194; **28**:295, 297, 299; **42**:226, 245, 253; **50**:44, 49

Sternheimer potential, 12:45

Sternheimer procedure, 10:6–7, 14

Steroids, 4:94, 95; **8**:163

infrared spectroscopy of, **7**:413

Stick boundary conditions, 49:29

Stick spectrum, 51:74, 77

Sticking probability, 10:306–308

Sticky collision, 28:29

Stilbene, 49:34

Stimulated emission, 12:119

cross section, **28**:441, 442

Stimulated photon echo, 47:Part 2:475–477

Stimulated Raman scattering, 23:51, 216; **42**:307; **50**:19

Stirling's approximation, 4:208

Stochastic aspects of nonequilibrium thermodynamics, 55:58–60

Stochastic components, Langevin equation, **53**:82

Mori theory, **53**:144, 145

primary system motion, **53**:76–77

velocity autocorrelation function, **53**:78–87

Stochastic differential equations, 38:313; **44**:440; **46**:234

Stochastic dynamics, 32:62

Stochastic equations, 1:142

Stochastic Fokker-Planck equation,

46:239

Stochastic geometry, 6:266

Stochastic Liouville equation, 15:103, 117 **35**:19, 20

Stochastic matrix, 51:132, 133

Stochastic model, 29:68; **41**:336

rotational relaxation, **44**:276

Stochastic process, 1:362, 368, 373; **17**:103; **26**:112, 142; **29**:41; **38**:69, 70, 77, 78, 92; **50**:415

antientropic, **38**:100

backward deterministic, **38**:94, 99, 101

backward nondeterministic, **38**:94, 99, 101

bidirectional deterministic, **38**:94

continuous parameter, **38**:69

decomposition, **38**:100

definition, **38**:69

directly perceptible, **38**:2, 72, 77, 78, 89

entropic, **38**:99

forward deterministic, **38**:93, 94, 99, 101

forward nondeterministic, **38**:93, 99, 101

Gaussian, **38**:86

induced by generic classical observable, **38**:70

molecular, **38**:9

nonlinear, **46**:225

one-sided, **38**:95

polymers, **51**:2

quantum, **38**:2, 5, 18

temporal symmetry of, **38**:92

Stochastic theory of chemical kinetics, 29:41, 66

Stochastic trajectory, 48:84

classical, **42**:310, 316, 346

classical equations, **42**:258, 291

recombination dynamics, **53**:108

Stochastic transition, classical dynamics and, **47**:Part 1:205–214

divergence of trajectories, **47**:Part 1:206

integrable systems, **47**:Part 1:206–208

K entropy, **47**:Part 1:212–214

mixing systems, **47**:Part 1:208–210

Poincaré surface of section, **47**:Part 1:205–206

typical systems, **47**:Part 1:210–212

exponentiating trajectories and, **47**:Part 1:231–233

models of, **47**:Part 1:220–231

critical point analysis, **47**:Part 1:225–226

Mo's method, **47**:Part 1:220–221

overlapping resonances, **47**:Part 1:226–230

variational equations, **47**:Part 1:221–225

numerical experiments on coupled oscillators, **47**:Part 1:214–219

Stochastic variables, 1:142; **44**:275; **46**:235

Stockmayer fluid, 48:231, 240, 267–268

Stockmayer potential, 12:297

modified, **44**:401

Stoeckly's theorem, 41:207, 208

Stoichiometric coefficients, 33:423

Stoichiometric matrix, 43:18

Stoichiometric network analysis, 43:12–13, 16

Stoichiometry, 49:457, 461–462, 474–480

bulk, **49**:474–477, 494, 517

layer, **49**:474, 477, 494, 517

Stokes-Einstein equation, 4:42

Stokes formula, 11:209, 254

Stokes law, 1:86; 11:262; 16:27; 35:73; 48:107

Stokes parameters, 33:186, 187

Stokes spectrum, 42:307

Stokes theorem, 19:350

Stokes vector, 33:186

Stoner theory, 24:245

Storage modulus, 35:97

Strain, in crystals, 40:474

Strain tensor, 35:96

Stratified medium, 37:34

Stratospheric chemistry, halogen compounds, 55:73–74

oxygen and nitrogen compounds, 55:72–73

Stratovarich type equations, 46:234

Streak camera, 49:19, 29

Streaming operators, 11:318, 321, 328, 356ff

Streamlines, 32:300

Stress tensor, 1:6, 7, 216, 220–223; 5:273, 292, 296, 297; 31:161; 33:407; 44:438; 49:373

Stretch modes, 36:34

Stripping, 10:199, 202, 321, 329, 358, 361–367; 19:55, 83; 28:29

recoil, 19:90, 93

see also Reactive scattering

Stripping mechanism, 19:101; 42:124

Strong collision, 35:19

Strong coupling, 4:149

correspondence principle, 36:66, 89

Strong coupling complex, 42:369, 384

Strong exchange limit, 35:20

Strong field deflection, 10:9–13

Strongly mixed states, 47:Part 2:366–368

Strontium, 41:141

autoionizing states, 47:Part 2:165–167

quenching of ^1P state, 50:353

quenching of ^3P state, 50:358, 383

Strontium bromide, Soret effect, 13:361

Strontium chloride, Soret effect, 13:361

Strontium titanate (SrTiO₃), 27:428, 505

Structural enumeration, 49:65–68

edge-sharing octahedra, 49:70–72, 97

Pólya's theorem, 49:68–70, 96

Structural indices, 7:31

Structural phase transition of solitons, 53:265

Structural sorting, 49:65–66, 70

A_2BO_4 minerals, 49:55

trigonal prismatic and octahedral coordination, 49:56

Structural transition, 46:365ff, 375ff, 400, 402

Structure, electronic, *see* Electronic structure

Structure-dependent theory, 33:203

Structure factor, 16:140, 145, 146; 17:261; 53:10, 11, 256

chlorine, 53:26, 27

cross section conversion, 53:48

Fourier transform, 53:54–55

heavy water, 53:42, 43

homonuclear diatomic fluids, 53:22

hydrogen chloride, 53:32

imaginary parts, 16:204–205, 207

inequality relations, 16:158

krypton, 53:17, 19, 20

liquid, 40:352

nitrogen, 53:54

normalized, 16:144, 149, 151, 154, 159

probability distribution, for centrosymmetric case, 16:139–145

for non-centrosymmetric case, 16:145–150

Toda lattice, 53:316–317

water, **53**:33, 36

Structure function, **17**:270; **31**:279, 280, 311; **34**:216, 228; **52**:368ff, 376, 392, 418

  diagrammatic representation, **31**:280

  long wavelength limit, **31**:312

Structure-independent equations, **33**:202, 203

Structure-independent Hamiltonian, **33**:208

Structure matrix, **5**:9ff

Structure of fluids, **53**:1–60

Stueckelberg oscillations, **30**:426; **42**:237, 579

Styrene, **54**:220–224

  polymerization, **2**:175

Styrene methyl methacrylate polymer, **2**:150

Su, Schreiffer, and Heeger model (SSH), **53**:287–291

Subadditivity of entropy, **51**:135, 137–139

Subclusters, **26**:119, 122

Subcritical instabilities, **26**:188, 205, 208

Subdynamics, **38**:145, 166

Subexponomial, **43**:181

Submillimeter detector technology, **44**:324

Submillimeter spectroscopy, **44**:259, 301, 309

Subspace, **41**:349–350; **44**:272, 292

  properties of, **43**:27

Substituents, **7**:30ff

  field effect of, **33**:392

  σ-core polarization by, **33**:392

Substituted hydrocarbons, **9**:65

Substitutional impurity, **52**:470

Substrate, **46**:375, 377, 378, 381

Subsurface states, **9**:156

  Shockley's, **9**:100ff, 141

Successive approximants, **44**:323

Successive partial orthogonalization, **14**:238

Succinic acid, **7**:543ff

Sudden approximation, **30**:78, 405; **36**:81, 82; **42**:289

Sugano's equations, **4**:176

Sugars, **7**:494

  absorption bands, **7**:494ff

  prebiotic synthesis, **55**:98–99

Suicide monomer, **2**:180

Sulfadiazine, spectra, **7**:528

Sulfonamides, NMR, **7**:528ff

Sulfur, Al (111) adsorption, **49**:602

  Ni (100) adsorption, **49**:594–595

Sulfur-6, **20**:28

Sulfur bridges, **7**:232

Sulfur-containing ligands, **8**:53

Sulfur dioxide, **42**:226

  clathrate, **2**:2

    in hydroquinone, **2**:7

  clusters, **52**:344–345

  collisional excitation, **30**:422

  hydrate, thermodynamic data and lattice constants, **2**:8

  radiative decay, **42**:245

Sulfur hexafluoride, **9**:181ff; **24**:42, 47

  hydrate, **2**:22, 47

    vapor pressure, **2**:22

  interatomic distances, **5**:156

  Langmuir's constants, **2**:47

  multiphoton dissociation of, **50**:128

  multiphoton excitation of, **50**:464

  rigid convex body constants, **5**:162

  spectroscopy of, **47**:Part 1:583–624

    coherent pulse propagation effects, field equations, **47**:Part 1:603–604

    polarization density, **47**:Part 1:603–617

radiated field, **47**:Part 1:609–611

sidebands, **47**:Part 1:611–613

spatial degeneracy, **47**:Part 1:613–617

thin-sample approximation, **47**:Part 1:606–609

v_3 band, **47**:Part 1:586–603

fitting of spectroscopic parameters, **47**:Part 1:596–598

transition moments, **47**:Part 1:598–601

vibration-rotation basis, **47**:Part 1:592–594

vibration-rotation Hamiltonian, **47**:Part 1:594–595

Sulfur methyl pentafluoride, barrier height of internal rotation, **2**:382

Sulfur trifluoromethyl pentafluoride, barrier height of internal rotation, **2**:382

Sulfuric acid, **34**:188

Sulfuryl chloride, **16**:181

Sum of angles formula, **16**:153

Sum-of-the-pairs theory, **48**:3

Sum rule, **1**:21; **4**:127, 136; **12**:150; **17**:108; **44**:309, 323, 514; **46**:318, 346

electronic intensities, **44**:577, 593, 607, 608, 616ff

oscillator strength, **4**:144

Summation convention, **46**:327

Sunspot, **32**:106

Super-hyperfine coupling, **5**:215ff

Superalkali atom, **50**:291

Superconductivity, **22**:319; **24**:233

microclusters, **40**:58

Supercontraction, **7**:266ff

Supercooled solution, **40**:137; **44**:258

Supercritical fluid, **2**:94

simple, **37**:155

Supercritical steam, **2**:94

Superdelocalizability, **7**:67

Superelastic scattering, **18**:15, 83, 84, 85; **28**:404, 405, 407, 433

Superexchange, **5**:223ff, 228ff

among monomers, **40**:459

Superfluid, **33**:3–4, 7, 8

component, **33**:3

dynamics, **33**:4, 5, 7

turbulence, **33**:8

Superfluid helium, **41**:178

^{3}He impurities, **33**:21–23, 28, 32

concentration limits, **33**:22, 23, 28

excitation spectrum, **33**:21

mean free path, **33**:23

polarization potential, **33**:32

volume excess, **33**:21

mobility, **33**:1–50

above 1.2K, **33**:7, 8, 39

definition, **33**:21

drag coefficient, **33**:28

formal expressions, **33**:24, 27, 54

in ^{4}He, **33**:17

negative charges, ^{3}He limited, **33**:18, 28–32

phonon limited, **33**:18, 32–38

roton limited, **33**:39–45

positive charges, ^{3}He limited, **33**:28–32

phonon limited, **33**:18, 32–38

roton limited, **33**:39–45

prefactors, **33**:39

pressure dependence, **33**:33, 38, 40, 41, 44

negative charge carriers, **33**:2ff, 13, 16–19, 30, 36

^{3}He condensation, **33**:30

bubble model, **33**:2, 3, 5, 13, 14

effective mass, **33**:5, 9, 17

in gas, **33**:2, 13

pressure dependence, **33**:14, 18, 33, 41

radius, **33**:4, 13, 14, 16–19

response coefficients, **33**:36, 37

surface energy, **33**:13

temperature dependence, **33**:14

variational calculation, **33**:15, 16

positive charge carriers, **33**:2, 10–12, 17–19, 36

   core radius, **33**:11, 12, 16–19

   core structure, **33**:12, 17

   effective mass, **33**:5, 9, 17

   electrostriction model, **33**:2, 5, 10–12

   in gas, **33**:2

   pressure dependence, **33**:11, 19, 41

   response coefficients, **33**:36, 38

   surface energy, **33**:11, 12, 38

   temperature dependence, **33**:12

scattering cross section, **33**:27ff

   ^{3}He impurities, **33**:27

   matrix elements, **33**:27, 43, 44

   momentum transfer, **33**:27

   phase shift analysis, **33**:30, 36

   phonons, **33**:27, 32, 33, 37

   rotons, **33**:41, 43, 44

transport, **33**:21–27, 41, 42

   Boltzmann equation, **33**:21–24

   dilute gas limit, **33**:22, 23

   Fokker-Planck equation, **33**:25

   linear approximation, **33**:25, 26

   momentum loss rate, **33**:24, 25, 41, 42

   variational principle, **33**:24

   viscosity of ^{4}He, **33**:22, 23

Superfluorescence, **50**:14

Supergraph, **43**:123

Superheating, **40**:138

Superimposed bases, split and double

shifts, **7**:134

Superionic conductor, **52**:488

Superlattice, **2**:120

Supermultiplet, **52**:35

Supermultiplicative function, **51**:8, 10, 31

Supernumerary rainbow, **12**:365; **30**:327, 342

Superoperator, **41**:345–348

Superparamagnetism, in microclusters, **40**:58

Superposition approximation, **11**:68; **37**:111, 142, 143, 145; **40**:300; **53**:196–201

   calculation of virial coefficients, **6**:269–270

   hard sphere fluids, **6**:235–237, 245–246, 265

   radial distribution function, **6**:185

Superposition of configurations, **2**:207, 250, 285, 295, 323; **9**:337

Superposition of phase equilibrium, **17**:14, 17

Superposition principle, **4**:88, 95; **25**:70–72, 90, 102, 115ff, 149, 167, 174; **30**:99, 103, 131

   distribution functions, **1**:153, 154

   relaxation processes, **1**:85

Superradiance, **25**:29; **34**:252; **46**:269

Supersaturation, **40**:137, 138, 140, 144, 149

   critical size, **40**:149

Superselection rule, **38**:34, 44

   Abelian, **38**:44, 47

Supersonic, *see also* Hypersonic

Supersonic beam, **47**:Part 1:5–6; **52**:265, 270–272

Supersonic jet, condensation in, **10**:290

   flow field of, **10**:285

   viscous effects in, **10**:284–287, 297–300

see also Nozzle beams

Supersonic nozzle, 40:54; **42**:360, 443

Supertrap, 40:455, 456

Supporting function in convex molecules, 5:149

Surf-riding condition, 27:448

Surface, 20:216, 228, 248

alloys, **20**:303

atomic motion, **27**:1, 30

current density, **27**:558

excess, **27**:558

diffuse, **27**:603, 604

electronic structure, **27**:1, 32

at equilibrium, **40**:190–197

free, **40**:137

groove growth on, **21**:669

homogeneous, **27**:557

ideal, **27**:142

inhomogeneous, **27**:557

magnetic structures, **20**:314

mercury salts, **9**:53

reconstruction, **20**:322

semiconductors, **27**:340, 345

sensitivity of electron spectroscopies, **27**:9

studied by photoemission spectroscopy, **49:49**:533–656

see also Dielectric surface; Gas-solid dynamics

Surface atom displacements, 20:290, 294, 300, 308

Surface cell, 7:311ff

Surface charge, 27:283, 284, 286, 367, 372

density, **27**:558

Surface-chemical potential, 6:146, 161

Surface compounds, 9:166

Surface crossing problem, 49:296–300

Surface diffusion, 40:166, 209, 224

Surface dynamics, 20:248

Surface energy, 1:203, 208–213, 231, 236; **26**:136, 137, 139; **27**:343, 546, 553; **41**:165, 166–168

liquid metal, **27**:553

Surface entropy, 1:228, 229

Surface excitons, 54:305–326, 427–428, 432–433

experimental evidence, **54**:319–320

fluorescence, **54**:320

reflection, **54**:319–320

transmission, **54**:319

simple theory, **54**:316–318

see also Excitons

Surface extended X-ray absorption fine structure measurement, 49:602

Surface fluctuations, 48:352–354

modes of, bending, **48**:350

squeezing, **48**:350

uncoupled, **48**:350

Surface fluorescence, 28:280, 282

Surface free energy, 1:229; **40**:142

Surface geometry, 9:146

Surface hopping, 30:59, 194; **42**:31, 104, 516, 524, 567

Surface impedance, 27:365, 379, 453

Surface ionization, 28:12, 29, 79; **42**:535

Surface lattice, 9:88

Surface layer, 7:311ff

Surface light scattering, 48:344–346

angular dependence of, **48**:346

by carbon dioxide, **48**:342

instrumental broadening effects, **48**:389

by water, **48**:352

Surface magnetism, 51:5

Surface migration, 40:187, 188

Surface modes, 27:268, 269, 357

extraordinary, **27**:528

generalized, **27**:431

mixed, **27**:467, 472, 497, 518

nonradiative, **27**:440, 441, 429, 527–528

optical phonons, **27**:357

ordinary, **27**:528

peeling off, **27**:473, 499

pseudosurface modes, **27**:467, 472, 497, 518

radiative, **27**:283, 292, 303ff, 341, 380–387, 408, 527

regular, **27**:431

resonant, **27**:518

strongly localized, **27**:490

transverse, **27**:450

Surface of section, 36:27

Surface phase, adsorption **6**:169–170

    Gibbs' law, **6**:151–153

        Langmuir's formula, **6**:170

    equation of state, **6**:167–168

    Gibbs' theory of capillarity, **6**:145

    ideal, **6**:163

    partial molar area, **6**:155, 159

    surface tension, *see* Surface tension

    Szyszkowski formula, **6**:166–167

    thermodynamics of, **6**:145–174

Surface photoelectric effect, 49:551

Surface plasma oscillation dielectric function, **27**:273–276

Surface plasma radiation, **27**:334

Surface plasmons, **20**:244; **27**:286, 309, 358, 416, 440, 454, 460, 578; **37**:1

    coupled with photons, **27**:294, 304, 328, 331, 339

    dispersion relation, **27**:582

    electron interaction, **27**:312, 315, 316, 321, 325

    metal-dielectric boundary, **27**:582

    nonradiative, **27**:580

        coupling to electromagnetic field, **27**:585, 586

        diffraction grating, **27**:586, 587

        dispersion, **27**:596

        diffuse surface, Epstein profile, **27**:611

        Lorentzian resonance, **27**:596

        second branch, **27**:603

    frustrated total reflection, **27**:590

    surface roughness, **27**:586, 587

    phase velocity, **27**:583

Surface polaritons, **27**:369; **37**:41

    coupled with photons, **27**:340

Surface pressure, **48**:335

Surface profile, **27**:346

Surface reactions, **27**:3

Surface reciprocal lattice vector, **49**:552

Surface resonance, **49**:571–574, 579–580

Surface roughness, **27**:327

Surface scattering, **24**:55

Surface sealing, **51**:38

Surface states, **9**:89, 98ff, 112, 131, 156, 164; **49**:571–593

    Al, **49**:568–569, 573, 577

    contamination, **49**:579–580

    crystal field methods, **9**:164

    Cu (111), **49**:571–573, 577

    GaAs, **49**:583–587

    localized, **9**:91ff, 98ff

    magnetic Ni (100), **49**:578

    Pd (111), **49**:580–581

    Si (111), **49**:583, 589–591, 617

    Ti (0001), **49**:608

Surface structure, **20**:223, 228, 264, 272, 296ff, 314–324; **27**:1, 2, 154

    adsorbed overlayers, **27**:145, 146, 148, 149, 159

    atomic geometry and symmetry, **27**:1ff, 25, 28, 142ff

    ideal surfaces, **27**:142, 143

        terrace-ledge-kink model, **27**:143

    nonideal surfaces, **27**:2, 150, 159,

181
classification of imperfections, 27:2, 142
disorder of the first kind, 27:183
disorder of the second kind, 27:183
disorder of the third kind, 27:183
multistep topographies, 27:2
single step imperfections, 27:2, 143
statistical mechanical model, 27:182
atomically flat, 27:8, 142
clean surfaces, 27:159
constant-momentum-transfer averaging method, 27:176
data reduction methods, 27:155, 156, 173
energy averaging methods, 27:180
Fourier transform methods, 27:157
kinematic methods, 27:173
microscopic model methods, 27:155
multistep topography, 27:143
Patterson function methods, 27:157
Surface tension, 1:203ff, 208–211, 218–220, 226–232; 6:282, 283; 9:23ff, 46ff, 80, 125; 24:55, 56; 26:179, 187, 192, 194; 27:552; 41:43–46, 47ff, 175
binary mixtures, 11:106
binary solutions, table of, 49:434
compressibility function, 9:72
correlation function form, 46:286, 295
corresponding states equation, 11:93
curvature dependence, 9:24, 46
density functional form, 46:286, 290
effect of pressure on, 6:172
Fowler form, 46:289, 295
halogen-containing carbons, 9:74
hard sphere fluids, 6:255, 261, 264, 268, 272, 280, 282
hydrocarbon liquids, 9:73
interfacial, 9:47
ionic systems, 11:93
k-dimensional, 46:303
Kirkwood-Buff formula, 46:288
liquid metals, 9:78; 27:553
copper, 27:554
under equilibrium conditions, 27:554
liquid mixtures, 6:163–168
mixed alkali nitrates, 11:107
molten salts, 9:52
non-metallic elements, 9:76
polar organic molecules, 9:75
potential form, 46:286
quantum mechanical, 46:289
reduced "thermodynamic", 11:93
Riezenberg-Zwanzig formula, 46:292
simple liquids, 9:52
van der Waals, 46:294, 295
water, 9:78
Surface waves, 20:230, 258; 46:317
Surface zone, 27:550
Surprisal analysis, 30:42, 307; 42:127; 47:Part 1:244–264; 50:482, 611; 53:343, 365–366
collisional activation and deactivation, 47:Part 1:257–261
distribution of rotational states, 47:Part 1:261–263
elimination reactions, 47:Part 1:245–248
F + RH, 47:Part 1:248–250
heavy ion transfer reactions, 47:Part 1:263–264
maximum entropy, 47:Part 1:250–254

multiphoton dissociation, **47**:Part 1:254–257

Surveillance, immune, **38**:394

Survival of fittest, **38**:235

Susceptibility, **12**:245; **17**:73; **33**:242, 248; **41**:116

electric, **1**:335

generalized, **31**:237

magnetic, *see* Magnetic susceptibility

perpendicular, **41**:165

spin, **41**:117

statistical theory of, **17**:75

tensor field, **33**:226

Swallowed shock, **10**:297

Swarm experiments, **28**:428

Swarm size, **44**:385

Swelling of heated macromolecules, 7:210, 228

Swihart's mode, **27**:324

Symbolic addition formula in phase determination, **16**:159, 161–197

Symmetric exchange reaction, **1**:24

Symmetric function, **8**:36

Symmetric random walk, **52**:437

Symmetric resonance in charge transfer, **10**:197–198, 239–240, 254

Symmetric stretching frequency, **1**:56

Symmetric top, **4**:230, 248, 250, 256, 263, 278, 299; **44**:11, 364, 449, 462;

photon catalysed dissociation, **50**:220

scattering from, **30**:377, 397

Symmetrized line shape, **51**:80

Symmetry, **20**:219, 229, 332, 334

broken, **43**:218

broken continuous, **49**:352

crystalline, **20**:8; **27**:470–471

group, **49**:185–186

molecular group, **49**:186–187

packing, **44**:478

permutation group, **44**:577

point group, **44**:576, 577

rotation-inversion, **49**:186, 342

spontaneous, **49**:312

translational, **49**:338, 350

two-dimensional structures, **20**:219

Symmetry breaking, **20**:72, 77; **29**:2, 3, 5, 64; **32**:4, 94; **38**:103, 272, 276; **55**:186, 189

chemotaxis, **29**:181

first primary branch, **38**:272

irreversibility and, **29**:13

Liesegang rings and, **29**:55

mitosis, **29**:54

photochemical reactions and, **29**:58

spatial, **46**:323

spontaneous, **31**:74

Symmetry breaking instability, **40**:240

Symmetry breaking variables, **31**:254

Symmetry constraint, **40**:363

Symmetry coordinates, **5**:255; **44**:10

Symmetry displacement, **5**:255

Symmetry-forbidden reaction, **42**:22

Symmetry-forbidden transition, **18**:19, 20, 39, 42, 63, 65

Symmetry number, **1**:20, 55; **16**:140

Symmetry operations, **5**:251, 253, 255; **20**:251, 262

in calculation of vibronic perturbations, **5**:251

Symmetry properties, atoms and molecules, **52**:1–111

wave functions, **1**:167

Symmetry-restoring, **49**:353

Symmetry-restoring variable, **46**:314

Symmetry-restricted covalency, **5**:68; **8**:53

Symmetry rules

photoemission spectroscopy, Al (100), **49**:575, 603

CO adsorption, **49**:542, 619–620,

625
Cu (100), **49**:556–558
excitation mechanism, **49**:534
magnetic Ni (100), **49**:578
N$_2$, **49**:544
TaS$_2$, **49**:548
Symmetry transformation, 38:16
Symmetry type, 8:48
Synapse, 19:340; **21**:604, 619
Synaptic vesicles, 21:622
Synchronization, 43:230
Synchronous amplification, 44:329
Synchronous detection, 44:329
Synchronously pumped laser, 49:2–10, 14–15, 19–21, 34–42
Synchrotron radiation, 42:227; **49**:534; **50**:258, 267; **52**:268
Syndiotactic placement, 2:166
Syneresis, 22:119, 125
Synergism of nonlinearity, 46:206
Synzyme, 39:109–176
  insolubilization, **39**:167
  minimum size, **39**:169
  specificity in catalysis, **39**:158, 168
  versus enzyme, **39**:157, 168, 175, 176
  see also Polyethylenimine
System, antidynamical, **38**:91
  classical, deterministic, **38**:47
    stochastic, **38**:48, 52
  closed, **38**:16
  definition of, **38**:11
  deterministic, **38**:92
  dynamical, **38**:91
  final, **38**:100
  illuminated, **29**:58
  nonanticipatory, **38**:91
  nondeterministic, **38**:92
  open, **29**:2, 31, 64
  quantal, **38**:49, 73
  size, **46**:196

System of equations of large scale fluctuations, 38:202
Systematic components, Brownian motion, **53**:82
  heatbath, **53**:90–91, 95
  primary system motion, **53**:76, 77
  solvent forces, **53**:78
Systematic uncertainty, 44:334
Szasz equation, 16:325–327
Szegö's theorem, 15:336

T

T$_1$, 25:9, 16
  double resonance, **25**:57
  longitudinal relaxation time, **40**:388ff, 418, 464, 469, 476
  power saturation, **25**:15
  pulse trains, **25**:47
  transient absorption, **25**:22
T$_2$, 25:9, 16
  double resonance, **25**:57
  pressure broadened, **25**:15
  pulse trains, **25**:47
  transient absorption, **25**:22
  transient emission, **25**:34
  transverse relaxation time, **40**:388ff, 419, 424ff, 464, 466ff, 472ff
T-matrix, 25:104; **31**:276, 277, 307; **42**:21
  center of mass representation, **33**:264
  didadic, **33**:251
  energy-shell elements of, **33**:252, 254
  L-space, two-body, **33**:265
  medium-adapted, **31**:307
  N-molecule, **33**:251
  pathological behavior, **31**:307
  tetradic, **33**:252
  two-body, **33**:237, 264, 265

T-operator, **25**:103

Tamm-Dancoff approximation, **36**:259; **44**:604, 610, 613, 622

Tamm's surface states, **9**:102ff, 106ff, 121, 139, 149, 156
energy, **9**:131
tetragonal lattice, **9**:119

Tanabe-Sugano determinants, **5**:72, 74

Tanabe-Sugano diagram, **8**:51, 59

Tandem oscillators, **43**:155

Tangent formula, **16**:152–153, 160–161, 187–191, 198–199, 204

Tangent mapping, **46**:108

Tangent reaction, **43**:129

Tangential pressure, **1**:206, 216, 221, 223, 226

Target theory, **7**:263

D-**Tartaric acid**, **16**:207

Tartrate ion, **7**:421

TaS$_2$, **49**:545, 547–548

Tauberian theorems, **52**:366, 401, 471
Laplace transforms, **52**:388–389
power series, **52**:396, 397

Tautomeric forms, **7**:103ff

Tautomeric rearrangement, **7**:115, 154

Taxonomy, of groups, **29**:199, 223
of proteins, **29**:192

Taylor number, **26**:207; **32**:119
subcritical, **32**:126

Taylor series, **40**:141

Taylor turbulent vortices, **32**:129

Taylor vortex flow, **32**:109, 122, 123, 131
wavelength, **32**:119, 131

Taylor's theorem, modified, **35**:64

Technetium, **41**:141

Technetium hexafluoride, **9**:285ff

TeH, electron resonance spectra, **18**:216, 228, 233

Telegraph equation, **52**:382, 383

Teleological explanation, **38**:91

Tellegen's theorem, **43**:27

Teller-Redlich product rule, **1**:20, 33, 34

Tellurium, quenching of, **50**:480

Temkin mean field theory, **40**:168

Temperature, **19**:332, 337, 342; **38**:6
boundary conditions for, **33**:429
characteristic quantum, **16**:35, 52–56
critical, **40**:138
dipole-dipole, **11**:303, 309, 315
generalized, **19**:327
primitive earth, **55**:87–88
reduced, **16**:225, 228, 233, 238ff, 248ff, 260ff, 278
rotational, **31**:160, 164, 179, 186–189, 196
spin, **11**:296
thermodynamic, **33**:410
translational, **31**:160, 164, 170, 186–189
Zeeman, **11**:309, 315

Temperature coefficient of resistivity, **51**:239ff, 249, 269, 270

Temperature-independent magnetic susceptibility, **5**:193

Temperature-independent paramagnetism, **7**:167

Temperature jump, **46**:398, 407, 410

Tendons, **7**:270, 277
reactions with aldehydes, **21**:559
shrinkage of, **21**:559

Tensor, irreducible, **31**:160

Tensor algebra, **9**:223

Tensor polarizability, **10**:3–4, 20–21, 25

Tensor polarization, **12**:563, 564

Tensor product, **44**:300

Term formula, **49**:153–155

Termination of polymerization, **2**:173

Ternary flames, **42**:15

o-Terphenyl, melting of, **6**:473

Tertiary process, **44**:399

Tesselation, **51**:21

Tetraaminopyrimidine, **7**:53

Tetraantimony hexaoxide, **20**:28

Tetraarsenic hexaoxide, **20**:28

Tetraboron tetrachloride, **20**:22

Tetracene, **47**:Part 1:6
  exciton bands, **23**:327

1, 2, 4, 5-Tetrachlorobenzene, **40**:408ff, 415ff, 424ff, 545ff, 466, 474ff
  spectrum of, **40**:406
  triplet excitons, **40**:453, 467

p-Tetrachlorohydroquinone, influence of radiation on, **2**:198

Tetracyanoethylene, **20**:17

Tetracyanoquinodimethane, electron affinity, **42**:564

n-Tetradecanoic acid, **41**:196

Tetrafluoromethane, **2**:32

Tetragonal field, **7**:371

Tetragonal lattice, **9**:119

Tetragonal symmetry, **5**:36, 43, 84ff

Tetrahedral complexes, **5**:36, 43, 80

Tetrahedral molecules, **12**:130
  fluid structure, **53**:45

Tetrahedral orbitals, **7**:18

Tetrahydrofuran, **47**:Part 1:641
  infrared spectra, **7**:494ff
  multiphoton dissociation, **50**:141

Tetrahydroquinone, quadrupole spectrum, **2**:195

Tetrahydroxy-*p*-benzoquinone dihydrate, **16**:174–175

Tetramer, water, **31**:27

Tetramethyl dioxetane, cleavage of, **42**:29, 30

2, 3, 4, 6-Tetranitroaniline, **16**:179

Tetraphenylcyclobutane, **16**:177

Tetrapyrroles, **7**:380, 399

Tetrazine, photodissociation of, **50**:314
  vapor, **49**:33

Thallium, thermodynamic functions of solutions of noble metals in, **2**:133

Thayer-Yardley plot, **42**:224; **47**:Part 2:294

Theile semi-invariants, **6**:423

Thermal conductivity, **1**:2, 134, 137, 139, 151; **5**:263, 292; **13**:350; **24**:272, 277, 283, 291, 310, 315, 329, 333; **31**:161, 164, 173, 174, 177, 186–189, 193, 196, 209; **37**:154, 186
  of gases in the critical region, **6**:223
  of reacting systems, **16**:103, 109, 123
  see also Conductivity, thermal; Heat flux

Thermal denaturation, **46**:365, 369, 377, 386, 391, 397

Thermal desorption, **41**:492

Thermal diffuse scattering, **20**:249, 289; **27**:64, 69

Thermal diffusion, **1**:135; **16**:101–130; **24**:38, 310; **32**:271, 274, 275, 281, 284
  chemical contribution, **16**:102, 109ff, 122–123
  flow type cell, **32**:273
  influence of chemical kinetic isotope effect on, **16**:103
  influence of reacting components on, **16**:103
  isotope separation, **16**:127
  liquid alcohol-water mixture, **16**:122
  physical contribution, **16**:112–113, 115–122
  presence of transformation, **16**:108ff
  Soret coefficient, **32**:269, 277
  static cell, **32**:269

Thermal diffusion coefficient, **16**:104, 128
  experimental values, **16**:113–121,

123, 126

polynary, **16**:104

reduced, **16**:126

total, **16**:103, 108, 115, 126

Thermal dissociation, 42:5

Thermal effects, ionic crystals, **27**:504–505

monatomic crystals, **27**:475–483, 523–524

proteins, nucleic acid, and viruses, **7**:201

Thermal energy beams, 12:364

Thermal energy, scattering at, **10**:119–120, 123, 125, 187

Thermal expansion of liquid, 6:281

Thermal fluctuations, 15:65ff

Thermal grooving, 21:673

Thermal inhomogeneous broadening, 47:Part 1:5

Thermal neutron scattering, 44:285

Thermal noise, 26:185, 206

Thermal relaxation, 24:9, 20, 21

Thermal stress, 5:292

Thermal wavelength, 20:43

Thermally activated hopping model, 31:468–471

Born-Haber cycle, **31**:471

Einstein mobility relation, **31**:468

electron exchange frequency, **31**:469, 470

electron mobility, **31**:469

ionization potential, **31**:471

Madelung energy, **31**:471

pair correlation function, **31**:468

probability density function, **31**:468

Thermally activated trap, 40:448

Thermochemical separation, 16:122

Thermochemistry, 52:353–354

Thermodiffusion, *see* Thermal diffusion

Thermodynamic analysis of transport, 39:328

Thermodynamic branch, 29:31, 64; **38**:266, 272, 273, 364; **39**:4, 8, 14

Thermodynamic buffering, 55:150–158

Thermodynamic constraint, 29:305, 344

Thermodynamic data, from diffusion measurements, **6**:300

Thermodynamic derivative, Mori equation, **53**:148

Thermodynamic equilibrium, 49:385–395, 425

Thermodynamic evolution criterion, 39:5

Thermodynamic factors in diffusion, 6:294, 301, 307, 312

Thermodynamic flux, 43:100

Thermodynamic force, 43:100

Thermodynamic functions, 17:12, 26ff; **40**:232

enthalpy, **17**:26, 27

entropy, **17**:27

excess enthalpy, **17**:26, 27

excess Gibbs free energy, **17**:27

excess volume, **17**:27

heat capacity, **17**:28, 31

Helmholtz free energy, **17**:32

mixing functions, **17**:29

volume, **17**:27

see also specific thermodynamic functions

Thermodynamic limit, 38:174, 178, 179, 181; **40**:291, 308; **41**:10–11

Thermodynamic parameters, 43:27

Thermodynamic potential, 17:44; **19**:217, 228, 297, 298

analytic continuation, **17**:54

conjectured for Ising model, **17**:54

expansion about point on coexistence curve, **17**:46

expansion about point on critical isotherm, **17**:46

Thermodynamic singularity, **40**:362

Thermodynamic stability criterion, **19**:219; **39**:5; **49**:399

Thermodynamic system, in contact with bath, **51**:130–135

Thermodynamic threshold, **38**:364, 406, 408

Thermodynamics, **32**:164; **43**:42, 94–100; **46**:165

  of chemical lasers, **47**:Part 2:72–80

  lasing criteria, **47**:Part 2:72–75

  maximal work and constraints, **47**:Part 2:75–76

  time dependence, **47**:Part 2:77–80

  generalized, **32**:1, 21

  irreversible, **1**:2; **16**:101, 102, 103; **29**:12, 30; **32**:13, 165

    linear, **3**:35; **29**:16, 30, 73

    nonlinear, **3**:39; **29**:17, 31, 73, 337

  quantum systems weakly coupled to thermal reservoirs, **38**:109–142

  nonlinear, **32**:61

  stellar systems, **26**:112, 115–118, 130, 139

Thermoelasticity, **32**:14

  nonlinear, **32**:15

Thermoelectric effects, **5**:305

Thermoelectric power, metal-ammonia solutions, **4**:322, 323, 333, 336, 370–371

  metallic glass, **51**:254–257, 269

Thermogravitation, interaction with thermal diffusion, **13**:350

Thermoluminescence, **41**:306, 307

  chlorophylls, **19**:353, 372, 373

Thermostatic availability function, **33**:417

Theta formalism, **9**:29, 32

Theta graph, **51**:11

Theta solvent, **51**:4

Theta temperature, **51**:4

Thin film, light scattering by, **48**:329–396

Thin film enhancement effect, **37**:23

Thiodiglycollic acid radical, **7**:553

Thionyl chloride, **16**:181

Thiophene, **40**:473

Third memory matrix, **44**:289

Thom classification theorem, **40**:244

Thomas-Fermi-Dirac treatment, **12**:348; **41**:63, 66

Thomas-Fermi theory, **41**:59–64, 68, 118

  relativistic, **41**:81

Thorium oxyperchlorate, **34**:198

Thorium perchlorate, **34**:198

THQ dihydrate, **16**:174–175

Three-body collision rate, **13**:106

Three-body correlation function, **53**:2

Three-body forces, **5**:186; **12**:160, 295, 299

Three-body potential, **12**:294, 300; **53**:13, 46, 47

Three-body reactive system, **49**:191–310

Three-body recombination and dissociation, **13**:102

Three capacitor model, **37**:141

Three-center elimination, **42**:139; **50**:143

Three-center integral, **36**:166

Three-center reactions, **1**:30

Three-level double resonance, **47**:Part 1:569–579

  inverted Raman, **47**:Part 1:574–575

  ladder, **47**:Part 1:570–574

Three phase equilibrium, **2**:18, 37

Three phase line, **2**:20, 22, 44; **17**:5, 8, 12, 15

  ice-hydrate-gas, **2**:23

Three-pulse echo, 40:467

Three-step model, 49:286

Threshold behavior, cross section, **36**:487

   endothermic charge transfer, **10**:256–257

   energy, **30**:422, 432

   excitation, **30**:17, 419, 450, 451, 454

   ionization, **30**:492, 529

   reactive systems, **10**:150, 153–154, 157, 162–164

   region of, **30**:121

Threshold-electron excitation, 36:422

   acetophenone, **36**:429

   benzaldehyde, **36**:429

   benzene, **36**:429

   benzoic acid, **36**:429

   fluorobenzene, **36**:429

   hexafluorobenzene, **36**:429

Threshold energy, 36:159; **41**:260, 265

Threshold excitation in electron impact spectroscopy, 18:19, 20, 43

Thymer E, 16:184

Thymidine, 16:184

Thymidylthymidine, 16:183

Thymine, 7:52ff, 85, 349

   internuclear distance, **7**:87

   molecular diagrams, **7**:94

   transition moment, **7**:95

THz spectroscopy, *see* Far infrared

Tidal interaction, 35:155–157, 171, 184

Tight-binding theory, 9:114ff; **53**:279

Tight complex, 42:121

Time, to absorption, **52**:410

   anisotropy of, **38**:8

   arrow of, **38**:6, 90

   complex, **25**:134ff, 139ff, 160ff; **51**:67

      branch points, **25**:136–140, 143, 153, 160, 174

to trapping, **52**:467

universal, **38**:8

Time autocorrelation matrix, 44:300

Time correlation functions, 17:65, 99, 113; **24**:311; **31**:253; **46**:265

   derivatives, **24**:312

   Gaussian approximation, **24**:312

   irreversibility theory, **53**:128–129

   long time decay, **24**:325

Time-dependent friction tensor, 44:266

Time-dependent Ginzburg-Landau model, 49:348, 353

Time-dependent Hartree-Fock theory, 44:608; **48**:15

Time-dependent Hartree theory, 44:616

Time-dependent Kirkwood *g* factor, 44:388

Time-dependent long-range order, 46:261

Time-dependent pair distribution function, 12:581; **51**:55, 68

Time-dependent perturbation theory, 5:243

Time-dependent projection operator, 42:292

Time-dependent thermodynamic functions, 47:Part 2:77–80

Time-dependent transformation, 46:215

   integral, **46**:238

   nonlinear, **46**:211

Time evolution, 35:52; **38**:10, 17, 49

   bidirectionally deterministic, **38**:10

   collisionally perturbed molecule, **47**:Part 2:348–350

   dynamical state, **35**:52

   ensemble averages, **35**:55

   semigroup, **38**:6

   singlet distribution function, **35**:66

   states, **38**:16

   stochastic, **38**:3

   unitary, **38**:3

universe of discourse, **38**:17

Time of flight technique, 30:285, 319, 456, 457; **42**:442, 446, 530; **44**:308; **47**:Part 2:134; **50**:257; **53**:52

Time ordered product, 41:123

Time ordering operator, 36:213

Time-resolved observable, 47:Part 1:13–14

Time-resolved spectroscopy, 49:21–34

Time reversal, 17:80; **43**:231, 260; **44**:11, 467; **52**:136–137, 179

Time reversal symmetry, 44:295, 345

Time scale, crossing, **35**:121, 136, 146, 182

fluid-dynamical, **35**:137

hybrid, **35**:175, 175, 178, 185

origin of life, **55**:86

relaxation, **35**:121, 136

secular, **35**:121

stellar systems, **26**:110, 113–115, 124, 133, 134, 141

Time symmetry breaking, 55:187

Tin, partial molar heats of solution in, **2**:138

Tin-lead solution, enthalpy of formation, **2**:143

TiO$_2$, 27:435, 447

Titanium, 41:141

(0001), **49**:206

Titanium complexes, vibronic perturbations, **5**:256

Titanium hexafluoride anion, vibronic perturbations, **5**:244

Titanium hexaquo cation, oscillator strength, **5**:252

vibronic perturbations, **5**:252

Titanium selenide, 41:160

Titanium sulfide, 41:159, 160

Tobacco mosaic virus, 7:222

photoreactivation, **7**:586

Toda lattice, 53:297–302, 311, 314, 321

phenomenological model, **53**:317

shock waves, **53**:308

structure factor, **53**:316–317

Toeplitz determinants, 15:333ff

Toeplitz matrices, 52:412

Toluene, 1:62, 99, 106, 114, 115

two-photon L_b spectrum, **54**:191–192

Tonks equation of state, 6:256, 257; **40**:325

Topological dynamics, 38:7

Topological soliton, 53:320

Topological structure of microclusters, 40:64–66

Toroid, 36:26, 31, 57

Torque, autocorrelation function, **44**:391, 429, 433

damping, **44**:362

derivative, **44**:280, 427

molecular, **34**:6, 20, 53, 99

Torque-dependent parameters, 44:470

mean, **44**:478

net force, **44**:402

term, **44**:350

Torsional energy, 22:48

Torsional motion, 40:430

Torsional oscillation, 2:375, 377, 380

Torus, 37:109

Total cross section, 42:396, 404, 453

Total internal reflection, 27:445

Toxvaerd's model, 41:209

Trace, theorem of, **1**:25

Traction vector, 49:359–360

Trajectory, 51:78, 95, 97

classical, **30**:77, 78, 87, 97, 201, 480; **42**:92, 93

complex, **30**:103, 132

divergence of, **47**:Part 1:206

free-radical, **53**:109

heatbath modeling, **53**:91, 99, 100

irregular, **46**:79

Monte Carlo, **30**:10, 489

properties of, **47**:Part 1:119–132

quantum, **53**:366

quasiperiodic, **46**:75, 76

recombination dynamics, **53**:108

separation of, **46**:82

solute-atom, **53**:77

space, **38**:69

spectral properties of, **46**:83

statistical behavior and, **47**:Part 1:231–233

statistical description, **38**:77

surface hopping, **30**:59, 175, 194; **49**:274

surface of section, **46**:81

vibrational relaxation of iodine molecules, **53**:210–211

Trajectory calculation, **41**:255–259; **50**:492; **53**:105–106

collisional ionization, **42**:524ff

Trans **effect**, **7**:374

Trans-**labilizing effects**, **7**:367

Trans **polymerization**, **2**:170

Trans **state of lipids**, **41**:196, 201, 202, 204, 205, 209, 210ff

Transconfiguration of dipoles, **1**:93, 95, 97, 105, 106

Transdetermination and homeosis phenomena, **55**:221–223

Transelectronases, electronic structure, **7**:71

Transfer Hamiltonian, **27**:231

Transfer-integral method, **53**:241, 242, 248, 256

Transfer ionization, **10**:199, 202

Transfer of charges, **1**:170, 262

Transfer spin correlation length, **53**:256

Transformable components, gradients of, **16**:108

Transformations, martensitic, **16**:353–354, 369, 376ff

deformation, **16**:353

spontaneous, **16**:353

nonlinear, **46**:211, 212

Transient absorption, **25**:21, 25, 28, 35

Transient behavior, **38**:346

Transient birefringence, **49**:32

Transient dichroism, **49**:32

Transient emission, **25**:29

after π, $\pi/2$ pulse train, **25**:42

after $\pi/2$, π pulse train, **25**:39

after pulse, **25**:35

after steady state absorption, **25**:31

Transient growth rate, **40**:187

Transient mutation, **25**:21; **35**:24

Transient phenomena, **46**:197, 235

Transition coherence, **40**:379, 381, 383

Transition complex, **1**:26, 64, 65

Transition kernels, **20**:139; **26**:222

approximate, **20**:150

detailed balance property, **20**:145

hard sphere gas, **20**:145

iterated, **20**:168, 208

matrix elements, **20**:150

moments, **20**:150

multistep probabilities, **20**:168

positive-definite character, **20**:142, 151

representation, degenerate kernels, **20**:152

orthogonal functions, **20**:151

separable, **20**:176

square integrability, **20**:147, 170

synthetic, **20**:150

transfer moments, **20**:150

Transition layer, **1**:204, 206, 207, 216, 220

Transition metal, **40**:146

magnetic susceptibility of, **3**:220

Transition metal complexes, 4:176

central field covalency, **5:**68

charge density, **5:**251

electron transfer spectra, **5:**75ff; *see also* Electron transfer spectra

electronegativity scale, **5:**126

electronic energy, **5:**251

forbidden transitions, **5:**241ff

halides, **34:**195

interelectronic repulsion parameters, **5:**68, 69

ligand field theory, **5:**42, 43, 44, 47, 48, 49

octahedral symmetry, **5:**50, 62, 72

Orgel diagrams, **5:**59

Pauli's hole equivalency theorem, **5:**44

principle of microsymmetry, **5:**127

square-planar symmetry, **5:**54, 66

stabilization, **5:**59

tetrahedral symmetry, **5:**80

Helmholz-Wolfsberg method, **5:**81

nephelauxetic series, **5:**68

optical spectra, **5:**75ff; *see also* Absorption; Electron transfer

orbital configuration, **5:**40, 62, 71

Racah's parameters, **5:**40, 71

Russell-Saunders coupling, **5:**39, 42, 46

spectrochemical series, **5:**65

symmetry-restricted covalency, **5:**68

Tanabe and Sugano's determinants, **5:**72, 74

vibrations of, **5:**251

vibronic perturbations, **5:**243, 253, 254

Transition metal ions, 4:174–188

paramagnetic, **53:**229–234

Transition moment, 7:93, 100; **8:**40;

44:551ff, 566, 573, 576, 591, 601ff, 611ff, 627

changes induced by an electric field, **13:**183

electric dipole, **4:**109, 163, 164; **44:**552ff, 563, 579, 582, 627, 632

electric quadrupole, **44:**552

fully retarded, **44:**551, 563, 589, 590, 592

magnetic dipole, **4:**109; **44:**552ff, 564, 579, 581, 595, 597, 627

origin dependence, **44:**584, 600, 608, 627, 628

origin independence, **44:**608

Transition moment operator, electric dipole, **44:**553

fully retarded, **44:**550

magnetic dipole, **44:**553

static magnetic field, **44:**598

Transition probability, 1:364, 368, 369; **10:**55–56, 126–127; **14:**162ff; **17:**88; **18:**17; **30:**91, 93, 126; **34:**248, 254, 260, 284; **36:**208, 211, 279, 283; **46:**400

collision theory, 1:354, 359ff, 363, 364, 367, 369, 370, 372

determined from oscillator strength, **5:**241, 243

forbidden transitions, **12:**75

irregular spectrum, **46:**122

random walk, **4:**2, 3–15, 16, 29, 45, 53

regular spectrum, **46:**122

Transition probability matrix, 1:377, 388; **34:**41, 49

Transition radiation, 27:318, 320

Transition rate, 43:247

between sites, **44:**397

Transition rate matrix, 33:78, 80, 102, 103, 114

cofactors, **33:**89, 90, 102, 115

eigenvalues of, **33:**102, 115

eigenvectors of, **33**:102, 115

Transition state, **33**:118, 134; **37**:288; **39**:109, 146, 150, 152, 161; **41**:65; **42**:64

alternate, **33**:34, 375, 377

biradical, **33**:379

coordinates, **25**:267

cycloaddition, **33**:375

Diels-Alder reaction, **33**:365, 375, 378

electrocyclization, **33**:374

H atom abstraction, **33**:372, 373, 378

hydrolysis, **33**:390

ion exchange, **33**:391

Möbius-like, **33**:376

as 'quasi-system', **33**:361

radical addition, **33**:365, 366, 368

sigmatropic reaction, **33**:376

Slater, **41**:90

solvolysis, **33**:391

symmetric, stability of, **33**:377

Transition state theory, **1**:16ff, 55, 66ff, 125, 179–182; **30**:16, 45; **33**:346, 347; **48**:72

applied to crystal growth, **21**:717

partition function of water, **21**:489, 490

of quenching, **50**:348

Transition temperature, proteins, **33**:105

two-state model, **33**:116, 132

Transition zone, **46**:281

inhomogeneous, **27**:547

liquid insulators, **27**:547

liquid metals, **27**:549

Transitional approximation, **30**:323

Transitions, **38**:327, 331

band-to-band, **40**:392, 414, 415

bound-to-free, **51**:60, 90

constant constraints, **38**:331

due to change in constraints, **38**:327

hard, **38**:327, 328

involving pseudostationary states, **38**:332

non-adiabatic, **30**:59, 132, 423, 465

soft, **38**:327

vibrational, **30**:49, 59, 92, 122

Translation group of surface lattice, **9**:88

Translation-rotation correlation function. *See* Rototranslational correlation function

Translation-rotation coupling, **17**:202; **44**:284

liquids, **34**:15

Translational absorption of rare gas mixtures

approximate dynamical methods, **51**:63–69

classical calculation, **51**:63

collision-induced, **51**:56–71

He-Ar, **51**:61

potential free model, **51**:64

quantum calculation, **51**:58–63

Translational diffusion coefficient, **17**:138

Translational invariance, **27**:356, 363, 396, 452; **40**:308

Translational motion, **34**:13

Translational phonon, **40**:472

Translational Raman scattering, **51**:84

Translational relaxation process, in collision theory, **5**:361–362, 368, 369

Translational space itinerant oscillator, **44**:373

Translational spectrum, classical and quantum calculations of, **51**:89–99

Translational temperature, **31**:160, 164, 170, 186–189

Translational velocity, **44**:441

Translationally equivalent molecules,

40:408, 415, 465

Transmission coefficient, 1:19, 24, 53, 61, 127; *see also* Absorption coefficient

Transmission probability, 25:78, 112ff, 131, 147, 156, 163

Transmission spectrum, 44:328

Transmittance, 27:436, 501

Transport, active, **29**:302, 303, 304; **39**:329–333

  amorphous insulator, **52**:388

  artificial membrane, **39**:287–315

    enzyme, **29**:117

    lipid, **29**:309

  biological membrane, **29**:301 **39**:324, 325, 328, 332

  carrier, **29**:302

  chemical, **39**:287

  chromatographic column, **52**:460

  dioleylphosphate membrane, **29**:325, 341

  disordered structure, **52**:475–484

  efficiency of, **39**:332

  electrical, **39**:287

  of excitation, **52**:373

  facilitated, **29**:302

  by gramicidin, **29**:315

  mechanism, **39**:306, 324, 325

  mediator, **29**:344

  by monactin, **29**:311

  nonequilibrium thermodynamics, **39**:328

  permeability ratio, **39**:295

  simple activated, **29**:302

  of sugar, **29**:304

  transference number, **39**:296, 304

  by valinomycin, **29**:311

Transport coefficients, 1:159; **11**:319, 335, 365; **12**:372; **24**:257, 283, 308; **36**:169, 170, 178, 190ff; **38**:173, 182, 183; **53**:132

  diffusion, **36**:190, 192, 194, 195

  relation to light scattering, **24**:17, 18

  second-order, **31**:168

  thermal conductivity, **36**:194

  thermal diffusion, **36**:194

  viscosity, **36**:190–192, 194, 195

Transport data, 51:269

Transport equations, 4:4, 8, 9, 14; **11**:186, 256, 283, 333

  coupled, **44**:298, 299

  disordered structure, **52**:477

  generalized, **11**:163, 174, 176, 182, 209, 225; **16**:15;

  macroscopic, **53**:82

  phenomenological, **53**:64, 139, 141

  relaxed motion, **15**:41

Transport matrix, 43:231

  generalized, **53**:148

Transport operator, 11:320, 336, 337, 356, 364

Transport processes, 1:135–139, 141, 147; **5**:261ff, 262–263, 272, 281, 294; **28**:221; **38**:175

  Boltzmann equation, **5**:263

  correlation functions in the calculation of, **5**:263, 302

  diffusion coefficients, **5**:263

  electrical conductivity, **5**:264, 267, 304

  entropy, **5**:285, 286–287

  Fourier's law, **5**:262, 290

  hydrodynamic equations, **5**:262–263, 305

  local equilibrium, **5**:278ff, 291, 303

  Markov process, **5**:263

  Navier-Stokes equation, **5**:262, 288

  steady state in non-isolated system, **5**:267

  stress tensors, **5**:292, 296, 273

  thermal conductivity, **5**:263, 292

  thermal stress, **5**:292

thermoelectric effect, **5**:305

viscoelastic medium, **5**:288

viscosity, **5**:263

Transport properties, 15:50ff; **21**:475; **42**:275, 354

  classical fluids, **37**:185, 197

  dense gases, **3**:289

  dilute gases, **3**:281

  electrolytes, **11**:6, 37, 38

  linear, of hard sphere gas, **20**:135ff

  particulate systems, **21**:753

Transverse correlation function, 53:248, 250

Transverse electric modes, 27:371, 390

Transverse electron polarization, 35:25

Transverse field, 41:84

Transverse magnetic modes, 27:370, 382, 383, 390, 528

Transverse optical frequency, 27:359, 377

Transverse pressure profile, 49:384–389

Transverse relaxation, 7:503

Transverse spin density correlation function, 44:438

Transverse spin wave, ferromagnet, **53**:263

Transvibronic reaction, 30:418

Trap-controlled hopping, 52:475

Trapezoid model, 8:150ff, 154

Trapped charges, 7:314

Trapped electron method, 18:19, 43, 48, 51, 60, 63, 71

Trapped soliton, 53:286–287

Trapping, average number of steps before, **52**:467

  models of, **52**:465–471

Trapping time, 52:409

Traps, 23:243; **40**:439; **52**:467, 471, 479

  concentration, **40**:457; **52**:468

  depth of, in mixed crystals, **8**:30ff

  in periodic sublattice, **52**:466

Traube's rule, 6:166

Tree function, 43:85

Tree networks, 43:10, 11, 77, 85–86

Trennduse, 10:314

Triangular kinetic scheme, 44:380

Triangular model, 8:154

Triatomic molecules, 9:65; **36**:20

  photodissociation of, **50**:279, 306

  as quenching agents, **50**:507

  Renner-Teller effect in, **52**:117–180

Triatomic van der Waals molecules. *See* van der Waals molecules

Tributyl phosphate, 47:Part 1:641

Trichlorotrifluoromethane, 55:76

Tricritical point, 41:33ff, 53

Trigonal hybrids, 7:19

Trimers, carbon monoxide, **52**:303

  ethylene, **52**:351

  nitrogen, **52**:303

  OCS, **52**:326–329

  water, **31**:26, 44

  see also Clusters

Trimethoxybenzoxy group, 16:189

Trimethylamine decahydrate, 34:198

Trimethylphosphate, 47:Part 1:641

Trimolecular model, 38:271; **49**:320–321

Trinucleotides, 7:135ff

Triphenylmethane dyes, 49:34

Triphenylmethyl in triphenylamine, optical absorption spectrum, **47**:Part 2:473

Triphenylmethyl sodium, as initiator, **2**:181

  polymerization, **2**:173

Triple beam experiment, 42:480

Triple dipole term, 53:13, 18

Triple point, 40:349

  carbon dioxide, **2**:97

density, **53**:45

Triple sample difference technique, **53**:38

Triplet, base pairs, **33**:100

closure at, **33**:82, 83

quenching process, **35**:9

spin relaxation time, **35**:9

T_1 measurements, **35**:26

Triplet exciton, 40:370, 392, 398, 411, 436, 445, 450ff, 465ff, 474ff

one-dimensional, **40**:401

Triplet model, 35:2

slow-motion limit, **35**:10

Triplet state, 4:198; **7**:553; **41**:145

chlorophyll, **7**:632

collisional energy transfer, **21**:169, 178

EPR studies, **7**:553ff

hyperfine structure, **7**:562ff

lifetime studies, **7**:564ff

photochemical processes, **7**:576ff

theory, **7**:287, 332

Triplet-triplet transfer, 7:568

Tris(3-dimethyl arsinopropyl)phosphine, 16:207

Tristate points, 43:37

Triticum vulgare, **7**:634ff

Tritium, 1:34, 49, 61

beam of, **30**:296

Tritium (T_2), ground state, dissociation energy, **12**:22

relativistic corrections, **12**:22

Tropone, 8:135ff

electronic transitions of, **13**:193

Troponin C, 39:83

Troposphere, hydrogen, nitrogen, and halogen compounds, **55**:74–76

Truesdell function, 22:173

Trypsin, 33:106; **39**:87, 88

Tryptophan, 7:59, 267, 579; **49**:34

fluorescence of, **33**:108

triplet state, **7**:579

Tryptophan-pteridine complex, 7:46

Tryptophan-riboflavin complex, 7:46

Tryptophan synthetase, 39:200

Tschebyscheff's polynomials, 4:12

Tube toroid, 36:22

Tumor cells, 29:51

Tumors, 8:177ff

development of, **7**:89, 148ff, 154

Tungsten hexachloride, 20:28

crystal structure, **5**:184

Tungsten hexafluoride, 9:192ff

Tunnel effect, 1:24, 25, 53

Tunnelling, 25:70, 72, 125, 128–140, 143, 144, 174; **30**:31, 99, 118, 453, 471; **36**:77, 154, 156, 161, 163; **49**:209, 219

DNA, **8**:178

Tunnelling center, 48:470

Tunnelling correction, 49:214

to reaction rates, **3**:131ff, 148

Tunnelling-orbiting phenomenon, 10:105, 125

Tunnelling state, 51:246, 251, 255

Turbulence, 32:30, 48

Turbulent diffusion, 52:365

Turbulent plasma, 38:176, 187, 188

Turing's instability, 49:321

Turn-over rule, 2:288–291, 309

Turning point, 10:77, 85, 89, 92

Two-bulb experiments, 49:461, 467

Two-color experiment, 50:134

Two degrees of freedom, 51:161–166

Two-dimensional crystal, 9:114ff

graphite lattice, **9**:116ff, 128ff

Two-electron integral transformations, MCSCF calculations, **54**:164–173

Two-electron system, 2:279, 300, 307

Two-hole single particle state, 49:537

Two-level atom, **38**:148

Two-level model, **44**:511, 513

Two-particle approximation, **9**:345

  Clausius-Mosotti, **48**:208

  Onsager, **48**:207

Two-particle wave function, **9**:333, 342, 353

Two-photon dissociation, **50**:195, 198, 278

Two-photon excitation, NO, **47**:Part 2:17–41

  quenching of fluorescence, by NO, **47**:Part 2:2–31

    by other molecules, **47**:Part 2:31–33

Two-photon ionization, **50**:235

Two-photon laser induced fluorescence, **50**:266, 306

Two-photon spectroscopy, **49**:39, 41–42

  perturbed benzenes, **54**:177–230; *see also* Benzenes

Two-point correlation function, **46**:253

Two-state approximation, **30**:468

Two-state case, **49**:282, 283

Two-state diffusion process, **52**:456

Two-state random walk, **52**:356, 376

Tyrosine, **7**:267

U

Uhlenbeck-Ornstein integral, **15**:357

Ultra high vacuum, **27**:3, 11

Ultrasonic attenuation, **48**:470

Ultrasonic interferometer, **28**:207, 208, 209

Ultrasonic velocity, **24**:21

Ultrasonic waves, absorption

  associated with longest time of Ising lattice, **33**:90, 92, 97

  calculated using two-state model,

    33:115–120

  experimental method, **33**:127

  oligomers, **33**:126, 131, 132

  polypeptides, **33**:89, 94, 95

  protein assemblies, **39**:242

  vinyl-polymers, **33**:126–135

  amplitude of, **33**:90

  velocity of propagation, **33**:126

Ultrasonics, **23**:43, 58, 77; **24**:20; **31**:159, 227

Ultraviolet, *see also* Vacuum ultraviolet

Ultraviolet absorption spectra, **7**:85; *see also* Absorption spectra

Ultraviolet photoelectron spectroscopy, **49**:466

Ultraviolet photoemission spectroscopy, **51**:257

Ultraviolet radiation, **7**:584

Umklapp process, **49**:551, 555, 557

  bulk, **49**:562–563

  surface, **49**:562–564

Uncertainty principle, **5**:305; **25**:88, 100, 133; **46**:154, 158

Uncorrelated ground state, **36**:242

Uncorrelated state, **26**:129

Undheim-McDonald theorem, **54**:63

Unfolding of singularity, **40**:244

Uniaxial crystals, **27**:434, 448, 527; **54**:339–340

Uniaxial ferromagnet, **53**:263–264

Unified model of metal-ammonia solutions, **4**:351

Uniform approximation, **30**:96, 322, 326; **36**:2, 57

Uniform positive background, **2**:255

Uniform states, **7**:6

Uniform system, chemical instability, **38**:318

Unilateral boundary value problem, **32**:54

Unimolecular decomposition, **13**:94;

50:158, 198; **52**:264, 270, 350

random walk model for, **15**:145

Unimolecular network, 43:32

Unimolecular reaction, 1:31, 367; **10**:176, 190; **15**:157; **21**:155; **30**:8; **42**:138; **47**:Part 1:56–58

induced by infrared radiation, **50**:396

producing detonations, **3**:104

producing flames, **3**:68

rate, **1**:29, 31–33

Unit meshes, 27:142, 152

Unitary equivalent, 49:160, 177

Unitary matrix, 4:130

Unitary transformation, 14:95; **25**:73–78, 91; **33**:155, 159, 204, 213, 228

canonical, **25**:73–75, 85

improper, **33**:206, 218

multiconfigurational Hartree-Fock state, **54**:6–8

United atom approximation, 2:276; **28**:117, 134

Univariant equilibrium, 2:19, 23

Univariant system, 2:91

Universality, critical phenomena and, **51**:20–25

Universality class, 51:24

Universality of critical exponents, 51:3, 5

Universe, age of, **26**:111, 113, 143

Universe of discourse, 38:8, 10, 23

Unlinked clusters, 22:121

Unpaired states electron density, 7:537

Unreacted pair probability, 53:111

Unrelaxed emission, 41:293

Unrestricted Hartree Fock operator, 36:219

Unrestricted Hartree-Fock method, 2:231; **14**:326

Unsaturated hydrocarbons, polarizability, **2**:76

Unsaturated ligand, 7:362, 366

Unsöld approximation, 12:48; **16**:71; **42**:175

Unstable eigenvalue, 38:163, 164

Unstable periodic trajectory, 36:37, 39–41

Unstable quantum state, 38:145, 167

Unstable region, 46:219

Unstable state, 46:197

Unzippering or unraveling, 7:230

Update methods, Broyden asymmetric rank-1, **54**:79–82

Broyden family (DFP and BFGS), **54**:82–86

Hessian, convergence characteristics, **54**:88–90

inverse, **54**:87–88

multiconfigurational self-consistent field state, **54**:78–101

numerical results, **54**:90–100

variational approach, **54**:86–87

Upright band, 8:34

Uracil, 7:52ff, 92

excitation energy, **7**:100ff

molecular diagrams, **7**:94, 102

oscillator strength, **7**:101

photodimerization, **15**:171ff

transition moment, **7**:95

Uranium hexachloride, 20:28

Uranium hexafluoride, 9:187ff; **20**:28

crystal structure, **5**:183

photodissociation of, **50**:260, 308

Uranyl compounds, multiphoton dissociation of, **50**:141

Urea, effect on protein, **39**:113

Urea radicals, 7:551

Urease, 29:120

pH effect, **39**:231

Uridine, infrared spectra, **7**:447ff, 453

Uridylic acid, 7:514

Urn model, 50:415

Ursell cluster functions, 11:321, 326, 328; **15**:202

V

Vacancy, 31:33
  clustering, **40**:151
  lattice site, **40**:151
Vaccinia, **7**:222
Vacuum microbalance, 7:247
Vacuum ultraviolet lasers, 47:Part 2:154–158
Vacuum ultraviolet monochromator, 52:268–269
  near normal incidence type, **52**:268
  Seya-Namioka type, **52**:268
Vacuum ultraviolet nonlinear optics and laser spectroscopy, 47:Part 2:153–181
Vacuum . ultraviolet photodetector, 52:269
Vacuum ultraviolet photolysis, 50:299
Vacuum ultraviolet radiation, 47:Part 2:153–181
  generation of harmonic radiation
    experimental, **47**:Part 2:164–165
    higher order nonlinear effects, **47**:Part 2:171–172
    magnesium vapor studies, **47**:Part 2:167–169
    saturation phenomena, **47**:Part 2:172–173
    strontium, **47**:Part 2:165–167
    third-order nonlinear phenomena, **47**:Part 2:160–164
Valance approximation, chemical reactivity problems, **33**:355
Valence band, 23:140; **51**:258
  characteristic function of, **5**:13
  positive holes, **7**:146
Valence bandwidth, 31:375

Valence bond method, 1:184, 198, 239; **2**:384, 386; **6**:24; **7**:21, 187ff; **30**:139; **36**:142–145, 149, 150, 157, 162ff; **42**:67, 74
  configurations, **36**:161, 166
  relation to electronegativity, **13**:58
Valence configuration, 45:238
Valence-core non-penetrability approximation, 16:319–320, 334–336
Valence electrons, pseudopotential type equation for motion of, **16**:325–327
Valence level, photoemission spectroscopy, **49**:594–638
Valence orbital, 23:144
Valence shell correlation in polyenes, 52:70–109
Valinomycin, 19:368; **29**:314
  ion transport, **29**:311
  oscillations and, **29**:148
Valinomycin-potassium complex, 21:584
van der Waals constant, 12:274
van der Waals contour, 44:371
van der Waals equation, 6:244, 257; **37**:172; **40**:230; **41**:22, 23, 26; **49**:375, 377, 413, 428–430, 436
  dense fluids, **34**:123
  exact for Kac model, **34**:123
van der Waals force, 1:267, 268, 273, 278, 292; **6**:398–406, 408; **7**:203; **9**:336; **12**:226ff, 249, 257ff, 276; **27**:343; **46**:384, 395, 397; **49**:104; **49**:545
  induction, **12**:226, 228, 249, 257, 258, 267, 276, 279
  liquids, **21**:485
  many-electron theory, **6**:408
  orientation, **12**:226, 228, 249, 257, 258, 267, 276, 279
  retardation effects, **12**:226, 227, 241, 249, 267, 279

van der Waals free energy model, **49**:423

van der Waals gas, coexistence curve, **6**:178, 187

van der Waals loop, **6**:244, 249, 276

van der Waals minimum, for H_2 $b^3\Sigma_g^+$ state, **12**:26

van der Waals modified model, **49**:413–416, 423

van der Waals molecules, **42**:353–420; **47**:Part 1:323–506; **52**:266
  Ar-I_2, **42**:363
  Ar-OCS, **42**:361
  binding energy, **47**:Part 1:349–352
  chemical synthesis and experimental probes, **47**:Part 1:325–330
  ClF-Ar, **42**:361
  ClF-Kr, **42**:415
  CO_2 dimer, **42**:359
  dimers, **47**:Part 1:427–453
    V → T process, **47**:Part 1:431–435
    V → V + T process, **47**:Part 1:435–439
    vibrational predissociation, **47**:Part 1:428–431, 439–453
      local mode basis, **47**:Part 1:440–442
      miniexciton basis, **47**:Part 1:442–443
      normal mode basis, **47**:Part 1:443–453
  electronic relaxation in, **47**:Part 2:370–372
  H_2 dimer, **42**:355, 362
  H_2-Ar, **42**:358, 382, 388, 394, 403, 408
  H_2-He, **42**:355, 392
  H_2-inert gases, 42–398, 399
  H_2-Kr, **42**:394, 407, 408
  H_2-Ne, **42**:392, 407
  H_2-Xe, **42**:394, 396
  HCl-Ar, **42**:355, 359, 361, 378, 383, 390, 411
  HD-Ar, **42**:358, 388
  He-I_2, **42**:363, 388, 389, 415
  He_2-I_2, **42**:354, 363
  He-NO_2, **42**:363
  He-tetrazine, **42**:354, 363
  heterodimers, V → T process, **47**:Part 1:431–435
    vibrational predissociation, **47**:Part 1:428–431, 439–453
  HF-Ar, **42**:361
  HF-ClF, **42**:361
  HF dimer, **42**:355, 359, 361, 415
  HF-HCl, **42**:361
  homodimers, dynamics of, **47**:Part 1:439–453
  infrared, microwave and radiofrequency spectra, **47**:Part 1:334–337
  iodine-argon, **47**:Part 1:358–360
  iodine-containing, **47**:Part 1:353–360, 467
  iodine-helium, **47**:Part 1:353–355
  iodine-neon, **47**:Part 1:355–358
  Kr dimer, **42**:355.
  Na-Ne, **42**:355
  Ne-Ar, **42**:355
  Ne dimer, **42**:355
  Ne-NH_3, **42**:281
  N_2O dimer, **42**:364
  photodissociation, **50**:316
    competing processes, **47**:Part 1:348–349
    product-state distribution, **47**:Part 1:352–360
  quasiclassical trajectory studies, **47**:Part 1:470
  spectroscopy and structure, diatomics, **47**:Part 1:331–334
    polyatomics, **47**:Part 1:334–343
  triatomics, collinear models of,

362   ADVANCES IN CHEMICAL PHYSICS

47:Part 1:372–413
   diabatic distorted wave treatment, 47:Part 1:390–395
   frozen diatomic model, 47:Part 1:372–388
   matrix elements, 47:Part 1:407–413
   Rosen's relative coordinates treatment, 47:Part 1:388–390
   vibrational predissociation, 47:Part 1:395–407
  numerical methods, 47:Part 1:413–427
   comparison with distorted wave calculations, 47:Part 1:417–420
  perpendicular vibrational predissociation, 47:Part 1:420–427
  scattering formalism, 47:Part 1:413–417
vibrational predissociation, 47:Part 1:23, 367–370, 453–464
  ArHCl, 47:Part 1:367
  electronically excited complexes, 47:Part 1:463–464
  hydrogen-bonded systems, 47:Part 1:336–337
  model calculations, 47:Part 1:454–463
  rotational effects on, 47:Part 1:470–482
  scaling theoretical analysis, 47:Part 1:467–470
  triatomics, 47:Part 1:370–427, 482–492
   collinear models, 47:Part 1:372–413
   numerical methods, 47:Part 1:413–427
vibrational relaxation in gases, 47:Part 1:367
visible and ultraviolet spectra,

47:Part 1:337–343
water dimer, 42:361
van der Waals potential, higher order, 12:228, 249
  nonadditive, 12:226, 249, 250, 254, 279
  pair, 12:228, 250, 257, 279
  rotational interaction, 12:267
  single-temperature, 12:278
  three-body, 12:228, 249, 250, 259
  two-body, 12:249, 250
  two-temperature, 12:228, 275, 276, 278
van der Waals radius, 2:25, 29, 31, 390; 44:478
van der Waals theory, 32:5; 34:107, 116; 41:47
  nonhydrogen bonding dense fluids, 34:107
  role of repulsive forces, 34:116
van der Waals well, 36:169, 176, 197; 50:564
van Hove correlation function, 44:284, 407, 416
  reorientation, 44:284
  self part, 44:404
  translational, 44:284
van Hove function, 31:159, 161
van Hove-Glauber scattering theory, 20:149, 163
van Hove self correlation function, 17:91, 191, 201
van Hove structure factor, 17:260; 53:10, 11, 54
van Hove weak coupling limit, 42:324
van Vleck theory of magnetic susceptibility, 3:181
van Vleck transformation, 44:23, 36
Vanadium, 41:141
  conductivity, 41:165
Vanadium complexes, 41:280

absorption spectra, **5**:256

Vapor deposition, **40**:204

Vapor phase, **46**:281

Vapor pressure, **16**:252–253, 275

alkali halides, **11**:92, 93

corresponding states expression for, **11**:88

effect of isotopic substitution, **4**:226–230, 286, 289–300

metal-ammonia solutions, **4**:306–308, 353, 354

pure carbon dioxide, **2**:96

Vaporization, cycle, **9**:63

enthalpy of, salts, **11**:90

Variable, extensive, **29**:86

macroscopic, **29**:32, 86

random, **29**:86

Variance, **46**:199

Variance-covariance matrix, **44**:268, 271

Variational collapse, **16**:288, 340

Variational functional, **25**:217, 251, 257, 259, 286

Variational methods, **1**:241ff, 251ff, 273; **2**:213, 318, 320, 326; **7**:9; **9**:345; **12**:22, 25; **13**:276; **14**:286; **23**:164; **25**:212–294; **26**:129, 132, 134; **30**:23, 27, 53; **34**:257; **35**:318, 319; **36**:31, 34, 144; **37**:11, 196, 197; **40**:260; **41**:114; **47**:Part 1:221–225

algebraic close coupling, **25**:214, 264ff, 278ff, 292

algebraic correlation, **25**:265, 268

anomaly-free, **25**:214, 243, 247, 260ff, 278ff

basis functions, **25**:212, 220ff, 238, 249ff, 264, 277ff, 287

close coupling, **25**:212, 213, 257, 264, 279

in continuum physics, **13**:289, 300

convergence, **25**:247, 248, 251, 257, 277, 278

correlation methods, **25**:258, 265, 266

derivative matrix, **25**:250, 251, 279, 281, 283, 292

Dirichlet problem, **32**:39, 40, 42, 52

Euler integral methods, **25**:213, 237, 243, 244, 247, 280

exchange approximation, **25**:213, 215, 282

Galerkin, **32**:40, 41

Garrett, **25**:282, 283

H_2, first excited state, **12**:26

ground state, **12**:22, 26

Harris and Michels trial function, **25**:214, 222, 243, 248, 259, 265ff, 280, 285

Harris's method, **25**:228, 243, 259, 277, 285

Huang, **25**:213, 231, 233

Hulthén, **25**:213, 218ff, 243, 255, 259, 271ff

Hulthén's second, **25**:213, 218, 221, 286, 287

Jacobi matrix (Heller and Yuman), **25**:283

Kato's method, **25**:219, 254

Kohn method, **25**:213, 218ff, 230ff, 265, 272, 280ff

Harris and Michels trial function, **25**:255–259

inverse, **25**:218

Kohn-Crawford, **25**:251, 253, 279, 282

Malik, **25**:235, 236, 254, 287

method of moments, **25**:243

minimum norm, **25**:214, 228, 230, 239ff, 259, 274ff

Harris and Michels, **25**:222–224

Moiseiwitsch, **25**:274, 281

Newton's, **25**:283

nonequilibrium, **46**:257

nonstationary, **25**:221, 224, 225, 239, 240–243, 273

numerical asymptotic function, **25**:253

optimized anomaly-free, **25**:243, 248, 266

optimized minimum norm, **25**:230, 239, 243

Payne, **25**:282

Percival, **25**:219ff, 243, 244, 286

perturbational, **6**:342, 344; **9**:335ff; **25**:255

Rayleigh-Ritz, **32**:40, 41

reaction rates, **13**:85, 89

Ritz, **32**:42

Ritz-Galerkin, **32**:42

Rubinow, **25**:214, 221, 230, 233ff, 254, 260ff, 291

Schwartz's method, **25**:254

Schwingers' method, **25**:254, 282, 283, 286, 288

in statistical mechanics, **13**:302; **24**:188, 189, 193

  Gibbs-Bogoliubov inequality, **24**:189

  Jordan-Golden inequality, **24**:194

  Peierls inequality, **24**:193

static approximation, **25**:213–215, 286, 287

static exchange approximation, **25**:213, 215, 287

stationary, **25**:225

time-dependent, **41**:120

vibration-rotation spectroscopy of polyatomic molecules, **37**:305–380

Variational parameter, **46**:262

Variational potential, **16**:293

Vector field, decomposition theorem, **33**:412

Vector model, **40**:378

Vector potential, **33**:155ff; **49**:536, 565

Vector quantum number, **36**:43

Vectorial dipolar correlation function, **44**:377

Vectorial photoeffect, **49**:551

Vedenov and Dykhne theory, **22**:167

Vehicular traffic, Prigogine-Herman theory, **32**:9

  probability of passing, **32**:10

Velocity, **31**:160

  angular, **31**:160, 168

  barycentric, **16**:105–106

  constant, **1**:29

  convective velocity field, **32**:92

  diffusion, **16**:104

  group, **46**:306

  peculiar, **31**:160

  phase, **46**:306

  potential, **46**:304

  propagation of ultrasonic waves, **33**:126

  reaction, **43**:17

  rotation of molecule, **33**:125

  soliton, **53**:275, 283, 284

  sound, in liquids, **21**:483

Velocity analysis, **30**:251

Velocity autocorrelation function, **17**:105, 167; **27**:484; **31**:85; **34**:148; **44**:286; **52**:374; **53**:77–87, 92, 101

  equivalent harmonic chain, **53**:97

  Langevin equation, **53**:82, 87

  linear response theory, **53**:135

  many-body dynamics, **53**:87

  molecular-time-scale generalized Langevin equation, **53**:215, 217

  normalized, **53**:102, 150–151, 156–159

  response function method, **53**:93, 139, 140

soliton, **53**:273

truncated chain model, **53**:100–101

Velocity change method, **42**:449, 465, 480

Velocity correlation function, **17**:130

long time tail, **24**:325

Velocity-dependent forces, **12**:12

Velocity distribution, **1**:131; **10**:10, 141, 166; **32**:10

beams of excited species, **10**:177–178

molecular beams, **12**:399

nozzle beams, **10**:277–279, 287–289, 311–315

products of reactive scattering, **10**:344–347

Velocity field, producing orientation or deformation, **33**:138, 139, 142

Velocity gradient, reduced, **33**:125

Velocity selection, **10**:15, 25, 139–141, 166, 177, 181, 326–327, 333, 344; **30**:250, 284, 319; **42**:442, 532

Velocity spectra, **44**:410

Vertebrate, limb development, **29**:263

Vertex approximation, **43**:177

Vertex function, **14**:230; **41**:112, 114, 125

Vertical electron affinity, **42**:515, 525; *see also* Electron affinity

Vestigial forces, **15**:50, 61

Vibration-rotation bands, **12**:351

Vibration-rotation coupling, **40**:17, 20, 26, 36; **49**:154, 158, 170–178

Vibration-rotation eigenvalue problem, **49**:184

Vibration-rotation energy transfer, **23**:46ff, 62ff, 70

Vibration-rotation excitation, **30**:223–227

Vibration-rotation perturbation kernel, **49**:176

Vibration-rotation unperturbed kernel, **49**:174–177

Vibration-vibration energy transfer, **50**:373

near resonance, **23**:64

spectrophone data, **23**:41, 42–46

Vibrational adiabaticity, **42**:116, 190

barrier height, **36**:160, 163, 165

Vibrational amplitude, anisotropy of, **27**:476

mean-square, **27**:475–482, 504–505

Vibrational and rotational collision processes, **42**:271–304

Vibrational basis functions, **37**:317, 319, 322

analytical, **37**:322

cubic spline, **37**:321, 323

harmonic oscillator, **37**:317, 319, 322, 323

Morse oscillator, **37**:323, 334

nonorthogonal Gaussian, **37**:323

numerical, **37**:321

sine functions, **37**:322, 323

Vibrational CI, **37**:332

Vibrational circular dichroism, **44**:597–598

Vibrational coordinates, **44**:447

cartesian, **37**:324, 326

curvilinear, **37**:338

dimensionless normal, **37**:329

internal, **37**:328

normal, **37**:326, 327

rectilinear, **37**:326, 327

reduced normal, **37**:314

Vibrational coupling constants, free, **36**:288

renormalized, **36**:280, 287, 310, 318

Vibrational deexcitation, **28**:173–185, 193–205, 224–236, 243–256, 391, 397, 437

BF_3, **28**:256

Br_2, **28**:205, 227, 232, 234

CCl_2F_2, **28**:227

CD_4, **28**:255

CF_3CN, **28**:249

CH_4, **28**:195, 202, 204, 205, 218, 220, 255

C_2H_2, **28**:249

C_2H_6, **28**:249

CH_2Br_2, **28**:249, 250

CH_2Cl_2, **28**:249, 250

Cl_2, **28**:202, 227, 232, 234, 247

CO, **28**:195, 197, 204, 215, 226, 232, 244ff, 254

CO_2, **28**:200ff, 216, 218, 250ff, 441, 453

C_3O_2, **28**:249

DCl, **28**:248

D_2S, **28**:254

F_2, **28**:227, 232

$GeCl_4$, **28**:256

H_2, **28**:182, 203, 218, 221, 234, 235, 240, 247, 249

HBr, **28**:205

HCl, **28**:245, 247, 248, 249

HF, **28**:249

H_2O, **28**:254

H_2S, **28**:254

I_2, **28**:227, 232

N_2, **28**:89, 196ff, 215, 227ff, 244, 388ff, 412, 426, 441

ND^+, **28**:223

ND_3, **28**:254

NH_3, **28**:254

NO_2, **28**:252, 253

O_2, **28**:181, 226ff, 243, 244, 255, 397ff, 412

OH, **28**:61, 413

PD_3, **28**:254

PH_3, **28**:254

SF_6, **28**:256

$SnCl_4$, **28**:256

Vibrational dephasing in liquids, **40**:1–48

Vibrational energy, **7**:203, 205

neutron diffraction studies, **53**:12

Placzek correction and, **53**:51

promoting reaction, **42**:28, 86, 88, 115

Vibrational energy transfer, **47**:Part 2:185–235, 528–529; **50**:49, 68

catastrophic cyclic paths, **47**:Part 2:220–222

CD_4, **47**:Part 2:192–193

CD_3H, **47**:Part 2:190–192

CH_4, **47**:Part 2:187–190

CH_2D_2, **47**:Part 2:193–198

CH_3F, **47**:Part 2:198–204

chemical reactivity, **47**:Part 2:222

collision induced, **42**:271ff; **47**:Part 2:245–258

collisional propensity rules, **47**:Part 2:229–231

continuous wave and discharge pumping, **47**:Part 2:222

degenerate mechanisms, **47**:Part 2:220–222

experimental techniques, **47**:Part 2:218–220

harmonic oscillator limit, **47**:Part 2:212–215

intermode coupling, **47**:Part 2:210–211

ladder-climbing effects, **47**:Part 2:211–212

mechanisms, **47**:Part 2:208–227

methane, deuteromethanes, and methyl fluoride, **47**:Part 2:187–227

polyatomic molecules, **47**:Part 2:12–13

quanta conservation, **47**:Part 2:223–227

temperatures, **47**:Part 2:209–222

vibrational temperatures, **47**:Part

2:218

see also Energy transfer; Intramolecular energy transfer; Vibrational deexcitation; Vibrational relaxation

Vibrational entropy in dilute solutions, 2:133

Vibrational excitation, 25:110, 130, 146, 147; 28:14, 39, 47–62, 81, 188, 296, 387, 437; 30:333; 42:142, 200

BaBr, 28:82

BaCl, 28:82

BaF, 28:82

BaI, 28:82

BaO, 28:90

C_2, 28:42, 43

CF_3, 28:63

CH_4, 28:255

CK, 28:43

CN, 28:42

CO, 28:53, 54, 61, 62, 63, 93, 94, 246

CO_2, 28:54, 55, 255, 432, 441, 454

CS, 28:50, 53

CsF, 28:81

D_2, 28:246

DBr, 28:83

DCl, 28:83, 246

DF, 28:54, 55, 92

DI, 28:83

in electron scattering, 18:15, 17, 21, 70ff, 80ff

angular dependence, 18:119, 122

C_2H_4, 18:21, 72

CO_2, 18:70, 71, 72

H_2, 28:223

HBr, 28:52, 58, 83, 93, 246

HCl, 28:52, 57, 58, 59, 63, 83, 93

H_2CO, 28:63

HD, 28:246

HF, 28:51, 60, 63, 92

HI, 28:52, 83, 93, 245, 246

IBr, 28:50

KBr, 28:26

KI, 28:68, 75

LiF, 28:81

N_2, 28:63, 64, 246, 388, 389, 432, 441, 454

NaCl, 28:4

NO, 28:61, 62, 94

N_2O, 28:62

O_2^+, 28:224, 225

O_2, 28:47, 48, 225, 400, 432

O_3, 28:87, 90

OH, 28:50, 51, 60, 61, 64

P_2, 28:62

S_2, 28:50

SiN, 28:42

SO, 28:50

Vibrational fluorescence, 28:216–221, 246, 247

Vibrational frequencies, $(NO)_2^+$, 52:291, 292

$CH_3COCH_3^+$, 52:546

CS_2^+, 52:309

NO^+, 52:290, 292

OCS^+, 52:325

one-dimensional random chain, 13:125, 126, 146, 149, 152, 156, 161, 162

SO_2^+, 52:335, 336, 339

Vibrational intensity distribution, NO^+, 45:104

Vibrational lifetime, 42:308

Vibrational predissociation, 42:354, 363, 389; 47:Part 1:24–26; 50:448; 52:306

ArHCl, 47:Part 1:367

hydrogen-bonded systems, 47:Part 1:366–367

quasiclassical trajectory studies, 47:Part 1:470

rotational effects on, **47**:Part 1:470–482

scaling theoretical analysis, **47**:Part 1:467–470

van der Waals molecules, **47**:Part 1:367–370

heteronuclear dimer, **47**:Part 1:428–431

iodine-rare gas, **47**:Part 1:467

triatomics, **47**:Part 1:470–427

see also van der Waals molecules

Vibrational relaxation, **23**:41, 44, 46, 53, 58; **34**:88; **41**:270, 288–289; **42**:209

cage effects, **53**:110

gas of diatoms, **15**:219

heating effects, **23**:46, 48, 53

hydrogen halides, **23**:45

infrared double resonance, **23**:48

intramolecular, **42**:221; **50**:118

iodine in simple solvents, **53**:209, 210–215

phase, **42**:306

populational, in liquids, **47**:Part 2:48–519

experimental studies, **47**:Part 2:507–513

laser studies, **47**:Part 2:508–513

ultrasonic absorptions studies, **47**:Part 2:507–508

non-Markovian effects, **47**:Part 2:513–516

theories of, **47**:Part 2:488–507

cell models, **47**:Part 2:496–497

Green's function methods, **47**:Part 2:502–503

hydrodynamic models, **47**:Part 2:497–499

isolated binary collision models, **47**:Part 2:489–491, 494–496

molecular dynamics simulation, **47**:Part 2:503–507

perturbation, **47**:Part 2:491–494

superposition approximations, **47**:Part 2:499–502

saturated absorption, **23**:45

of small molecules in dense media, **42**:305–352

Cl_2 in Ar, **42**:311, 318

CO in Ar, **42**:330, 339,

HCl in Ar, **42**:339, 349

hydrodynamical model, **42**:348

N_2 in Ar, **42**:322, 346, 349

N_2 liquid, **42**:322

NH in Ar, **42**:335, 339

OH in Ne, **42**:339

one-dimensional models, **42**:313

solids, **42**:305

see also Vibrational deexcitation; Vibrational energy transfer

Vibrational SCF, **37**:326

Vibrational states of two-body systems, **10**:124–126

Vibrational step, **52**:288

Vibrational structure in photoelectron spectra, **36**:273, 279, 310, 324, 334

Vibrational studies, **49**:21–28

liquids, **49**:26–28

mixed crystals, **49**:21–25

semiconductors, **49**:25–26

Vibrational temperature, **42**:23, 34; **52**:265, 272

Vibrational wave function, **37**:314, 324, 354; **44**:22

assignment of quantum number, **37**:353

plot of, **37**:354–358, 361

Vibrations, hexafluoride molecules, **9**:169ff

lattice, **4**:162

molecular, **2**:12; **4**:248, 300

Vibronic coupling, **9**:208ff; **44**:44

experimental spectra, **9**:264ff

osmium hexafluoride, **9**:301

Vibronic eigenfunction, **44**:45

Vibronic energy levels, Renner-Teller effect, **52**:117–180

Vibronic mechanisms

two-photon spectra of perturbed benzenes, **54**:198–201

B_{1u} state, **4**:200–201

B_{2u} state, **54**:198–201

Vibronic perturbations, **5**:242–243, 253, 295

benzene, **5**:244

nuclear gradients, **5**:244, 252

strength of, **5**:242, 252, 256, 258

titanium hexafluoride anion, **5**:244

Vibronic states, **42**:3, 211

NO, low resolution emission spectra, **47**:Part 2:29

Vibrons, **44**:489

Vinyl polymers, **2**:154, 172; **33**:75, 115, 147; *see also* Polymers

Vinylidene dicyamide, **2**:155

Vinylidene monomer, polymerization of, **2**:164

Violent collisions, **10**:197–207, 212–218

Virial coefficients, **2**:105ff, 112; **6**:235, 265, 269, 276; **9**:2, 26, 40; **12**:370; **20**:41, 43, 47

asymptotic behavior, **6**:272

classical theory, **20**:43

convex molecules, **5**:164ff

fourth and fifth, hypernetted chain approximation, **6**:265

netted chain approximation, **6**:265

scaled particle theory, **6**:265

hard sphere fluid, **6**:242, 272; **20**:39, 49–51, 57–59, 63

mixture, **6**:269–270

helium, **20**:58, 59

mixture of hard cubes, **6**:272

nonspherical molecules, **5**:168

quantum theory, **20**:58, 59, 65

second, **20**:39, 49, 50; **24**:223

direct, **20**:60, 62

exchange, **20**:60, 62

hard spheres, **20**:39, 49, 50, 51, 58, 65

helium, **20**:58, 59

nuclear spin, **20**:61

square well, **20**:50

second and third, **31**:38

up to third by Born-Green theory, **6**:269

van der Waals virial expansion, **6**:234

Virial expansion, **2**:104; **22**:80; **41**:26; **51**:25; **53**:16, 28, 39, 40, 58

compared with molecular dynamics calculations, **48**:446–449

generalized Boltzmann equation, **11**:337, 356

hard sphere fluid, **6**:242, 245–246, 263

second virial dielectric coefficient, **46**:13

second virial Kerr coefficient, **46**:16–17

Virial lines, **6**:277–278

Virial ratio, **26**:134, 138

Virial theorem, **1**:166; **6**:232, 244; **7**:25; **12**:55, 61; **35**:133, 136; **41**:62, 65; **44**:402, 514

Virtual modes, **27**:383, 436–439, 442

Virtual orbital, **26**:250; **42**:172

Virtual state, **18**:98

Viruses, **7**:414, 433

attachment and entry, **7**:223

heat effects on, 7:201ff, 222

infectivity loss, 7:233

infrared spectroscopy, 7:414, 433

multiplication of, 7:223

sterilization of, 7:211

Viscoelastic medium, 5:288

Viscoelasticity, 32:14

Viscometric functions, 35:96

Viscosity, 1:2, 135–139, 147ff, 158, 159–163; 5:263, 297; 31:4, 161; 35:96, 100, 106; 46:329

anomalous, 6:199, 471–472

bulk, 5:297

colloidal solutions, 6:472

critical region, 6:197–207, 223

dense media, 21:475

dependence of the polarization, 35:21

dynamic, 35:106

high frequency limit, 35:106, 108, 110, 111

rigid body, 35:110

fluids near glass transition, 48:456, 493, 503, 509

Arrhenius behavior, 48:459, 468

Doolittle equation, 48:469, 476, 494, 495, 509

experimental results, 48:460, 495, 496

non-Arrhenius behavior, 48:468

pressure dependence of, 48:496

Vogel-Fulcher equation, 48:461, 468

glass, 51:228

internal, 33:77, 78, 138–146

matrix of, 33:78, 124, 136, 145

intrinsic, 33:125

macromolecules, 35:96

metal-ammonia solutions, 4:308, 309, 354, 355

Newtonian flow, 21:273

non-Newtonian flow, 21:282

polymers, 39:111

Robinson's equation, 21:282

shear, 1:148, 150–152, 158, 160; 5:297

polymer solutions, 33:131–144

solvent, 33:123, 124, 128, 133–147

volume, 1:148, 150, 158

see also Shear viscosity; Bulk viscosity

Viscosmetry, 1:78

Viscous fields, 32:14

multilayered, 32:14

Viscous flow, 1:162

Viscous liquid, 44:397

Visualization techniques, Schlieren, 32:139

shadowgraph, 32:139

Vitamin B$_{12}$, 7;369, 399; 16:212

Vitamins, 7:433

Vlasov equation, 26:133; 46:334

Void, in crystals, 40:151

Volterra equation, 17:106, 217

Volterra-Lotka model, 19:249, 268, 269, 305

thermodynamic properties of, 19:252

Volterra noise, 19:333, 334, 336

Volume, 48:456, 464, 502, 505

configurational, 48:480

cutoff, 26:139, 140

effect of internal pressure on, 13:190

effective interaction, 13:115

excess, 16:229, 253–260, 270–272, 276, 278; 17:27

excluded, 48:81

free, see Free volume

reduced, 16:225, 227–228, 238, 240, 276

relative, 16:240

Volume change, conformation change,

33:87, 97

on mixing, **16**:247, 252–253, 270

see also Polymers

Volume expansion of metal-ammonia solutions, **4**:309–311, 355–357

Volume fractions, related to multicomponent diffusion, **6**:294, 296, 297

Volume ratio, **33**:406

Volume states, energy of, **9**:135, 155

von Mises distribution, **44**:282

Von Neumann algebras, **22**:330

Von Neumann equation, **16**:3, 5, 21

Von Neumann-Liouville equation, **16**:4–9, 39–40

Von Neumann-Liouville operators, **16**:6, 16, 20

Voronoi polyhedra, **40**:64–65; **48**:443, 472, 473, 476

Vorticity, **32**:79, 91

quantized, **33**:6, 7, 18

Vulcanization, stochastic model of, **15**:166

W

Wade's rules, **49**:51, 92

Waiting time density, **52**:370

Wald's identity, **52**:414–416

Walden's rule, **31**:342

Wallace's model, **9**:128

Walsh correlation diagram, **42**:121

Wangsness-Bloch equations, **11**:295

Wannier model, **41**:275

Wannier series, **31**:373

Warburg apparatus, **7**:637

Water, **7**:239; **34**:184; **47**:Part 1:641

adsorption, conductivity and, **7**:251ff

entropy of, **7**:244

heat of, **7**:244

hysteresis effects, **7**:243ff

isotherms, **7**:240ff

kinetics of, **7**:247, 249

on metals, **21**:363

on solid protein, **7**:238

BNS potential, **31**:63

bound in polypeptide chains, **7**:203

cell models, **31**:59

clathrates, **31**:25

coordination number, **31**:34

critical point, **31**:43

diatomics-in-molecules potential energy surface, **42**:72, 81, 99

dielectric relaxation time, **31**:4

dimer, **42**:361

interaction, **31**:15

dipole moment, **31**:6, 22

dissociation, **31**:13, 88

$E \rightarrow V$ energy transfer, **47**:Part 2:399, 406

effect of small ion on structure of, **13**:365

electron correlation of, **2**:324

electron impact spectra, **18**:70, 71

equation of state, **31**:36, 42, 92

equivalent transformation orbital of, **6**:388

fluid structure, **53**:33–44

heat capacity, **31**:4

hydrogen bonding, **31**:17, 18, 42, 58, 79, 80

ice, Ih, **31**:24

polymorphs, **31**:81

iceberg model, **31**:80

interactions with proteins, **21**:567

lattice energy, Bell, **31**:53

Fleming-Gibbs, **31**:49

liquid, partition function, **21**:489

mean field approximation, **31**:51

-methane-propane ternary system, **2**:23

microclusters of, **40**:55, 130
nearest-neighbor polyhedron, **31**:45
normal modes, **31**:8, 35
pair potentials, **31**:15, 30, 38; **53**:46–47
quadrupole moment, **31**:8
self-diffusion constant, **31**:84
structure, **31**:5; **46**:385, 396, 397
surface tension and compressibility, **9**:78
tetramer, **31**:27
thermodynamic energy, **31**:35, 45, 48
thermodynamic properties, **21**:494
trimer, **31**:26, 44
x-ray scattering, **31**:52, 65, 75
see also Ice
Watson-Crick model, **7**:54, 116, 342ff; **8**:177
Watts potential, **53**:39, 40
Wave equation, **27**:389, 396, 425; **36**:18
scalar, **27**:389
vector, **27**:389, 396, 425
Wave function, **4**:115, 123, 130, 164, 174, 184, 253, 265, **25**:212
antisymmetrical, **2**:260, 261; **4**:238
asymptotic form, **25**:216
coordinate, **36**:16
electronic, **4**:116, 128
geminal, **13**:59
ground state, **41**:62
determinantal, **41**:85
molecular, **4**:68
momentum, **36**:16
one-electron, **4**:105
paired-electron, **9**:336ff
polymer, **4**:168
relativistic, **41**:83
symmetrical, **4**:238
two-particle, **9**:333
Wave operator, **8**:3; **14**:300, 316, 335;

36:349, 371; **52**:235
adiabatic, **14**:302
Wave packet, **27**:407, 410
Wave vector, **27**:407
complex, **27**:407, 410, 429, 525–526
three-dimensional, **27**:365, 394, 489
two-dimensional, **27**:365, 436, 445, 452, 465, 527
Waves, of activation, **29**:270
chemical, **38**:334; *see also* Chemical waves
in chemical reactions, **29**:36, 55, 152, 156, 164
in chemotaxis, **29**:153, 180
of concentration, **39**:15–21, 30
in embryonic development, **29**:58, 269
in glycolysis, **29**:146; **39**:32
longitudinal, **27**:450, 460
phase diffusion theory, **29**:55
propagation, **43**:244
stability, **38**:346
transverse, **27**:450
Weak collision approximation, **12**:565, 566
Weak coupling, **4**:147; **31**:199, 201–205, 210–219; **48**:92
collision operator, **35**:147, 152, 159, 175
dispersion interaction and, **53**:284–286
effective potential theory, **31**:210–219
Prigogine-Nicolis-Misguich-Davis theory, **31**:202–205
Rice-Allnatt theory, **31**:199–201
Weak coupling complex, **42**:369, 384
Weak coupling condition, **35**:135
Weak coupling equation, **35**:161, 183
Weak coupling limit, **38**:116, 144
physical significance, **38**:116

van Hove limit, **38**:116

Weak electrolyte model, 31:87

Weak field deflection, 10:9, 13–15

Weak instability, 43:49

Weak quantization, 18:105

Weakly adsorbed chain, 52:453

Weeks-Chandler-Andersen theory, 49:424–425

Weight function, 49:119–120, 140, 143, 172

   internal, **49**:180

   unperturbed, **49**:184

Weighted rectangular function, 44:329

Weighting coefficients, 53:279

Weiss approximation, 22:351; **41**:22

Weiss constant, 7:611

Weiss mean field theory, 17:44

   algebraic function for, **17**:44

   from cluster expansions, **17**:47

   ferromagnetism, **40**:321, 361

Weissenberg hypothesis, 35:98

Werfelmeirer packing sequence, 40:75–76

Werner-Meyer method, 54:136–138

Wertheim continuum, 48:212

Wertheim-Fröhlich expression, Felderhof rederivation of, **48**:279

Weyl group, 38:97

Weyl operator, 38:82

Weyl problem, 40:101

Weyl system, 38:96

Wheeler equation, 49:125–126, 182–184

Wheeler integral kernel, 49:142, 187

White noise, 38:86

Whitten-Rabinovitch formula, 50:115

Whole-grain heating effect, 40:57

Wick's theorem, 6:94; **14**:44, 133, 138; **36**:215, 216; **46**:213; **52**:196

Wide angle scattering, 42:468

Wien filter, 45:25

Wiener-Feynman-Kac formula, 46:247

Wiener integral, 15:356

   polymer statistics, **22**:1, 23ff, 35ff, 65, 72, 76

Wiener-Khinchin theorem, 7:507; **15**:66; **17**:74

Wiener measure, in polymer statistics, **22**:24, 25, 50, 62

Wiener process, 44:279; **46**:247

Wiener statistics, 44:401

Wiener theorem, 49:157

Wigner 3j coefficients, 14:497

Wigner crystallization, 41:188

Wigner distribution function, 1:314, 326, 332, 342; **16**:7–8; **38**:84; **46**:124

   spatially dependent, **16**:23

Wigner-Eckart theorem, 9:222ff; **14**:104, 117, 400; **44**:41, 122

Wigner function, 46:127ff

   classical limit for irregular states, **46**:129

   classical limit for regular state, **46**:128

   relationship to spatial autocorrelation function, **46**:130

   semiclassical limit for regular states, **46**:128

Wigner rate, 13:108

Wigner-Seitz cell, 44:533

Wigner-Seitz potential, 31:394, 409, 415, 429

Wigner-Seitz radius, 41:102

Wigner-Weisskopf approximation, 33:213; **38**:147; **40**:181; **42**:326

Wigner-Weisskopf-Dicke model, 53:364

Wigner-Wilkins equation, 20:177, 179, 194

Wigner-Wilkins kernel, 20:145

square integrability of, **20**:170

Wigner-Witmer rules, 28:116, 135; **30**:58; **42**:20

Wigner's rule, 30:446

Wigner's theory, cellular method, **2**:252, 304, 306, 318

Wilkins model, 7:344

Wilson-Frenkel theory, 40:167, 210

Winding number, continued fraction representation, **46**:107

irrational, **46**:78, 106

rational, **46**:77, 105

WKB approximation, 20:156, 176, 180; **22**:56, 70; **25**:70, 91, 102, 105, 111, 132, 138, 171; **28**:180, 233; **29**:87; **30**:82, 88, 102, 129, 321, 338, 377, 471, 480; **36**:73; **46**:250; **50**:200; **53**:301, 315

Wood's anomalies, 27:337

Woodward-Hoffman rules, 30:58; **33**:348, 374, 375, 375, 377, 379, 381; **42**:20.29; **50**:145

Wool, 7:266, 277

Work function, 31:342, 403; **41**:165, 491

determination of, **27**:230

hydrocarbon mixture, **31**:403, 404

non-polar fluid, **31**:342, 361, 381

Workhorse approximation, 10:379

Wormlike chain, 52:365, 433, 441, 445–448

Wrapped distribution, 44:288

Wrapped normal distribution, 44:268, 282, 288

Wrapped variate, 44:288

Wurtzite lattice, 27:448

X

Xα method, 41:64–66

Xα statistical exchange approximation, 26:258

X-ray absorption fine structure, extended (EXAFS), 49:469–472, 480, 510, 519, 643

X-ray analysis, 2:4, 10, 24, 28, 29

X-ray appearance potential spectroscopy, 41:491

X-ray crystallography, 39:55, 59, 62–64

X-ray detectors, 34:164

X-ray diffraction, 1:210

argon, **53**:13

critical region, **6**:194–197

energy dispersive, **51**:234

fluid structure studies, **53**:3–6

liquids, **34**:157

metal-ammonia solutions, **4**:311, 355–357

metallic glass, **51**:231, 248

metallic solution, **2**:128

natural fibers, **7**:264

nitrogen, **53**:26

water, **53**:35

X-ray fluorescence, 27:18

X-ray photoelectron spectroscopy, 27:28; **49**:466

X-ray photoemission spectroscopy, 51:257

X-ray production of color centers, 4:194

X-ray scattering, 27:342

X-ray single crystal studies, 49:484

X-ray source, 34:160

stability, **34**:161

X-ray spectroscopy, soft, **51**:259

X-ray structure factor, 53:41, 44, 56

x-trap, 40:454, 455, 467

Xanthine, 7:47ff, 50ff

Xanthine oxidase, 7:77, 620

XeAr, 52:284

XeKr, 52:284

Xenon, 16:127; **27**:476, 479, 524; **34**:180; **47**:Part 2:179; **49**:395

adsorption on Pd (100), **49**:636

chains, semiclassical quantization **53**:315–316

clathrates, **2**:30, 41

dimer, **46**:62; **52**:280

hydrate, **2**:3, 32, 34

    thermodynamic data and lattice constants, **2**:8

light scattering, **24**:42, 56

London coefficient for, **6**:399

polarizability, **46**:62

    magnetic susceptibility, and ionization energy, **2**:71

potential energy, **2**:71

Rydberg states of, **42**:580

structural and thermodynamic data, **16**:346–348

see also Collisional ionization

Xenon-helium system, **2**:95

XY model, **40**:196; **49**:348, 352, 354

    planar, **40**:201

Y

Yang-Lee theory of phase transitions, **37**:144

Yeast, **29**:51, 138

    glycolysis in *S. carlsbergensis*, **29**:138

Yohimene, **7**:414

Young diagram, **14**:94; **51**:116

Young-Laplace equation, **46**:283, 305; **49**:365, 409

Yttrium, **41**:141

Yvon-Born-Green hierarchy, **37**:11, 140, 145, 155, 181, 184; **40**:290, 294, 314–318, 339–349; **49**:358ff, 371, 379, 383, 386, 392ff, 404, 411, 417

    sum rules, **49**:392–395

Z

z-axis ligands, **7**:371, 376, 393, 400

Z-cut crystalline quartz, **44**:334

Zeeman cross section, **28**:352

Zeeman effect, **30**:395; **42**:245; **44**:33, 89, 116

    diamagnetic, **44**:33

    nuclear, **44**:34

    orbital, **44**:93

    paramagnetic, **44**:33, 90

    rotational, **44**:94

    second order, **44**:34, 98

    translational, **44**:101

    vibrational, **44**:34, 96

Zeeman energy, **7**:177

Zeeman Hamiltonian, **18**:176

Zeeman interaction, **18**:209

Zeeman splitting, **7**:166

Zeolites, **44**:397

Zermolo's paradox, **22**:360

Zero deficiency networks, **43**:11

Zero deficiency theorem, **38**:322

Zero differential overlap approximation, **1**:249, 250, 254, 256, 257, 259; **7**:36; **18**:269, 271, 291; **26**:255

    atomic orbitals, **18**:269

    SCF matrix elements, **18**:271

Zero-dimensional regime, **49**:314, 319–320, 345, 347, 351, 354

Zero-field splitting, **7**:554; **40**:401, 409, 410, 412

Zero-field transition, **40**:411, 413

Zero of energy, **36**:145

Zero-order function, **7**:12

Zero phonon line, **41**:273

Zero-point energy, **1**:25, 33, 64, 65, 69, 233–236; **5**:173; **27**:344

Zero-point motion, **53**:39

Zero-point vibration, **2**:235; **27**:479

Zero temperature limit, of electron gas, 6:112–113

Zeta potential of slime mold, 39:33

Zhabotinsky reaction, 29:156, 157; 39:16, 30, 35; 43:222; *see also* Belousov-Zhabotinsky reaction

Ziegler type catalyst, 2:162

Ziman theory, 51:244–250, 254ff, 261, 270ff

Zimm-Bragg theory, 15:301

Zinc, 41:141, 166
  collisional excitation, 30:445
  quenching of ^1P state, 50:351, 379
  quenching of ^3P state, 50:352

Zinc dimethyldithiocarbamate, 16:181, 183

Zinc oxide, 27:448–449, 510

Zinc proteins, 39:61, 84, 85

Zinc sulfide, 27:443

Zinc sulfide structure, 49:50, 54, 61

Zinc-tin alloy (Zn_5Sn_5), calculation of thermodynamic quantities, 2:136

Zirconium, 41:141

Zunger method, 49:79

Zustandsummme, 7:206

Zwanzig expansion, 37:142

Zwanzig-Mori transport theory, 33:256

Zwanzig projection operator method, 42:324

Zwolinski-Eyring model, 21:93